Advances in Transition-Metal Mediated Heterocyclic Synthesis

Advances in Transition-Metal Mediated Heterocyclic Synthesis

Edited by

Daniel Solé
Universitat de Barcelona, Barcelona, Spain

Israel Fernández
Universidad Complutense de Madrid, Madrid, Spain

ELSEVIER

ACADEMIC PRESS
An imprint of Elsevier

Academic Press is an imprint of Elsevier
125 London Wall, London EC2Y 5AS, United Kingdom
525 B Street, Suite 1800, San Diego, CA 92101-4495, United States
50 Hampshire Street, 5th Floor, Cambridge, MA 02139, United States
The Boulevard, Langford Lane, Kidlington, Oxford OX5 1GB, United Kingdom

British Library Cataloguing-in-Publication Data
A catalogue record for this book is available from the British Library

Library of Congress Cataloging-in-Publication Data
A catalog record for this book is available from the Library of Congress

ISBN: 978-0-12-811651-7

For Information on all Academic Press publications
visit our website at https://www.elsevier.com/books-and-journals

Publisher: John Fedor
Acquisition Editor: Emily McCloskey
Editorial Project Manager: Amy Clark
Production Project Manager: Nilesh Kumar Shah
Cover Designer: Christian Bilbow

Typeset by MPS Limited, Chennai, India

Contents

List of Contributors

Benito Alcaide Universidad Complutense de Madrid, Madrid, Spain

Pedro Almendros Instituto de Química Orgánica General-CSIC, Madrid, Spain

Daniel A. Cruz Instituto de Productos Naturales y Agrobiología-CSIC, La Laguna, Santa Cruz de Tenerife, Spain; Instituto Universitario de Bio-Orgánica Antonio González (CIBICAN), Departamento de Química Orgánica, Universidad de La Laguna, Avda. Astrofísico Fco., La Laguna, Tenerife, Spain

Israel Fernández Universidad Complutense de Madrid, Madrid, Spain

Bartolo Gabriele University of Calabria, Arcavacata di Rende, CS, Italy

Fangdong Hu Linyi University, Linyi City, China

Lu Liu University of Wisconsin, Madison, WI, United States

Juan M. López-Soria Instituto de Productos Naturales y Agrobiología-CSIC, La Laguna, Santa Cruz de Tenerife, Spain; Instituto Universitario de Bio-Orgánica Antonio González (CIBICAN), Departamento de Química Orgánica, Universidad de La Laguna, Avda. Astrofísico Fco., La Laguna, Tenerife, Spain

Víctor S. Martín Instituto de Productos Naturales y Agrobiología-CSIC, La Laguna, Santa Cruz de Tenerife, Spain; Instituto Universitario de Bio-Orgánica Antonio González (CIBICAN), Departamento de Química Orgánica, Universidad de La Laguna, Avda. Astrofísico Fco., La Laguna, Tenerife, Spain

Teresa Martínez del Campo Universidad Complutense de Madrid, Madrid, Spain

Sergio Mata Universidad de Oviedo, Oviedo, Spain

Pedro O. Miranda Instituto de Productos Naturales y Agrobiología-CSIC, La Laguna, Santa Cruz de Tenerife, Spain; Instituto Universitario de Bio-Orgánica Antonio González (CIBICAN), Departamento de Química Orgánica, Universidad de La Laguna, Avda. Astrofísico Fco., La Laguna, Tenerife, Spain

Kilian Muñiz Institute of Chemical Research of Catalonia (ICIQ), Tarragona, Spain and ICREA, Barcelona, Spain

Juan I. Padrón Instituto de Productos Naturales y Agrobiología-CSIC, La Laguna, Santa Cruz de Tenerife, Spain; Instituto Universitario de Bio-Orgánica Antonio González (CIBICAN), Departamento de Química Orgánica, Universidad de La Laguna, Avda. Astrofísico Fco., La Laguna, Tenerife, Spain

Jennifer M. Schomaker University of Wisconsin, Madison, WI, United States

Victoria Sinka Instituto de Productos Naturales y Agrobiología-CSIC, La Laguna, Santa Cruz de Tenerife, Spain; Instituto Universitario de Bio-Orgánica Antonio González (CIBICAN), Departamento de Química Orgánica, Universidad de La Laguna, Avda. Astrofísico Fco., La Laguna, Tenerife, Spain

Daniel Solé Universitat de Barcelona, Barcelona, Spain

Rubén Vicente Universidad de Oviedo, Oviedo, Spain

Jianbo Wang Peking University, Beijing, China

Foreword

It is not rare to hear these days that chemistry is a "mature science." Such a statement is oftentimes used to emphasize the lack of big, or maybe one should rather say "mass media newsworthy" discoveries, in our discipline over the past decades. For most chemistry practitioners though, this perception could not be further from the truth. Chemistry continues to bloom with the development of novel, more efficient, and sustainable methods towards the preparation of basic commodities, intermediate building blocks, and high-value end products, at faster paces than ever. Further, chemistry's centrality to other disciplines such as medicine, biology, material science, and nanotechnology, to name just a few, is currently undeniable and will likely continue to grow in the years to come.

In this vibrant context, heterocyclic chemistry plays a prominent role. A large fraction of drugs and pharmaceuticals in the market are primarily based on heterocyclic scaffolds. Further, more than 90% of molecules occurring naturally contain heterocycles in their core structure. Cutting-edge materials, finely tuned to deliver specific photophysical and electronic properties, cannot be devised in the absence of heterocyclic frameworks. As such the impact and ramifications that heterocycles have in both basic physical and life sciences, as well as in applied disciplines, is unparalleled by any other chemical motif. Still, while heterocyclic synthesis might be regarded as one of the most venerable areas of chemistry, and its well-established foundations can be found in the chemistry textbooks used in the formative years of the undergraduate chemistry curriculum, the number of new synthetic tools targeting heterocycles is exponentially growing. Arguably the most sustainable and efficient methods in the numerous present and upcoming developments stem from transition-metal catalysis.

In this book, selected aspects highlighting the impact of transition metals on heterocyclic synthesis have been compiled. Starting with a chapter focused on the synthesis of nonaromatic oxygen-containing cycles from allenols by Pedro Almendros and Benito Alcaide, the reader will transition to recent developments in the synthesis of N-containing heterocycles via the intramolecular diamination of alkenes summarized by Kilian Muñiz in Chapter 2, Intramolecular Diamination of Alkenes. Palladium, arguably the most synthetically useful transition metal in the periodic table, is the focus of Chapter 3, Synthesis of Heterocycles by Palladium-Catalyzed Carbonylative Reactions, in which

Bartolo Gabriele summarizes its impact in carbonylative reactions towards the construction of heterocyclic frameworks. Jianbo Wang offers in Chapter 4, Synthesis of Heterocyclic Compounds Based on Transition-Metal-Catalyzed Carbene Coupling Reactions, a compelling overview on creative applications of carbene couplings towards heterocyclic synthesis. Earth abundant metals are also a source of inspiration towards the construction of heteroatom-containing cycles as illustrated by Chapters 5, Synthesis of Heterocycles With Iron Salts as Sustainable Metal Catalysts, and Chapter 7, Zinc-Mediated Synthesis of Heterocycles, by Juan I. Padrón and Rubén Vicente, respectively. Novel tools towards the synthesis of aziridines and their application towards the efficient generation of complex amines are described by Jennifer M. Schomaker in Chapter 6, Allene Aziridination as a Tool for the Synthesis of Complex Amines. Finally, Israel Fernández and Daniel Solé focus Chapter 8, Controlling Selectivities in Palladium Catalyzed Cyclization Reactions Leading to Heterocycles: From Ambiphilic Reactions of Arylpalladium Species to Carbene Insertions, on mechanistic aspects governing Pd-catalyzed synthesis of heterocycles, highlighting also the power of modern computational tools in organic chemistry.

While none of these contributions aims to be comprehensive (what would be an almost impossible endeavor given the breath and extension of the field), I am positive that this book will offer the reader insightful perspectives on selected aspects of this lively topic. In addition, I am confident the body of work summarized here will serve as a source of inspiration for future developments in the both highly demanding and highly demanded arena of heterocyclic synthesis.

Cristina Nevado

Department of Chemistry, University of Zurich, Zürich, Switzerland

Chapter 1

Metal-Mediated Synthesis of Nonaromatic Oxacycles From Allenols

Benito Alcaide[1], Pedro Almendros[2] and Teresa Martínez del Campo[1]
[1]Universidad Complutense de Madrid, Madrid, Spain, [2]Instituto de Química Orgánica General-CSIC, Madrid, Spain

1.1 INTRODUCTION

Nonaromatic oxacycles are ubiquitous structural units that are extensively encountered in a number of biologically active natural products and functional molecules, and therefore their stereocontrolled synthesis remains an intensive research area.[1–7] Transition-metal catalyzed nucleophile addition to allenes is a very powerful tool for the synthesis of functionalized molecules containing heteroatoms in an atom efficient manner.[8–13] The reaction on allenic systems has an added value as the regiochemistry can be controlled by using different metals or by changing the substituents or the length of the chain between the allene and the heteroatom in the intramolecular case. The mechanisms for the cyclization of allenols in the presence of transition metal complexes are summarized in Scheme 1.1. The metal can coordinate to one or the other double bonds of the allene in a Z2-coordination mode, promoting the *exo*- or *endo*-attack to the terminal or internal carbon atoms to form different-sized heterocycles. Five- and six-membered rings are often preferred, but larger rings have been obtained by modulating the substitution in the starting material and the metal employed. The purpose of this chapter is to highlight the metal-mediated processes leading to the formation of at least one C–O bond through an intramolecular reaction with an allenyl unit to obtain nonaromatic oxacycles. This chapter has been organized taking into account the nature of the metal catalyst.

Advances in Transition-Metal Mediated Heterocyclic Synthesis.
DOI: https://doi.org/10.1016/B978-0-12-811651-7.00001-7

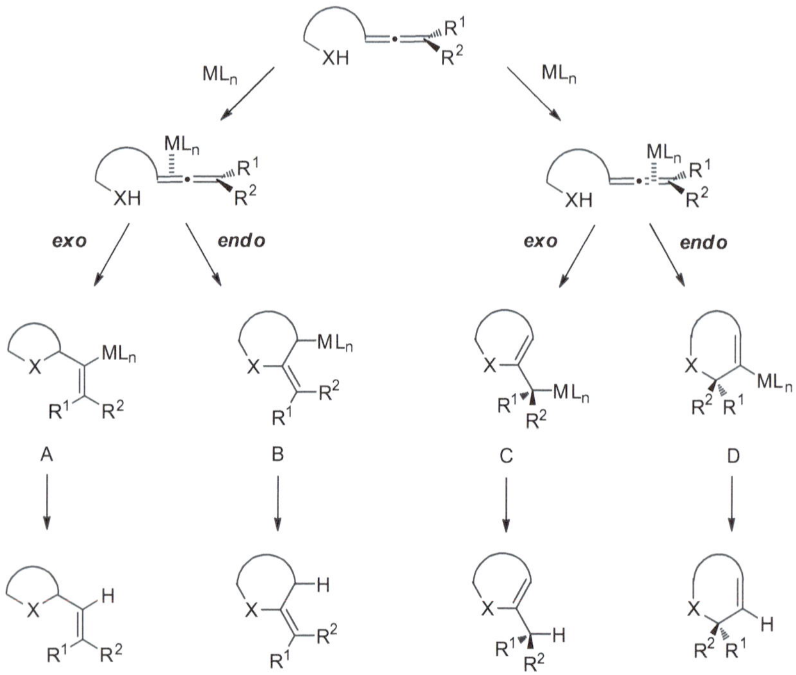

SCHEME 1.1 Possible paths for the metal-catalyzed cyclization of allenols.

1.2 SILVER-CATALYZED REACTIONS

Olsson and Claesson reported in 1979 the first reaction of α-allenols in the presence of catalytic amounts of silver salts to give dihydrofurans[14] (Scheme 1.2).

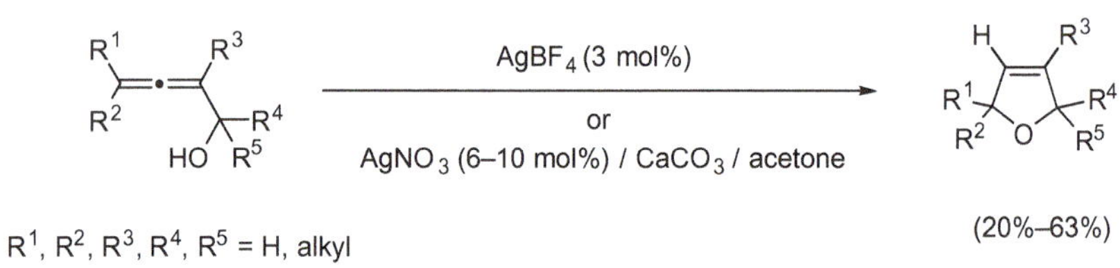

SCHEME 1.2 First reported silver-catalyzed reaction of α-allenols.

Silver nitrate has been the salt of choice in many transformations of α-allenols, but stoichiometric amounts have to be used in some challenging cases. Alcaide, Almendros, et al. reported the synthesis of spirocyclic dihydrofuran derivatives in the presence of 1 equivalent of $AgNO_3$[15,16] (Scheme 1.3).

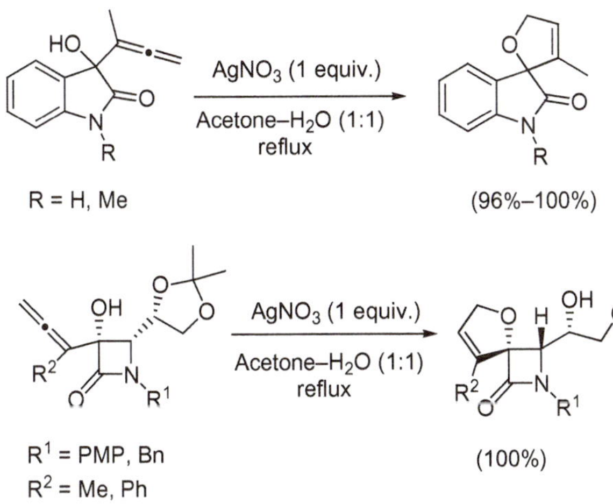

SCHEME 1.3 Stoichiometric silver-promoted formation of spirocyclic dihydrofurans.

The catalytic version of this reaction has been carried out by Marshall et al.[17–19] They reported that the reaction with enantioenriched α-allenols with $AgNO_3$ (20 mol%) led to stereospecific cyclization to the corresponding dihydrofurans, with a presumed attack on the allenic double bond *syn* to the side chain containing the OH group (Scheme 1.4).

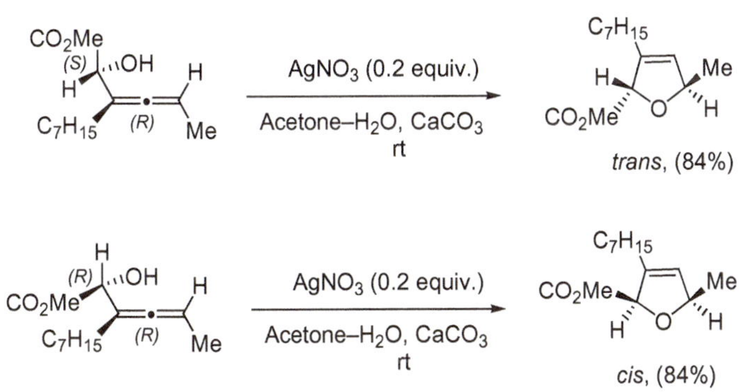

SCHEME 1.4 $AgNO_3$-catalyzed cyclization of enantioenriched α-allenols.

The reaction also seems to be chemoselective. Ma et al. reported that bis-allyl substituted α-allenols can be converted to the 2,5-dihydrofuran derivatives by the attack of the oxygen atom onto the allene, without the involvement of the extra double bonds in the reaction[20] (Scheme 1.5).

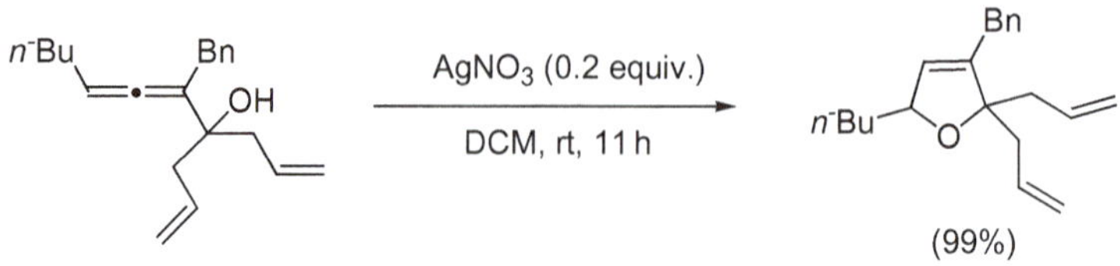

SCHEME 1.5 Preferred mode of cyclization of dienyl α-allenols: allene versus alkene attack.

However, Lee et al. found that when an alkyne is present, the silver salt tends to be more alkynophilic than allenophilic, and nucleophilic attack of the alcohol occurs onto the activated triple bond to give substituted furans[21] (Scheme 1.6).

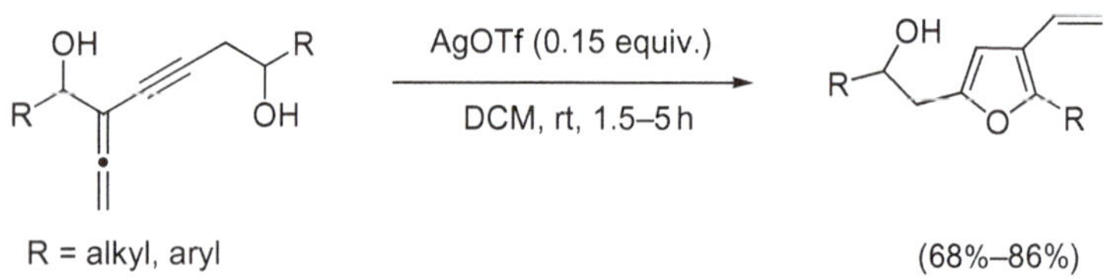

SCHEME 1.6 Preferred mode of cyclization of alkynyl α-allenols: alkyne versus allene.

The regioselectivity of the reaction has been studied by Marshall and Aurrecoechea with α,α′-allendiols and Krause with bis(α-allenols). Marshall et al.[22] and Aurrecoechea et al.[23] reported independently that α,α′-allendiols with primary and secondary hydroxyl groups led to the corresponding dihydrofurans by cyclization with the more substituted alcohol by a complexation of the Ag^+ to the less congested double bond of the allenic system (Scheme 1.7).

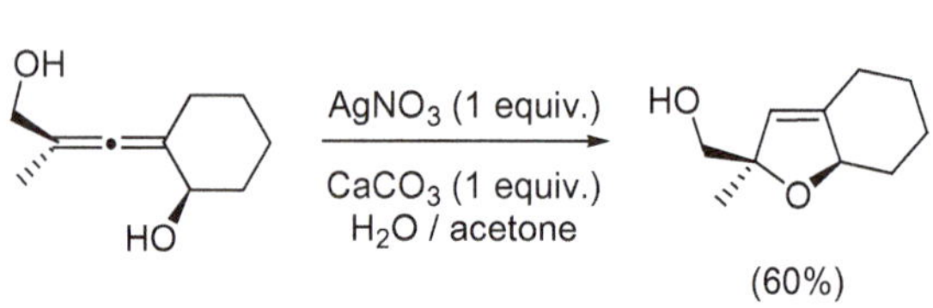

SCHEME 1.7 Regioselectivity in the cyclization of α,α′-allendiols.

Krause et al. reported the silver-catalyzed bis-cyclization of bis (α-allenols), but 30 mol% of $AgNO_3$ and prolonged reaction times were required[24] (Scheme 1.8). The steric effects have a big influence and the bis-cyclization was only achieved effectively for the less substituted derivatives.

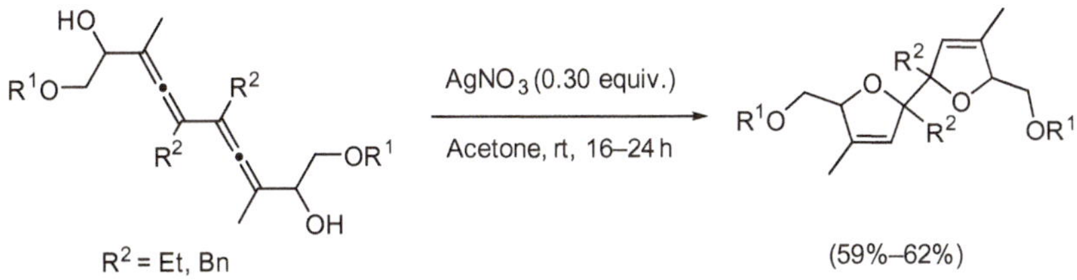

SCHEME 1.8 Bis-cyclization of bis(α-allenols).

Cyclization of α-allenols with silver salts as a catalyst has been widely used as the key step in many total syntheses of natural products. Standaert et al. reported the synthesis of (+)-furanomycin 1 in seven steps,[25] where one of the steps is the cyclization of an α-allenol mediated by silver (Scheme 1.9).

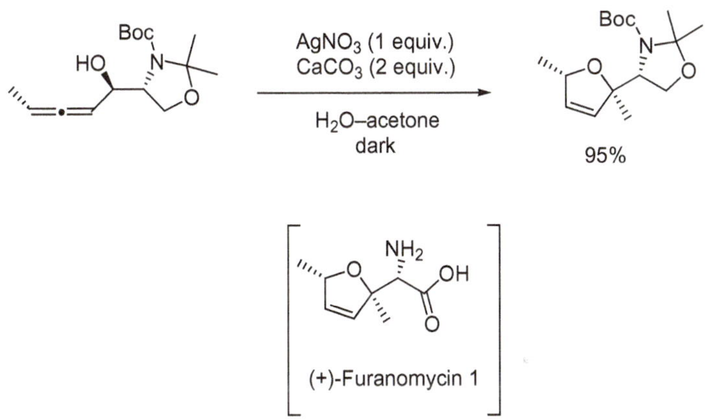

SCHEME 1.9 Cyclization of α-allenol as the key step in the total synthesis of (+)-furanomycin 1.

Fürstner et al. carried out the synthesis of several natural products such as amphidinolide X and Y[26,27] (Scheme 1.10) and leiodolide B[28] (Scheme 1.11) containing furan units that were obtained by this methodology.

There are a few examples of silver-catalyzed reaction of β-allenols. The first reaction was reported by Claesson in 1979.[14] β-Allenols needed more active conditions to afford 2*H*-pyrans by the attack of the oxygen atom to the terminal carbon atom of the distal double bond of the allenic system by a 6-*endo* cyclization (Scheme 1.12).

Gore et al. reported a favorable reaction of the β-hydroxy group versus the γ-hydroxy group in the silver-mediated cyclizations of β,β- and β,γ-allene-diols.[29] β,β-Allene-diols in the presence of 10 mol% of silver

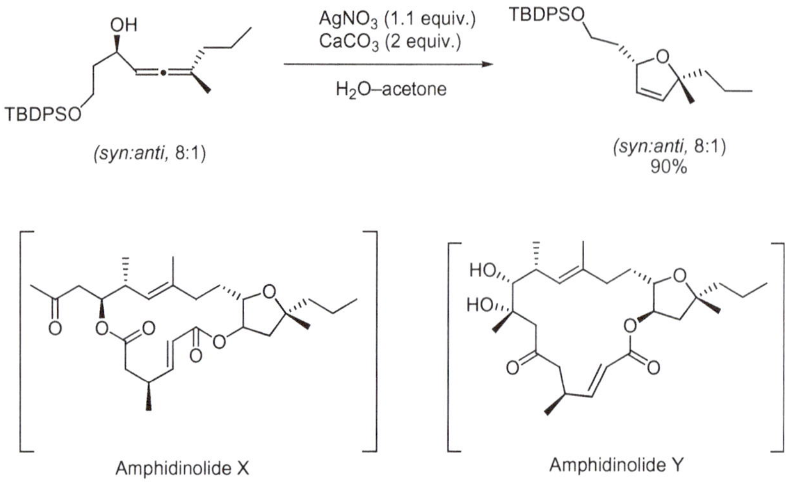

SCHEME 1.10 Cyclization of α-allenol moieties in the total synthesis of amphidinolide X and Y.

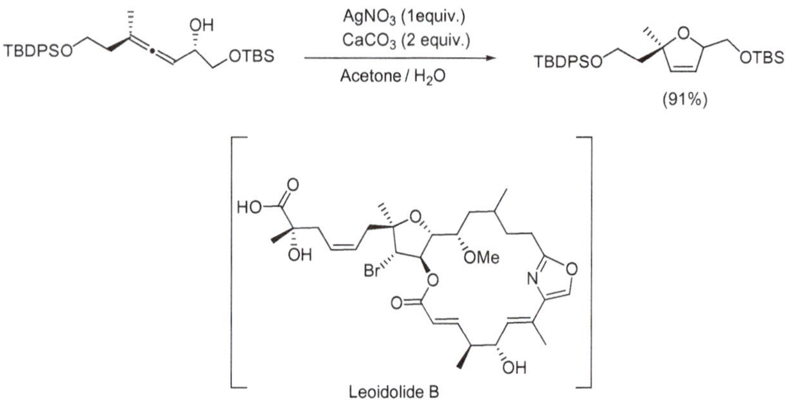

SCHEME 1.11 Cyclization of an α-allenol in the total synthesis of leiodolide B.

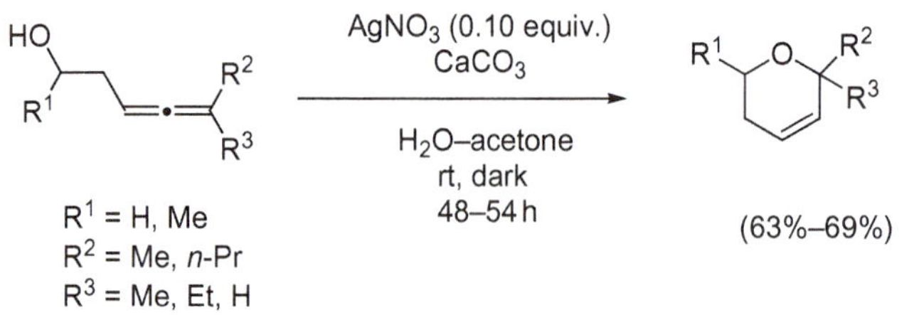

SCHEME 1.12 Silver-catalyzed cyclization of β-allenols.

nitrate gave bicyclic acetals by the attack to the central carbon atom of the allene, together with the expected dihydropyrans. Unsubstituted β,γ-allene-diols also gave the bicyclic acetals and the dihydropyran resulted from the attack of the β-hydroxyl group. Surprisingly, when the β,γ-allene-diol had a methyl group attached to the terminal carbon atom of the allene, only the dihydropyran from the attack of the β-hydroxy group was observed (Scheme 1.13).

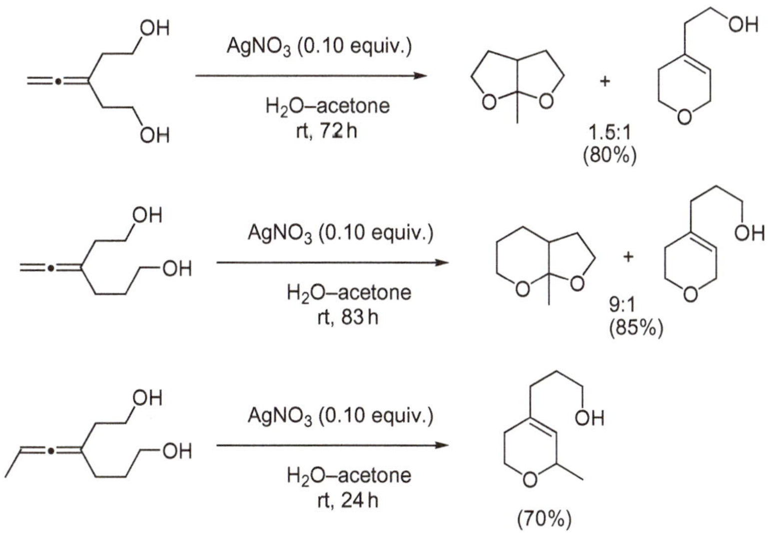

SCHEME 1.13 Selectivity on the silver-catalyzed cyclization of β,β- and β,γ-allene-diols.

Hii et al.[30] reported the few examples in the literature of silver-catalyzed reaction of γ-allenols. They reported the 5-*exo-trig* cyclization of γ-allenols with silver salts to give vinyl-substituted tetrahydrofurans (Scheme 1.14).

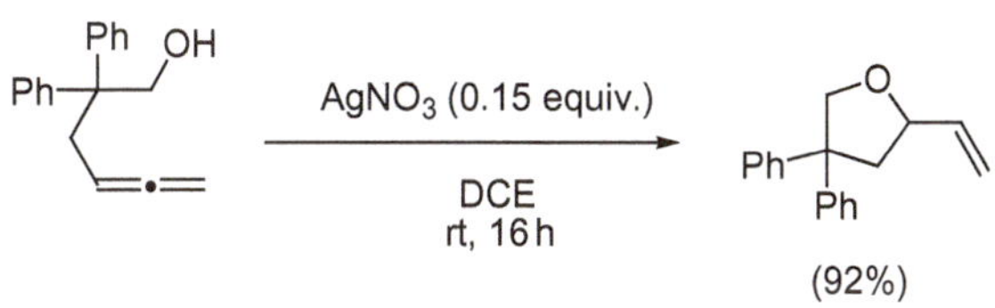

SCHEME 1.14 Silver-catalyzed cyclization of γ-allenols.

The same group[31] reported the enantioselective version of the reaction using silver salts of chiral acids. The tetrahydrofurans were obtained with enantioselectivities up to 73% ee using as catalysts oxophosphorus(V) phosphinate and phosphate (TADDOL-POOH) complexes of silver (Scheme 1.15).

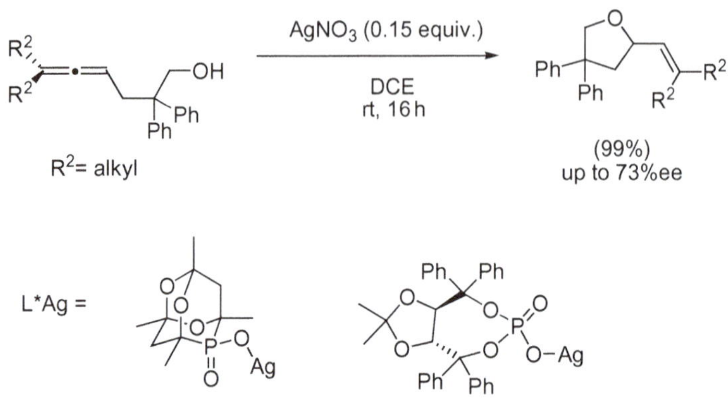

SCHEME 1.15 Enantioselective silver-catalyzed cyclization of γ-allenols.

Reaction with δ-allenols needed stoichiometric amounts of silver salts to give tetrahydropyrans by 6-*exo* cyclization. Gore et al.[32] reported that up to 6 equivalents of $AgNO_3$ were needed when the allenic system was fully substituted (Scheme 1.16).

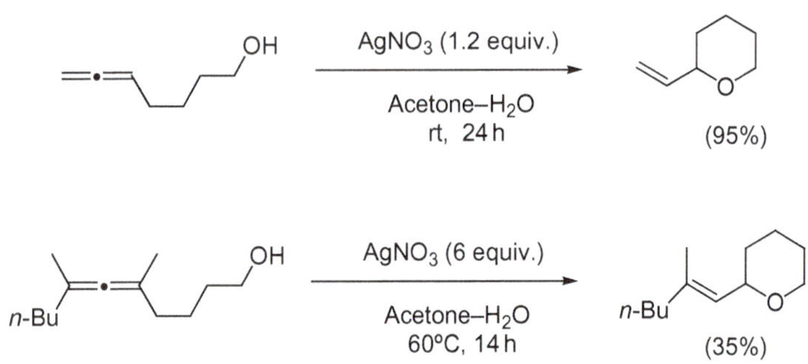

SCHEME 1.16 Stoichiometric silver-promoted reaction of δ-allenols.

Gallagher et al.[33] reported that the reaction is stereoselective because of the obtention of *cis*-2,6-disubstituted tetrahydropyrans from secondary δ-allenols (Scheme 1.17).

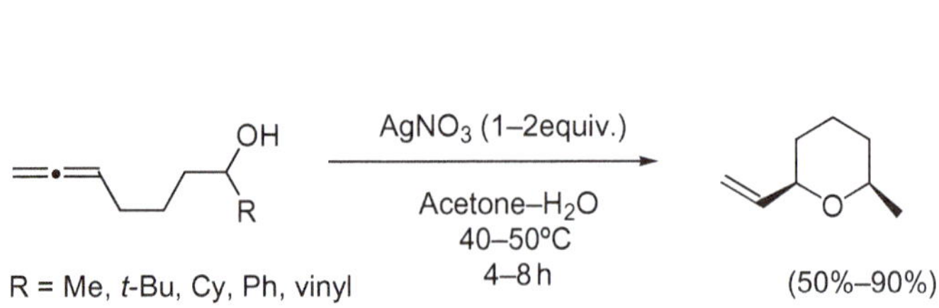

SCHEME 1.17 Stereoselectivity on the silver-promoted cyclization of δ-allenols.

1.3 GOLD-CATALYZED REACTIONS

The use of gold entails a number of advantages such as shorter reaction times, milder conditions, and lower catalyst loading than silver. The first gold-catalyzed cyclization of α-allenols was accomplished by Krause et al.[34,35] They reported the synthesis of chiral 2,5-dihydrofurans by treatment of α-allenols with catalytic amounts of $AuCl_3$ in nonpolar solvents (Scheme 1.18). Many functionalities are tolerated under these conditions. In the case of chiral allenes bearing alkyl substituents, the stereochemical information of the axial chirality is completely transferred to the newly formed stereogenic center.[36] In contrast, substrates with phenyl or electron-rich aromatic substituents undergo epimerization when treated with gold catalysts.

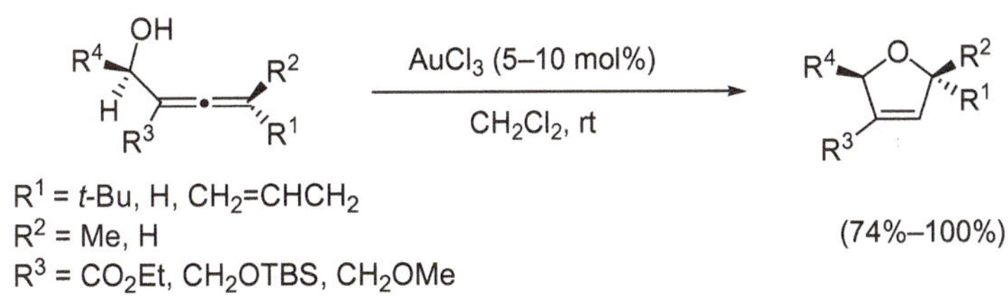

SCHEME 1.18 Gold-catalyzed cycloisomerization of α-allenols.

The cyclization of α-allenols was carried out in the presence of different gold catalysts such as $AuBr_3$ and AuCl, carbophilic Au(I) species (e.g., Ph_3PAuCl in the presence of additives). Hegedus et al. reported an efficient gold-catalyzed cyclization of chiral γ-substituted allenamides to highly functionalized dihydrofurans (Scheme 1.19). The reactions proceed rapidly and without erosion of the stereocenters.[37]

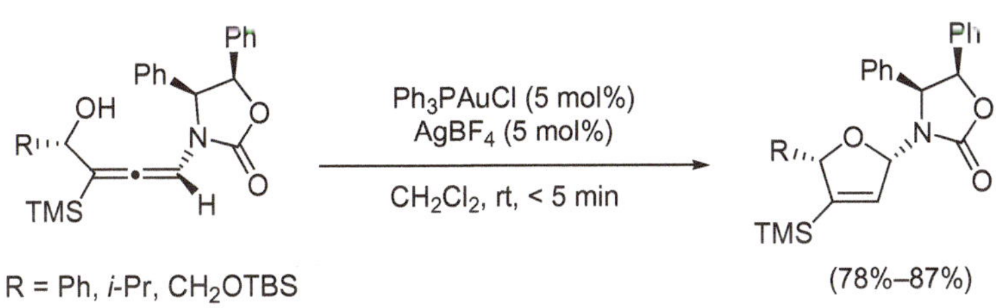

SCHEME 1.19 Gold-catalyzed cyclization of chiral α-allenols.

Krause et al. carried out the cyclization of an α,α-dihydroxyallene with catalytic amounts of Au(III) in the presence of pyridine to give a spiro-dihydrofuran by a chemoselective cycloisomerization; of the two secondary hydroxy groups, the one in the benzylic position participated in the cyclization[38] (Scheme 1.20). The use of pyridine increased the yield of the cyclization.

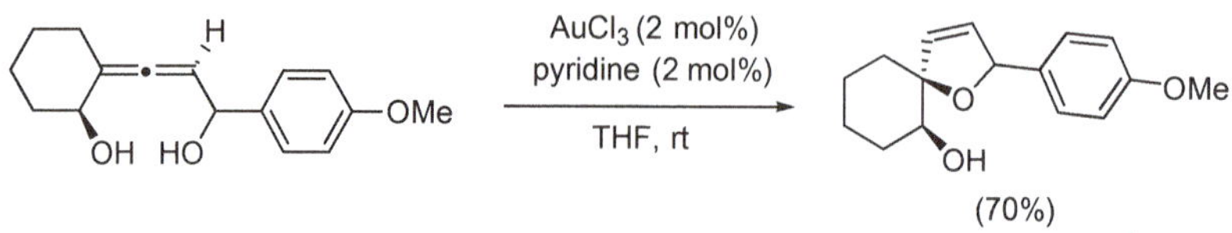

SCHEME 1.20 Chemoselective gold-catalyzed cyclization of an α,α-dihydroxyallene.

Volz and Krause[39,40] reported the first total synthesis of the β-carboline alkaloids (−)-isocyclocapitellin and (−)-isochrysotricine by Pictet-Spengler reaction of a chiral tetrahydrofuran with tryptamine (Scheme 1.21). Key tetrahydrofuran intermediate was obtained from the corresponding α,β-dihydroxyallene with complete axis-to-center chirality transfer by the use of gold(III) chloride as the catalyst. An analogous gold-catalyzed cycloisomerization of dihydroxyallene to 2,5-dihydrofuran was employed recently by Kocienski et al.[41] in their synthesis of ionomycin−calcium complex (Scheme 1.22). The gold-catalyzed cycloisomerization of those α,β-dihydroxyallenes is not only stereoselective but also highly chemoselective, because no product resulting from nucleophilic attack of the β-hydroxy group was observed.

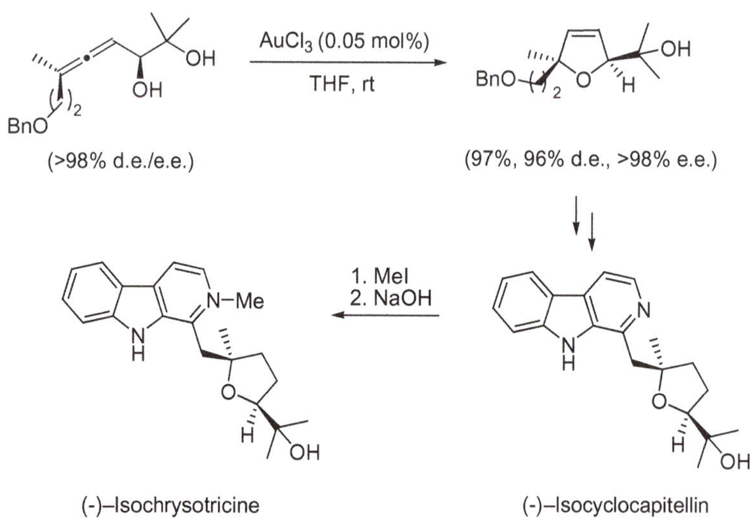

SCHEME 1.21 Stereo- and regioselective gold-catalyzed cyclization of α,β-dihydroxyallene: synthesis of β-carboline alkaloids.

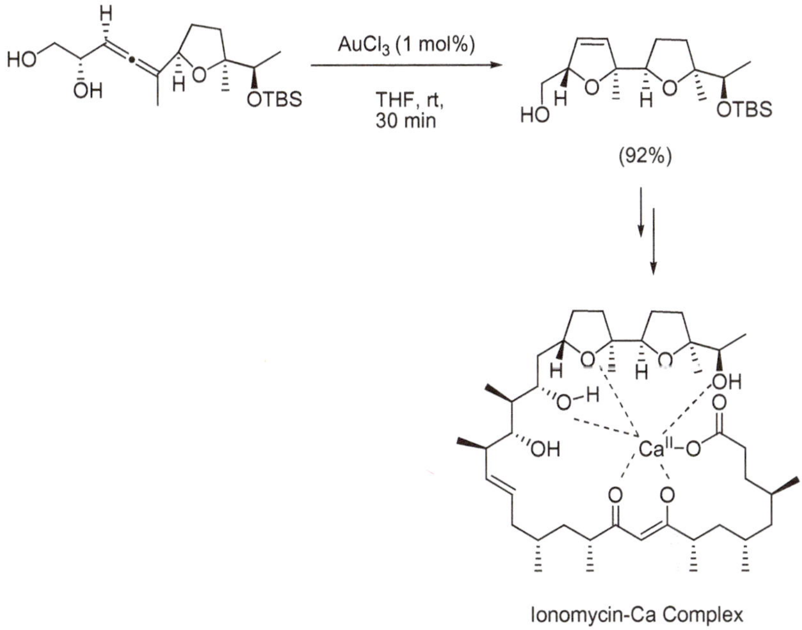

SCHEME 1.22 Synthesis of ionomycin–calcium complex.

Both gold (I) and gold (III) salts are shown to be active (pre)catalysts for the cycloisomerization of allenols. But, the question is: What is the oxidation state of the catalytically active species? In this way, Hashmi et al.[42] suggest that gold (III) chloride is reduced in situ to gold (I) in the presence of the α-allenol. This is concluded from the formation of the oxidative coupling product together with the dihydrofuran (Scheme 1.23). An increase of the amount of the gold (III) salt caused a raised yield of the oxidative product, which indicates that $AuCl_3$ is indeed acting as oxidizing agent in this transformation, and is therefore reduced to gold (I). On the other hand, it is known that gold (I) chloride can undergo dismutation to gold (III) and gold(0) in the presence of an allene.[43]

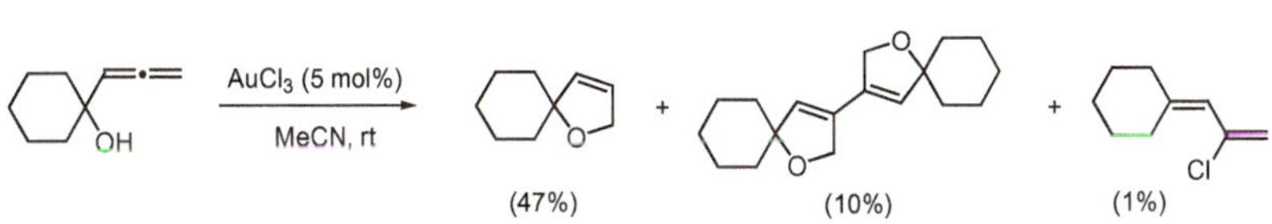

SCHEME 1.23 Oxidative coupling by gold (III) chloride.

The chemoselectivity of the reaction has been studied by Alcaide, Almendros, et al.[44] The chemodifferentiation of an allene versus an alkene moiety using Au(III) catalysis was reported. Gold-catalyzed oxycyclizations of enallenols did totally control the reaction toward the corresponding dihydrofurans (Scheme 1.24).

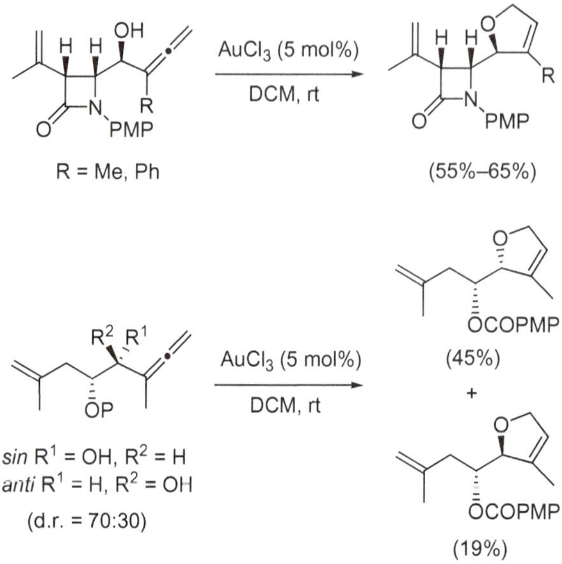

SCHEME 1.24 Preferred mode of cyclization of enallenols: allene versus alkene.

Lee et al.[45] reported an interesting example of a chemoselective gold-catalyzed transformation. A diol bearing both an alkyne and an allene moiety can afford the corresponding dihydrofuran when gold(III) chloride was used as the catalyst for the reaction. The gold salt activates preferentially the allene moiety in the presence of an alkyne (Scheme 1.25).

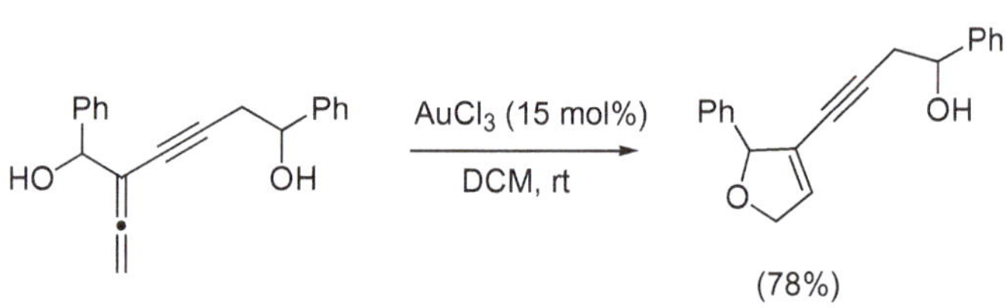

SCHEME 1.25 Preferred mode of cyclization of allenynol: allene versus alkyne.

A serious drawback of homogeneous gold catalysis is that the catalyst is usually reduced to metallic gold after the reaction or upon workup, so that it cannot be recycled. This problem can be overcome by using ionic liquids as the reaction medium. Thus, Krause et al.[46] reported the cycloisomerization of an α-allenol to dihydrofuran using $AuBr_3$ in the imidazolium-derived medium [BMIM][PF_6] (Scheme 1.26). This catalytic system is not only stable to water and air but can also be recycled easily without loss of efficiency.

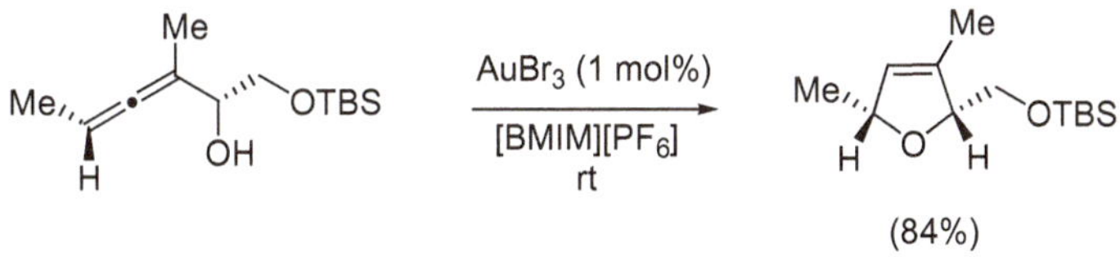

SCHEME 1.26 Gold-catalyzed cycloisomerization of α-allenols in an ionic liquid.

The cyclization of these same α-allenols can also be carried out in water using tetrachloroauric acid as a catalyst.[47] Compared to traditional gold catalysts in organic solvents, this new catalytic system is more environmentally friendly, and the gold catalyst can be reused after complete conversion of the substrate.

A novel reactivity in gold catalysis, an unusual preference for the 4-*exo-dig* cyclization in α-allenols as well as the rare β-hydride elimination reaction, was described by Alcaide, Almendros, et al.[48] Starting from salicylaldehyde-derived phenyl-substituted α-allenols, the gold-catalyzed synthesis of oxetenes was achieved (Scheme 1.27).

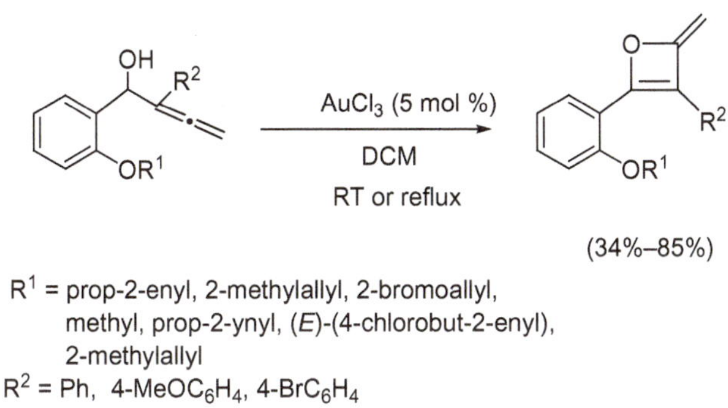

SCHEME 1.27 Synthesis of oxetenes through 4-*exo*-oxycyclization reaction of α-allenols under gold catalysis.

The gold-catalyzed cyclizations of β-allenols are often very slow, resulting in reaction times of several days. Krause et al.[49] reported the rapid 6-*endo* cyclization of β-hydroallenes to give the corresponding 5,6-dihydro-2*H*-pyrans in the presence of a cationic gold catalyst formed in situ from Ph_3PAuCl and $AgBF_4$ (Scheme 1.28). The reaction of some of these β-allenols with 5 mol% of AuCl in dichloromethane at room temperature typically requires 5 days for complete conversion to the corresponding dihydropyrans. However, the addition of *N*-iodosuccinimide to the reaction accelerated the process; the complete conversion to the iodinated dihydropyran was afforded within 1 min at room temperature.[50,51] This effect is probably caused by a rapid iododeauration of a σ-gold intermediate by N-iodosuccinimide (NIS), which is activated by the gold catalyst (Scheme 1.29).

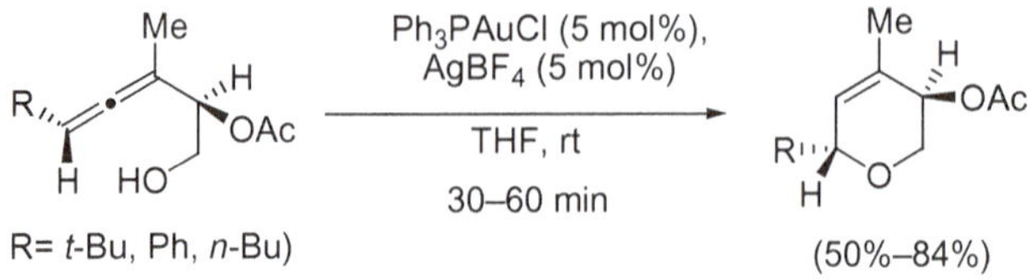

SCHEME 1.28 Gold-catalyzed cycloisomerization of β-allenols.

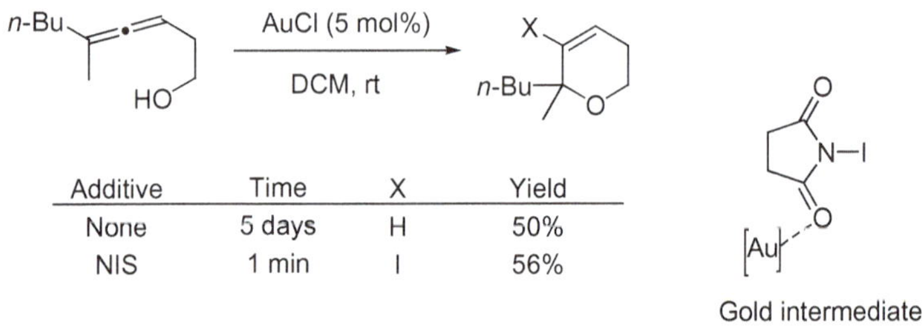

Additive	Time	X	Yield
None	5 days	H	50%
NIS	1 min	I	56%

SCHEME 1.29 Gold-catalyzed cycloisomerization of β-allenols with AuCl and NIS as additive.

The above reaction has already found application in target-oriented synthesis.[52] The gold catalyzed cycloisomerization of a β-allenol to the corresponding dihydropyran was the key step in the total synthesis of the sesquiterpenoid (*R, R, R*)-bejarol (Scheme 1.30).

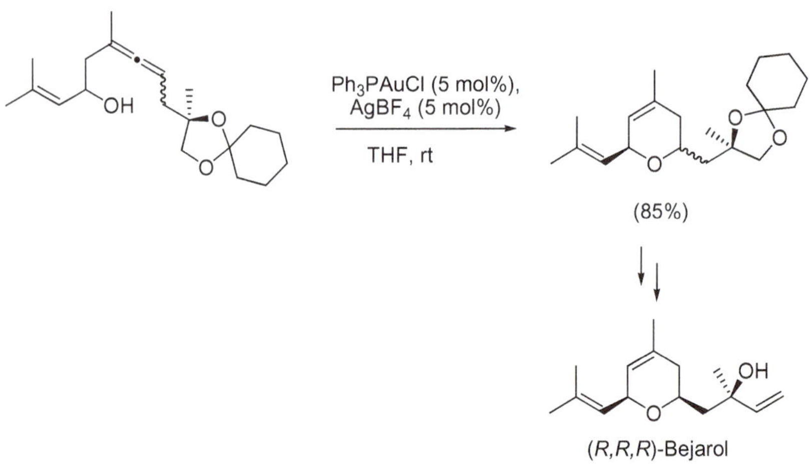

SCHEME 1.30 Total synthesis of the sesquiterpenoid (*R,R,R*)-bejarol.

As it has been shown, the preferred mode of cyclization for α- and β-allenols is *endo*. By contrast, γ-allenols preferred the *exo*-selective attack in gold-catalyzed cyclizations. Alcaide, Almendros, et al.[53] reported the reaction of γ-allenol tethered β-lactams with $AuCl_3$ to exclusively afford tetrahydrofuran derivatives resulting from a 5-*exo*-attack (Scheme 1.31).

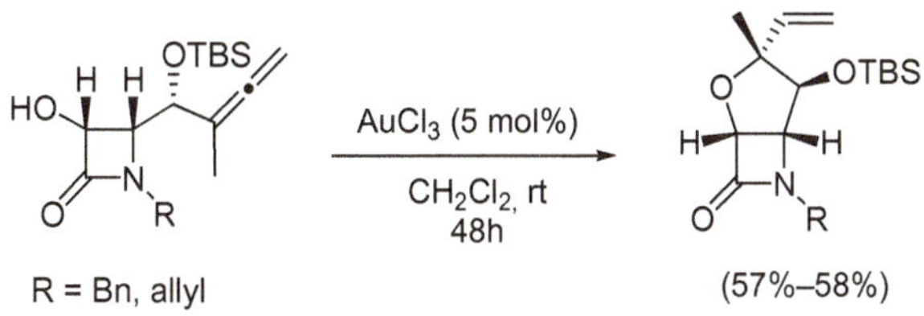

SCHEME 1.31 Cyclization mode of γ-allenols.

Another representative gold-catalyzed *exo*-cyclization is the transformation of a γ-allenol into the corresponding tetrahydrofuran, which was reported by Widenhoefer et al.[54] (Scheme 1.32).

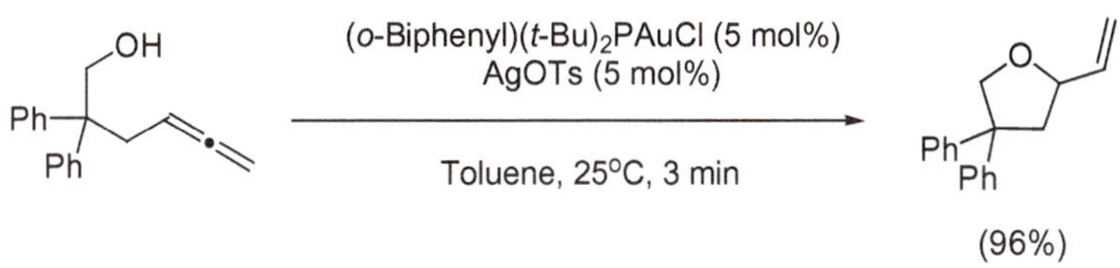

SCHEME 1.32 *exo*-Selective cyclization of a γ-allenol.

The same research group extended[55] this cyclization to the synthesis of enantiomerically enriched tetrahydrofurans by using chiral phosphine ligands with the gold catalyst (Scheme 1.33).

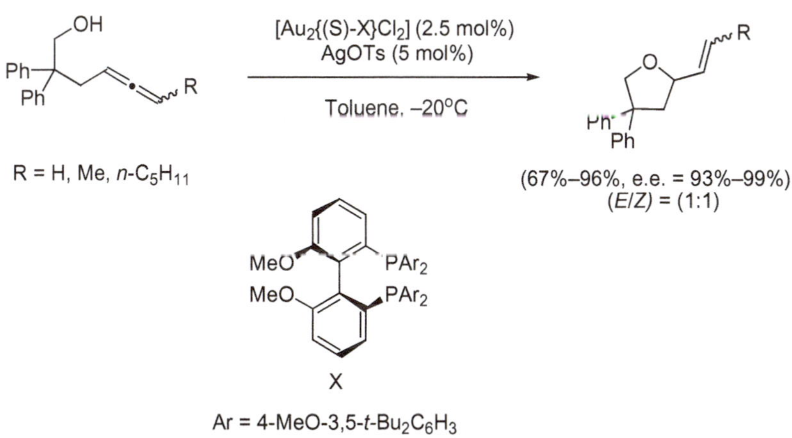

SCHEME 1.33 Enantioselective cycloisomerization of γ-allenols.

An alternative catalytic system reported by Toste et al.[56] takes advantage of a chiral counterion, which was introduced into the catalyst by a silver salt (Scheme 1.34).

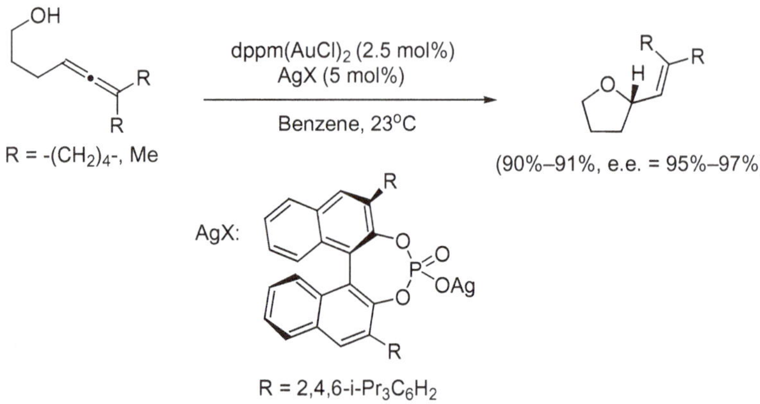

SCHEME 1.34 Chiral counterion effect in the enantioselective cyclization of γ-allenols.

In contrast with the *exo*-cyclization mode, Alcaide, Almendros, et al.[57] reported the preferential regioselective 7-*endo* cyclization reaction by the treatment of an acyclic γ-allenol with $AuCl_3$ to furnish the corresponding tetrahydrooxepine (Scheme 1.35).

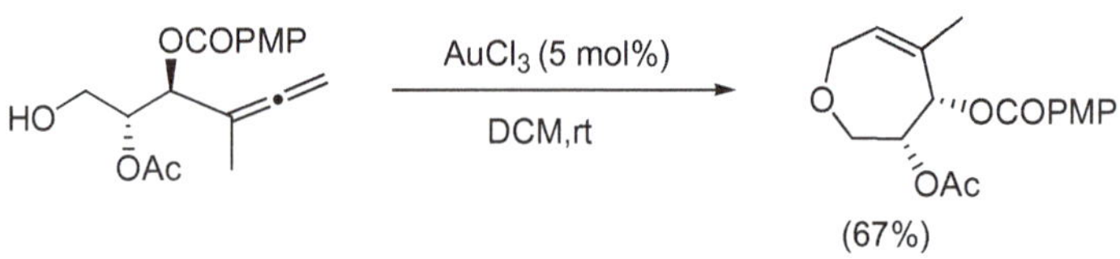

SCHEME 1.35 7-*endo*-Cyclization of γ-allenols.

This same research group[58] found that $AuCl_3$ was a selective catalyst to perform a 6-*endo* cycloisomerization from β,γ-allendiols by a chemo- and regiospecific attack of the secondary hydroxy group at the terminal allene carbon atom (Scheme 1.36).

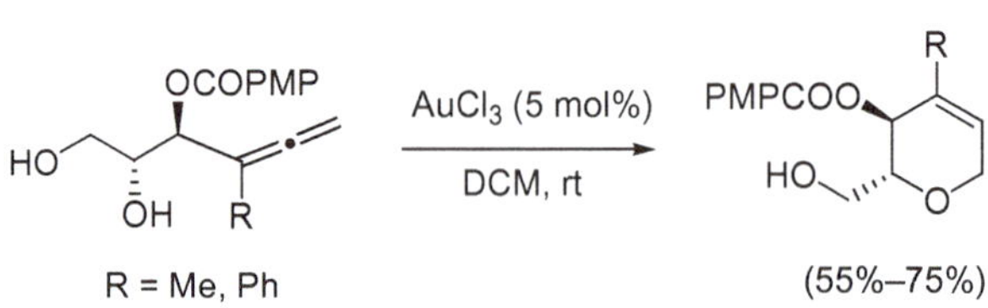

SCHEME 1.36 Chemo- and regiospecific cycloisomerization of β,γ-allendiols.

Widenhoefer et al.[54] reported the gold-catalyzed hydroalkoxylation of the axially chiral δ-allenol to form the tetrahydropyran with 20:1 *E*:*Z* selectivity (Scheme 1.37).

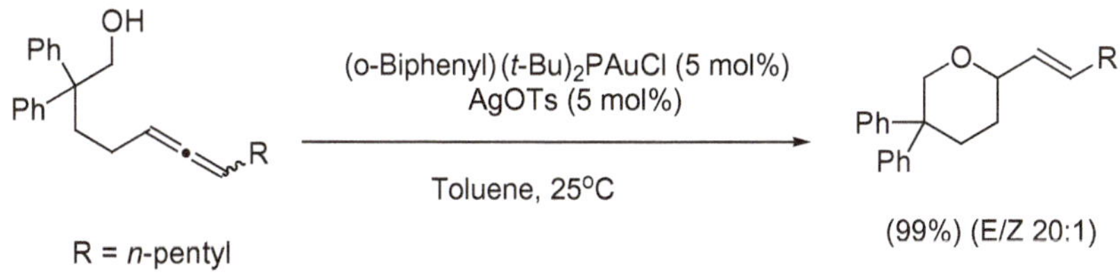

SCHEME 1.37 Cycloisomerization of a δ-allenol.

1.4 PLATINUM-CATALYZED REACTIONS

Intramolecular addition of oxygen nucleophiles to allenes catalyzed by platinum is another tool for the preparation of oxygen-heterocycles. The main advantage of platinum is the small charge of catalyst that is needed in the reaction. Alcaide, Almendros, et al.[59–61] reported some examples of the reaction of α-allenols to give preferentially dihydrofurans by 5-*endo* attack of the oxygen atom to the distant double bond (Scheme 1.38).

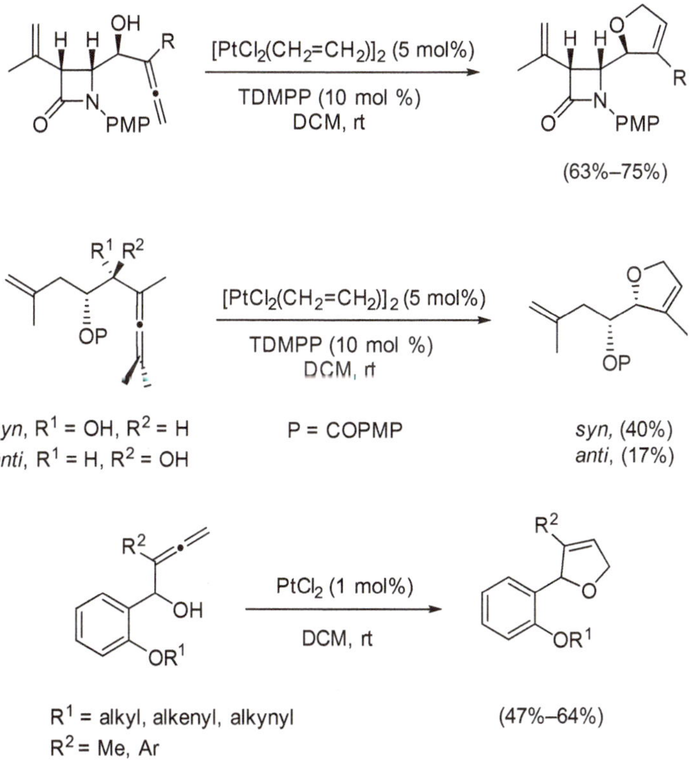

SCHEME 1.38 Platinum-catalyzed reaction of α-allenols.

The cyclization of β-allenols catalyzed by platinum has been scarcely explored. Alcaide, Almendros, et al.[57] reported the unexpected 5-*exo*-cyclization reaction of β-allenols catalyzed by $[PtCl_2(CH_2{=}CH_2)]_2$ and TDMPP [tris(2,6-dimethoxyphenyl)phosphine]. The reaction seems to proceed by the attack of the oxygen atom to the central allene carbon atom to give an intermediate dihydrofuran that aromatizes under the reaction conditions (Scheme 1.39).

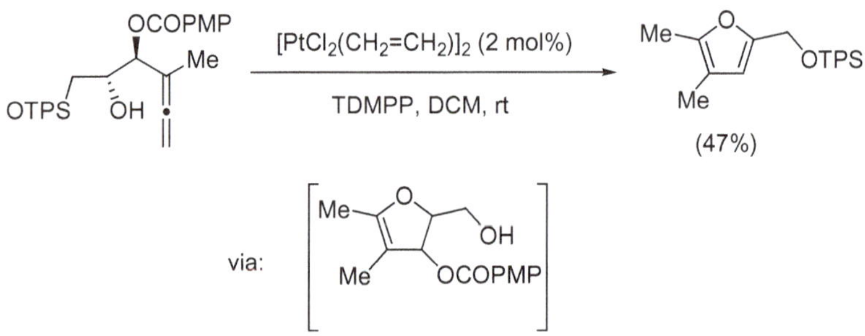

SCHEME 1.39 Platinum-catalyzed 5-*exo*-cyclization of a β-allenol.

The reaction of the analogous β,γ-allenediol,[62] obtained by removal of the tert-butyldiphenylsilyl (TPS)-protecting group, led to the chemo- and regioselective 6-*endo* cyclization of the secondary hydroxyl group at the distal double bond, with the concurrent oxidation to give the dihydropyran aldehyde, which can be trapped by a Wittig reaction (Scheme 1.40).

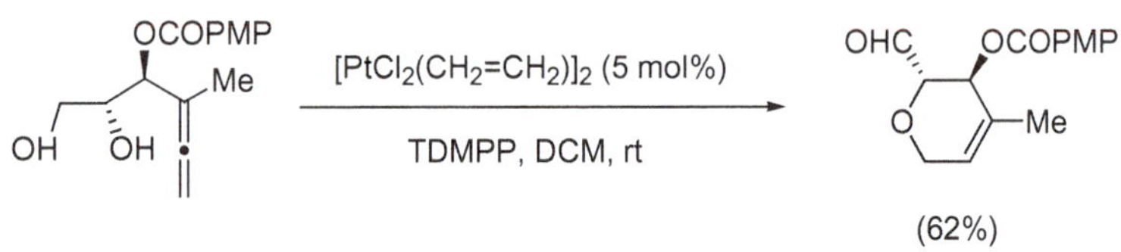

SCHEME 1.40 Platinum-catalyzed 6-*endo* cyclization of a β,γ-allenediol.

The γ-allenol with the β-hydroxy group protected[57] gave a tetrahydrooxepine. Its formation can be explained by the 7-*endo* cyclization of the hydroxyl group to the distal double bond of the allene with concomitant isomerization of the newly formed endocyclic double bond (Scheme 1.41).

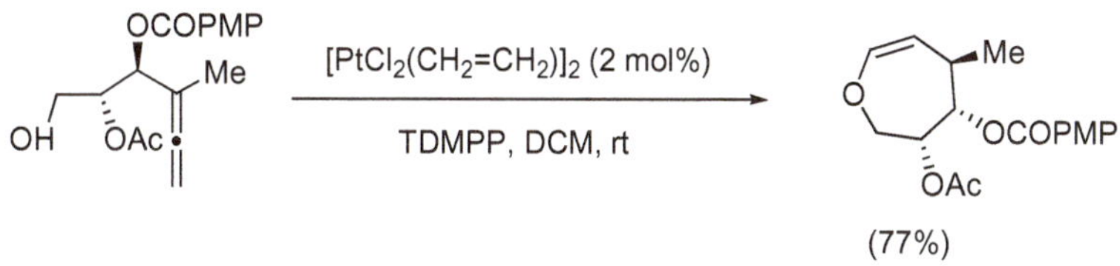

SCHEME 1.41 Platinum-catalyzed 7-*endo* cyclization of a γ-allenol.

In contrast, the reaction of a β-lactam γ-allenol[53,63] gave the 5-*exo* cyclization to afford the tetrahydrofuran derivative in low yield (Scheme 1.42).

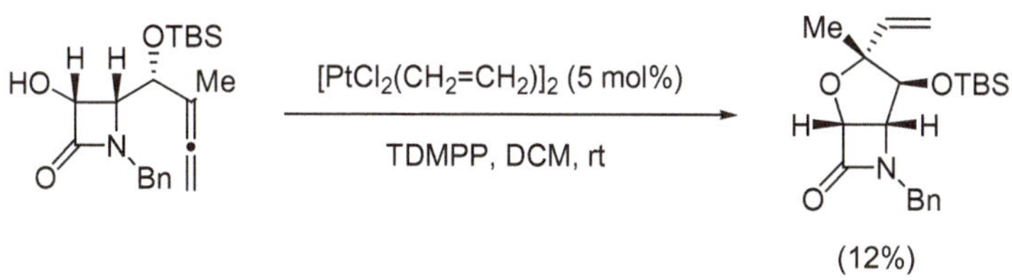

SCHEME 1.42 Platinum-catalyzed 5-*exo* cyclization of a γ-allenol.

Similarly, a γ,δ-allenic diol[62] reacted through the γ-hydroxy group to give the 5-*exo* cyclization product to obtain the corresponding tetrahydrofuran, albeit in low yield (Scheme 1.43).

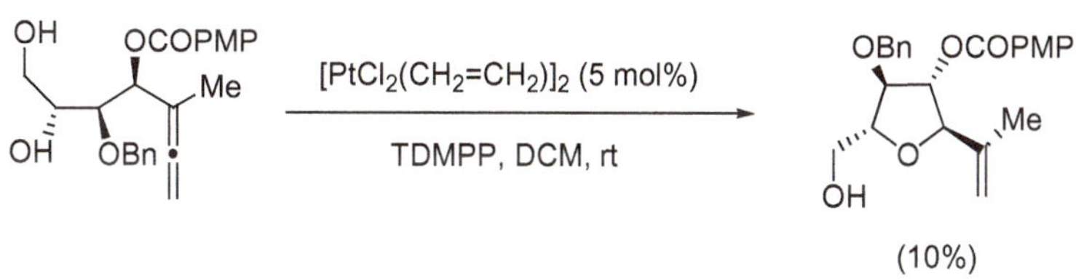

SCHEME 1.43 Platinum-catalyzed 5-*exo* cyclization of a γ,δ-allenic diol.

Widenhoefer et al. reported[64] the 6-*endo* cyclization of simple γ-allenols in the presence of platinum catalysts in moderate yields (Scheme 1.44). Altogether, these examples have shown the low reactivity of γ-allenols with platinum catalysts.

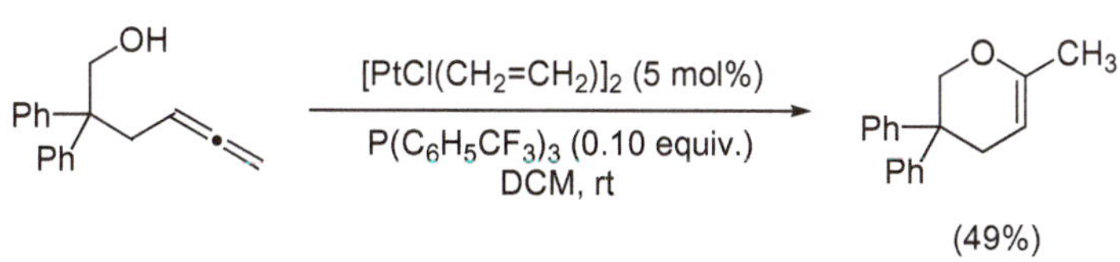

SCHEME 1.44 Platinum-catalyzed 6-*endo* cyclization of a γ-allenol.

1.5 PALLADIUM-CATALYZED REACTIONS

The palladium-catalyzed cyclizations from allenols mainly occur in a domino fashion in which the initial cyclization is followed by an additional coupling allowing to synthesize highly functionalized oxacycles.

Tsuji et al.[65] reported in 1985 the first coupling-oxycyclization sequence of α-allenols with aryliodides to give (1-arylvinyl)oxiranes (Scheme 1.45). A plausible mechanism implies the arylpalladation of the allenic group and the subsequent nucleophilic attack of the hydroxyl group to the resulting allylpalladium intermediate.

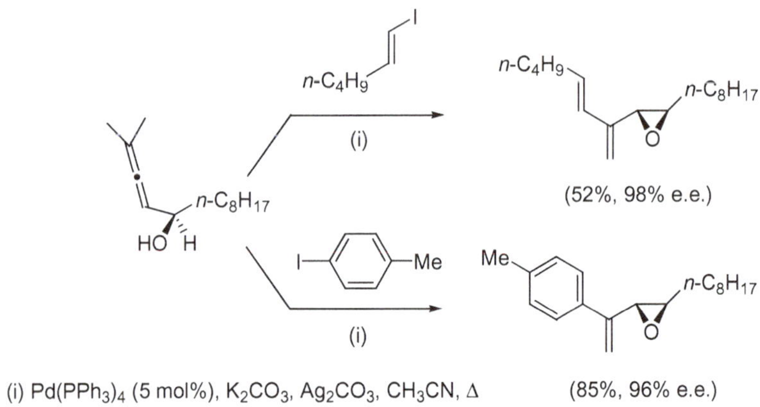

SCHEME 1.45 Palladium-catalyzed coupling-oxycyclization of α-allenols.

Ma et al.[66,67] reported the same reactivity in absence of silver salts and DMF as the solvent. Under these palladium(0) catalytic conditions, α-allenols underwent reaction to afford the *trans*-(1-arylvinyl)oxiranes (Scheme 1.46).

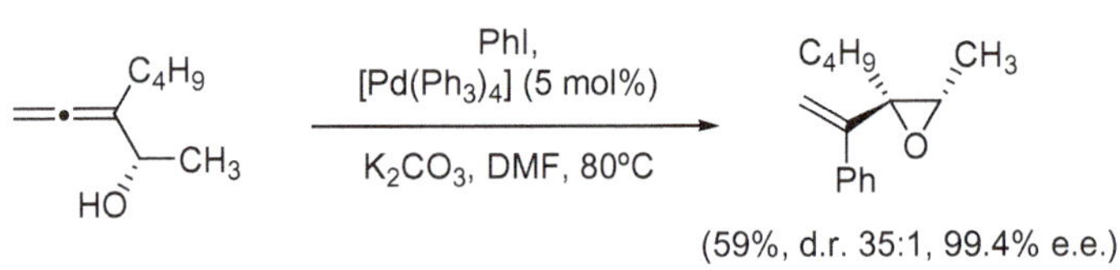

SCHEME 1.46 Palladium(0) coupling-oxycyclization of α-allenols.

Ma et al.[68] first disclosed the synthesis of dihydrofurans from the palladium (II) catalyzed cyclizative coupling of α-allenols with allylic halides (Scheme 1.47).

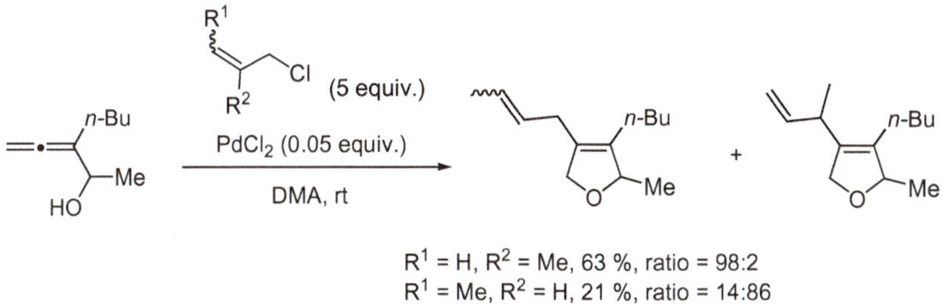

SCHEME 1.47 Palladium (II)-catalyzed coupling-cyclization of α-allenols.

Alcaide, Almendros, et al.[59,69] have generously exploited the oxycyclization-Heck addition reactivity using as substrates β-lactam-tethered α-allenols. The palladium (II)-catalyzed cyclization led to the corresponding dihydrofurans through a favored 5-*endo* cyclization mode (Scheme 1.48).

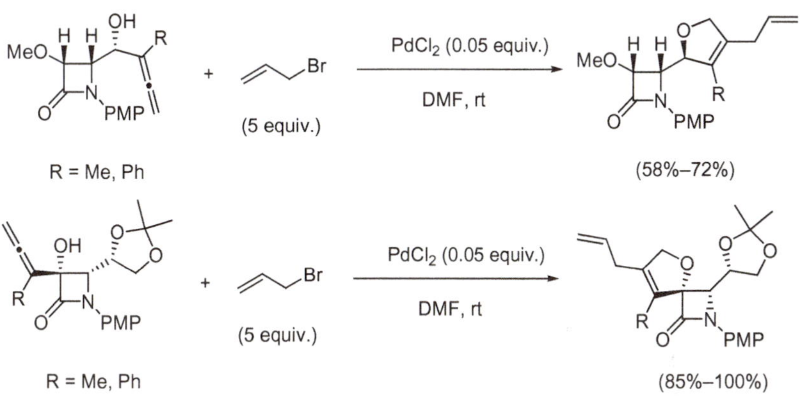

SCHEME 1.48 Palladium (II)-catalyzed oxycyclization-Heck addition of α-allenols.

The same research group[70] has also used Baylis-Hillman acetates instead of allyl halides to obtain spirocyclic lactams by a 5-*endo* cyclization-coupling reaction (Scheme 1.49).

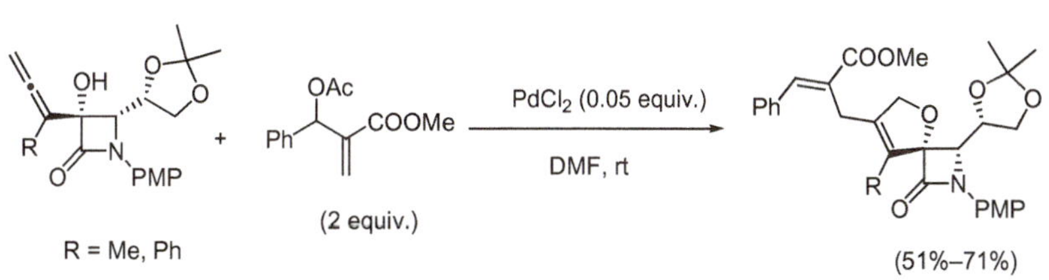

SCHEME 1.49 Palladium (II)-catalyzed oxycyclization-coupling of α-allenols with Baylis-Hillman acetates.

Ma et al.[71] reported the synthesis of dihydropyrans from β-allenols and allylic halides through a 6-*endo* cyclization-coupling reaction catalyzed by palladium (II) (Scheme 1.50).

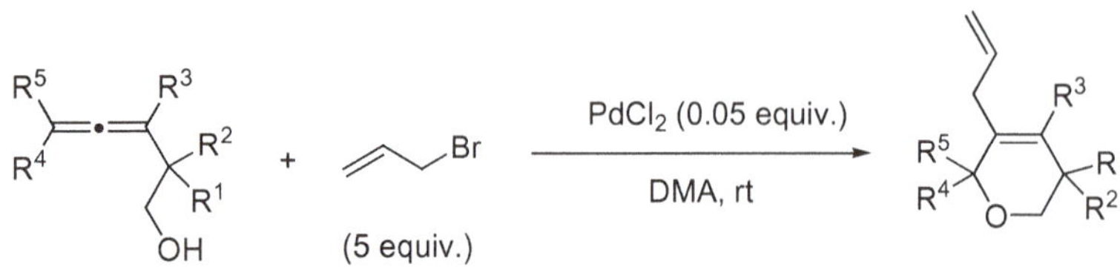

R^1 = H, Me, Ph, *n*-Bu
R^2 = H, Me
R^3 = H, Me, Bn, *n*-Bu
R^4 = H, *n*-Hex
R^5 = H, $(CH_2)_5$

(34%–87%)

SCHEME 1.50 Palladium (II)-catalyzed cyclization-coupling of β-allenols.

Recently, the Ma's protocol, using DMF as solvent instead DMA, has been used by Bäckvall et al.[72] to synthesize a bicyclic compound (Scheme 1.51).

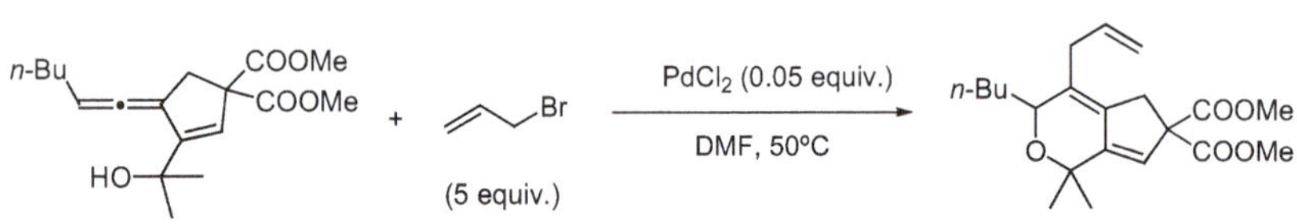

(82%)

SCHEME 1.51 Palladium (II)-catalyzed cyclization-coupling of a β-allenol linked monocycle.

Walkup et al.[73] first exploited this reactivity within γ-allenols. The treatment of γ-allenols with aryl or alkenyl halides catalyzed by $(Ph_3P)_4Pd$ and base afforded vinyl-tetrahydrofurans by a cyclization-coupling protocol but with a very poor stereoselectivity to the *trans* product (Scheme 1.52).

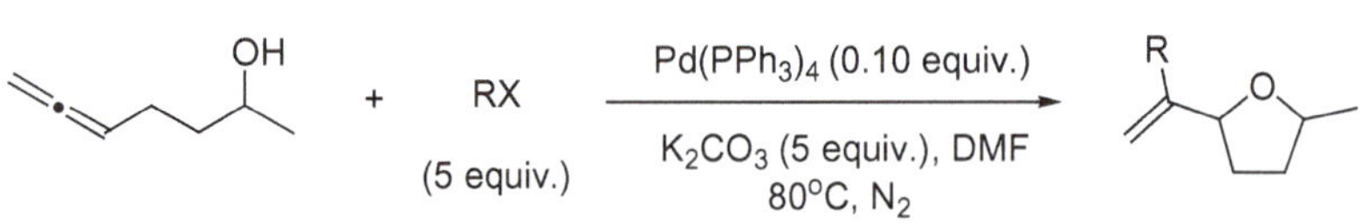

R = aryl, alkenyl
X = Br, I

(14%–76%)

SCHEME 1.52 Palladium-catalyzed cyclization-coupling of γ-allenols with aryl or alkenyl halides.

Recently, Ma et al.[74] reported the palladium(0)-catalyzed asymmetric coupling cyclization of terminal γ-allenols with aryl iodides to afford enantioenriched tetrahydrofurans (Scheme 1.53).

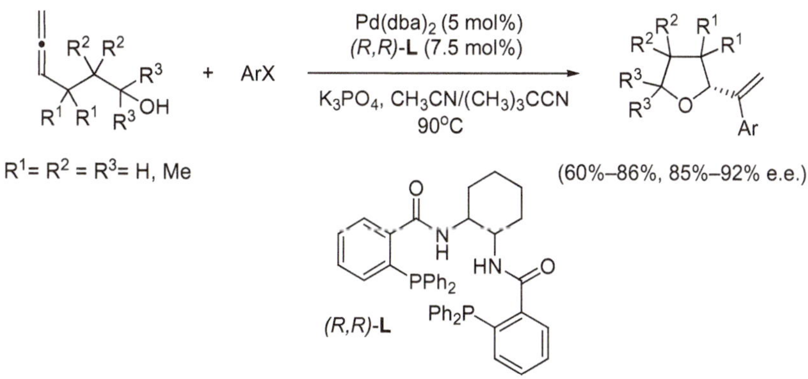

SCHEME 1.53 Palladium-catalyzed asymmetric cyclization-coupling of terminal γ-allenols with aryl iodides.

Alcaide, Almendros, et al.[53,57,63,75] reported the synthesis of allylated seven-membered heterocycles by palladium (II)-catalyzed cyclization processes through the reaction of allylic bromides with both β-lactam-tethered γ-allenols (Scheme 1.54) and acyclic γ-allenols (Scheme 1.55).

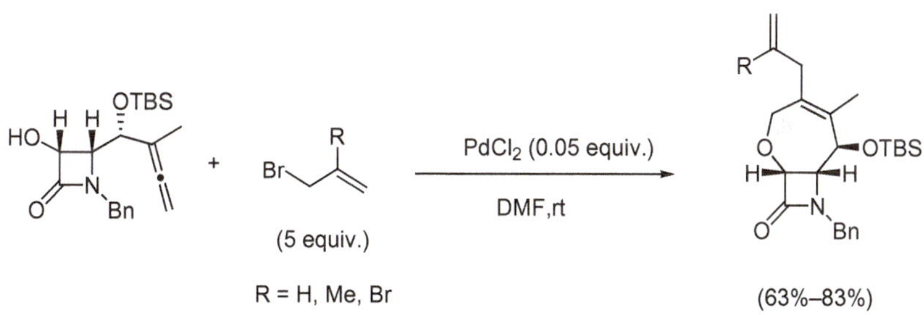

SCHEME 1.54 Palladium-catalyzed cyclization-coupling of β-lactam tethered γ-allenols.

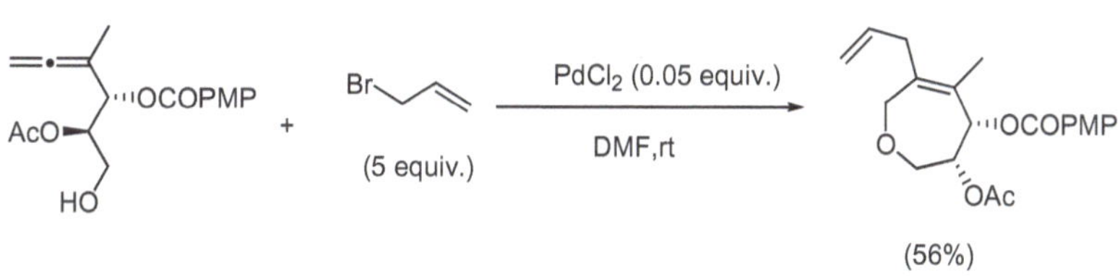

SCHEME 1.55 Synthesis of allylated tetrahydrooxepine by palladium (II)-catalyzed cyclization-coupling processes from allylic bromides and γ-allenols.

Alcaide, Almendros, et al.[62] reported the reaction of γ,δ-allendiols with allylic bromides catalyzed by palladium. The acyclic γ,δ-allendiols afforded the corresponding tetrahydrooxepines, while related γ,δ-allendiols bearing a β-lactam ring reacted through an 8-*endo* oxycyclization to yield oxocines (Scheme 1.56).

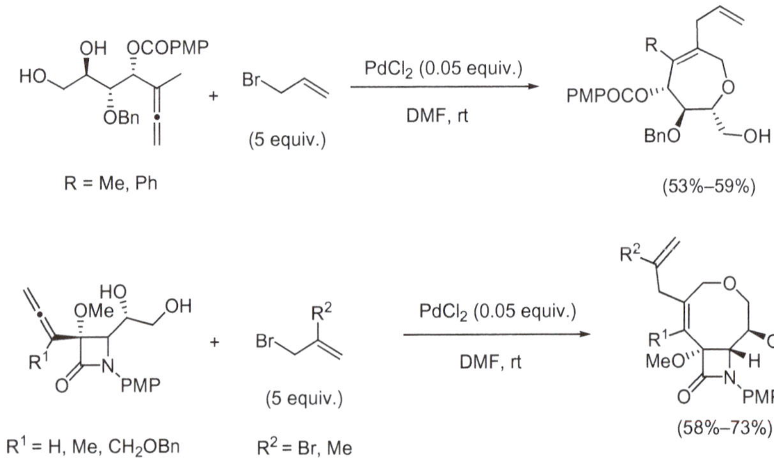

SCHEME 1.56 Palladium-catalyzed cyclization-coupling of γ,δ-allendiols with allyl bromides.

Recently, Alcaide, Almendros, et al.[76] disclosed the activity of palladium nanoparticles (PdNPs) as a convenient catalyst for the chemoselective cycloisomerization reaction of various α-allenol derivatives in water under ligandless conditions. The heterogeneous catalyst also offers the advantages of recyclability, ease removal, and high turnover frequency that the homogeneous catalyst (Scheme 1.57).

1.6 LANTHANUM-CATALYZED REACTIONS

Organolanthanide complexes are known to be highly active hydrofunctionalization catalysts. Their versatility in hydroamination and hydrophosphination raises the intriguing question of whether or not these efficient catalysts could mediate hydroalkoxylation. In this way, Marks et al.[77] reported the effective and selective intramolecular HO/cyclization of allenyl alcohols mediated by homogeneous lanthanide catalysts (Scheme 1.58). Besides, the mechanism of precious metals did not apply for this catalyst. The lanthanum-based mechanism invokes terminal C = C insertion into the Ln–O bond and η^3-allyl lanthanide complex formation.

Alcaide, Almendros, et al.[63,75,78] described regiocontrolled lanthanum-catalyzed heterocyclization reactions of γ-allenol derivatives leading to a variety of fused enantiopure oxacycles. The use of $La[N(SiMe_3)_2]_3$ usually

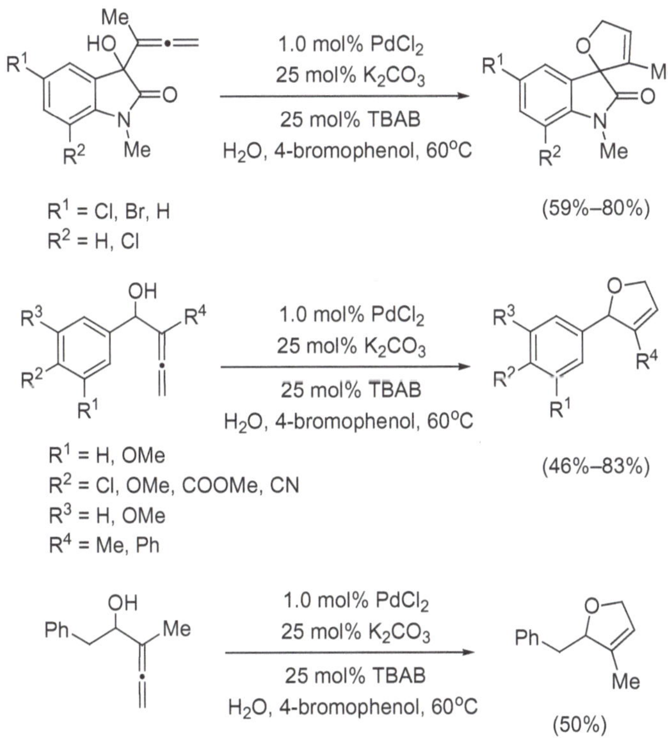

SCHEME 1.57 Synthesis of 2,5-dihydrofurans through PdNPs-catalyzed oxycyclization reaction of α-allenols.

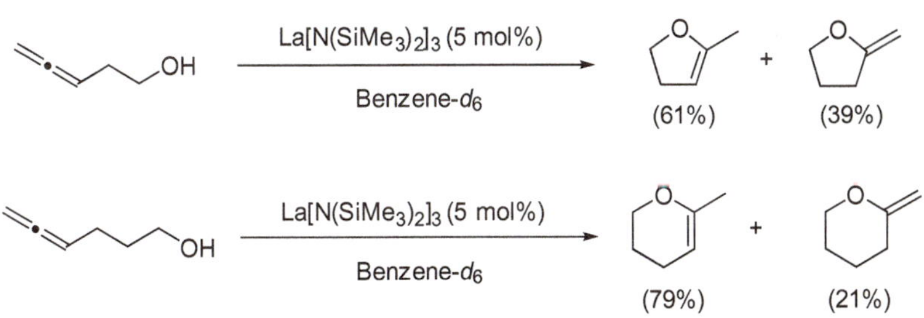

SCHEME 1.58 Lanthanum-catalyzed cyclization of allenyl alcohols.

favors the formation of dihydropyrans by a 6-*endo* oxycyclization. In addition, they observed that the presence of a methoxymethyl-protecting group not only masks a hydroxy functionality, but also exerts directing effects as a controlling unit in a regioselectivity reversal (7-*endo* vs 6-*endo* cyclization). Besides, the regioselectivity of the La-catalyzed cycloetherification can be tuned (5-*exo* vs 7-*endo*) simply through subtle variation in the substitution pattern of the allene component (Ph vs Me) (Scheme 1.59).

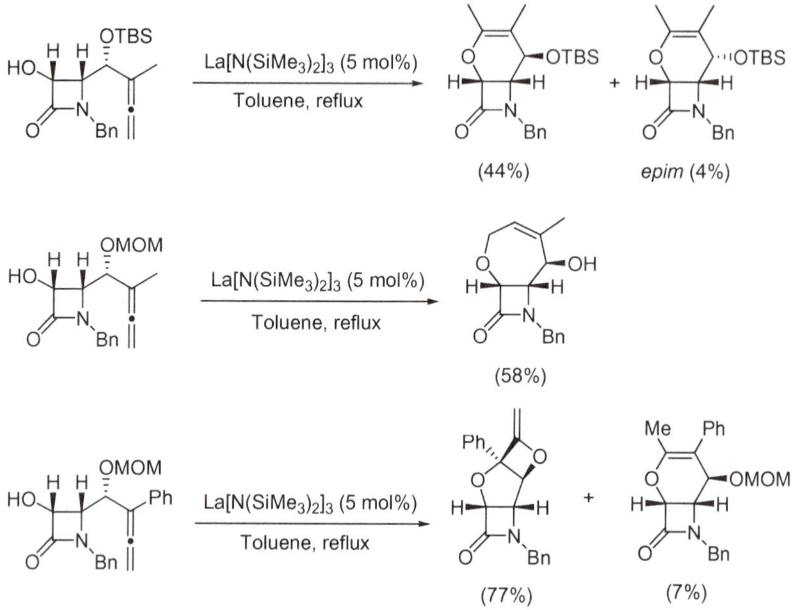

SCHEME 1.59 Lanthanum-catalyzed cyclization of γ-allenyl alcohols.

The same research group[58] reported the preparation of a furan derivative through the reaction of a β,γ-allenic diol by using the lanthanide–amide-catalyzed protocol. This outcome could be explained through a selective 5-*exo* cyclization by the attack of the secondary hydroxy group at the central allene carbon atom to give the nonisolable dihydrofuran, which under the reaction conditions suffers aromatization to form the furan. Interestingly, the reaction of a γ,δ-allenic diol under identical conditions to those given earlier could proceed smoothly with complete product selectivity to give a tetrahydrofuran through 5-*exo* cycloisomerization of the secondary hydroxy group toward the internal allene carbon atom (Scheme 1.60).

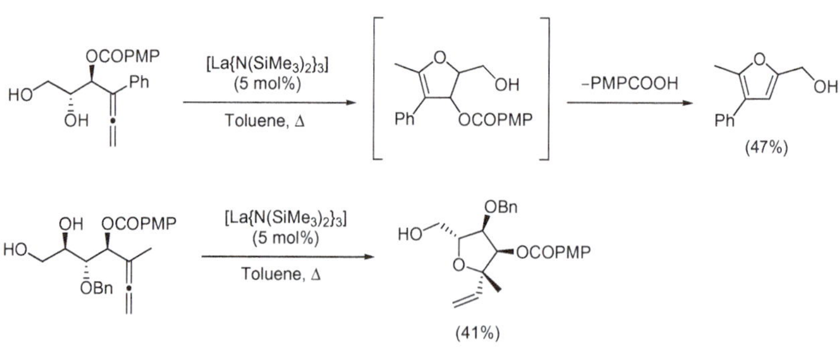

SCHEME 1.60 Lanthanum-catalyzed cyclization of β,γ- and γ,δ-allenic diols.

1.7 GALLIUM-CATALYZED REACTIONS

Alcaide, Almendros, et al.[79] explored a gallium-catalyzed coupling–cyclization between indolinone-tethered α-allenols and phenols toward the preparation of dihydrobenzofuran-linked oxindoles. It may be presumed that the gallium salt merges allene and phenol groups together via a bidentate mode, with the hydroxyl group in the allenol being eliminated (Scheme 1.61).

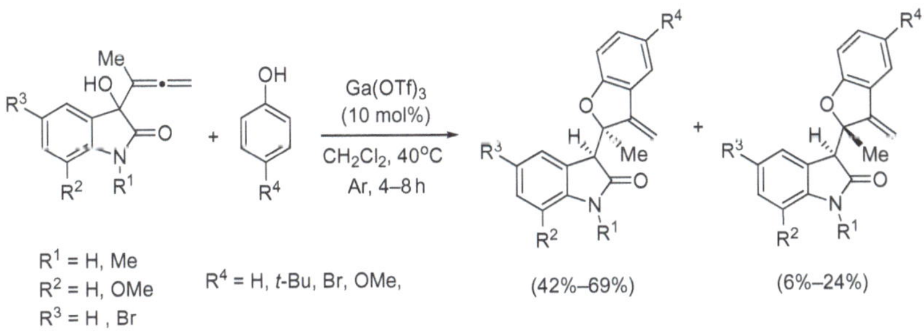

SCHEME 1.61 Gallium-catalyzed coupling-oxycyclization of α-allenols and phenols.

1.8 CONCLUSION

Metal-mediated oxygen nucleophile addition to allenols is a very powerful tool for the synthesis of functionalized oxacycles. The reaction has been explored using a wide variety of metals, and gold is one of the most effective. Although silver, platinum, gallium, or lanthanum have been less studied, interesting reactivities or regioselectivities have been observed, showing the great potential and versatility of this methodology. This chapter has highlighted the most important reactions of allenols with all these metals.

REFERENCES

1. Larrosa, I.; Romea, P.; Urpí, F. Synthesis of Six-Membered Oxygenated Heterocycles Through Carbon-Oxygen Bond-Forming Reactions. *Tetrahedron* **2008,** *64*, 2683–2723.
2. Wolfe, J. P.; Hay, M. B. Recent Advances in the Stereoselective Synthesis of Tetrahydrofurans. *Tetrahedron* **2007,** *63*, 261–290.
3. Clarke, P. A.; Santos, S. Strategies for the Formation of Tetrahydropyran Rings in the Synthesis of Natural Products. *Eur. J. Org. Chem.* **2006,** 2045–2053.
4. Snyder, N. L.; Haines, H. M.; Peczuh, M. W. Recent Developments in the Synthesis of Oxepines. *Tetrahedron* **2006,** *62*, 9301–9320.
5. Hou, X. L.; Yang, Z.; Yeung, K. S.; Wong, H. N. C. In *Progress in Heterocyclic Chemistry;* Gribble, G. W., Joule, J. A., Eds.; Vol. 17; Elsevier: Oxford, 2005; pp 142–171.
6. Eicher, T.; Hauptmann, J. S., Eds. *The Chemistry of Heterocycles: Structure, Reactions, Syntheses and Applications.;* Wiley-VCH: Weinheim, 2003.
7. Álvarez-Corral, M.; Muñoz-Dorado, M.; Rodríguez-García, I. Silver-Mediated Synthesis of Heterocycles. *Chem. Rev.* **2008,** *108*, 3174–3198.

8. Ma, S. Some Typical Advances in the Synthetic Applications of Allenes. *Chem. Rev.* **2005,** *105*, 2829–2872.
9. Krause, N.; Hashmi, A. S. K., Eds. *Modern Allene Chemistry;* Wiley-VCH: Weinheim, 2004.
10. Zimmer, R.; Dinesh, C. U.; Nandanan, E.; Khan, F. A. Palladium-Catalyzed Reactions of Allenes. *Chem. Rev.* **2000,** *100*, 3067–3126.
11. Krause, N.; Winter, C. Gold-Catalyzed Nucleophilic Cyclization of Functionalized Allenes: A Powerful Access to Carbo- and Heterocycles. *Chem. Rev.* **2011,** *111*, 1994–2009.
12. Muñoz, M. P. Silver and Platinum-Catalysed Addition of O–H and N–H Bonds to Allenes. *Chem. Soc. Rev* **2014,** *43*, 3164–3183.
13. Le Bras, J.; Muzart, J. Palladium-Catalysed Inter- and Intramolecular Formation of C–O Bonds From Allenes. *Chem. Soc. Rev.* **2014,** *43*, 3003–3040.
14. Olsson, L.-I.; Claesson, A. Synthesis of 2,5-Dihydrofurans and 5,6-Dihydro-2*H*-Pyrans by Silver (I)-Catalyzed Cyclization of Allenic Alcohols. *Synthesis* **1979,** 743–745.
15. Alcaide, B.; Almendros, P.; Rodriguez-Acebes, R. Efficient Entry to Diversely Functionalized Spirocyclic Oxindoles From Isatins Through Carbonyl-Addition/Cyclization Reaction Sequences. *J. Org. Chem.* **2006,** *71*, 2346–2351.
16. Alcaide, B.; Almendros, P.; Martínez del Campo, T.; Rodriguez-Acebes, R. Diversity-Oriented Preparation of Enantiopure Spirocyclic 2-Azetidinones From α-Oxo-β-Lactams Through Barbier-Type Reactions Followed by Metal-Catalyzed Cyclizations. *Adv. Synth. Catal.* **2007,** *349*, 749–758.
17. Marshall, J. A.; Wang, X.-j Synthesis of Enantioenriched α-Hydroxy-α-Allenylacetic Acids by [2,3] Wittig Rearrangement of α-(Propargyloxy)Acetates. *J. Org. Chem.* **1991,** *56*, 4913–4918.
18. Marshall, J. A.; Yu, R. H.; Perkins, J. F. Diastereo- and Enantioselective Synthesis of Allenylcarbinols Through S_E2' Addition of Transient Nonracemic Propargylic Stannanes to Aldehydes. *J. Org. Chem.* **1995,** *60*, 5550–5555.
19. Marshall, J. A.; Sehon, C. A. Synthesis of Furans and 2,5-Dihydrofurans by Ag(I)-Catalyzed Isomerization of Allenones, Alkynyl Allylic Alcohols, and Allenylcarbinols. *J. Org. Chem.* **1995,** *60*, 5966–5968.
20. Chen, B.; Lu, Z.; Chai, G.; Fu, C.; Ma, S. An Efficient Double 1,2-Addition Reaction of 2,3-Allenoates With Allyl Magnesium Chloride. *J. Org. Chem.* **2008,** *73*, 9486–9489.
21. Kim, S.; Lee, P. H. Cyclization of Allenyne-1,6-Diols Catalyzed by Gold and Silver Salts: An Efficient Selective Synthesis of Dihydrofuran and Furan Derivatives. *Adv. Synth. Catal.* **2008,** *350*, 547–551.
22. Marshall, J. A.; Pinney, K. G. Stereoselective Synthesis of 2,5-Dihydrofurans by Sequential S_N2' Cleavage of Alkynyloxiranes and Silver(I)-Catalyzed Cyclization of the Allenylcarbinol Products. *J. Org. Chem.* **1993,** *58*, 7180–7184.
23. Aurrecoechea, J. M.; Solay, M. Stereoselective Samarium Diiodide-Promoted Intermolecular Coupling of Alkynyloxiranes With Ketones. Synthesis of 2,3-Pentadiene-1,5-Diols. *Tetrahedron Lett.* **1995,** *36*, 2501–2504.
24. Poonoth, M.; Krause, N. Stereoselective Synthesis of Conjugated Bisallenols as Precursors of Novel bis(2,5-Dihydrofuran) Derivatives. *Adv. Synth. Catal.* **2009,** *351*, 117–122.
25. VanBrunt, M. P.; Standaert, R. F. A Short Total Synthesis of (+)-Furanomycin. *Org. Lett.* **2000,** *2*, 705–708.
26. Lepage, O.; Fürstner, A. Total Synthesis of Amphidinolide X. *J. Am. Chem. Soc.* **2004,** *126*, 15970–15971.
27. Fürstner, A.; Kattnig, E.; Lepage, O. Total Syntheses of Amphidinolide X and Y. *J. Am. Chem. Soc.* **2006,** *128*, 9194–9204.

28. Larivée, A.; Unger, J. B.; Thomas, M.; Wirtz, C.; Dubost, C.; Handa, S.; Fürstner, A. The Leiodolide-B puzzle. *Angew. Chem. Int. Ed.* **2011,** *50*, 318–323.
29. Hu, Y.; Yi, R.; Wu, F.; Wan, B. Synthesis of Functionalized Oxazoles Via Silver-Catalyzed Cyclization of Propargylamides and Allenylamides. *J. Org. Chem.* **2013,** *78*, 7714–7726.
30. Arbour, J. L.; Rzepa, H. S.; White, A. J. P.; Hii, K. K. Unusual Regiodivergence in Metal-Catalysed Intramolecular Cyclisation of γ-Allenols. *Chem. Commun.* **2009,** 7125–7127.
31. Arbour, J. L.; Rzepa, H. S.; Contreras-García, J.; Adrio, L. A.; Barreiro, E. M.; Hii, K. K. Silver-Catalysed Enantioselective Addition of O – H and N – H Bonds to Allenes: A New Model for Stereoselectivity Based on Noncovalent Interactions. *Chem. Eur. J.* **2012,** *18*, 11317–11324.
32. Audin, P.; Doutheau, A.; Gore, J. Preparation d'α-Alkenyl-2 Tetrahydropyrannes par Cyclisation d'alcohols δ-Alleniques. *Tetrahedron Lett.* **1982,** *23*, 4337–4340.
33. Gallagher, T. The Cyclisation of Secondary Allenic Alcohols. *J. Chem. Soc., Chem. Commun.* **1984,** 1554–1555.
34. Hoffmann-Röder, A.; Krause, N. Gold (III) Chloride Catalyzed Cyclization of α-Hydroxyallenes to 2,5-Dihydrofurans. *Org. Lett.* **2001,** *3*, 2537–2538.
35. Krause, N.; Hoffmann-Röder, A.; Cansius, J. From Amino Acids to Dihydrofurans: Functionalized Allenes in Modern Organic Synthesis. *Synthesis* **2002,** 1759–1774.
36. Deutsch, C.; Gockel, B.; Hoffmann-Röder, A.; Krause, N. Golden Opportunities in Stereoselective Catalysis: Optimization of Chirality Transfer and Catalyst Efficiency in the Gold-Catalyzed Cycloisomerization of α-Hydroxyallenes to 2,5-Dihydrofurans. *Synlett* **2007,** 1790–1794.
37. Hyland, C. J. T.; Hegedus, L. S. Gold-Catalyzed and *N*-Iodosuccinimide-Mediated Cyclization of γ-Substituted Allenamides. *J. Org. Chem.* **2006,** *71*, 8658–8660.
38. Deutsch, C.; Lipshutz, B.; Krause, N. Small But Effective: Copper Hydride Catalyzed Synthesis of Alpha-Hydroxyallenes. *Angew. Chem. Int. Ed.* **2007,** *46*, 1650–1653.
39. Volz, F.; Krause, N. Golden Opportunities in Natural Product Synthesis: First Total Synthesis of (–)-Isocyclocapitelline and (–)-Isochrysotricine by Gold-Catalyzed Allene Cycloisomerization. *Org. Biomol. Chem.* **2007,** *5*, 1519–1521.
40. Volz, F.; Wadmam, S. H.; Hoffmann-Röder, A.; Krause, N. Gold Catalysis in Stereoselective Natural Product Synthesis: (+)-Linalool Oxide, (–)-Isocyclocapitelline, and (–)-Isochrysotricine. *Tetrahedron* **2009,** *65*, 1902–1910.
41. Gao, Z.; Li, Y.; Cooksey, J. P.; Snaddon, T. N.; Schunk, S.; Viseux, E. M. E.; McAteer, S. M.; Kocienski, P. J. A Synthesis of an Ionomycin Calcium Complex. *Angew. Chem. Int. Ed.* **2009,** *48*, 5022–5025.
42. Hashmi, A. S. K.; Blanco, M. C.; Fischer, D.; Bats, J. W. Gold Catalysis: Evidence for the In-Situ Reduction of Gold(III) During the Cyclization of Allenyl Carbinols. *Eur. J. Org. Chem.* **2006,** 1387–1389.
43. Lemière, G.; Gandon, V.; Agenet, N.; Goddard, J.-P.; de Kozak, A.; Aubert, C.; Fensterbank, L.; Malacria, M. Gold(I)- and Gold(III)-Catalyzed Cycloisomerization of Allenynes: a Remarkable Halide Effect. *Angew. Chem. Int. Ed.* **2006,** *45*, 7596–7599.
44. Alcaide, B.; Almendros, P.; Martínez del Campo, T. Chemodivergence in Alkene/Allene Cycloetherification of Enallenols: Iron Versus Noble Metal Catalysis. *Chem. Eur. J.* **2008,** *14*, 7756–7759.
45. Kim, S.; Lee, P. H. Cyclization of Allenyne-1,6-Diols Catalyzed by Gold and Silver Salts: An Efficient Selective Synthesis of Dihydrofuran and Furan Derivatives. *Adv. Synth. Catal.* **2008,** *350*, 547–551.

46. Aksin, Ö.; Krause, N. Gold-Catalyzed Synthesis of 2,5-Dihydrofurans in Ionic Liquids. *Adv. Synth. Catal.* **2008,** *350,* 1106–1112.
47. Winter, C.; Krause, N. Towards Sustainable Homogeneous Gold Catalysis: Cycloisomerization of Functionalized Allenes in Water. *Green Chem.* **2009,** 1309–1312.
48. Alcaide, B.; Almendros, P.; Martínez del Campo, T.; Fernández, I. Fascinating Reactivity in Gold Catalysis: Synthesis of Oxetenes Through Rare 4-*exo*-Dig Allene Cyclization and Infrequent β-Hydride Elimination. *Chem. Commun.* **2011,** *47,* 9054–9056.
49. Gockel, B.; Krause, N. Golden Times for Allenes: Gold-Catalyzed Cycloisomerization of β-Hydroxyallenes to Dihydropyrans. *Org. Lett.* **2006,** *8,* 4485–4488.
50. Gockel, B.; Krause, N. Synthesis of Bicyclic Ethers by a Gold/Palladium/Gold-Catalyzed Cyclization/Cross Coupling Sequence. *Eur. J. Org. Chem.* **2010,** 311–316.
51. Poonoth, M.; Krause, N. Stereoselective Synthesis of Conjugated Bisallenols as Precursors of Novel bis(2,5-Dihydrofuran) Derivatives. *Adv. Synth. Catal.* **2009,** *351,* 117–122.
52. Sawama, Y.; Sawama, Y.; Krause, N. First Total Synthesis of (*R*,*R*,*R*)- and (3*R*,5*S*,9*R*)-Bejarol by Gold-Catalyzed Allene Cycloisomerization and Determination of Absolute Configuration of the Natural Product. *Org. Biomol. Chem.* **2008,** *6,* 3573–3579.
53. Alcaide, B.; Almendros, P.; Martínez del Campo, T. Metal-Catalyzed Regiodivergent Cyclization of γ-Allenols: Tetrahydrofurans Versus Oxepanes. *Angew. Chem. Int. Ed.* **2007,** *46,* 6684–6687.
54. Zhang, Z.; Liu, C.; Zinder, R. E.; Han, X.; Qian, H.; Widenhoefer, R. A. Highly Active Au (I) Catalyst for the Intramolecular *exo*-Hydrofunctionalization of Allenes With Carbon, Nitrogen, and Oxygen Nucleophiles. *J. Am. Chem. Soc.* **2006,** *128,* 9066–9073.
55. Zhang, Z.; Widenhoefer, R. A. Gold(I)-Catalyzed Intramolecular Enantioselective Hydroalkoxylation of Allenes. *Angew. Chem. Int. Ed.* **2007,** *46,* 283–285.
56. Hamilton, G. L.; Kang, E. J.; Mba, M.; Toste, F. D. A Powerful Chiral Counterion Strategy for Asymmetric Transition Metal Catalysis. *Science* **2007,** *317,* 496–499.
57. Alcaide, B.; Almendros, P.; Martínez del Campo, T.; Soriano, E.; Marco-Contelles, J. L. Metal-Catalyzed Cyclization of β- and γ-Allenols Derived From D-Glyceraldehyde—Synthesis of Enantiopure Dihydropyrans and Tetrahydrooxepines: An Experimental and Theoretical Study. *Chem. Eur. J.* **2009,** *15,* 9127–9138.
58. Alcaide, B.; Almendros, P.; Carrascosa, R.; Martínez del Campo, T. Metal-Catalyzed Cycloetherification Reactions of β,γ- and γ,δ-Allendiols: Chemo-, Regio-, and Stereocontrol in the Synthesis of Oxacycles. *Chem. Eur. J.* **2010,** *16,* 13243–13252.
59. Alcaide, B.; Almendros, P.; Martínez del Campo, T.; Redondo, M. C.; Fernández, I. Striking Alkenol Versus Allenol Reactivity: Metal-Catalyzed Chemodifferentiating Oxycyclization of Enallenols. *Chem. Eur. J.* **2011,** *17,* 15005–15013.
60. Alcaide, B.; Almendros, P.; Martínez del Campo, T. Chemodivergence in Alkene/Allene Cycloetherification of Enallenols: Iron Versus Noble Metal Catalysis. *Chem. Eur. J.* **2008,** *14,* 7756–7759.
61. Alcaide, B.; Almendros, P.; Fernández, I.; Martínez del Campo, T.; Naranjo, T. Platinum-Catalyzed Divergent Reactivity of α-Hydroxyallenes: Synthesis of Dihydrofurans and α,β-Unsaturated Ketones. *Adv. Synth. Catal.* **2013,** *355,* 2681–2685.
62. Alcaide, B.; Almendros, P.; Carrascosa, R.; Martínez del Campo, T. Chemo- and Regioselective Palladium-Catalyzed Oxycyclization Reactions of Allendiols: Preparation of Five-, Six-, and Eight-Membered Cycles. *Chem. Eur. J.* **2009,** *15,* 2496–2499.
63. Alcaide, B.; Almendros, P.; Martínez del Campo, T.; Soriano, E.; Marco-Contelles, J. L. Regioselectivity Control in the Metal-Catalyzed O – C Functionalization of γ-Allenols, Part 1: Experimental Study. *Chem. Eur. J.* **2009,** *15,* 1901–1908.

64. Qian, H.; Han, X.; Widenhoefer, R. A. Platinum-Catalyzed Intramolecular Hydroalkoxylation of γ- and δ-Hydroxy Olefins to Form Cyclic Ethers. *J. Am. Chem. Soc.* **2004,** *126*, 9536–9537.
65. Shimizu, I.; Sugiura, T.; Tsuji, J. Facile Synthesis of Beta.- Aryl- or. Beta.- Alkenyl-. Beta.- Methyl-. Alpha., Beta.-Unsaturated Carbonyl Compounds by Palladium Catalyzed Reaction of 1,2-Dien-4-Ols With Aryl or Alkenyl Halides. *J. Org. Chem.* **1985,** *50*, 537–539.
66. Ma, S.; Zhao, S. Pd(0)-Catalyzed Insertion-Cyclization Reaction of 2,3-Allenols With Aryl or Alkenyl Halides. Diastereoselective Synthesis of Highly Optically Active Trans-2,3-Disubstituted Vinylic Oxiranes. *J. Am. Chem. Soc.* **1999,** *121*, 7943–7944.
67. Xu, D.; Xu, Y.; Li, L.; Ma, S. Efficient Preparation of Highly Optically Active (*S*)-(−)-2,3-Allenols and (*R*)-(+)-2,3-Allenyl Acetates by a Clean Novozym-435-Catalyzed Enzymatic Separation of Racemic 2,3-Allenols. *Chem. Eur. J* **2002,** *8*, 5012–5018.
68. Ma, S.; Gao, W. Efficient Synthesis of 4-(2′-Alkenyl)-2,5-Dihydrofurans via $PdCl_2$-Catalyzed Coupling–Cyclization Reaction of 2,3-Allenols With Allylic Halides. *Tetrahedron Lett.* **2000,** *41*, 8933–8936.
69. Alcaide, B.; Almendros, P.; Martínez del Campo, T.; Rodríguez-Acebes, R. Metal-Assisted Synthesis of Enantiopure Spirocyclic β-Lactams From Azetidine-2,3-Diones. *Tetrahedron Lett.* **2004,** *45*, 6429–6431.
70. Alcaide, B.; Almendros, P.; Martínez del Campo, T.; Quirós, M. T. Synthesis of Spiroheterocycles by Palladium-Catalyzed Domino Cycloisomerization/Cross-Coupling of α-Allenols and Baylis–Hillman Acetates. *Chem. Eur. J.* **2009,** *15*, 3344–3346.
71. Ma, S.; Gao, W. Efficient Synthesis of 4-(2’-Alkenyl)-2,5-Dihydrofurans and 5,6-Dihydro-2*H*-Pyrans via the Pd-Catalyzed Cyclizative Coupling Reaction of 2,3- or 3,4-Allenols With Allylic Halides. *J. Org. Chem.* **2002,** *67*, 6104–6112.
72. Deng, Y.; Bäckvall, J.-E. Palladium-Catalyzed Oxidative Acyloxylation/Carbocyclization of Allenynes. *Angew. Chem. Int. Ed.* **2013,** *52*, 3217–3221.
73. Walkup, R. D.; Guan, L.; Mosher, D.; Kim, S. W.; Kim, Y. S. Cyclizations of γ-Hydroxyallenes Mediated by Organopalladium Halides: A Convergent Synthetic Route to Substituted 5-Alkenyl-2-Furanones and 2-Alkenyltetrahydrofurans. *Synlett* **1993,** 88–90.
74. Xie, X.; Ma, S. Palladium-Catalyzed Asymmetric Coupling Cyclization of Terminal γ-Allenols With aryl Iodides. *Chem. Commun.* **2013,** *49*, 5693–5695.
75. Alcaide, B.; Almendros, P.; Martínez del Campo, T.; Soriano, E.; Marco-Contelles, J. L. Regioselectivity Control in the Metal-Catalyzed Functionalization of γ-Allenols, Part 2: Theoretical Study. *Chem. Eur. J.* **2009,** *15*, 1909–1928.
76. Alcaide, B.; Almendros, P.; González, A. M.; Luna, A.; Martínez-Ramírez, S. Palladium Nanoparticles in Water: A Reusable Catalytic System for the Cycloetherification or Benzannulation of α-Allenols. *Adv. Synth. Catal.* **2016,** *358*, 2000–2006.
77. Yu, X.; Seo, S.-Y.; Marks, T. J. Effective, Selective Hydroalkoxylation/Cyclization of Alkynyl and Allenyl Alcohols Mediated by Lanthanide Catalysts. *J. Am. Chem. Soc.* **2007,** *129*, 7244–7245.
78. Alcaide, B.; Almendros, P.; Martínez del Campo, T.; Soriano, E.; Marco-Contelles, J. L. Heterocyclization of Allenes Catalyzed by Late Transition Metals: Mechanisms and Regioselectivity. *Top. Curr. Chem.* **2011,** *302*, 183–224.
79. Alcaide, B.; Almendros, P.; Herrera, F.; Luna, A.; de Orbe, M. E.; Torres, M. R. Gallium-Catalyzed Domino Arylation/Oxycyclization of Allenes With Phenols. *J. Org. Chem.* **2015,** *80*, 4157–4163.

Chapter 2

Intramolecular Diamination of Alkenes

Kilian Muñiz
Institute of Chemical Research of Catalonia (ICIQ), Tarragona, Spain and ICREA, Barcelona, Spain

2.1 INTRODUCTION

Vicinal diamines constitute significant core units in wide areas of chemical sciences reaching from natural products and pharmaceuticals to ligands in transition metal complexes and material sciences.[1–7] A subgroup of particular interest refers to heterocycles as one of the predominating families in organic chemistry. They may incorporate vicinal diamines as an important subunit. These entities can be commonly generated from an intramolecular diamination of alkenes, in which the two transferable nitrogen atoms are linked within a tether (Scheme 2.1). This tether can either constitute a useful functional group

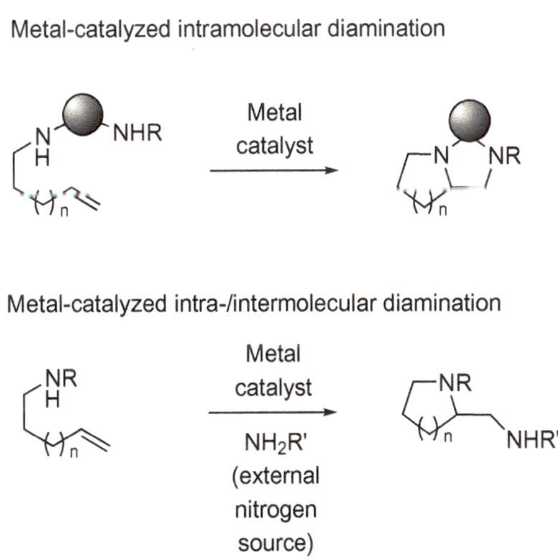

SCHEME 2.1 Conceptual pathways for heterocycle synthesis within diamination reactions of alkenes.

Advances in Transition-Metal Mediated Heterocyclic Synthesis.
DOI: https://doi.org/10.1016/B978-0-12-811651-7.00002-9

(as in the case of, for example, guanidines or ureas) or can be removed to provide free saturated nitrogen heterocycles with an exocyclic methylenylamine.

A second approach relies on the use of intramolecular aminometallation of alkenes to directly generate aminomethylenyl-substituted five- and six-membered heterocycles. This approach is prominently based on an initial step comprising the most robust aminopalladation chemistry, but has also been accomplished for alternative transition metal catalysts. It enables the formation of higher-functionalized derivatives of the ubiquitous pyrrolidine and piperidine cores.

The vicinal diamination of alkenes has matured into a versatile synthetic tool.[8] The large body of different reactions is the result of the continuous synthetic efforts by a growing number of experimental groups. A comprehensive review of the topic is not the general aim of this chapter. Instead, the reader is referred to several seminal reviews on the topic.[6–10] In the following, major reactions are highlighted that provide synthetically useful access to the class of diamine-incorporating heterocycles.

Despite a long-standing occupation with the development of transition metal–mediated diamination of alkenes that dates back to fundamental work by Barluenga, Sharpless, and Bäckvall in the 1970s,[11–16] catalysis solutions had remained elusive for this particular transformation. The major reason may be assigned to reside in the excellent ligation properties of the diamine products to transition metals. This context is illustrated by diaminations using bis and trisimido osmium species, which form stable osmaimidazolidine products.[3,13–16] While these compounds may be considered metallaheterocycles, their high cost has rendered further interest in the exploration of this particular class of compounds rather low.

2.2 METAL-CATALYZED SYNTHESIS OF HETEROCYCLES THROUGH THE OXIDATIVE INTRAMOLECULAR VICINAL DIAMINATION OF ALKENES

2.2.1 Palladium Catalysis

To overcome potential catalyst poisoning by the diamine products, starting in 2005 the Muñiz group developed intramolecular reactions using tethered nitrogen sources.[17] This chemistry employs a conventional iodine(III) reagent as the terminal oxidant (Scheme 2.2). It provides bicyclic urea

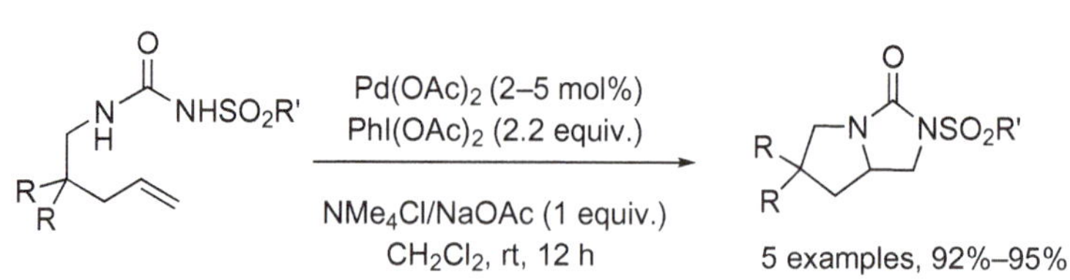

SCHEME 2.2 Palladium-catalyzed intramolecular diamination of alkenes using ureas as nitrogen sources.

molecules, which form selectively over the potentially competing isourea products.

The reaction proceeds through a sequence of base-assisted aminopalladation followed by the selective oxidation of the metal center in the resulting σ-alkyl palladium(II) intermediate. This results in the formation of the corresponding σ-alkyl palladium(IV) complex, in which the carbon–palladium bond is highly polarized and gives rise to nucleophilic displacement of the palladium(IV) by given nucleophiles.[18,19] In the present case, this role is carried out by the remaining amido group in the urea entity, and in order to provide suitable nucleophilicity, sulfonamide components are usually preferred, although the reactions are not restricted to them. The second step of alkyl-nitrogen bond formation constituted a rare case of a catalytic transformation of this type. It derives directly from the inherent power of the palladium(II/IV) catalysis to enable nucleophilic bond formation at the metalated carbon center. This topic has recently been discussed in detail.[20]

The full mechanistic details of this transformation were elucidated by NMR titration studies, deuterium labeling, and Hammett correlation studies, and a catalytic cycle was proposed (Scheme 2.3).[21]

The reaction starts with the base-assisted formation of a coordination complex between the palladium catalyst and the urea substrate. The upcoming intramolecular alkene oxidation proceeds through *syn*-aminopalladation[22] to give the intermediary six-membered palladacycle, which undergoes metal oxidation with iodine(III). As proven by isotope-labeling experiments, the following second C–N bond formation warrants an inversion of configuration, which is mechanistically accomplished through dissociation of the sulfonamide from the metal coordination sphere followed by an S_N2-type displacement of the electrophilic palladium(IV). This step closes the cyclic urea and liberates the reduced palladium(II) catalyst. Its course was further confirmed through a combined experimental and theoretical investigation by Álvarez and Muñiz.[23]

High oxidation state palladium(IV) catalysis[18,19] has proven a successful guiding concept for the successful development of further examples of this type of reactions. It is particular value within the stereochemical control in diamination reactions of internal alkenes. It was employed by Muñiz for the selective diamination of symmetric and unsymmetric stilbene derivatives (Scheme 2.4).[24] Again, sulfonamides are employed as nitrogen sources and the reaction proceeds with complete diastereoselectivity. In case of the symmetric substrates, only the C2-symmetric products are formed, while the corresponding meso-derivatives are not observed. This example again underlines the power of palladium(II/IV) catalysis for selective alkene oxidation. For the present case of heterocycle synthesis, it provides an effective entry into bisindoline products, and the potential over-oxidation to the corresponding bisindoles was not observed.

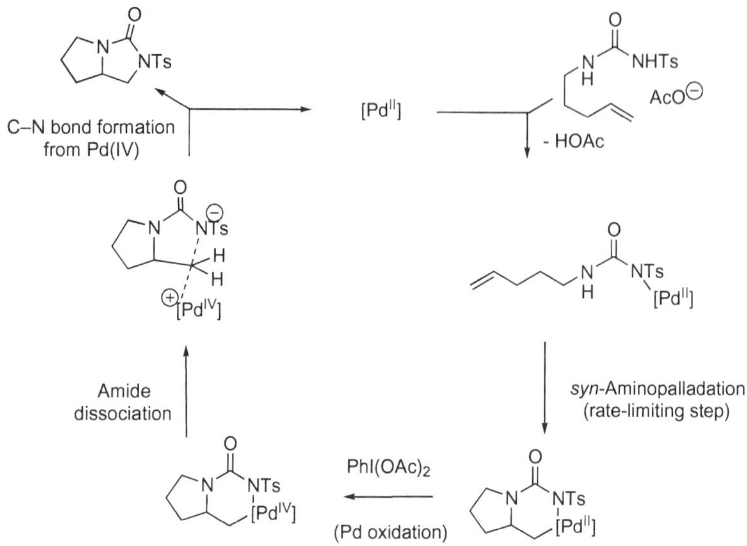

SCHEME 2.3 Mechanism of the palladium(II/IV)-catalyzed urea-based intramolecular diamination of alkenes.

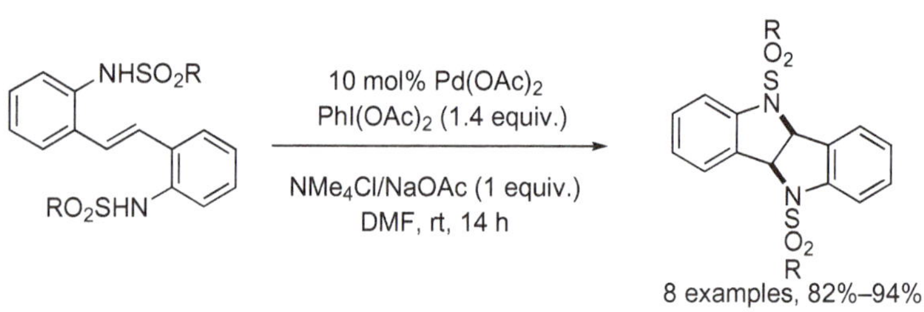

SCHEME 2.4 Synthesis of bisindolines by palladium(II/IV)-catalyzed diamination of stilbenes.

Since the palladium high oxidation state catalysis within the iodine(III) reoxidation system displayed some important substrate limitations, subsequent exploration focused on the development of reaction conditions that included copper(II) salts as terminal oxidants. Their stoichiometric use is explained by the fact that the aerobic conditions of the well-established Wacker oxidation require an overall acidic medium,[25,26] under which the second C–N bond formation is usually prevented. Instead, aminoxygenation products are obtained selectively as demonstrated for an aerobic platinum-catalyzed reaction.[27]

Overall, the copper(II)-based catalyst systems greatly enhance the scope of the intramolecular diamination reaction as demonstrated for the case of ureas as nitrogen sources.[28] For example, a palladium/copper(II) bromide system works with yields comparable to the palladium/iodine(III) reaction. The new conditions were found to be general. In fact, they are not restricted to the formation of five-membered heterocyclic products, but can also be employed to generate the corresponding six-membered ring annulation products (Scheme 2.5). They were also employed in the synthesis of diamines from higher-functionalized spacers given more advanced heterocycles such as pyrazines.[8]

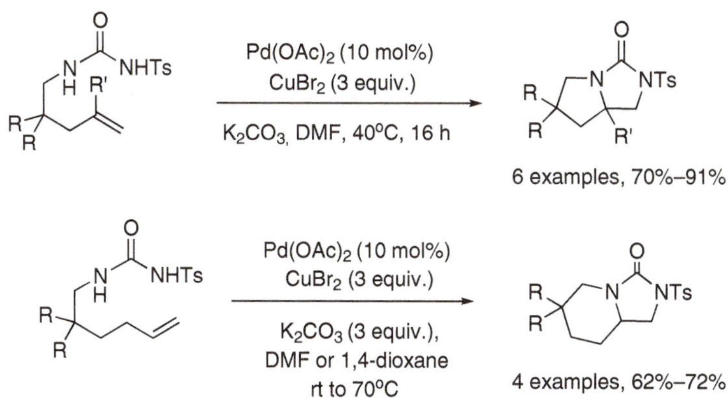

SCHEME 2.5 Diamination of urea using the palladium/copper bromide system.

Moreover, internal alkenes also perform well under the present conditions and provide the corresponding cyclic ureas as major products (Scheme 2.6), although the formation of the corresponding isourea derivatives as minor by-products (5%–31%, not depicted) cannot be entirely suppressed.

Based on the final *trans*-stereochemistry of the product, the reaction must follow pathways that are mechanistically different from the usual ones of the terminal alkenes. Based on stoichiometric control experiments, the involvement of halogenated intermediate was postulated. After the initial aminopalladation, an interaction between the palladium(II) and the copper bromide oxidant initiates a transient oxidation,[25,29] which weakens the α-carbon bond to palladium and promotes a nucleophilic bromination at this position. This additional S_N2 step provides an explanation for the overall stereochemical outcome. Assisted by copper salt, the subsequent nucleophilic substitution by

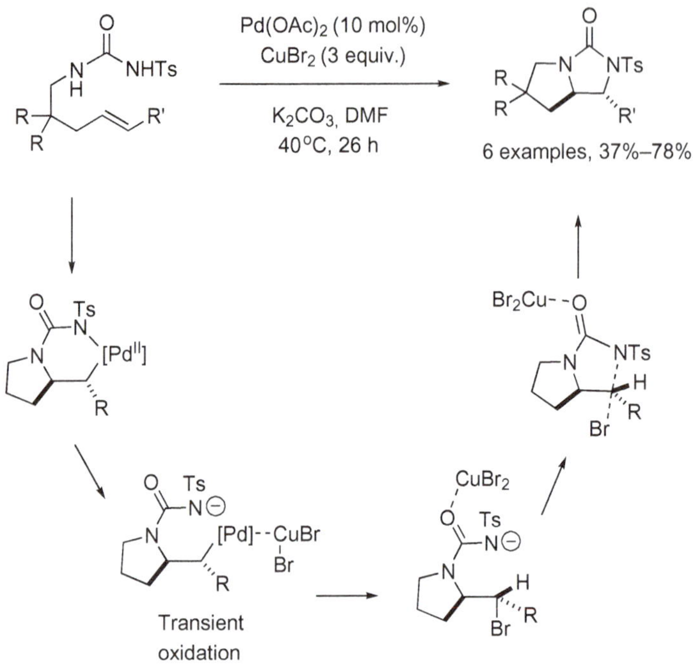

SCHEME 2.6 Palladium-catalyzed intramolecular diamination of internal alkenes with copper oxidants.

the urea amide leads to additional inversion of configuration and to an overall *trans*-stereochemistry in the cyclized diamination products.

Additional reactions within the palladium/copper manifold include sulfamides and, in particularly, guanidines. Given the importance of cyclic guanidine alkaloids in natural products, the direct implementation of this heterocyclic unit through a diamination of alkenes represents a striking synthetic alternative to the common construction from up to three individual C–N bond–forming events. Indeed, such an oxidative cycloguanidination proceeds smoothly with copper(II) chloride as terminal oxidant (Scheme 2.7).[30] It employs standard Boc- and Cbz-protection at the guanidine units, which can be readily removed at a later stage. The diamination reaction proceeds under unprecedentedly mild conditions and tolerates both five- and six-membered ring annulations. Selective deuterium labeling at the alkene terminus provides the mechanistic information that the cycloguanidination reactions also proceed involving *syn*-aminopalladation and a direct C–N bond formation within a transition state from transient oxidation.

As to another important substrate class, acrylates and their derivatives are usually beyond the scope regarding the palladium/iodine(III) systems.

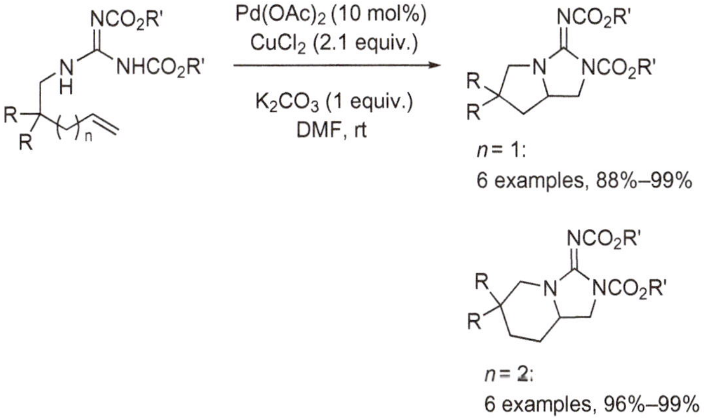

SCHEME 2.7 Cycloguanidination through palladium-catalyzed intramolecular diamination of alkenes.

In contrast, they can be conveniently cyclized under palladium catalysis in the presence of copper bromide as terminal oxidant. They provide a straightforward access to *syn*-configured ester-substituted urea products (Scheme 2.8).[31] The diastereomeric excesses depend on the size of the ester group, and smaller alkyl groups represent the preferred esters. Following the initial *syn*-aminopalladation, this importance on sterics is explained by a transition state of transient oxidation for the second C–N bond formation. This step thus constitutes a direct C–N bond formation in α-position to carboxylic groups, which represents an alternative approach to amino acid synthesis.

In addition to ureas, sulfamides were explored, which lead to the opposite *anti*-configured diamination products (Scheme 2.8).[32] This overall stereochemistry can be understood by the participation of a α-brominated intermediate arising from transient oxidation, which could indeed be isolated in one of the cases. The final C–N bond formation to the α-amino ester thus provides the second inversion of configuration at this particular center. As a result, both *syn*- and *anti*-configured diamino carboxylic ester derivatives can be accessed through intramolecular diamination reactions of acrylates depending on the chosen carbonyl or thionyl spacer, respectively.

Demonstrating the synthetic use of this transformation, Muñiz employed one of the diamination products as a 1,2,3-trisubstituted synthon for a short six-step synthesis of the alkaloid absouline (Scheme 2.9).[31]

This palladium catalysis for diamination was recently explored further by Broggini and Chiacchio (Scheme 2.10).[33,34] They employed the palladium/copper catalysis system for the construction of bicyclic piperazinones employing either ureas or sulfamates as the starting material. The reactions are based on the established conditions. For the urea version, the reactions employed N-arylated ureas and proceeded in very good yields in favor of diamination over aminooxygenation using the depicted conditions.[33]

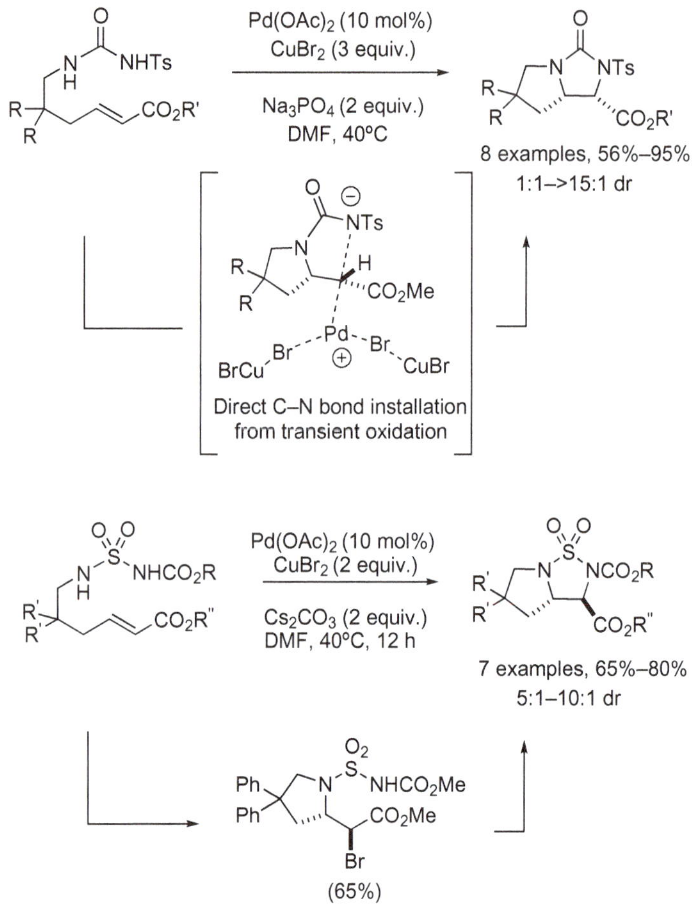

SCHEME 2.8 Diastereoselective intramolecular diamination reactions of acrylates.

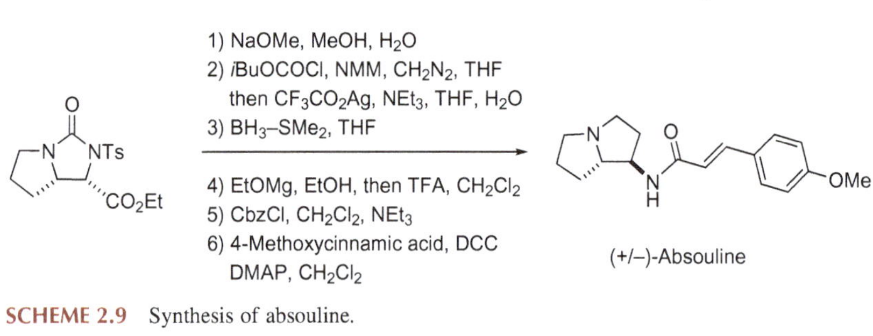

SCHEME 2.9 Synthesis of absouline.

Through the isolation of an intermediary alkyl chloride, the authors could demonstrate the role of the copper, which was postulated to engage in transient oxidation of the σ-alkyl palladium intermediate from the initial aminopalladation of the alkene. The competing formation of an alkyl chloride may

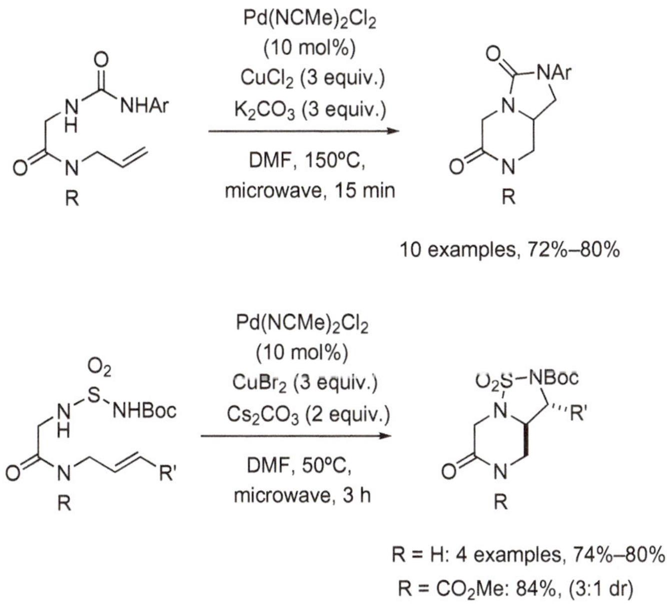

SCHEME 2.10 Palladium-catalyzed piperazinone syntheses.

be the consequence of the use of less nucleophilic aniline components for the second C–N bond formation. In any case, the successful use of these urea substrates constitutes a noteworthy addition to the field.

In a second approach, the related sulfamides were explored.[34] In these cases, slight modifications to copper bromide as terminal oxidant and cesium carbonate as base were established as optimum conditions. While the reaction worked well for terminal alkenes, the corresponding methyl acrylate derivative furnished a mixture of two diastereosiomers, with the *trans*-isomer representing the major one. This outcome was explained by the assumption of a halogenated intermediate following the related chemistry from the Muñiz group.

2.2.2 Other Metal Catalysis Using Iodine(III) Oxidants

The concept of high oxidation catalysis is obviously not restricted to palladium. In fact, the mechanistic constraints of *syn*-aminopalladation/*anti*-C–N bond formation of the palladium catalyses are not compatible with the requirements of cyclic substrates. This led to the inspection of alternative metal catalysis manifolds that would initiate diamination reactions through an *anti*-aminometallation pathway. Due to their linear coordination geometry, gold complexes were considered as good candidates and indeed provide the corresponding diamination reactions within a gold(I/III) catalytic cycle (Scheme 2.11).[35] The gold(III) oxidation state generated from oxidation with

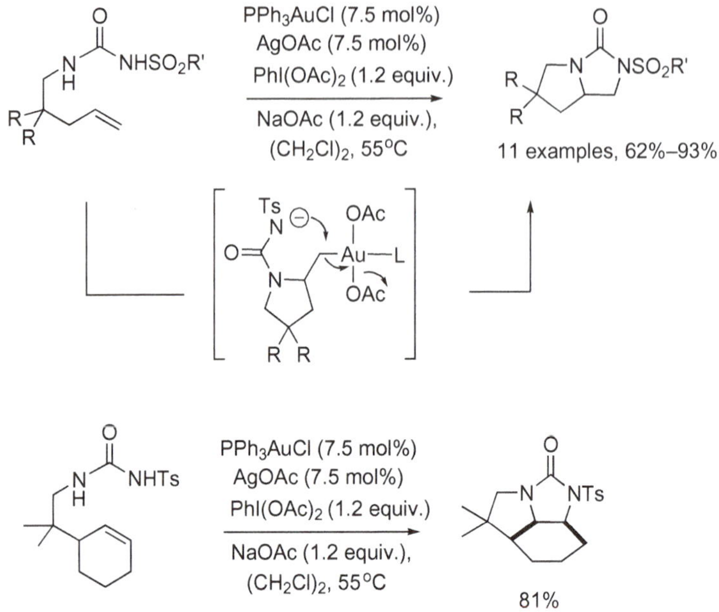

SCHEME 2.11 Homogeneous gold(I/III)-catalyzed diamination of alkenes.

iodine(III) was confirmed from a control experiment using model complex PPh_3-AuMe and lead to direct nucleophilic displacement of the high oxidation state gold catalyst. Such a scenario should enable to tolerance of cyclic double bond, and this could in fact be successfully demonstrated for a cyclohexyl derivative.

In another exploration, sulfamides were explored. These substrates are considered interesting since the resulting diamination products can usually be converted into the free diamines under more convenient conditions than it is the case for the urea derivatives. Unfortunately, the palladium(II)/iodine (III) system only promotes aminooxygenation reactions, while palladium/copper combinations lead to significant sulfamide degradation. In contrast to these palladium catalyses, nickel does promote the corresponding diamination reaction (Scheme 2.12), although it is limited to terminal alkenes and to five-membered ring annulation.[36]

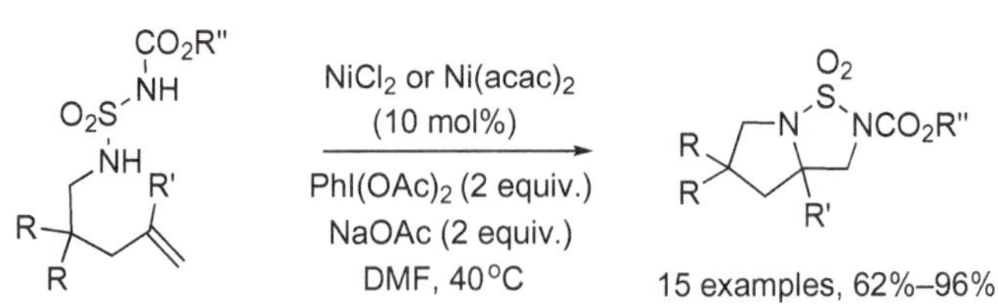

SCHEME 2.12 Nickel(II)-catalyzed intramolecular diamination of alkenes.

2.2.3 Copper Catalysis

Copper was identified as an excellent promoter for intramolecular diamination of alkenes in 2005, when Chemler reported a stoichiometric reaction based on copper(II) acetate.[37–39] The subsequent development of catalytic variants was mainly made for intra-/intermolecular reactions that are summarized in Section 2.3.2.

With respect to intramolecular diamination under copper catalysis, Chiba developed an aerobic reaction using amidines as nitrogen sources. These substrates can be accessed from condensation of primary amines and nitriles (Scheme 2.13).[40]

The copper catalyst is crucially supported by a 2,2′-bipyridine ligand, and the reaction works well for five-membered ring annulation. The authors proposed a mechanism involving a higher oxidation state copper(II) superoxo or peroxo species that is arising from aerobic oxidation. Sequential single-electron oxidation would finally provide a nitrene as the key intermediate, and its corresponding 1,3-dipolar resonance form undergoes intramolecular [3 + 2]-cycloaddition with the alkene. The concerted mechanism of this cyclization leads to complete diastereoselectivity in the diamination of internal alkenes. The copper(II) catalyst is not involved in the actual diamination process, but rather generates the reactive dipolar species, and is regenerated by reoxidation of its reduced Cu(I) precursor under aerobic conditions.

In a more recent application, Zeng and coworkers could demonstrate that the standard urea tethers can be employed in an enantioselective transformation under chiral copper catalysis.[41] The reaction conditions are widely based on a related example by Chemler for intra-/intermolecular reactivity, but could be nicely elaborated for a significant number of examples. The reaction was usually developed and tested for mono-substituted terminal alkenes, although a methallyl derivative also provided excellent enantioselectivity.

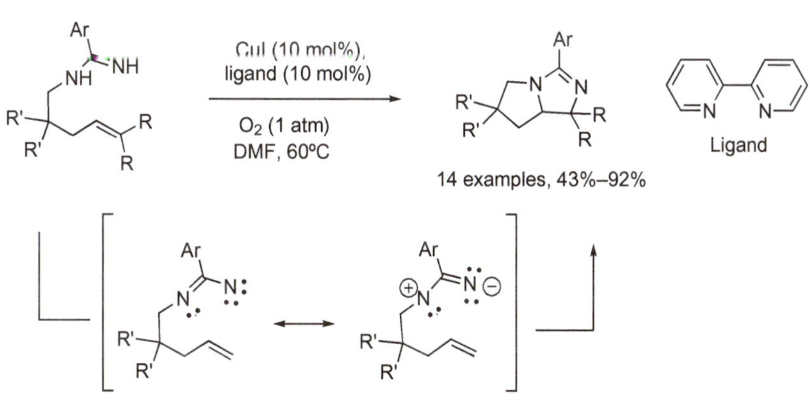

SCHEME 2.13 Copper-catalyzed aerobic intramolecular diamination with amidines as nitrogen sources.

These examples ended the challenging search for an enantioselective intramolecular diamination of the original urea systems (Scheme 2.14).

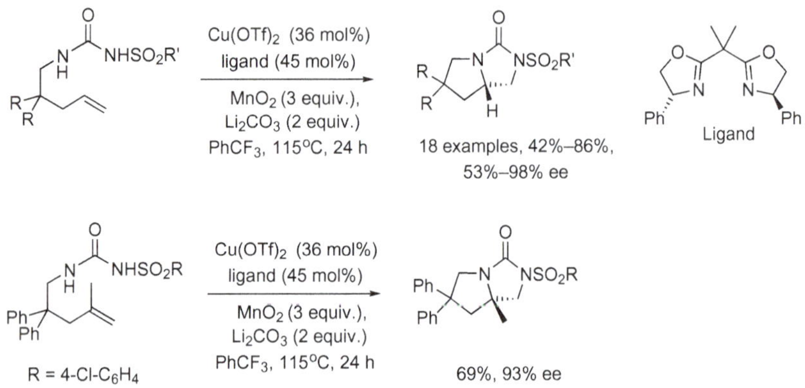

SCHEME 2.14 Copper-catalyzed enantioselective intramolecular diamination.

2.3 METAL-CATALYZED SYNTHESIS OF HETEROCYCLES THROUGH THE OXIDATIVE INTRA-/INTERMOLECULAR VICINAL DIAMINATION OF ALKENES

2.3.1 Palladium Catalysis

The discovery that the well-established aminopalladation products from intramolecular aminative alkene functionalization can be intercepted with NFSI enabled Michael to devise synthetically useful palladium catalysis for the intra-/intermolecular diamination (Scheme 2.15).[42,43]

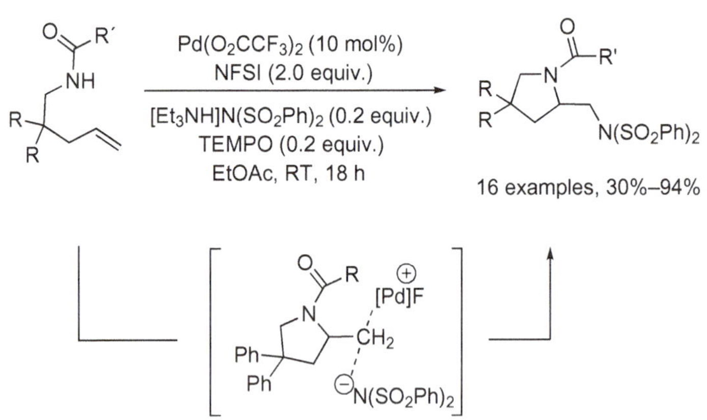

SCHEME 2.15 Palladium(II/IV)-catalyzed intra-/intermolecular diamination.

This new diamination reaction proceeds under robust synthetic conditions and is a very useful entry into the aminomethylenyl pyrrolidine unit. The reaction is limited to terminal alkenes, but proceeds well for a range of different nitrogen substituents. Mechanistic studies suggest a sequence of *anti*-aminopalladation followed by palladium oxidation with the electrophilic fluorine of NFSI. The resulting Pd(IV) intermediate undergoes nucleophilic attack at the α-carbon leading to reductive C−N bond formation. This step is reminiscent of the S_N2 reaction pathway in related intramolecular diamination reactions.

In a subsequent application, Zhang developed a related synthesis of isoindolinones.[44] Starting from 2-vinyl benzamides, a robust oxidative diamination to the corresponding isoindolinone products could be accomplished with a catalyst from palladium acetate and bipyridine as ligand. NFSI served as the oxidant and second nitrogen source. The products can be transformed into the free lactams by treatment with samarium diiodide (Scheme 2.16).

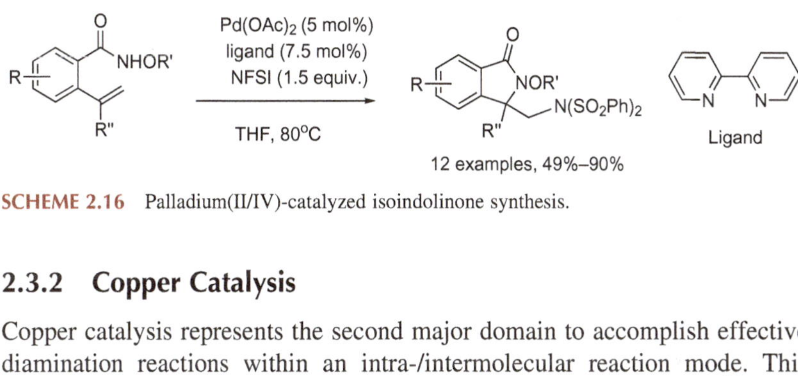

SCHEME 2.16 Palladium(II/IV)-catalyzed isoindolinone synthesis.

2.3.2 Copper Catalysis

Copper catalysis represents the second major domain to accomplish effective diamination reactions within an intra-/intermolecular reaction mode. This work is largely based on developments from Chemler and her group.[45]

Chemler developed the corresponding version of an intra-/intermolecular diamination of a series of *N*-allyl ureas and *N*-mesyl 2-allylanilines (Scheme 2.17). This reaction proceeds with excellent selectivity and employs a tuned copper 2-ethylhexanoate catalyst. It employs unactivated sulfonamides as the nitrogen source in the second C−N bond-forming step. Chemler could show that a catalyst derived from copper triflate and a chiral bisoxazoline ligand can perform this diamination in an enantioselective manner with 71% ee. By that time, this transformation represented the proof of principle for the asymmetric diamination of alkenes under metal catalysis.

Building on this work, Chemler further explored the diamination of 2-allyl aniline derivatives. After extensive screening, the final optimized protocol included application of a bulky sulfonamide group and provided a general diamination reaction with high enantiomeric excesses (Scheme 2.18). Under related conditions, simple sulfonamides can also undergo this reaction and provide useful 2-aminomethylyl pyrrolidines with up to 72% ee.[46]

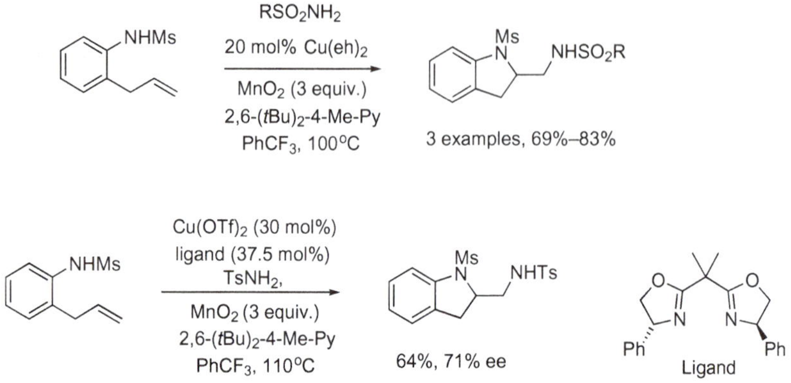

SCHEME 2.17 Copper-catalyzed diamination under intra-/intermolecular reaction control.

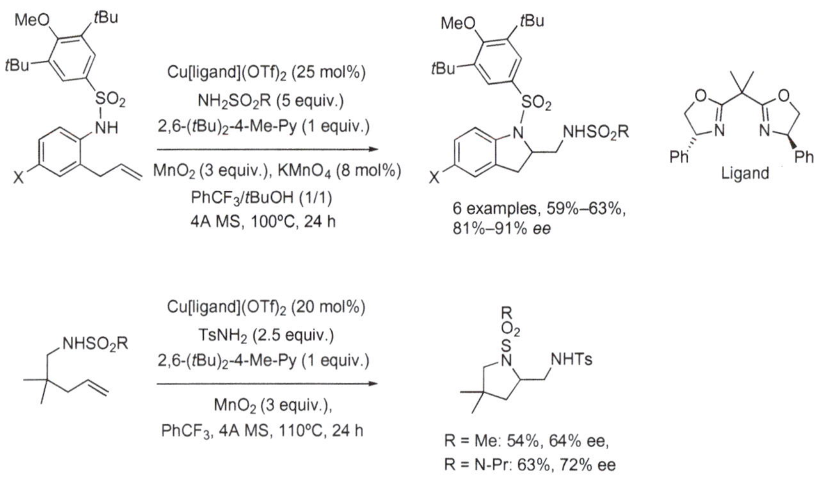

SCHEME 2.18 Copper-catalyzed enantioselective diamination under intra-/intermolecular reaction control.

In a subsequent application of their achiral copper catalyst system, the same group presented a stereoselective synthesis of isoxazolidines (Scheme 2.19).[47] This protocol provided a large toleration of different substituents at nitrogen and in the carbon framework leading to a great number of divers products, which could be formed in up to 92% yield and with >20:1 diastereomeric ratio. An attempt to carry out an enantioselective transformation met with limited success, but demonstrated that it should generally be possible to carry out the initial aminocupration under enantiocontrol.

The isoxazolidine heterocycles are versatile building blocks, in particular after reductive O–N bond cleavage, which provides 1,3,5-trisubstituted alkyl derivatives.

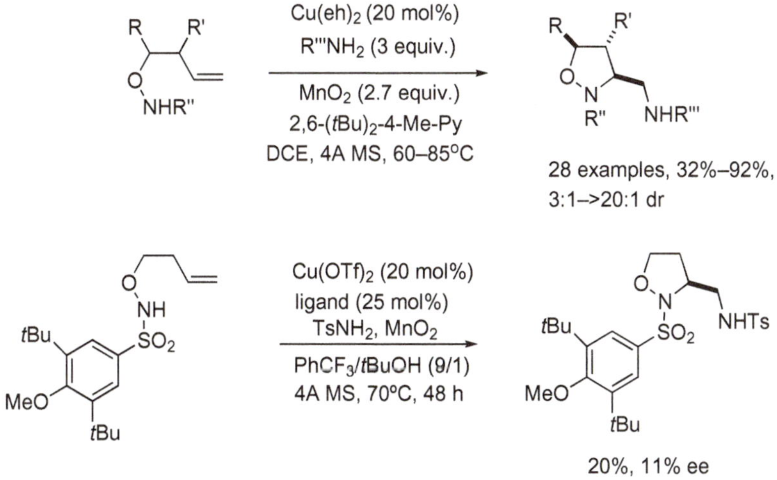

SCHEME 2.19 Copper-catalyzed synthesis of isoxazolidine heterocycles.

2.4 METAL-CATALYZED SYNTHESIS OF HETEROCYCLES THROUGH THE OXIDATIVE INTERMOLECULAR VICINAL DIAMINATION OF ALKENES

Although generally not covered in this chapter, diamination reactions starting from an intermolecular C–N bond formation can be of interest to the synthesis of heterocycles as well, presuming that the second C–N bond formation proceeds in an intramolecular reaction. This requires that both nitrogen sources be included into a single reagent. A larger number of reports in the area are based on chemistry developed by Shi and make use of small strained three-membered ring reagents such as thiadiaziridines, diaziridinones, and their imine derivatives. The basic transformations can be conducted with homogeneous copper catalysts bearing tertiary phosphine ligands. These reactions were suggested to involve an oxidative insertion of the copper(I) catalyst into the N–N bond followed by generation of a nitrogen radical, which adds to the double bond of the alkene. The resulting carbon-centered radical undergoes C–N bond formation with the remaining nitrogen atom and regenerates the copper catalyst. Successful protocols include di-*tert*-butylthiadiaziridine, di-*tert*-butyldiaziridinylidene cyanamide, and di-*tert*-butyldiaziridinone as reagents in these diamination reactions providing cyclic sulfamides, guanidines, and ureas, respectively.[9,48–50] The power of this methodology was demonstrated in a short synthesis of the pharmaceutical SCH 425078, which consisted of a diamination of a preformed chiral, nonracemic allylic ether. This reaction gave a mixture of two diastereomers (35% of the desired and 30% of the undesired one), which could be separated. Deprotection by trifluoroacetic acid (TFA) directly provides the target compound (Scheme 2.20).

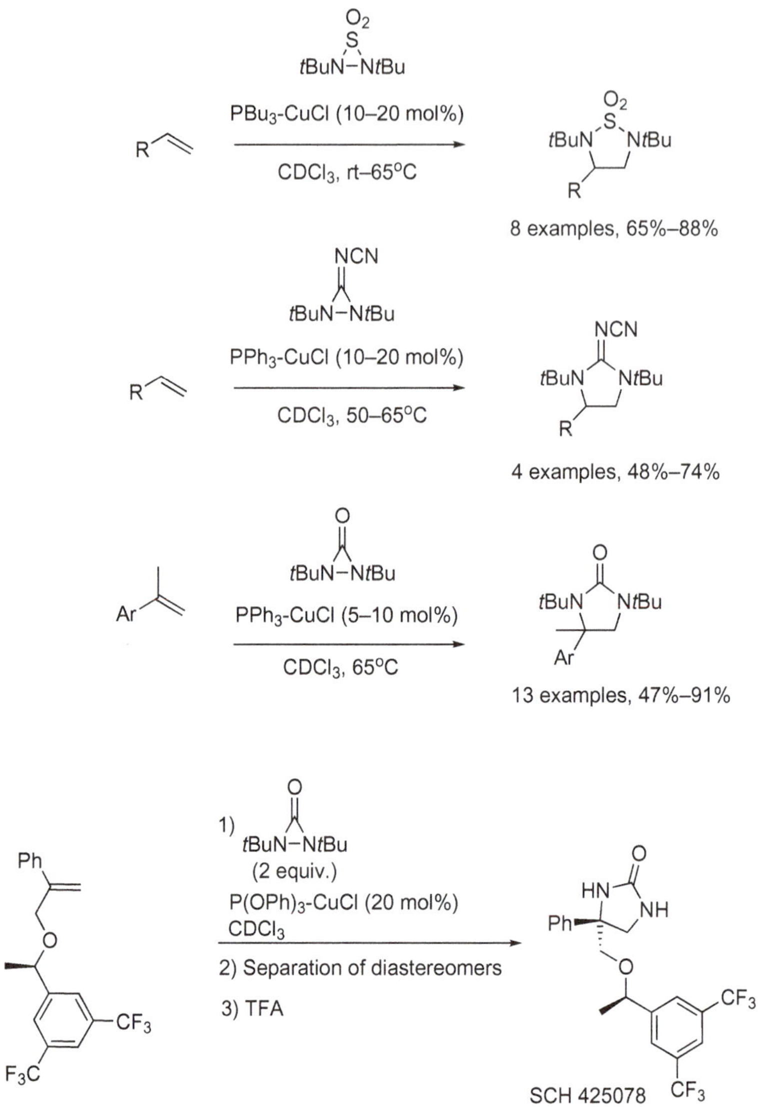

SCHEME 2.20 Copper-catalyzed intermolecular diamination reactions.

A related reaction was used by Miao and coworkers to install a tethered diamine moiety on a fullerene core (Scheme 2.21).[51] To this end, an iodine (I) promoter needed to be chosen, which was generated in situ from an iodine(III) source and molecular iodine. Free sulfamides or a phosphoryldiamide were employed as nitrogen sources and the reaction is believed to proceed through an iodine-initiated radical pathway. While the sulfamide

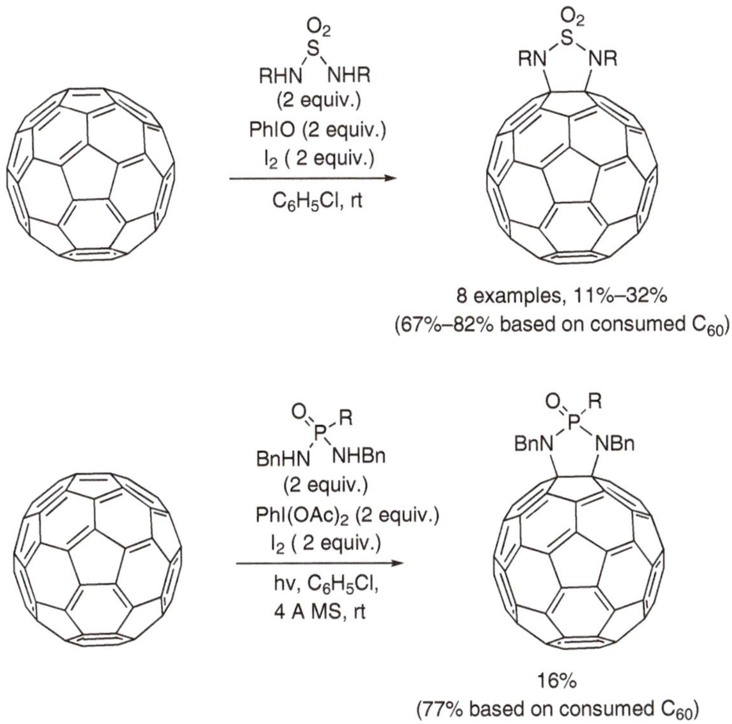

SCHEME 2.21 Iodine-mediated intermolecular diamination reactions of fullerene.

motive could be transferred with relatively high selectivity, the phosphoryldiamide gave an N-benzylated aziridine as the main product. After extensive experimentation, the reaction could be carried out with preferential selectivity for diamination (Scheme 2.22).

Within a conceptually alternative, Du Bois and coworkers developed a sequential two-step synthesis of diamines, which consisting of an initial aziridination followed by rearrangement to the corresponding cyclic sulfamides.[52] The reaction employs Boc-sulfamide as the nitrogen source for an initial rhodium-catalyzed aziridination. This step is followed by the addition of sodium iodide in DMF, which promotes aziridine opening followed by nucleophilic substitution to the final diamine derivative. The reaction was developed mostly for terminal alkenes, but (*Z*)-alkenes can also be employed successfully, while an (*E*)-alkene provided low yield only. Since the aziridine ring opening and the subsequent nucleophilic substitution proceed under complete stereocontrol, internal alkenes provide single diastereomeric products.

For racemic N-Boc allyl glycine, the reaction generates two diasteromeric products in a 1:1-ratio, which can be deprotected and then separated at the stage of the free sulfamides. Their individual guanidination, followed by final protecting group removal, provides a concise racemic synthesis of the

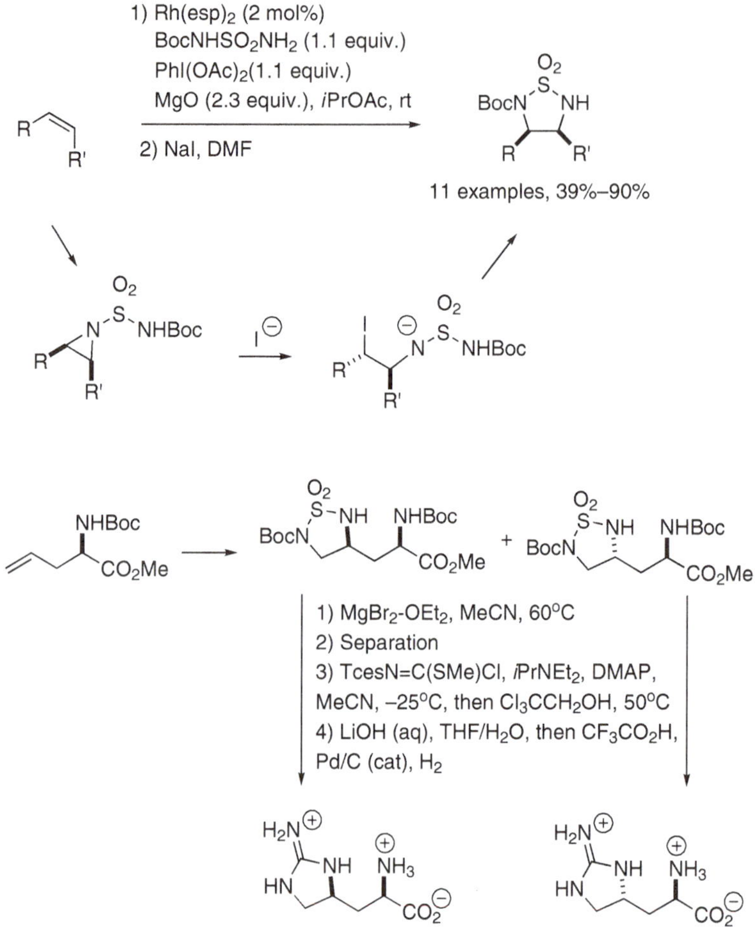

(+/–)-Enduracididine (+/–)-Allo-enduracididine

SCHEME 2.22 Rhodium-catalyzed intermolecular aziridination/diamination reactions and application to the synthesis of enduracididine and allo-enduracididine.

two guanidinium natural products enduracididine and allo-enduracididine. It again demonstrates the synthetic power of diamination reactions with cyclic products as synthetically reliable precursors in alkaloid synthesis.

2.5 SUMMARY AND OUTLOOK

The two main variants for intramolecular diamination of alkenes have so far enabled unprecedented access to a number of heterocycle systems. These reactions are characterized by the application of defined molecular catalysts on the basis of transition metals. For some of the reactions, suitable chiral ligands have emerged that have enabled the first examples of enantioselective diamination reactions under intramolecular reaction control. Several of the reactions have provided advanced building blocks for the application in total synthesis of natural products or pharmaceuticals. The future asymmetric diamination of alkenes looks more and more promising and one can expect that continuous developments will ultimately provide a larger number of robust and reliable protocols including highly enantioselective versions of general applicability.

REFERENCES

1. Lucet, D.; Le Gall, T.; Mioskowski, C. The Chemistry of Vicinal Diamines. *Angew. Chem. Int. Ed.* **1998,** *37*, 2580–2627.
2. Kolb, H. C.; VanNieuwenhze, M. S.; Sharpless, K. B. Catalytic Asymmetric Dihydroxylation. *Chem. Rev.* **1994,** *94*, 2483–2547.
3. Muñiz, K. Imido-Osmium(VIII) Compounds in Organic Synthesis: Aminohydroxylation and Diamination Reactions. *Chem. Soc. Rev.* **2004,** *33*, 166–174.
4. Muñiz, K. The Development of Asymmetric Diamination of Alkenes With Imido-Osmium Reagents. *New J. Chem.* **2005,** *29*, 1371–1385.
5. De Jong, S.; Nosal, D. G.; Wardrop, D. J. Methods for Direct Alkene Diamination, New & Old. *Tetrahedron* **2012,** *68*, 4067–4105.
6. Cardona, F.; Goti, A. Metal-Catalysed 1,2-Diamination Reactions. *Nat. Chem.* **2009,** *1*, 269–275.
7. de Figueiredo, R. M. Transition-Metal-Catalyzed Diamination of Olefins. *Angew. Chem. Int. Ed.* **2009,** *48*, 1190–1193.
8. Martínez, C.; Muñiz, K. Development of Intramolecular Vicinal Diamination of Alkenes: From Palladium to Bromine Catalysis. *J. Org. Chem.* **2013,** *78*, 2168–2174.
9. Zhu, Y.; Cornwall, R. G.; Du, H.; Zhao, B.; Shi, Y. Catalytic Diamination of Olefins via N–N Bond Activation. *Acc. Chem. Res.* **2014,** *47*, 3665–3678.
10 Gómez Aranda, V.; Barluenga, J.; Aznar, F. The Addition of Aromatic Amines to Alkenes in the Presence of Thallium(III) Acetate. *Synthesis* **1974,** 504–505.
11. Barluenga, J.; Alonso-Cires, L.; Asensio, G. Mercury(II) Oxide/Tetrafluoroboric Acid—A New Reagent in Organic Synthesis; A Convenient Diamination of Olefins. *Synthesis* **1979,** 962–964.
12. Bäckvall, J.-E. Stereospecific Palladium-Promoted Vicinal Diamination of Olefins. *Tetrahedron Lett.* **1978,** 163–166.
13. Chong, A. O.; Oshima, K.; Sharpless, K. B. Synthesis of Dioxobis(*tert*-Alkylimido) Osmium(VIII) and Oxotris(*tert*-Alkylimido)Osmium(VIII) Complexes. Stereospecific Vicinal Diamination of Olefins. *J. Am. Chem. Soc.* **1977,** *99*, 3420–3426.
14. Anhaus, J. T.; Kee, T. P.; Schofield, M. H.; Schrock, R. R. Planar "20-Electron" Osmium Imido Complexes. A Linear Imido Ligand Does Not Necessarily Donate Its Lone Pair of Electrons to the Metal. *J. Am. Chem. Soc.* **1990,** *112*, 1642–1643.
15. Muñiz, K.; Iesato, A.; Nieger, M. Diamination of Olefins: Synthesis, Structures and Reactivity of Osmaimidazolidines. *Chem. Eur. J.* **2003,** *9*, 5581–5596.
16. Muñiz, K.; Nieger, M.; Mansikkamäki, H. Stereochemically Defined Osmium Centers From Asymmetric Diamination of Olefins: Mechanistic Implications for Osmium-Mediated Acrylate Functionalization. *Angew. Chem. Int. Ed.* **2003,** *42*, 5958–5961.

17. Streuff, J.; Hövelmann, C. H.; Nieger, M.; Muñiz, K. Palladium(II)-Catalyzed Intramolecular Diamination of Unfunctionalized Alkenes. *J. Am. Chem. Soc.* **2005,** *127*, 14586–14587.
18. Muñiz, K. High-Oxidation-State Palladium Catalysis: New Reactivity for Organic Synthesis. *Angew. Chem. Int. Ed.* **2009,** *48*, 9412–9423.
19. Hickman, A. J.; Sanford, M. S. High-Valent Organometallic Copper and Palladium in Catalysis. *Nature* **2012,** *484*, 177–185.
20. Martínez, C.; Iglesias, A.; Muñiz, K. The Quest for Palladium-Catalysed Alkyl-Nitrogen Bond Formation. *Chem. Rec.* **2016,** *16*, 2561–2572.
21. Muñiz, K.; Hövelmann, C. H.; Streuff, J. Oxidative Diamination of Alkenes With Ureas as Nitrogen Sources: Mechanistic Pathways in the Presence of a High Oxidation State Palladium Catalyst. *J. Am. Chem. Soc.* **2008,** *130*, 763–773.
22. Minatti, A.; Muñiz, K. Intramolecular Aminopalladation of Alkenes as a Key Step to Pyrrolidines and Related Heterocycles. *Chem. Soc. Rev.* **2007,** *36*, 1142–1152.
23. Iglesias, A.; Álvarez, R.; de Lera, A. R.; Muñiz, K. Palladium-Catalyzed Intermolecular C (sp3)-H Amidation. *Angew. Chem. Int. Ed.* **2012,** *51*, 2225–2228.
24. Muñiz, K. Advancing Palladium-Catalyzed C – N Bond Formation: Bisindoline Construction From Successive Amide Transfer to Internal Alkenes. *J. Am. Chem. Soc.* **2007,** *129*, 14542–14543.
25. Henry, P. M., Ed. *Palladium Catalyzed Oxidation of Hydrocarbons;* Reidel: Dordrecht, 1980.
26. Stangl, H.; Jira, R. Die Durch Palladium(II)Chlorid und Kupfer(II)Chlorid Katalysierte Oxychlorierung von Äthylen zu Äthylenchlorhydrin. *Tetrahedron Lett.* **1970,** *11*, 3589.
27. Muñiz, K.; Iglesias, A.; Fang, Y. Platinum-Catalysed Aerobic 1,2-Aminooxygenation of Alkenes. *Chem. Commun.* **2009,** 5591–5593.
28. Muñiz, K.; Hövelmann, C. H.; Campos-Gómez, E.; Barluenga, J.; González, J. M.; Streuff, J.; Nieger, M. Intramolecular Diamination of Alkenes With Palladium(II)/Copper(II) Bromide and IPy_2BF_4: The Role of Halogenated Intermediates. *Chem. Asian J.* **2008,** *3*, 776–788.
29. Hamed, O.; Henry, P. M. Oxidation of Olefins by Palladium(II). 16. A New Palladium(II)-Catalyzed Asymmetric Chlorohydrin Synthesis. *Organometallics* **1998,** *17*, 5184–5189.
30. Hövelmann, C. H.; Streuff, J.; Brelot, L.; Muñiz, K. Direct Synthesis of Bicyclic Guanidines Through Unprecedented Palladium(II) Catalysed Diamination With Copper Chloride as Oxidant. *Chem. Commun.* **2008,** 2334–2336.
31. Muñiz, K.; Streuff, J.; Chávez, P.; Hövelmann, C. H. Synthesis of Diamino Carboxylic Esters by Palladium-Catalyzed Oxidative Intramolecular Diamination of Acrylates. *Chem. Asian J.* **2008,** *3*, 1248–1255.
32. Chavez, P.; Kirsch, J.; Streuff, J.; Muñiz, K. Palladium-Catalyzed Intramolecular Diamination of Acrylic Esters Using Sulfamates as Nitrogen Source. *J. Org. Chem.* **2012,** *77*, 1922–1930.
33. Broggini, G.; Barbera, V.; Beccalli, E. M.; Chiacchio, U.; Fasana, A.; Galli, S.; Gazzola, S. Selective Intramolecular Palladium(II)-Catalyzed Aminooxygenation vs. Diamination of Alkenylureas: Efficient Microwave-Assisted Reactions to Bicyclic Piperazinones. *Adv. Synth. Catal.* **2013,** *355*, 1640–1648.
34. Chiacchio, U.; Broggini, G.; Romeo, R.; Gazzola, S.; Chiacchio, M. A.; Giofrè, S. V.; Gabriele, B.; Mancuso, R.; Floresta, G.; Zagni, C. Intramolecular Oxidative Palladium-Catalyzed Diamination Reactions of Alkenyl Sulfamates: An Efficient Synthesis of 1,2,5Thiadiazolo-Fused Piperazinones. *RSC Adv.* **2016,** *6*, 57521–57529.

35. Iglesias, A.; Muñiz, K. Oxidative Interception of the Hydroamination Pathway: A Gold-Catalyzed Diamination of Alkenes. *Chem. Eur. J.* **2009,** *15*, 10563–10569.

36. Muñiz, K.; Streuff, J.; Hövelmann, C. H.; Nuñez, A. Exploring the Nickel-Catalyzed Oxidation of Alkenes: A Diamination by Sulfamide Transfer. *Angew. Chem. Int. Ed.* **2007,** *469*, 7125–7127.

37. Zabawa, T. P.; Kasi, D.; Chemler, S. R. Copper(II) Acetate Promoted Intramolecular Diamination of Unactivated Olefins. *J. Am. Chem. Soc.* **2005,** *127*, 11250–11251.

38. Zabawa, T. P.; Chemler, S. R. Copper(II) Carboxylate Promoted Intramolecular Diamination of Terminal Alkenes: Improved Reaction Conditions and Expanded Substrate Scope. *Org. Lett.* **2007,** *9*, 2035–2038.

39. Chemler, S. R. Evolution of Copper(II) as a New Alkene Amination Promoter and Catalyst. *J. Organomet. Chem.* **2012,** *696*, 150–158.

40. Wang, Y.-F.; Zhu, X.; Chiba, S. Copper-Catalyzed Aerobic 3 + 2-Annulation of *N*-Alkenyl Amidines. *J. Am. Chem. Soc.* **2012,** *134*, 3679–3682.

41. Fu, S.; Yang, H.; Li, G.; Deng, Y.; Jiang, H.; Zeng, W. Copper(II)-Catalyzed Enantioselective Intramolecular Cyclization of *N*-Alkenylureas. *Org. Lett.* **2015,** *17*, 1018–1021.

42. Sibbald, P. A.; Michael, F. E. Palladium-Catalyzed Diamination of Unactivated Alkenes Using *N*-Fluorobenzenesulfonimide as Source of Electrophilic Nitrogen. *Org. Lett.* **2009,** *11*, 1147–1149.

43. Sibbald, P. A.; Rosewall, C. F.; Swartz, R. D.; Michael, F. E. Mechanism of *N*-Fluorobenzenesulfonimide Promoted Diamination and Carboamination Reactions: Divergent Reactivity of a Pd(IV) Species. *J. Am. Chem. Soc.* **2009,** *131*, 15945–15951.

44. Li, Y.; Kou, X.; Ye, C.; Zhang, X.; Yang, G.; Zhang, W. Preparation of Isoindolinones via a Palladium-Catalyzed Diamination. *Tetrahedron Lett.* **2017,** *58*, 285–288.

45. Sequeira, F. C.; Turnpenny, B. W.; Chemler, S. R. Copper-Promoted and Copper-Catalyzed Intermolecular Alkene Diamination. *Angew. Chem. Int. Ed.* **2010,** *49*, 6365–6368.

46. Turnpenny, B. W.; Chemler, S. R. Copper-Catalyzed Alkene Diamination: Synthesis of Chiral 2-Aminomethyl Indolines and Pyrrolidines. *Chem. Sci.* **2014,** *5*, 1786–1793.

47. Khoder, Z. M.; Wong, C. E.; Chemler, S. R. Stereoselective Synthesis of Isoxazolidines via Copper-Catalyzed Alkene Diamination. *ACS Catal.* **2017,** *7*, 4775–4779.

48. Zhao, B.; Yuan, W.; Shi, Y. Cu(I)-Catalyzed Intermolecular Diamination of Activated Terminal Olefins. *Org. Lett.* **2007,** *9*, 4943–4945.

49. Wen, Y.; Zhao, B.; Shi, Y. Cu(I)-Catalyzed Diamination of Disubstituted Terminal Olefins: An Approach to Potent NK_1 Antagonist. *Org. Lett.* **2009,** *11*, 2365–2368.

50. Zhao, B.; Du, H.; Dhi, Y. Cu(I)-Catalyzed Cycloguanidination of Olefins. *Org. Lett.* **2008,** *10*, 1087–1090.

51. Yang, H.-T.; Lu, X.-W.; Xing, M.-L.; Sun, X.-Q.; Miao, C.-B. Hypervalent Iodine Reagent Mediated Diamination of 60Fullerene With Sulfamides or Phosphoryl Diamides. *Org. Lett.* **2014,** *16*, 5882–5885.

52. Olson, D. E.; Su, J. Y.; Roberts, D. A.; Du Bois, J. Vicinal Diamination of Alkenes Under Rh-Catalysis. *J. Am. Chem. Soc.* **2014,** *136*, 13506–13509.

Chapter 3

Synthesis of Heterocycles by Palladium-Catalyzed Carbonylative Reactions

Bartolo Gabriele
University of Calabria, Arcavacata di Rende, CS, Italy

3.1 INTRODUCTION

In the course of the years, palladium catalysis has acquired an increasing importance in the synthesis of heterocycles. This is mainly due to the versatility of palladium-based catalysts in promoting a plethora of functional group transformations and interconversions, under mild reaction conditions and with high efficiency and selectivity. Another peculiar aspect of palladium complexes is related to the possibility to efficiently catalyze the direct incorporation of carbon monoxide into an organic substrate. A process like this, resulting in the direct synthesis of carbonyl compounds from the simplest and readily available C-1 unit, is called carbonylation.[1–3] Carbonylation reactions are now recognized as the most valuable and sustainable approach for the introduction of the carbonyl group into a substrate, in view of the large availability of carbon monoxide, the high atom economy of the process, and the possibility to synthesize in one step high value–added compounds by the catalytic activation of simple building blocks.[1–43] Recently, efforts have also been made aimed at using CO surrogates (such as formic acid or formaldehyde) instead of gaseous carbon monoxide.[44–50] Although these processes are attractive in a laboratory scale for the production of fine chemicals, particularly in academia, it should be considered that CO still remains the best choice in terms of availability and atom economy.

Considering that palladium-based catalysts may be able to catalyze both heterocyclization[51–87] and carbonylation reactions,[1–50] it is not surprising that many important processes have been developed during the past years for the direct synthesis of carbonylated heterocycles through palladium-catalyzed carbonylative heterocyclization reactions. In these processes a

Advances in Transition-Metal Mediated Heterocyclic Synthesis.
DOI: https://doi.org/10.1016/B978-0-12-811651-7.00003-0

palladium-based species is able to promote both the heterocyclization of a suitably functionalized substrate and the incorporation of one or more CO molecules, with formation of carbonylated heterocycles in a single catalytic process. Since some important reviews and books were recently published on the Pd-catalyzed carbonylative synthesis of heterocycles,[16–18] this chapter is mainly devoted to give an overview of the most recent advancements in this field (from 2012 to April 2017).[88]

3.2 PALLADIUM-CATALYZED CYCLOCARBONYLATIONS LEADING TO HETEROCYCLES

A cyclocarbonylation is a carbonylative annulation process in which carbon monoxide is inserted into the final cycle.[89] This section highlights the recent developments in the field of palladium-catalyzed cyclocarbonylation processes leading to heterocyclic derivatives.

3.2.1 Under Oxidative Conditions

A palladium-mediated oxidative carbonylation is a process in which carbon monoxide is inserted into an organic substrate SH_2 with the formation of a carbonylated product S(CO) and simultaneous reduction of the palladium species that has promoted the reaction.[90] The most common case is related to the reduction of an L_2PdX_2 species (where X is an anionic counterion, e.g., a halide, and L is a ligand) to Pd(0) with concomitant formation of 2 mol of HX. Clearly, to achieve a catalytic process an external oxidant is necessary to reoxidize Pd(0) to the original Pd(II) species, as shown in Scheme 3.1 (in this and in the following mechanistic schemes of this chapter, all the unreactive ligands are omitted for clarity). When the oxidative carbonylation occurs with annulation and incorporation of CO into the final cycle, the process is called oxidative cyclocarbonylation.

$$SH_2 + CO + PdX_2 \longrightarrow S(CO) + Pd(0) + 2\,HX$$
$$Pd(0) + 2\,HX + [OX] \longrightarrow PdX_2 + [OXH_2]$$
$$\overline{}$$
$$SH_2 + CO + [OX] \xrightarrow{PdX_2} S(CO) + [OXH_2]$$

SCHEME 3.1 A generic Pd(II)-catalyzed oxidative carbonylation process. *SH_2*, organic substrate; *X*, anionic counterion; *[OX]*, oxidative agent; *[OXH_2]*, reduced oxidative agent.

3.2.1.1 With Alkyne and Alkene Derivatives (Without Direct C–H Activation)

Of particular importance is the Pd(II)-catalyzed oxidative cyclocarbonylation of functionalized acetylenic substrates, which may take place with high catalytic efficiency and under particularly mild conditions owing to the elevated coordination ability of the triple bond to Pd(II).[1–3,16–18,23,90–97]

The Pd(II)-catalyzed oxidative carbonylation of simple alkynes may lead to different products depending on reaction conditions.[1–3,39] Usually the reaction is carried out in the presence of a nucleophilic species NuH (such as an alcohol), and the main products obtained are maleic derivatives, deriving from an oxidative dicarbonylation process (i.e., a process in which two molecules of CO are incorporated into the final product) (Scheme 3.2A). In the case of terminal alkynes, *sp* C–H activation is also possible under suitable conditions, with formation of propiolic derivatives, from an oxidative monocarbonylation process (Scheme 3.2B).[98,99] Both kind of reactivity can be exploited for the carbonylative synthesis of heterocycles, when applied to suitably functionalized acetylenic substrates[17]; however, the C–H cyclocarbonylative acetylenic activation will be treated in Section 3.2.1.2.

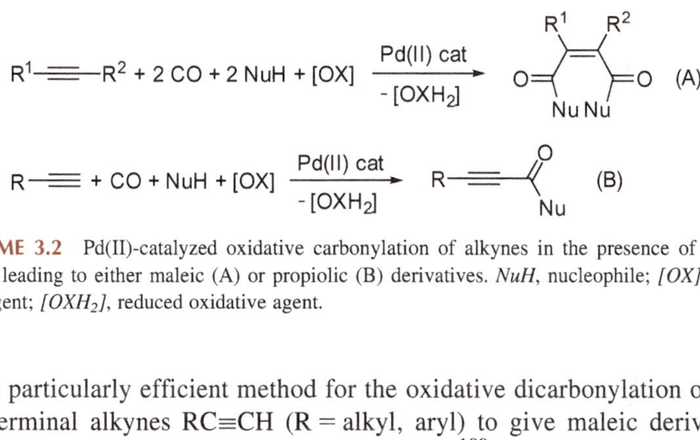

SCHEME 3.2 Pd(II)-catalyzed oxidative carbonylation of alkynes in the presence of nucleophiles leading to either maleic (A) or propiolic (B) derivatives. *NuH*, nucleophile; *[OX]*, oxidative agent; *[OXH$_2$]*, reduced oxidative agent.

A particularly efficient method for the oxidative dicarbonylation of simple terminal alkynes RC≡CH (R = alkyl, aryl) to give maleic derivatives was disclosed by our research group in 1992.[100] It involved the use of a very simple catalytic system, consisting of the PdI_4^{2-} anion (formed in situ from PdI_2 in the presence of an excess of KI) with O_2 as the external oxidant ([OX] = (1/2)O_2, [OXH_2] = H_2O). Depending on the reaction conditions[101] the process could afford maleic diesters (NuH = R′OH, R = alkyl),[100,102] maleic acids (NuH = H_2O),[102,103] or maleic anhydrides.[103] Maleic diesters or acids were formed through the *syn* insertion of the triple bond of the substrate into an I-Pd-CO_2R' species (obtained from PdI_2, CO, and R′OH, R′ = H or alkyl) to give a vinylpalladium intermediate, followed by CO insertion to afford an acylpalladium species and nucleophilic displacement by R′OH (Scheme 3.3). The efficacy of the process was related to a particularly efficient reoxidation of Pd(0) to Pd(II), which took place through the oxidation of HI (ensuing from the carbonylation process) by O_2 to give I_2 followed by oxidative addition of the latter to PdI_2 (Scheme 3.3).[102–104]

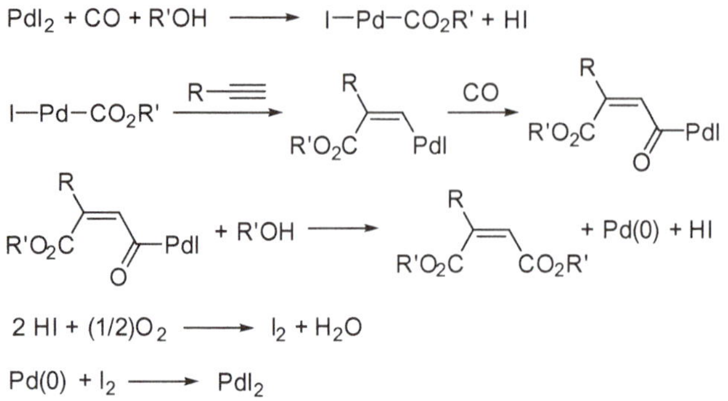

SCHEME 3.3 Mechanism of the PdI_2/KI-catalyzed oxidative dicarbonylation of terminal alkynes leading to maleic diesters or acids (R′ = H or alkyl; anionic iodide ligands are omitted for clarity).

This kind of reactivity (with R′ = alkyl, in particular) has led to the synthesis of a variety of heterocyclic ester derivatives, when applied to acetylenic substrates bearing a suitably placed heteronucleophilic group (YH, Y = O, or NR"). A first mechanistic possibility involves the intramolecular trapping of the initially formed acylpalladium intermediate (possibly, via the formation of a palladacycle followed by reductive elimination) (Scheme 3.3, pathway *a*). Alternatively, intramolecular triple bond insertion into the Y-(CO)PdI species formed by the reaction between the YH group, CO, and PdI_2, followed by alkoxycarbonylation (i.e., carbon monoxide insertion followed by nucleophilic displacement by the alcohol R′OH), may take place (Scheme 3.4, pathway *b*). Functionalized β-lactones,[91,92] γ-lactones,[92] β-lactams,[93] oxindoles,[94] and pyridinones[95] were synthesized by this methodology.

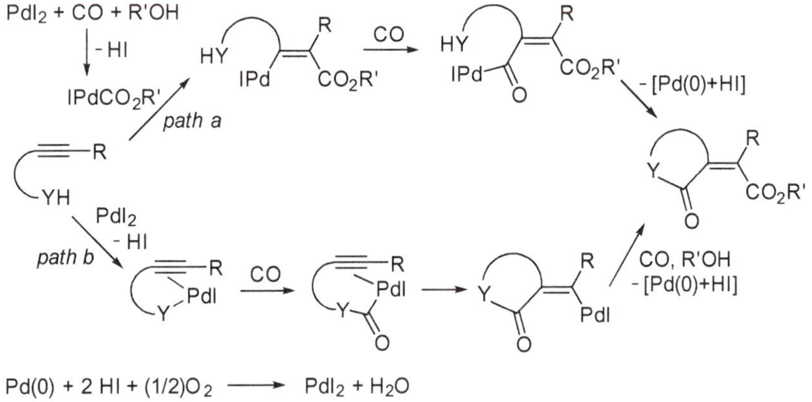

SCHEME 3.4 Mechanistic pathways leading to heterocycles by PdI_2/KI-catalyzed oxidative cyclocarbonylation of suitably functionalized acetylenic derivatives (YH = heteronucleophilic group, Y = O or NR"; R′ = alkyl).

More recently the PdI_2/KI-catalyzed oxidative cyclocarbonylation approach was used for synthesizing novel ζ-lactam derivatives with antitumor activity against MCF-7 and MDA-MB-231 breast cancer cell lines.[105] 2-(2-Ethynylphenoxy)anilines were allowed to react with CO (32 atm), O_2 (from air, 8 atm), and alcohols (such as MeOH or EtOH, also used as solvent) at 100°C for 15–24 h in the presence of PdI_2 (2 mol%) and KI (20 mol%) to afford alkyl (*Z*)-(6-oxo-5,6-dihydro-12-oxa-5-azabenzo[*a*,*d*]cycloocten-7-ylidene)acetates in fair to good yields (53%–75%) (Scheme 3.5). Density functional theory (DFT) calculations were in agreement with the cyclocarbonylation–alkoxycarbonylation mechanism shown in Scheme 3.4, pathway *b*.[105]

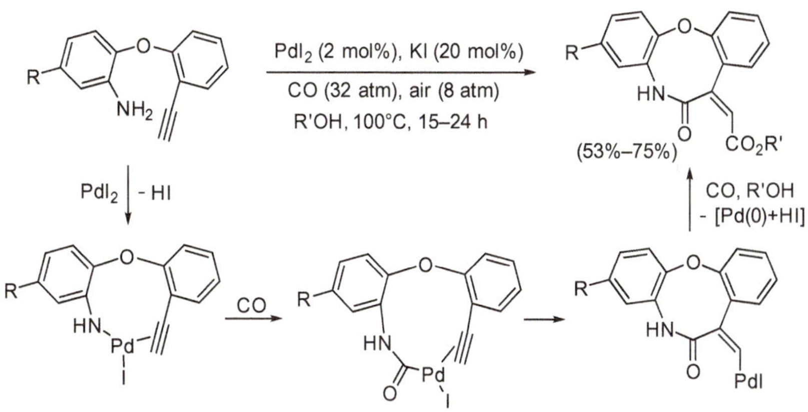

SCHEME 3.5 Synthesis of ζ-lactam derivatives by PdI_2/KI-catalyzed oxidative cyclocarbonylation of 2-(2-ethynylphenoxy)anilines (R = H, Me, OMe, COMe, CO_2Me, CN, CF_3; R′ = Me, Et).

Olefinic substrates have also been successfully used in oxidative cyclocarbonylation reactions to give heterocycles.[1–3,16–18] A particularly important recent example was published by You and Lei. β-Lactam derivatives were synthesized in moderate to excellent yields (27%–98%) by treating allyl amines with 1 atm of CO in *N*,*N*-dimethylformamide (DMF) at 100°C in the presence of $PdCl_2/PCy_3$ (Cy = cyclohexyl) as the catalytic system and $Cu(OPiv)_2$ (Piv = pivaloyl) as the oxidizing agent, through a mechanism involving *N*-palladation, CO insertion, double bond insertion and β-H elimination (Scheme 3.6).[106]

In a similar way the formation of coumarins from 2-vinylphenols was reported by Alper and coworkers, using $Pd(OAc)_2$/dppb [dppb = 1,4-bis (diphenylphosphino)butane] as catalyst and benzoquinone or O_2 as oxidant (Scheme 3.7),[107] and 3-methyleneindolin-2-ones were synthesized from 2-vinylanilines using $Pd(OAc)_2$ as catalyst (5 mol%) and Ag_2CO_3 as oxidant (Scheme 3.8).[108]

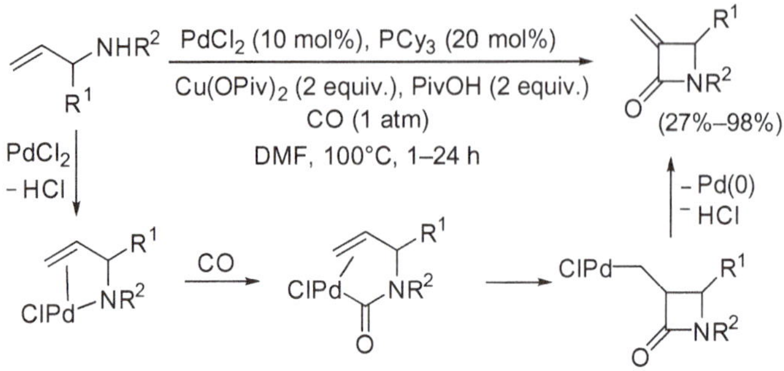

SCHEME 3.6 Synthesis of β-lactam derivatives by $PdCl_2$-catalyzed oxidative cyclocarbonylation of allyl amines (R^1 = H, alkyl; R^2 = alkyl, aryl).

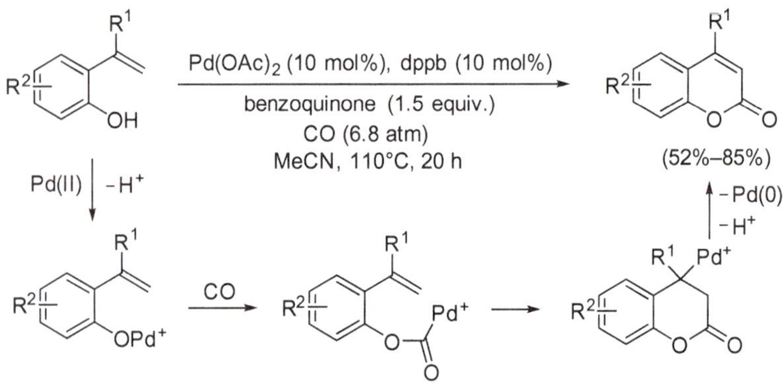

SCHEME 3.7 Synthesis of coumarins by $Pd(OAc)_2$-catalyzed oxidative cyclocarbonylation of 2-vinylphenols (R^1 = H, Me, Ph; R^2 = H, Cl, Br, Me, OEt).

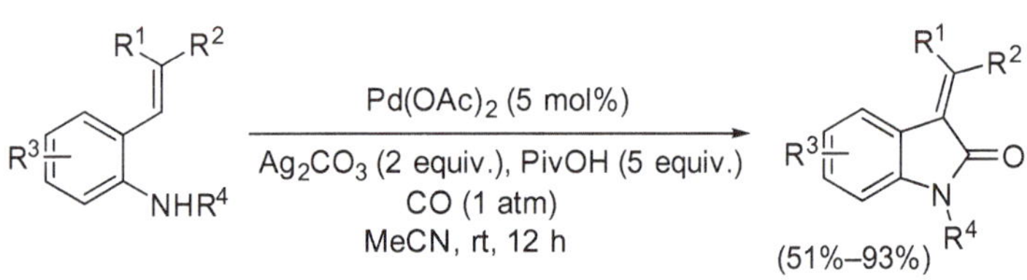

SCHEME 3.8 Synthesis of 3-methyleneindolin-2-ones by $Pd(OAc)_2$-catalyzed oxidative cyclocarbonylation of 2-vinylanilines (R^1, R^2 = H, alkyl, aryl, CO_2Bu, CN; R^3 = H, Me, Cl, COMe; R^4 = alkyl, benzyl, allyl, Ac).

An interesting $PdCl_2$-catalyzed multicomponent oxidative cyclocarbonylation process was recently developed by Yang and coworkers.[109] It involved the reaction between homoallylic alcohol and 3-arylpropiolates in an ionic liquid (IL) as the solvent (1-carboxymethyl-3-methylimidazolium chloride, $[C_2O_2mim]Cl$) to afford γ-lactone derivatives through a proposed mechanism involving *anti*-chloropalladation of the triple bond followed by double bond insertion, CO insertion, and intramolecular nucleophilic displacement

(Scheme 3.9). The ensuing Pd(0) species was then reoxidized by the Wacker-type $CuCl_2/O_2$ oxidation system.[109] It should be noted, however, that as found relatively often in the oxidative carbonylation literature employing O_2 as the oxidizing agent,[17] the conditions employed (CO: O_2 = 3:1) were within the explosion limits for a CO–O_2 mixture.[110]

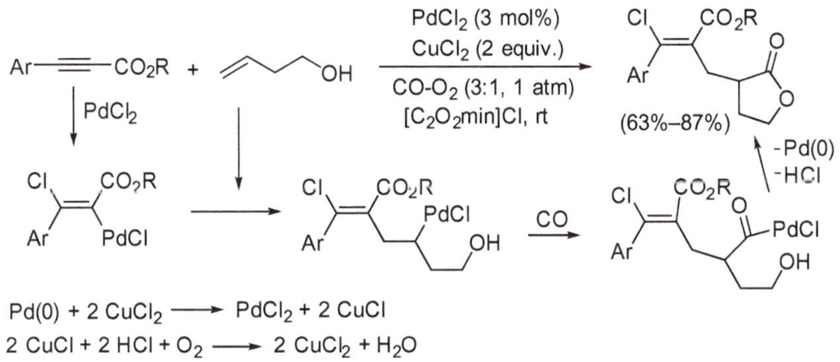

SCHEME 3.9 Synthesis of γ-lactone derivatives by $PdCl_2$-catalyzed oxidative cyclocarbonylation of homoallylic alcohol with 3-arylpropiolates (Ar = aryl, R = ethyl, allyl, phenyl).

3.2.1.2 *With Acetylenic C–H Activation*

As shown in Scheme 3.2B, under suitable oxidative carbonylation conditions, a direct *sp* C–H activation of the terminal triple bond of 1-alkynes may take place, with formation of propiolate derivatives.[90,98,99] In particular, our research group reported the first example of a catalytic oxidative monoaminocarbonylation of alkyl- and arylacetylenes to give 2-ynamides in 2001.[98] The reaction was performed using secondary amines as nucleophiles, at 100°C in dioxane as the solvent and under 20 atm of a 4:1 mixture of CO–air in the presence of the PdI_2/KI catalytic system, already successfully employed for the oxidative dicarbonylation (see Section 3.2.1.1). The basic conditions favored the direct ≡CH palladation, with formation of an alkynylpalladium species, from which the final product was obtained via CO insertion and nucleophilic displacement by the amine (Scheme 3.10).[98]

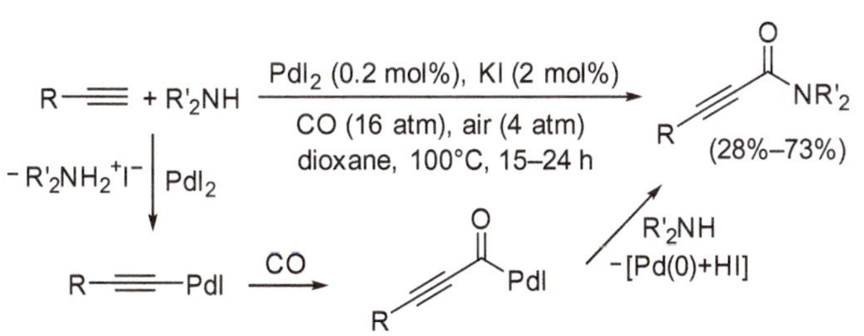

SCHEME 3.10 Synthesis of propiolic amides by PdI_2/KI-catalyzed oxidative monoaminocarbonylation of terminal alkynes (R = alkyl, aryl).

With suitably substituted alkyne substrates, this kind of reactivity may result in a cyclocarbonylation process leading to heterocyclic derivatives, according to the general concepts shown in Scheme 3.11. In pathway *a* the initially formed 2-ynamide intermediate undergoes external conjugate addition by an amine, with formation of an intermediate from which the final product is formed by lactonization or lactamization; 4-dialkylamino-5*H*-furan-2-ones[111] and 4-dialkylamino-1,5-dihydropyrrol-2-ones[112] were synthesized by this approach. In pathway *b* the overall process consists of two concatenated catalytic cycles (oxidative monocarbonylation of the triple bond followed by cyclocarbonylation), both catalyzed by the same catalyst, and it is therefore an example of "auto-tandem catalysis."[113] 5-(Carbamoylmethylene)oxazolidinones, starting from α,α-disubstituted 2-ynylamines, were obtained by this latter approach (with water promoting the cyclocarbonylation step, Scheme 3.12), in conventional organic solvents (such as 1,2-dimethoxyethane, DME)[114] or, more recently, in the IL 1-ethyl-3-methylimidazolium ethyl sulfate ($EmimEtSO_4$) as unconventional solvent (Scheme 3.13).[115] In this latter case the catalyst/IL system could be successfully recycled.[115]

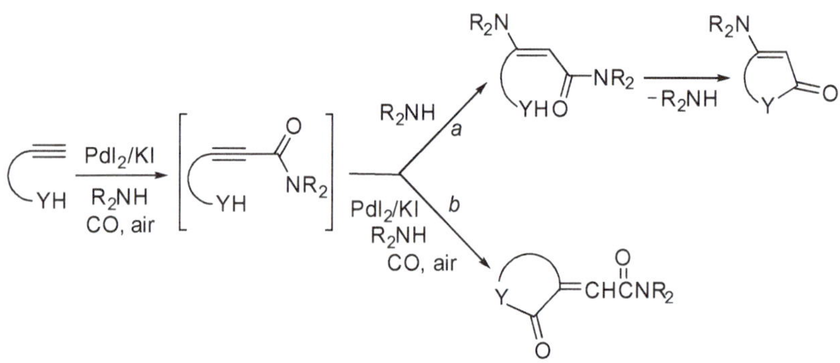

SCHEME 3.11 Synthesis of heterocycles by PdI_2/KI-catalyzed oxidative monoaminocarbonylation followed by either intermolecular conjugate addition and cyclization (pathway *a*) or oxidative cyclocarbonylation (pathway *b*) (Y = heteronucleophile).

Imidazothiazinones were also recently successfully synthesized from 2-(prop-2-ynylthio)imidazoles.[116] In this case, as shown in Scheme 3.14, in the cyclocarbonylation cycle the imidazole nitrogen acted as heteronucleophile, eventually leading to the final bicyclic products, through CO insertion, triple bond insertion, protonolysis, and isomerization.[116] In a similar way, benzimidazothiazinones were obtained from 2-prop-2-ynylsulfanyl-1*H*-benzimidazolium salts.[117]

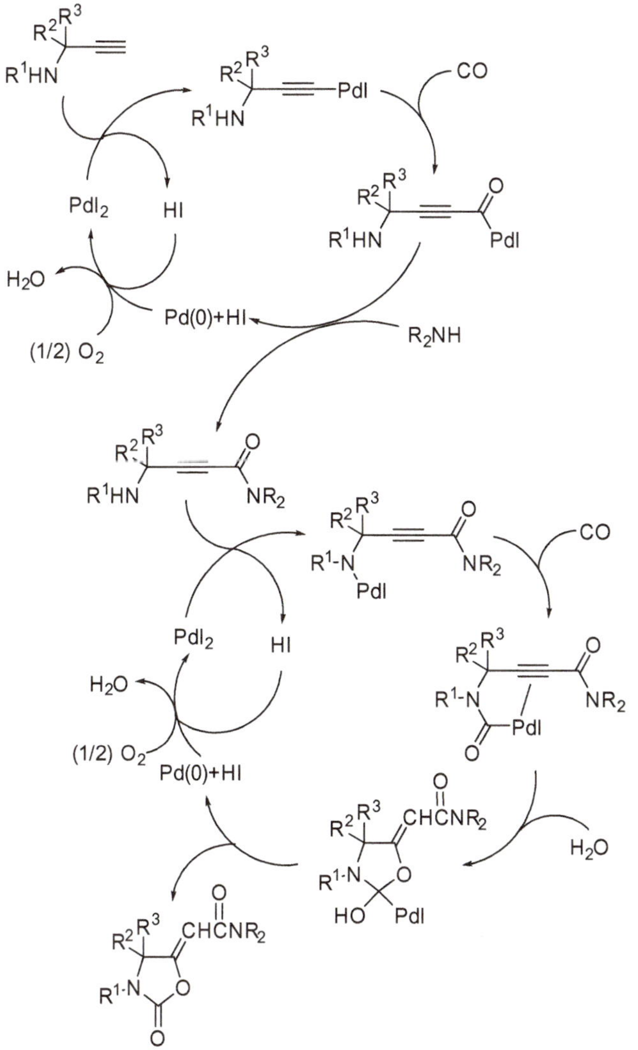

SCHEME 3.12 Formation of functionalized oxazolidinones by PdI_2/KI-promoted auto-tandem catalysis: oxidative monoaminocarbonylation of α,α-disubstituted 2-ynylamines followed by oxidative cyclocarbonylation (R^1, R^2, R^3, R = alkyl).

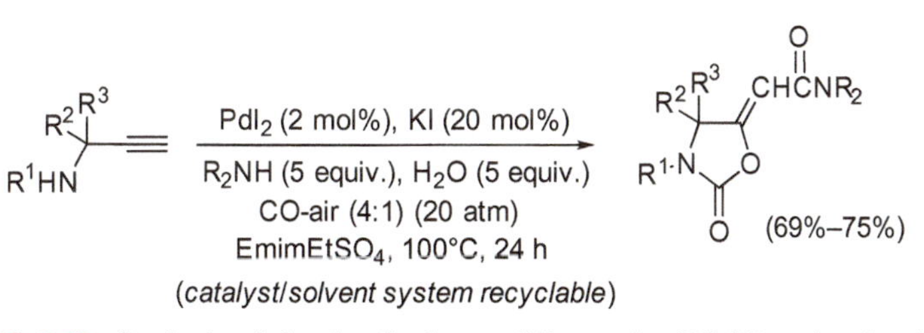

SCHEME 3.13 Synthesis of functionalized oxazolidinones by PdI_2/KI-catalyzed oxidative monoaminocarbonylation followed by oxidative cyclocarbonylation in $EmimEtSO_4$ (R^1, R^2, R^3, R = alkyl).

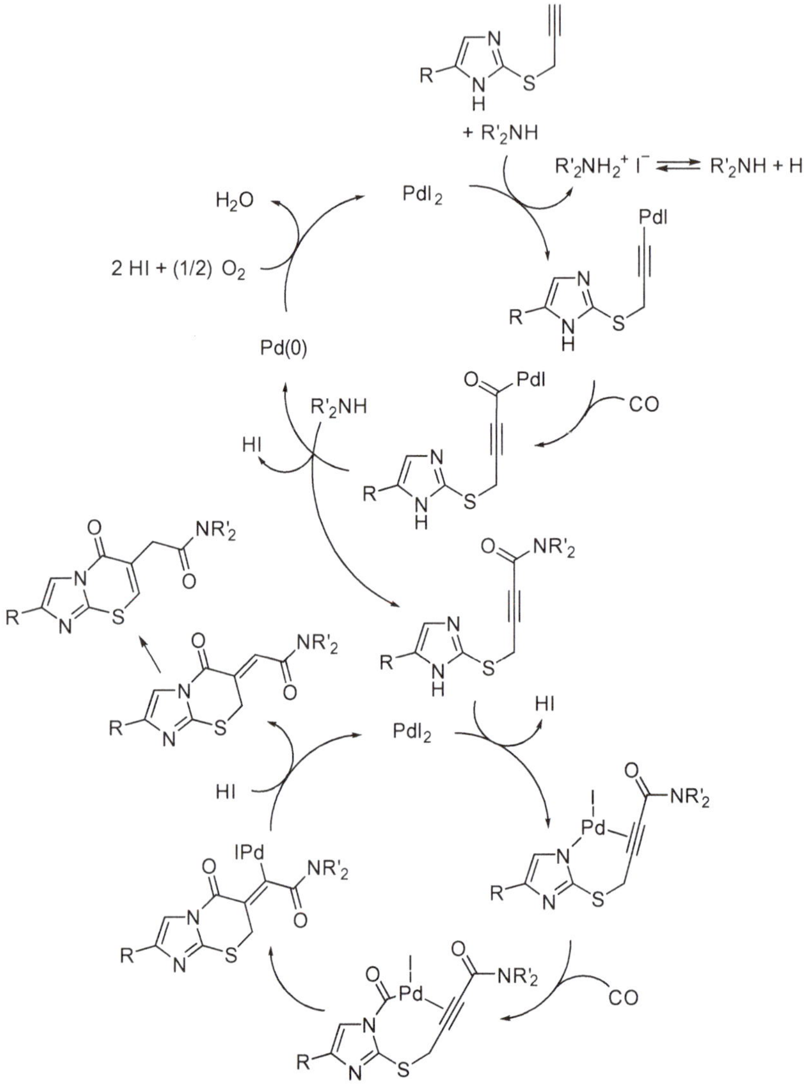

SCHEME 3.14 Synthesis of imidazothiazinones by PdI_2/KI-promoted auto-tandem catalysis: oxidative monoaminocarbonylation of 2-(prop-2-ynylthio)imidazoles followed by oxidative cyclocarbonylation (R = H, alkyl, aryl; R′ = alkyl).

3.2.1.3 With Alkenylic C–H Activation

Pd(II)-promoted olefinic C–H activation may also take place under suitable conditions, leading to heterocycles by cyclocarbonylation. Thus 1,3-oxazin-6-ones were obtained by the reaction of *N*-(1-arylvinyl)amides with CO (1 atm) in DMF at 80°C with $Pd(OAc)_2$ as catalyst and in the presence of KI, Ac_2O, 1,4-diazabicyclo[2.2.2]octane (DABCO) as base, and Cu $(OAc)_2$ as oxidant.[118] As shown in Scheme 3.15 the proposed mechanism starts with sp^2 C–H palladation followed by CO insertion, nitrogen deprotonation, and reductive elimination.[118] The same process was recently also performed under photoredox catalysis using O_2 as oxidant and $Ru(bpy)_2Cl_2$ [(bpy) = 2,2'-bipyridine] as photosensitizer in the presence of 4,5-bis(diphenylphosphino)-9,9-dimethylxanthene (xantphos) as ligand and under irradiation of 3 W blue LEDs for 6–26 h.[119]

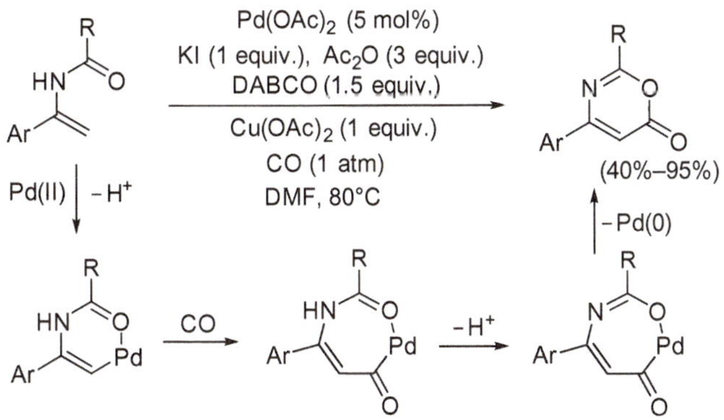

SCHEME 3.15 Synthesis of 1,3-oxazin-6-ones by $Pd(OAc)_2$-catalyzed oxidative cyclocarbonylation of *N*-(1-arylvinyl)amides (R = Me, Et, Ph; Ar = aryl).

In a related work, and under similar conditions, pyrido[1,2-*a*]pyrimidin-4-ones were synthesized in moderate to high yields starting from *N*-(1-arylethylidene)pyridin-2-amines (Scheme 3.16).[120]

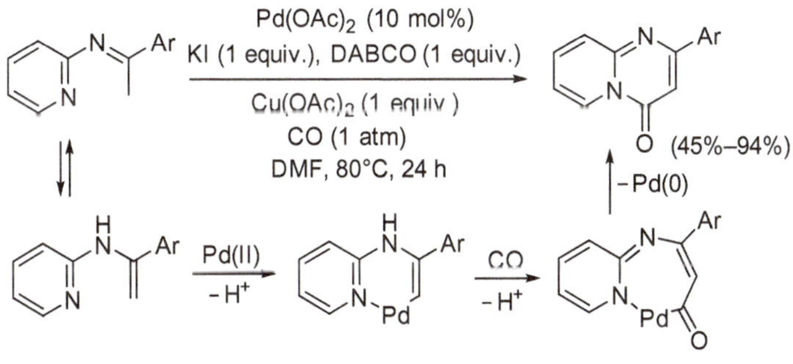

SCHEME 3.16 Synthesis of pyrido[1,2-*a*]pyrimidin-4-ones by $Pd(OAc)_2$-catalyzed oxidative cyclocarbonylation of *N*-(1-arylethylidene)pyridin-2-amines (Ar = aryl).

3.2.1.4 With Aromatic C–H Activation

Pd(II)-mediated aromatic C–H activation, combined with oxidative cyclocarbonylation, is a powerful tool for the synthesis of carbonylated heterocycles, and impressive progress has been made in this field over the past years.[16–18] When starting from an aromatic substrate bearing a heteronucleophilic substituent YH, the process may occur through either *ortho*-palladation followed by CO insertion and reductive elimination (Scheme 3.17, pathway *a*) or by palladation of the heteroatom Y followed by CO insertion, *ortho*-palladation and reductive elimination (Scheme 3.17, pathway *b*) (in this latter case, intramolecular *ortho*-palladation by the $-YPd^+$ species may also precede CO insertion). Sequential double aromatic palladation may also take place with substrates in which two aryl groups are bonded to a heteroaromatic group (Scheme 3.18).

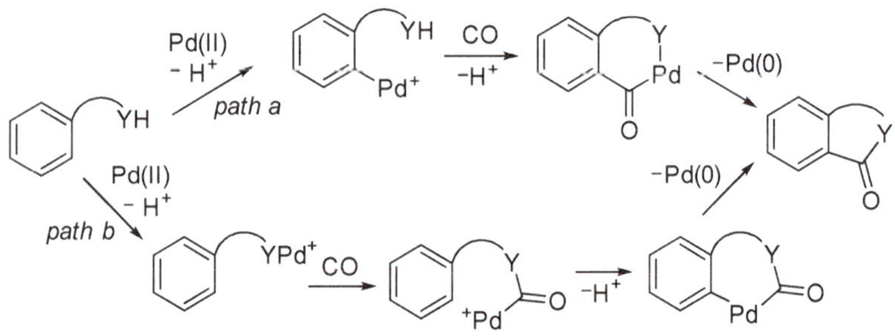

SCHEME 3.17 Different pathways leading to heterocycles by Pd(II)-catalyzed oxidative cyclocarbonylation with aromatic C–H activation (YH = heteronucleophilic group).

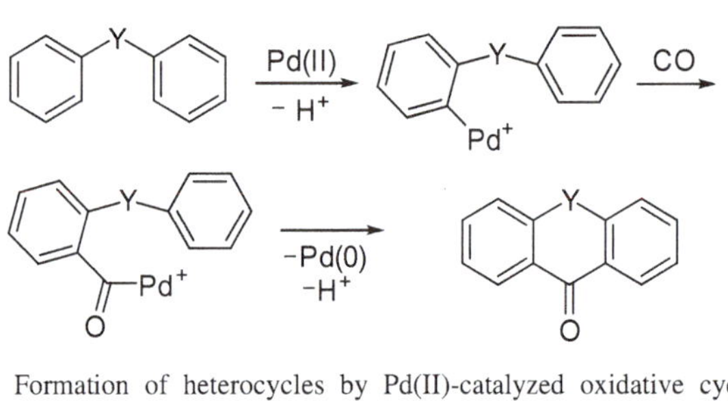

SCHEME 3.18 Formation of heterocycles by Pd(II)-catalyzed oxidative cyclocarbonylation with double aromatic C–H activation (Y = heteroatom).

Some benzo[*c*]phenanthridinones were synthesized by $Pd(OAc)_2$-catalyzed oxidative cyclocarbonylation of alkoxy-substituted 1-amino-2-phenyltetralins and 1-amino-2-phenylnaphthalenes, according to the mechanism shown in Scheme 3.17, pathway *a*, as exemplified in Scheme 3.19 (DMSO = dimethylsulphoxide).[121]

In a similar way, oxaphosphorinanones (Scheme 3.20A) (DCE = 1,2-dichloroethane),[122] isochromenoquinolinones (Scheme 3.20B),[123] benzoxazinones (Scheme 3.20C) (dba = dibenzylideneacetone; TsOH = tosic acid),[124]

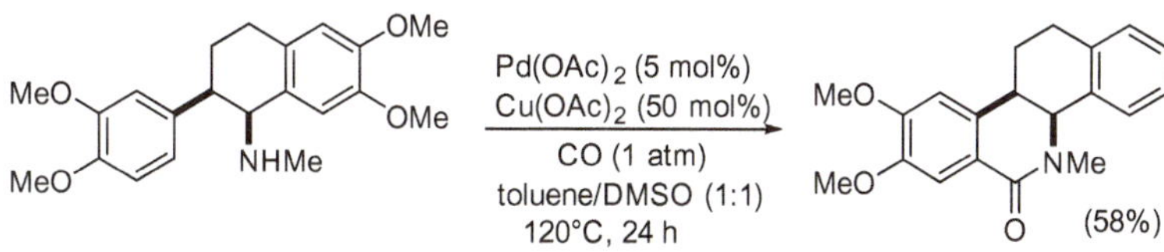

SCHEME 3.19 Synthesis of 2,3,8,9-tetramethoxy-5-methyl-4*b*,5,11,12-tetrahydrobenzo[*c*]phenanthridin-6(10b*H*)-one by Pd(OAc)$_2$-catalyzed cyclocarbonylation of 2-(3,4-dimethoxyphenyl)-6,7-dimethoxy-*N*-methyl-1,2,3,4-tetrahydronaphthalen-1-amine with aromatic C–H activation.

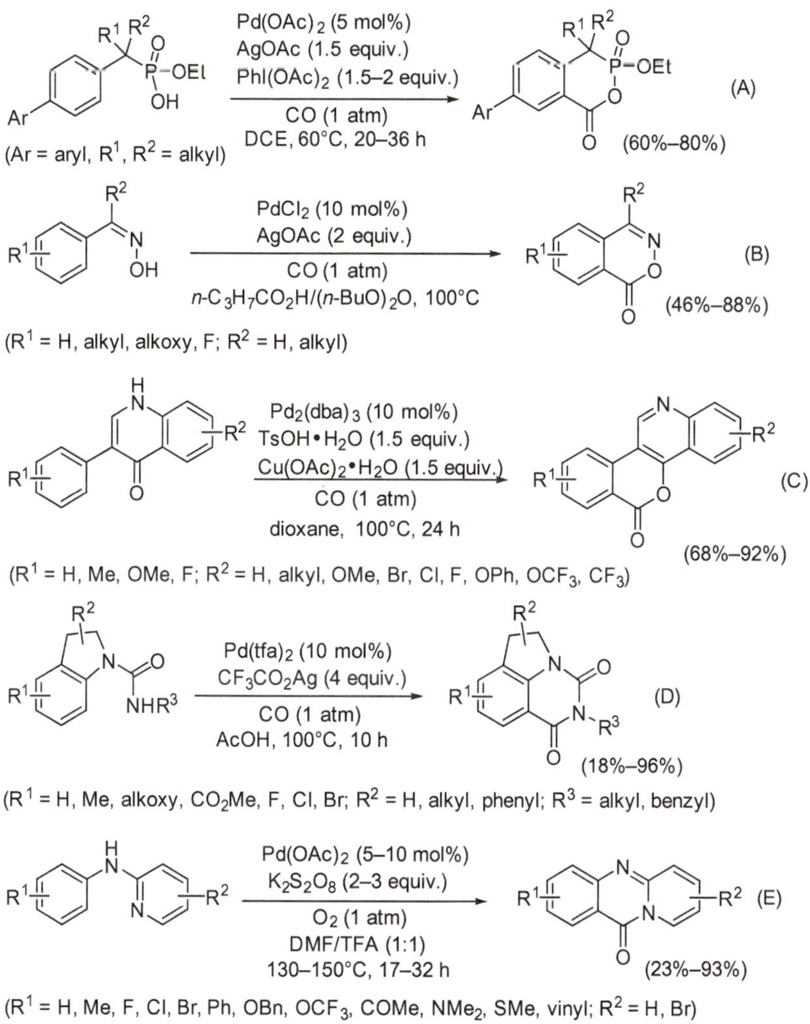

SCHEME 3.20 Synthesis of miscellaneous heterocycles by Pd(II)-catalyzed cyclocarbonylation with aromatic C–H activation.

pyrroloquinazolinediones (Scheme 3.20D) (tfa = trifluoroacetate),[125] and pyridoquinazolinones (Scheme 3.20E) (TFA = trifluoroacetic acid)[126,127] were recently synthesized; in this latter case, DMF could be used as the in situ CO source. The synthesis of isatoic anhydrides from *N*-methyl anilines has also been realized, as shown in Scheme 3.21.[128] The proposed mechanism involves the initial Pd(II)-catalyzed oxidative carbonylation of the substrates to give an *N*-methylanthanilic acid followed by Pd(II)-catalyzed oxidative cyclocarbonylation of the latter (Scheme 3.21).[128,129]

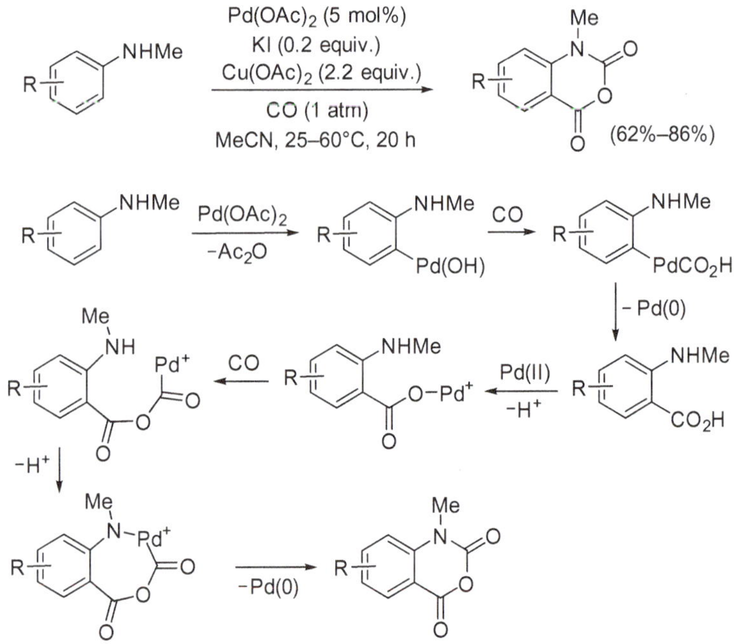

SCHEME 3.21 Formation of isatoic anhydrides by Pd(II)-catalyzed oxidative cyclocarbonylation of *N*-methyl anilines with aromatic C–H activation (R = H, Me, OMe, F, Cl, Br, NO_2, CHO, COMe, CO_2Et).

The mechanistic pathway starting with palladation of the heteroatom Y (Scheme 3.17, pathway *b*), either by deprotonation or Y-alkyl cleavage, has been proposed in the formation of a variety of heterocycles, such as dibenzopyranones (Scheme 3.22A),[130,131] benzo[*cd*]indolones (Scheme 3.22B) (DMA = *N,N*-dimethylacetamide),[132] and phenanthridinones (Scheme 3.22C).[133] The possibility for a double CO insertion was also recently successfully verified by the Lei's group (Scheme 3.22D).[134] Note, however, that in some of the processes shown in Scheme 3.19 the reactions were carried out within the explosion limits for a CO/O_2 mixture.[110]

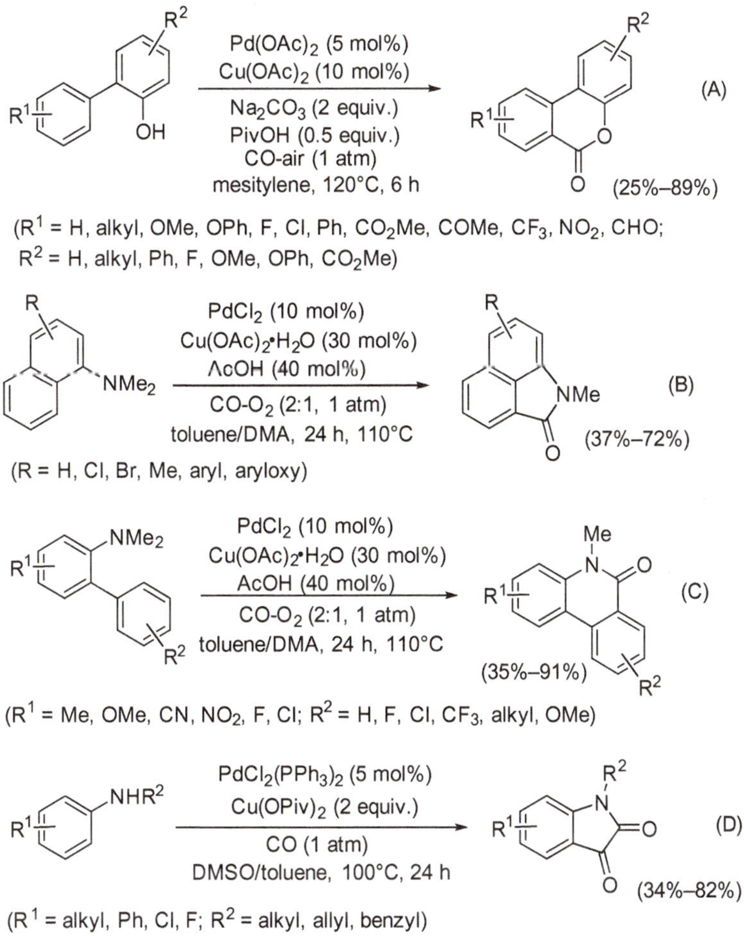

SCHEME 3.22 Synthesis of miscellaneous heterocycles by Pd(II)-catalyzed cyclocarbonylation with aromatic C–H activation.

In some cases the aromatic substrate can also be coupled with another unsaturated compound. Thus 3-methyleneisoindolin-2-ones were obtained by the oxidative carbonylation of tertiary anilines with α-olefins, through a proposed mechanism involving *ortho*-palladation, double bond insertion, β-H elimination to give 2-vinyl-*N*,*N*-dimethylanilines, followed by *N*-dealkylative palladation, CO insertion, intramolecular double bond insertion and β-H elimination (Scheme 3.23) [(dppp = 1,3-bis(diphenylphosphino)propane].[135] In a similar way the coupling of *N*-aryl-pyridine-2-amines with norbornene[136] or with internal alkynes to give 2-quinolinones (Scheme 3.24A)[137] and of anilines with arynes [formed in situ from 2-(trimethylsilyl)aryl triflates] to give phenanthridinones (Scheme 3.24B) (TMS = trimethylsilyl)[138] were also recently reported. In the first two cases, $[Mo(CO)_6]$ was used as a solid CO source.[136,137]

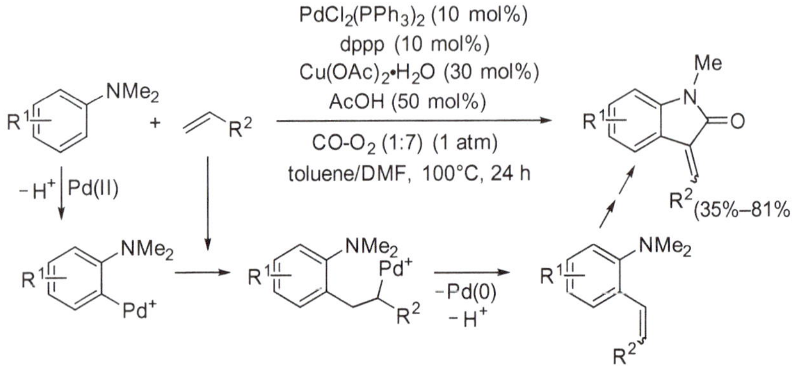

SCHEME 3.23 Synthesis of 3-methyleneisoindolin-2-ones by Pd(II)-catalyzed oxidative cyclocarbonylation of tertiary anilines with α-olefins with aromatic C–H activation (R^1 = H, Me, F, Cl, Br, CO_2Et, CN, CF_3; R^2 = aryl, CO_2Bu).

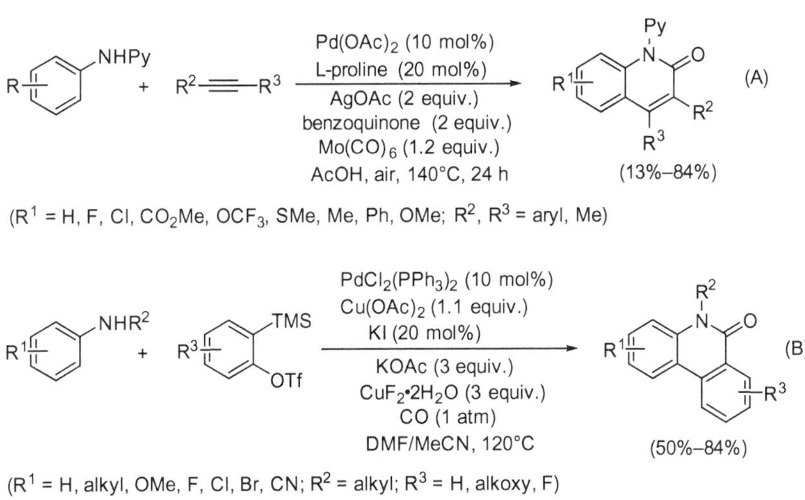

SCHEME 3.24 Synthesis of 2-quinolinones (A) and phenanthridinones (B) by Pd(II)-catalyzed multicomponent oxidative cyclocarbonylation with aromatic C–H activation.

Cyclocarbonylation with double aromatic C–H activation of Ar-Y-Ar′ (Y = heteroatom, Scheme 3.18) is a very interesting and innovative approach for the synthesis of polycyclic heterocycles. The group of Lei reported the synthesis of xanthones (Scheme 3.25A)[139] and acridones (Scheme 3.25B)[140] (DTBP = di-*tert*-butyl peroxide) by employing this methodology.

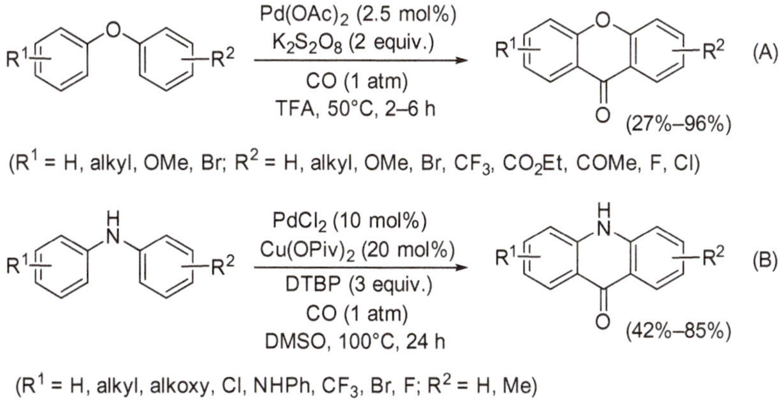

SCHEME 3.25 Synthesis of xanthones (A) and acridones (B) by Pd(II)-catalyzed oxidative cyclocarbonylation with double aromatic C–H activation.

3.2.1.5 *With Aliphatic C–H Activation*

The more challenging aliphatic sp^3 C–H activation has also been achieved. As exemplified in Scheme 3.26, β-lactams were obtained by $Pd(OAc)_2$-catalyzed oxidative cyclocarbonylation of suitably substituted cyclic aliphatic amines under 1 atm of CO–air and in the presence of $Cu(OAc)_2$, through a 5-membered ring cyclopalladation pathway.[141]

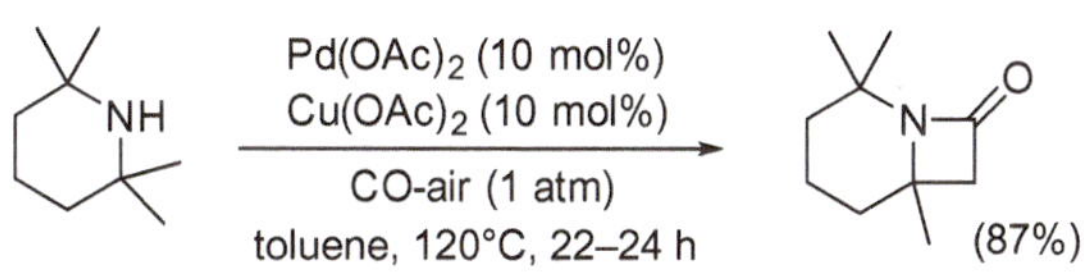

SCHEME 3.26 Synthesis of β-lactams through Pd(II)-catalyzed oxidative cyclocarbonylation of cyclic aliphatic amines with sp^3 C–H activation.

3.2.1.6 *With Diols, β-Amino Alcohols, Diamines, and Related Compounds*

The Pd(II)-catalyzed oxidative cyclocarbonylation of diols, β-amino alcohols, and diamines is an excellent method for the direct synthesis of very important classes of heterocycles, that are, cyclic carbonates, oxazolidin-2-ones, and cyclic ureas, respectively.[17,142–148] The cyclocarbonylation process occurs through the general mechanism outlined in Scheme 3.27, involving palladation of the YH group (Y = NR or OR) followed by carbon monoxide insertion and intramolecular nucleophilic displacement (possibly, via the formation of a palladacycle followed by reductive elimination).

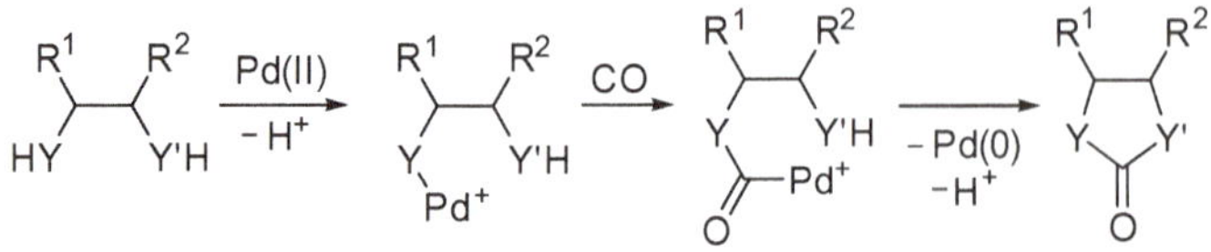

SCHEME 3.27 Formation of cyclic carbonates, oxazolidin-2-ones, or cyclic ureas by Pd(II)-catalyzed oxidative cyclocarbonylation of diols, β-amino alcohols, and diamines, respectively (Y, Y′ = NR, O).

The PdI_2/KI catalytic system, successfully employed by our research group for the oxidative carbonylation of alkynes (see Section 3.2.1.1), also proved valuable for the oxidative cyclocarbonylation of β-amino alcohols to oxazolidin-2-ones,[142,143] diamines to give cyclic ureas[144] and 1,2- and 1,3-diols to give 5- and 6-membered cyclic carbonates (including glycerol carbonate from glycerol).[145] More recently the oxidative cyclocarbonylation of β-amino alcohols and 2-aminophenols to oxazolidin-2-ones and benzoxazolones in the IL 1-butyl-3-methylimidazolium tetrafluoroborate ($BmimBF_4$) as nonconventional solvent, with the possibility to recycle the catalyst, was also achieved (Scheme 3.28).[149]

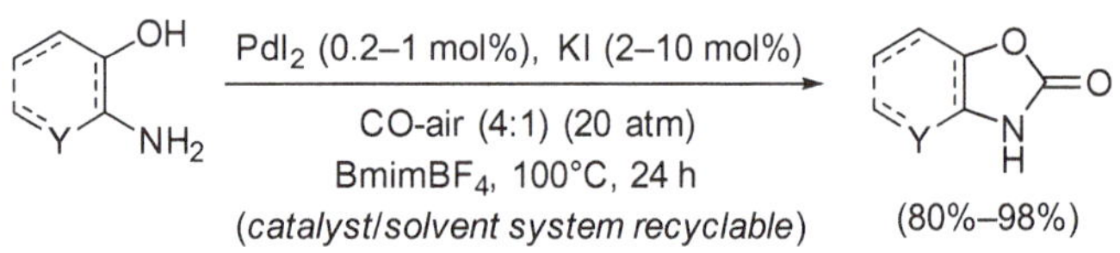

SCHEME 3.28 Synthesis of oxazolidinones and benzoxazolones by PdI_2/KI-catalyzed oxidative cyclocarbonylation of β-amino alcohols and 2-aminophenols in $BmimBF_4$ (Y = CH, N).

The oxidative cyclocarbonylation of aryl hydrazides to give oxadiazolones was also recently reported, by the groups of Chen and coworkers[150] and Jiang and coworkers[151] independently, using different reaction conditions (conditions A[150] or B[151] of Scheme 3.29) [dppf = 1,1-bis(diphenylphosphino)ferrocene].

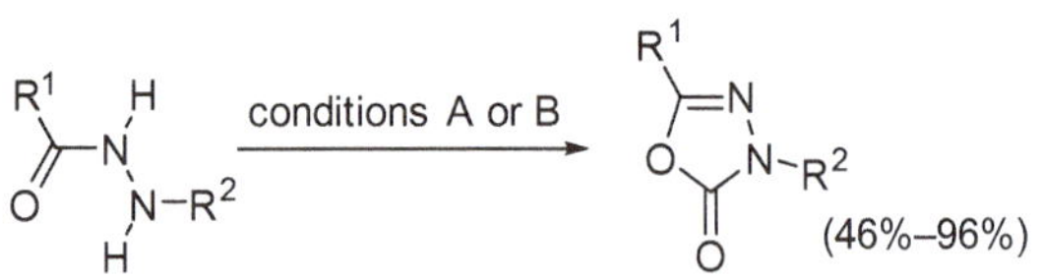

conditions A: $Pd(OAc)_2$ (10 mol%), dppf (20 mol%), $Cu(OAc)_2 \cdot H_2O$ (1 equiv.), NaOAc, CO (1 atm), toluene, 80°C (R^1, R^2 = aryl)

conditions B: $Pd(tfa)_2$ (5 mol%), CuO (1.5 equiv.) or Ag_2CO_3 (2 equiv.) CO (1 atm), TFA (3 equiv.), MeCN, 60°C, 12 h (R^1 = aryl, R^2 = H, aryl)

SCHEME 3.29 Synthesis of oxadiazolones by Pd(II)-catalyzed oxidative cyclocarbonylation of aryl hydrazides.

3.2.2 Under Nonoxidative Conditions

3.2.2.1 With Functionalized Halides

Aryl and vinyl halides, when suitably functionalized with a heteronucleophilic group YH (Y = O, NR), may lead to carbonylated heterocycles by palladium-catalyzed cyclocarbonylation.[1–3,16,18,152–155] As shown in Scheme 3.30 the process is catalyzed by Pd(0), so it does not require the presence of any external oxidant. The mechanism starts with the oxidative addition of the C–X bond (usually, X = I or Br) to Pd(0) followed by CO insertion and intramolecular nucleophilic displacement by YH (possibly, via the formation of a palladacycle followed by reductive elimination). Since HX is formally obtained from the process, the reaction is usually carried out in the presence of a base. The process is essentially the intramolecular version of the classical Heck alkoxy- and aminocarbonylation of halides.[1–3]

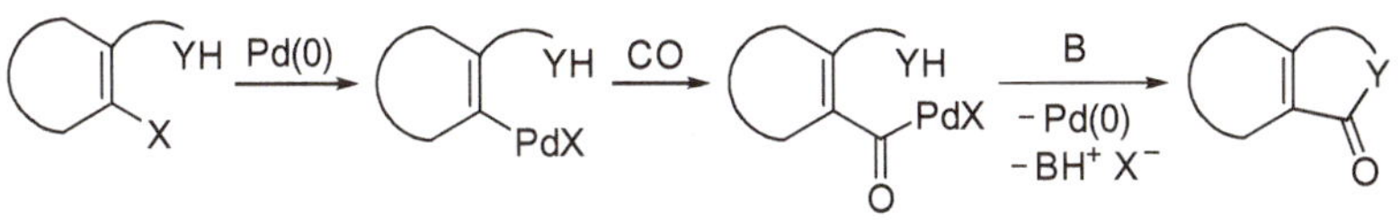

SCHEME 3.30 General mechanism for the Pd(0)-catalyzed cyclocarbonylation of functionalized aryl or vinyl halides (Y = O, NR; X = I, Br; B = base).

Several important examples of cyclocarbonylation of functionalized halides have been reported in the recent literature. Thus maleic anhydrides were obtained by cyclocarbonylation of β-bromo-α,β-unsaturated acids, using $PdCl_2$/dppp as catalytic precursor and Et_3N as base under 10 atm of CO[156] (Scheme 3.31A), and phthalimides were synthesized from 2-halobenzoic acids or esters in the presence of a nucleophilic primary amine.[157,158] In a similar way, *N*-protected α-fluoroalkyl 2-iodobenzylamines were converted into the corresponding isoindolinones using $Pd(PPh_3)_4$ as catalyst under atmospheric pressure of CO[159] (Scheme 3.31B) (PG = protecting group; Boc = *tert*-butoxycarbonyl; Cbz = carbobenzyloxy). The cyclocarbonylation approach also proved valuable for the carbonylative construction of polycyclic heterocycles, as exemplified in Scheme 3.31C.[160] Carbon monoxide surrogates were also successfully employed in place of gaseous CO, as demonstrated by the synthesis of phthalimides from 2-iodobenzamides with phenyl formate as CO source[161] (Scheme 3.31D) (TMEDA = tetramethylethylenediamine) or by the synthesis of benzoxazinones from 2-haloanilides with phenyl formate[161] (Scheme 3.31E, conditions A) or paraformaldehyde[162] (Scheme 3.31E, conditions B) as CO source.

Aryl halides bearing a heterocyclic nucleophile in a suitable position have also been effectively cyclocarbonylated, as exemplified in Scheme 3.32 (dibpp = diisobutylphosphinopropane; Ad = adamantyl).[163–165]

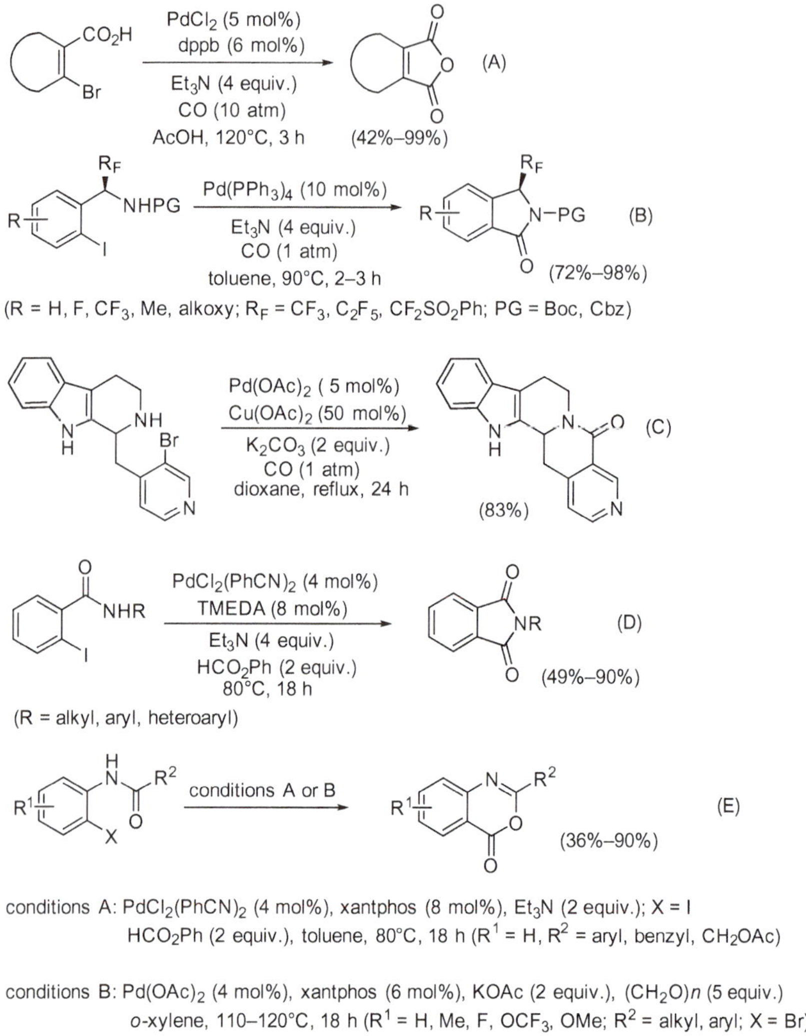

SCHEME 3.31 Synthesis of miscellaneous heterocycles by Pd(0)-catalyzed cyclocarbonylation of suitably functionalized aryl halides.

Interestingly, sequential double cyclocarbonylation has also been recently reported, as shown in Scheme 3.33 for the synthesis of isochromenoquinolindiones starting from *N*-(2-bromophenyl)-2-(2-iodophenyl)acetamides.[166]

In another interesting cascade process, α-chloroketones were converted into 4-hydroxy-2-pyranones through dimerization of the acylketene ensuing from oxidative addition to Pd(0), CO insertion, and β-H elimination (Scheme 3.34) (THF = tetrahydrofuran).[167]

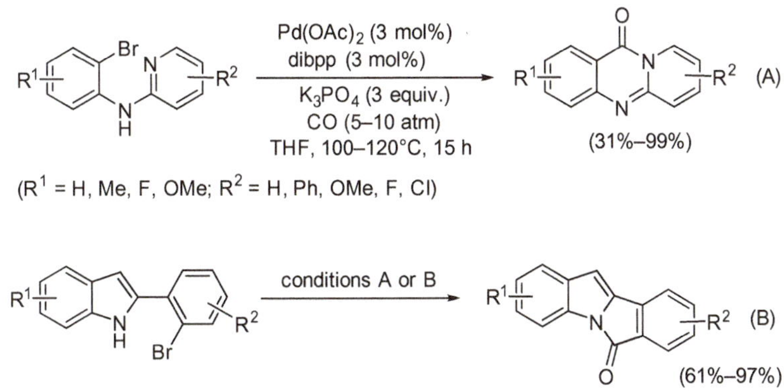

conditions A: $PdCl_2$ (5 mol%), PPh_3 (10 mol%), Et_3N (6 equiv.), CO (10 atm) toluene, 110°C, 5 h (R^1 = H, alkyl, Cl; R^2 = H, OMe)

conditions B: $Pd(OAc)_2$ (5–10 mol%), $BuPAd_2$ (15–30 mol%), DABCO (3 equiv.), CO (1 atm) DMSO, 120°C,12 h (R^1 = H, Me, Cl; R^2 = H, Me, OMe, F, Cl)

SCHEME 3.32 Synthesis of pyridoquinazolinones (A) and isoindoloindolones (B) by Pd(0)-catalyzed cyclocarbonylations.

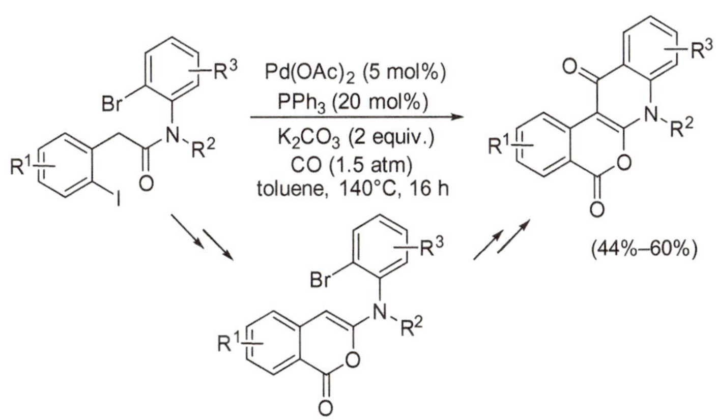

SCHEME 3.33 Synthesis of isochromenoquinolindiones by Pd(0)-catalyzed sequential double cyclocarbonylation (R^1 = H, OMe; R^2 = Me, Bn; R^3 = H, Me, F).

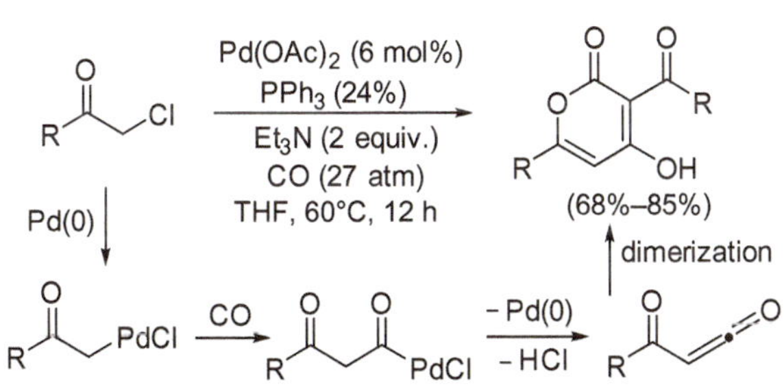

SCHEME 3.34 Synthesis of 4-hydroxy-2-pyranones by Pd(0)-catalyzed sequential carbonylation—acylketene dimerization (R = alkyl, aryl).

3.2.2.2 With Halides Together With Nucleophilic Partners

Of particular interest is the possibility to directly synthesize a carbonylated heterocyclic system starting from organic halides in conjunction with another building block, in a multicomponent fashion, and considerable efforts have been made in the recent literature for developing such procedures in an efficient and selective manner.[16,18] The additional component besides the halide may be a nucleophile, an electrophile, an alkene, or an alkyne. We will consider the case of a nucleophilic component first.

Simple aryl bromides have been cyclocarbonylated in the presence of suitably functionalized nucleophiles.[168] Thus with 2-hydroxyacetophenones a variety of flavones were efficiently synthesized, through a possible mechanism involving oxidative addition to Pd(0), CO insertion, nucleophilic displacement by the phenolic group of the 2-hydroxyacetophenone and intramolecular aldol condensation (Scheme 3.35) (DBU = 1,8-diazabicyclo [5.4.0]undec-7-ene).[169] In a similar way, quinazolinones were prepared from aryl bromides and 2-aminobenzamides[170] or 2-aminobenzonitrile (with the cyano group being hydrated in situ to the benzamide group).[171]

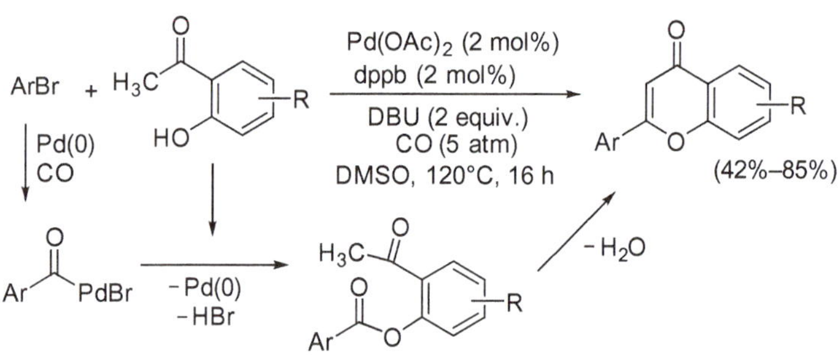

SCHEME 3.35 Synthesis of flavones by Pd(0)-catalyzed cyclocarbonylation of aryl bromides with 2-hydroxyacetophenones (Ar = aryl; R = H, Me, OMe, Cl, F).

Functionalized aryl halides can also serve as substrates for the cyclocarbonylation leading to heterocycles in the presence of external nucleophilic partners.[172–176] In particular, 1,2-dihaloarenes may undergo double carbonylative annulation in the presence of nitrogen nucleophiles.[177,178] As exemplified in Scheme 3.36, phthalimides[179,180] and thienopyrrolodiones[181] could be synthesized by this approach. With 2-aminobenzyl amine as nucleophile the double carbonylation led to the formation of isoindoloquinazolinones (Scheme 3.37).[182]

The groups of Beller and Wu found that when 2-bromofluorobenzenes are allowed to react in the presence of a suitable bidentate nucleophile, the initial aminocarbonylation of the C–Br bond is followed by intramolecular aromatic nucleophilic substitution, eventually leading to formation of a

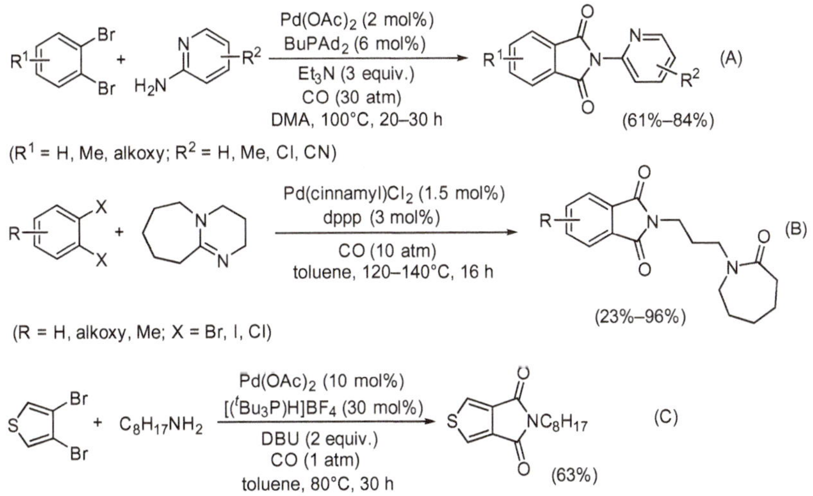

SCHEME 3.36 Synthesis of phthalimides (A and B) and thienopyrrolodiones (C) by Pd(0)-catalyzed double cyclocarbonylation.

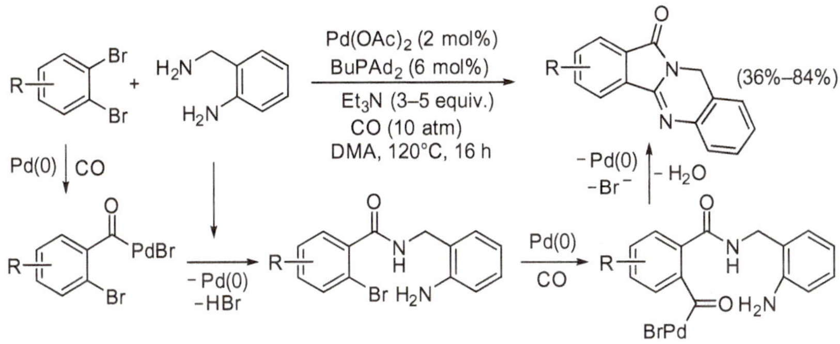

SCHEME 3.37 Synthesis of isoindoloquinazolinones by Pd(0)-catalyzed double cyclocarbonylation (R = H, Me, OMe).

cyclocarbonylated heterocycle. Accordingly, fused quinazolinones were obtained using 2-aminopyridine as nucleophilic partner, as exemplified in Scheme 3.38.[183] Interestingly the regiochemistry of the process toward the formation of the linear or the angular fused quinazolinone derivative could be modulated by the nature of the base employed: while DBU favored the formation of the linear isomer, Et_3N promoted the formation of the angular product.[183] In a similar way, dibenzodiazepinones (Scheme 3.39A),[184] pyridobenzoxazepinones (Scheme 3.39B),[185] and chromones (Scheme 3.39C)[186,187] were obtained from 2-fluorobromobenzenes or 3-bromo-2-chloropyridines with 2-aminophenols and 1,2-diarylethan-1-ones.

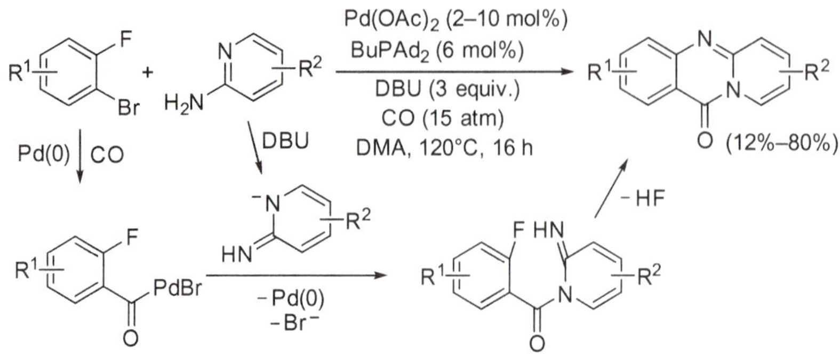

SCHEME 3.38 Synthesis of linear fused quinazolinones by Pd(0)-catalyzed carbonylation/intramolecular aromatic nucleophilic substitution (R^1 = H, Me, F, CHF_2, Cl, COMe; R^2 = H, Me, F).

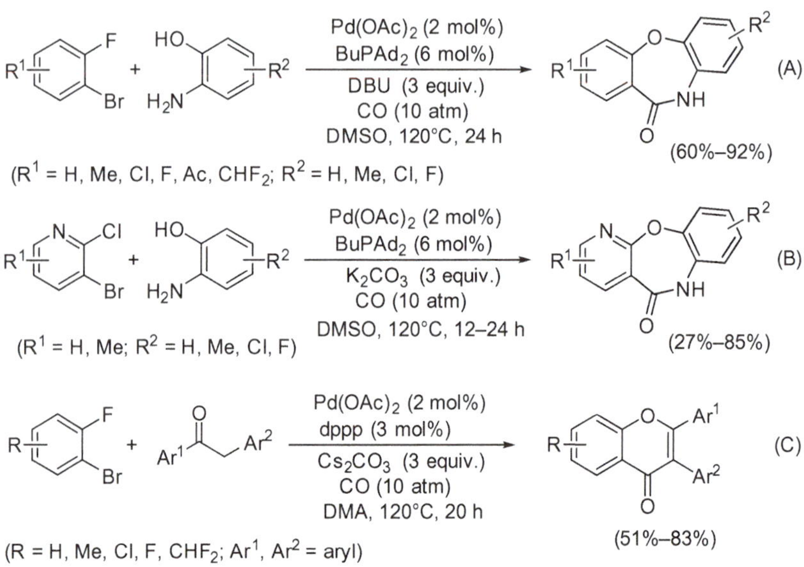

SCHEME 3.39 Synthesis of miscellaneous heterocycles by Pd(0)-catalyzed carbonylation/intramolecular aromatic nucleophilic substitution reactions.

Quinazolinone derivatives were conveniently synthesized from *N*-(*o*-bromophenyl)imidates and primary amines.[188] The possible reaction mechanism involves initial aminocarbonylation of the C–Br bond followed by intramolecular nucleophilic substitution on the iminoether moiety (Scheme 3.40).

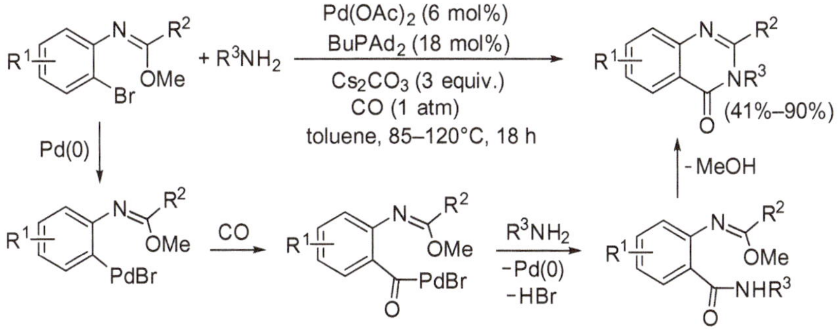

SCHEME 3.40 Synthesis of 4(3*H*)-quinazolinones from *N*-(*o*-bromophenyl)imidates and primary amines (R^1 = H, Me, OMe, CO_2Me, F, Cl; R^2 = phenyl, *t*-Bu, benzyl; R^3 = aryl, alkyl, benzyl, allyl).

In a similar way, quinazolinones were also obtained from 2-bromoanilines and orthoformates through the formation of 2-formamidobenzamides as intermediates (Scheme 3.41).[189] The reaction could also be conveniently performed using Pd/C as catalyst in the absence of phoshine ligands and with catalyst recovery.[190] 2-Bromoformanilides have also been directly used as substrates together with organic nitro compounds, with $Mo(CO)_6$ serving both as reducing agent for nitro compounds to give the corresponding amines and as carbon monoxide source.[191]

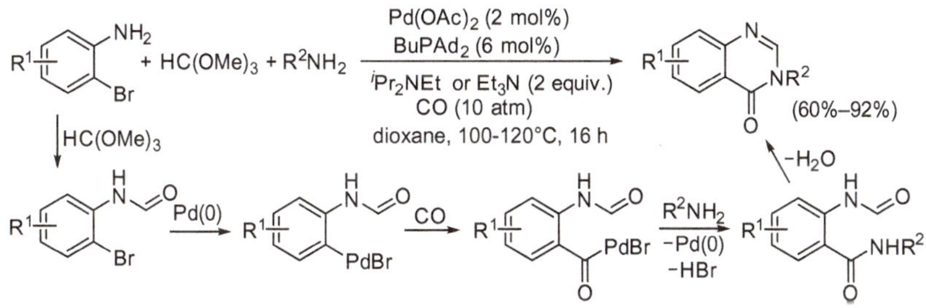

SCHEME 3.41 Synthesis of 4(3*H*)-quinazolinones from 2-bromoanilines by formylation/Pd(0)-catalyzed cyclocarbonylation (R^1 = H, Me, Cl, F, CF_3, OCF_3; R^2 = aryl, alkyl, benzyl).

When allowed to react with 2-bromobenzyl amines, 2-bromoanilines were converted into isoindoloquinazolinones.[192] In this case, in a possible mechanistic hypothesis the nucleophilic displacement by the 2-bromobenzyl amine was followed by intramolecular aminocarbonylation and intramolecular condensation, as shown in Scheme 3.42.[192]

2-Bromobenzaldehydes have also been successfully used as substrates in conjunction with a nucleophilic partner. For example, phthalazinones were synthesized using hydrazines as nucleophiles, through a mechanistic pathway involving condensation of the aldehyde with the amino group of the hydrazine followed by intramolecular aminocarbonylation[193] (Scheme 3.43).

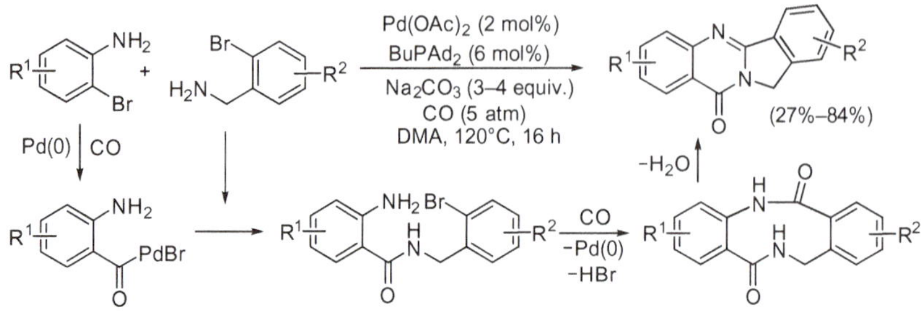

SCHEME 3.42 Synthesis of isoindolo[1,2-*b*]quinazolin-10(12*H*)-ones from 2-bromoanilines and 2-bromobenzyl amines (R^1 = H, Me, F, Cl, Ac, CF_3; R^2 = H, Me, F).

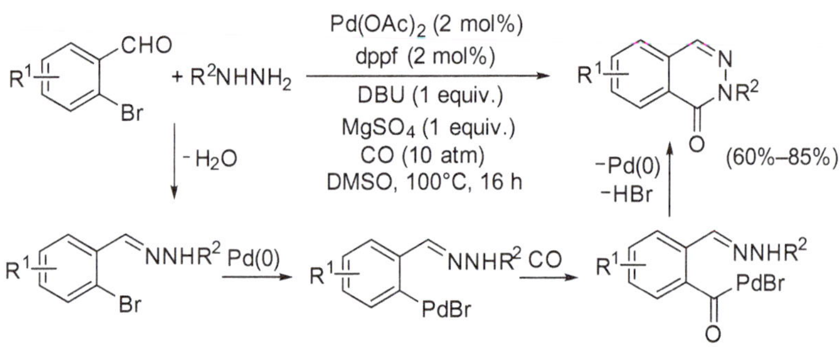

SCHEME 3.43 Synthesis of phthalazinones from 2-bromobenzaldehydes and hydrazines (R^1 = H, CF_3, F, Me, alkoxy; R^2 = Me, aryl).

On the other hand, functionalized isoindolinones were obtained from 2-bromobenzaldehyde and 2-bromoanilines, possibly via aminocarbonylation followed by cyclization and intramolecular alkoxycarbonylation[194] (Scheme 3.44). Interestingly the same products could also be obtained starting from

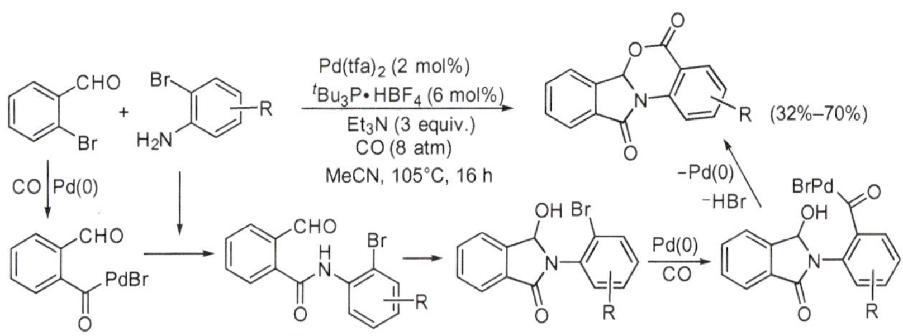

SCHEME 3.44 Synthesis of functionalized isoindolinones from 2-bromobenzaldehyde and 2-bromoanilines (R = H, Cl, Me, Ac).

2-formylbenzoic acids instead of 2-bromobenzaldehydes.[194] In a similar way, quinazolinediones were synthesized from 2-bromobenzonitriles and anilines, through aminocarbonylation, cyclization, imine isomerization, and further intramolecular aminocarbonylation (Scheme 3.45).[195]

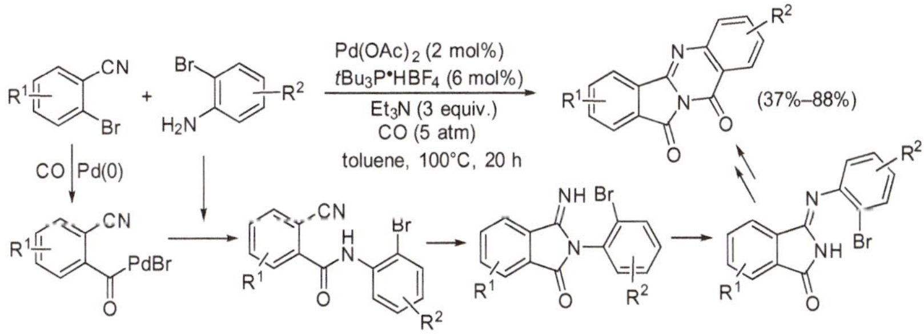

SCHEME 3.45 Synthesis of quinazolinediones from 2-bromobenzonitriles and 2-bromoanilines (R^1 = H, Me, OMe, F; R^2 = H, Cl, F, Ac, CN, CF_3).

As reported by the group of Wu and Beller, benzylic chlorides can be converted into chromenones when allowed to react with CO and salicylic aldehydes.[196] Oxidative addition to Pd(0) is followed by CO insertion, nucleophilic displacement by the phenolic group, and intramolecular aldol condensation (Scheme 3.46).[196] A similar sequence of steps (oxidative addition, CO insertion, nucleophilic displacement, and intramolecular condensation) would also account for the formation of pyrazol-3-one and pyrazol-5-one derivatives from α-chloroketones and hydrazines (Scheme 3.47).[197] An alternative pathway, starting from the condensation of the oxo group of α-chloroketones with hydrazines followed by cyclocarbonylation, could also explain the formation of pyrazol-5-ones.[197]

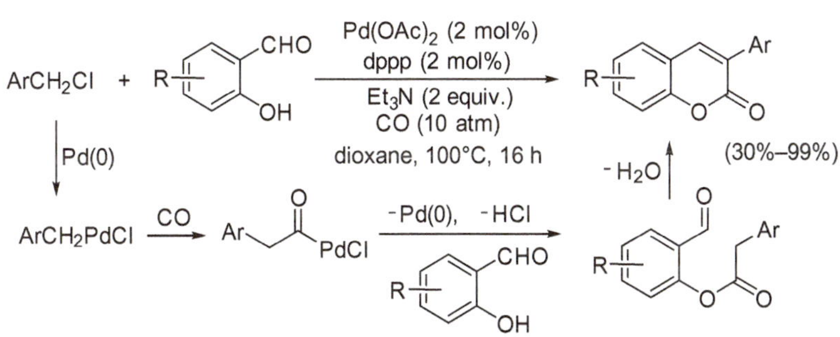

SCHEME 3.46 Synthesis of chromenones from benzyl chlorides and salicylic aldehydes (Ar = aryl; R = H, Me, CO_2Me, *t*-Bu, Cl, OMe).

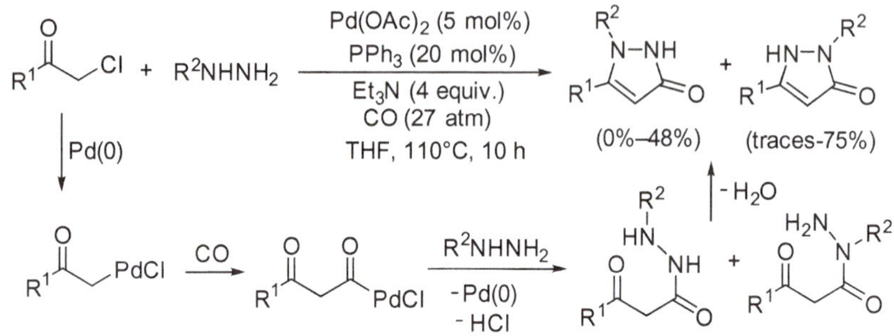

SCHEME 3.47 Formation of pyrazolones from α-chloroketones and hydrazines (R^1 = alkyl, aryl; R^2 = *t*-Bu, aryl).

3.2.2.3 With Halides Together With Electrophilic Partners

The carbonylation reaction of halides bearing a suitably placed nucleophilic group, carried out in the presence of an appropriate electrophilic molecule, may lead to cyclocarbonylated heterocycles when the reaction between the nucleophilic group and the electrophile leads to an intermediate which can undergo intramolecular nucleophilic displacement.[16,18,198–200] An interesting recent example was reported by the group of Alper, and consisted in the reaction between 2-iodoanilines with a Michael acceptor, such as diethyl ethoxycarbonylbutenedioate. The initial conjugate addition followed by oxidative addition, CO insertion, and intramolecular attack by the enolate function led to the final dihydroquinolinone derivatives (Scheme 3.48).[201]

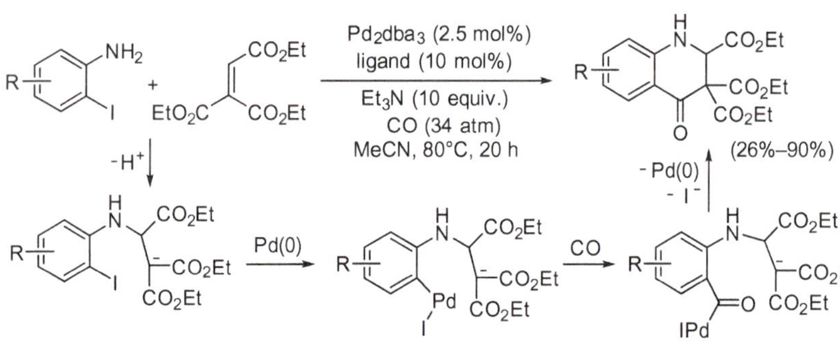

SCHEME 3.48 Synthesis of dihydroquinolinones from 2-iodoanilines and diethyl ethoxycarbonylbutenedioate [R = H, Cl, Me, OMe; ligand = 2-(di-*tert*-butylphosphino)biphenyl].

Wu and Beller reported the reaction between 2-bromoanilines and carboxylic acid anhydrides. Through a possible mechanism involving acylation followed by intramolecular alkoxycarbonylation, benzoxazinone derivatives were synthesized in fair to high yields (Scheme 3.49).[202] In a similar way the reaction of 2-bromoanilines with isocyanates, with $Mo(CO)_6$ as the carbon monoxide source, led to the formation of 2-aminobenzoxazinones

(Scheme 3.50A) [binap = (2,2'-bis(diphenylphosphino)-1,1'-binaphthyl)],[203] while dihydrobenzodioxepinones were synthesized from 2-bromophenols and epoxides (Scheme 3.50B).[204]

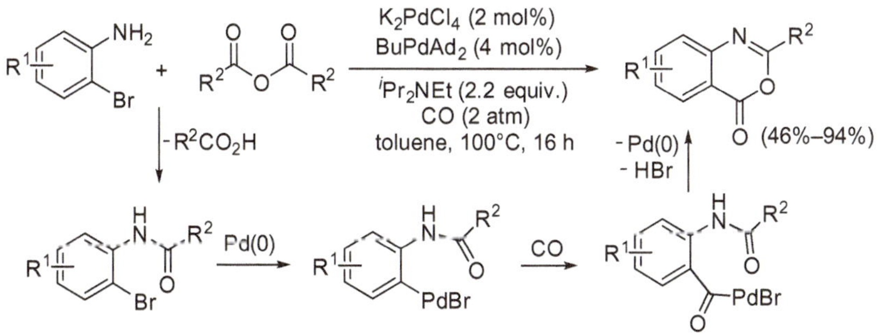

SCHEME 3.49 Synthesis of benzoxazinones from 2-bromoanilines and carboxylic acid anhydrides (R^1 = H, Me, F, Cl, Ac; R^2 = alkyl, alkenyl, CF_3, Ph).

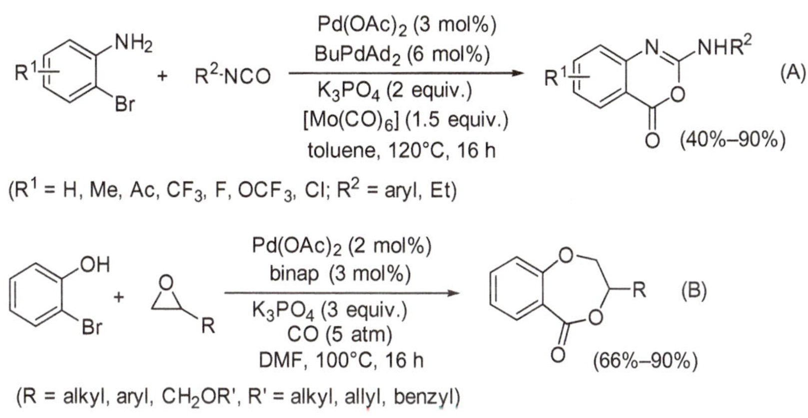

SCHEME 3.50 Synthesis of 2-aminobenzoxazinones from 2-bromoanilines and isocyanates (A) and of dihydrobenzodioxepinones from 2-bromophenols and epoxides (B).

3.2.2.4 With Halides Together With Alkenes and Imines

Functionalized halides can, under certain circumstances, be cyclocarbonylated together with alkenes and related compounds (such as allenes and imines) to give heterocycles.[16,18,23,205,206] An interesting double carbonylative Heck reaction of aryl halides and α-olefins was reported by the Beller's group, and afforded 5-arylfuranones (Scheme 3.51).[207] The proposed mechanism starts with the oxidative addition to Pd(0), followed by CO insertion, alkene insertion, deprotonation to form an enolate ligand (thus hindering the β-H elimination), further CO insertion, intramolecular nucleophilic displacement, and isomerization (Scheme 3.51).[207]

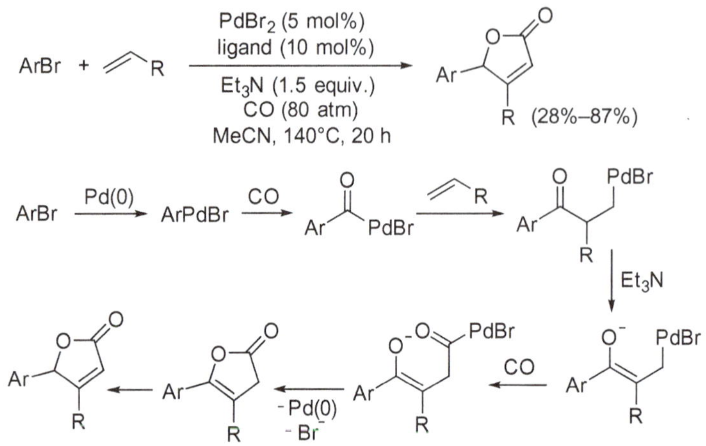

SCHEME 3.51 Synthesis of 5-arylfuranones from aryl bromides and α-olefins (Ar = aryl, heteroaryl; R = alkyl, $SiEt_3$; ligand = 2-(dicyclohexylphosphino)-1-(2-methoxyphenyl)-1*H*-pyrrole).

β-Lactams were synthesized by the reaction between benzyl chlorides and imines (Scheme 3.52).[208] For this process the authors proposed a mechanism involving the formation of a ketene as intermediate (from oxidative addition to Pd(0), CO insertion, and β-H elimination), from which the final product was obtained by [2 + 2] cycloaddition with the imine (Scheme 3.52).[208]

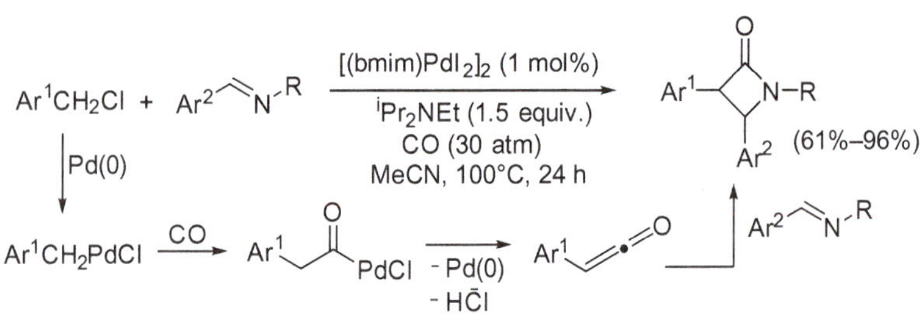

SCHEME 3.52 Synthesis of β-lactam derivatives from benzyl chlorides and imines (Ar^1, Ar^2 = aryl; R = benzyl, Ph, Pr, CH_2CO_2Me).

3.2.2.5 *With Halides Together With Alkynes and Arynes*

Alkynes are excellent partners in the palladium-catalyzed cyclocarbonylation of simple and suitably functionalized halides, which may lead to important heterocyclic derivatives.[16,18,23,209–219] In the case of aryl halides *ortho*-functionalized with a heteronucleophile and terminal alkynes, an initial carbonylative Sonogashira coupling may take place followed by intramolecular conjugate addition (Scheme 3.53, pathway *a*). On the other hand, for internal alkynes the oxidative addition to Pd(0) is followed by triple bond insertion, CO insertion, and intramolecular nucleophilic displacement (Scheme 3.53, pathway *b*).

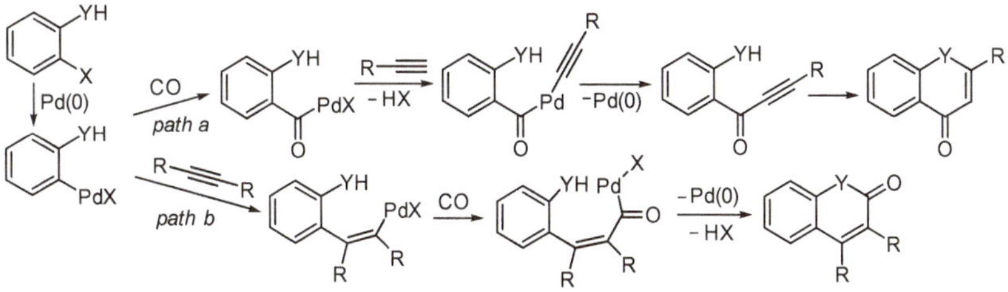

SCHEME 3.53 Different pathways for the formation of heterocycles from *ortho*-functionalized aryl halides (Y = heteroatom).

In a recent example, 2-iodophenol was efficiently coupled with terminal acetylenes to give flavones, in the presence of Pd/C as reusable catalyst and in the absence of added ligands, probably through pathway *a* of Scheme 3.53 (Scheme 3.54A).[220] In a similar way, thiochromenones were obtained from 1-fluoro-2-iodobenzenes, which were converted in situ to 2-iodothiophenates (Scheme 3.54B).[221] In this process the use of a reagent capsule was important to avoid the catalyst poisoning and to reduce the manual operations.[221] On the other hand a similar sequence of steps, but involving the conversion of the alkynylpalladium complex to a palladium enolate followed by CO insertion and reductive elimination, allowed the synthesis of 3-alkylidenefuran-2-ones from aryl iodides and benzyl acetylenes (Scheme 3.55).[222]

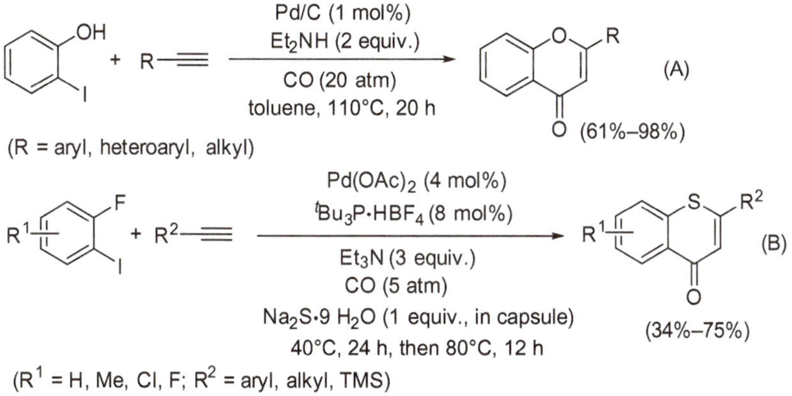

SCHEME 3.54 Synthesis of flavones from 2-iodophenols and terminal alkynes (A) and of thiochromenones from 1-fluoro-2-iodobenzenes, Na_2S, and terminal alkynes (B).

Interestingly, even arynes, obtained in situ from suitable precursors, have been successfully used as partner together with 2-iodoanilines in a cyclocarbonylation reaction leading to either phenanthridinones or acridones depending on reaction conditions.[223] As exemplified in Scheme 3.56 for the formation of phenanthridinones, aryne insertion took place after oxidative addition to Pd(0), followed by CO insertion and reductive elimination (TBAI = tetrabutylammonium iodide).[223]

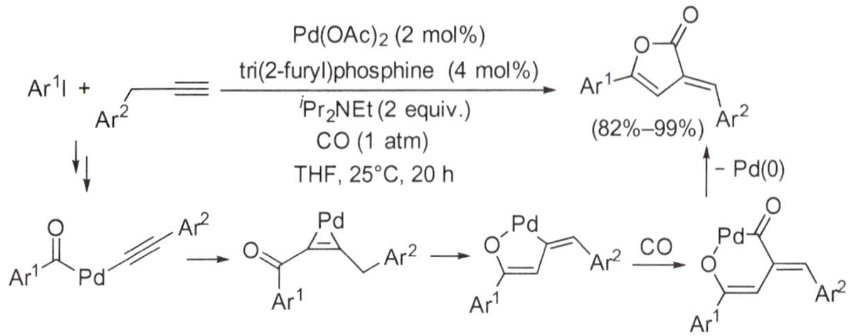

SCHEME 3.55 Synthesis of 3-alkylidenefuran-2-ones from aryl iodides and benzyl acetylenes (Ar^1 = aryl, heteroaryl; Ar^2 = aryl).

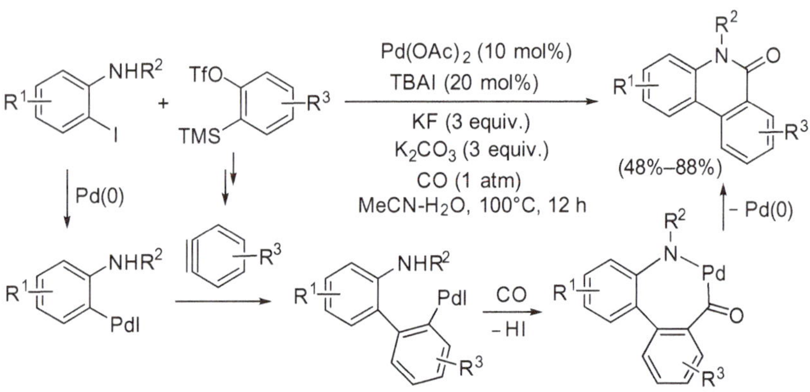

SCHEME 3.56 Synthesis of phenanthridinones from 2-iodoanilines and 2-(trimethylsilyl)phenyl trifluoromethanesulfonates as aryne precursors (R^1 = H, alkyl, CN, F, Cl, Br; R^2 = alkyl, benzyl; R^3 = H, alkoxy).

3.2.2.6 *With Other Substrates*

Other substrates besides halides can be cyclocarbonylated to afford heterocycles.[1–3,16,18,23,89,224–239] When carbonylated under reductive conditions, even simple alkynes can lead to important heterocyclic derivatives, such as furan-2(5*H*)-ones, through rather complicated mechanistic pathways.[224–228]

Alkenes and alkynes functionalized with a suitably placed nucleophilic group are excellent substrates for the synthesis of heterocycles by cyclocarbonylation reactions.[1–3,16,18,23,89,230–239] These processes can be catalyzed either by Pd(0) or Pd(II) species, through different mechanistic pathways, as exemplified in Scheme 3.57 in the case of unsaturated alcohols. In pathway *a*, referring to catalysis with Pd(0), the oxidative addition of the –OH group to Pd(0) takes place, followed by CO insertion, intramolecular double or triple bond insertion (only the *endo* cyclization mode is shown), and reductive elimination.[232] On the other hand, with Pd(II)-based catalysts, the formation of an alkoxycarbonylpalladium species may occur from the reaction between

PdX_2 (X = anionic counterion to Pd(II), e.g., a halide) and the alcoholic function of the substrate, followed by CO insertion; the alkoxycarbonylpalladium species then undergoes intramolecular double or triple bond insertion (only the *exo* cyclization mode is shown) and protonolysis to give the final product with regeneration of the Pd(II) catalyst (Scheme 3.57, pathway *b*).[233] The latter pathway can also be operative with unsaturated amines to give lactams.[234]

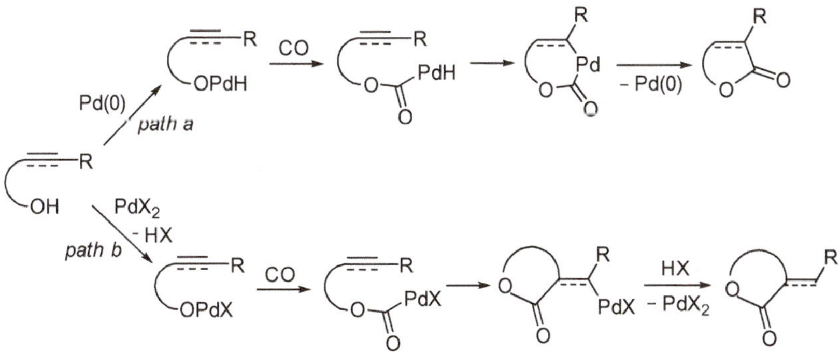

SCHEME 3.57 Possible mechanistic pathways leading to oxygen heterocycles by Pd(0)- or Pd (II)-catalyzed (pathways *a* and *b*, respectively) cyclocarbonylation of unsaturated alcohols.

Of particular interest is the cyclocarbonylation of allylic and propargylic alcohols and of suitably functionalized phenolic derivatives, which can be smoothly converted into lactones.[1–3,16,18,23,89] In particular the formation of seven-membered lactones was reported to occur from 2-allyl phenols under conditions of Pd(0) catalysis.[235–239] More recently, suitable conditions were found for the selective synthesis of dihydrocoumarins.[240] As shown in Scheme 3.58 the plausible mechanism starts with the oxidative addition of

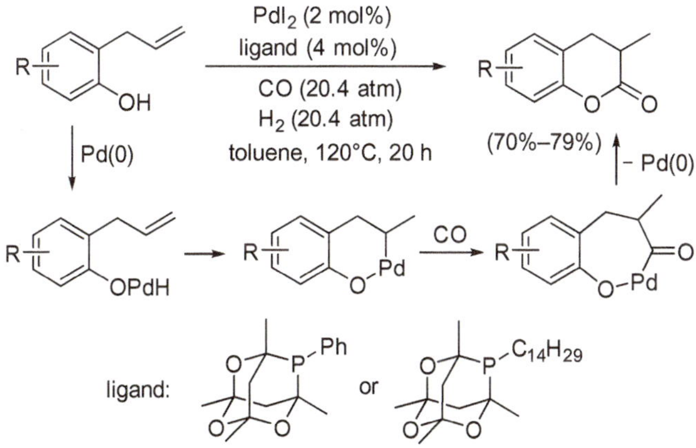

SCHEME 3.58 Synthesis of dihydrocoumarins by cyclocarbonylation of 2-allyl phenols (R = H, OMe, Cl, alkyl, Br).

the phenolic group to Pd(0) (readily formed in situ owing to the presence of hydrogen in the gaseous mixture), followed by *exo* double bond insertion, CO insertion, and reductive elimination.[240]

Recently, benzofuranones were obtained from 2-hydroxybenzyl alcohols using formic acid as the CO source, through a proposed mechanism involving benzylic palladation with elimination of water followed by CO insertion and nucleophilic displacement[241] (Scheme 3.59). Interestingly, 2-hydroxybenzyl alcohols could also be formed in situ from phenols and aldehydes under acidic conditions (CF_3CO_2H).[242]

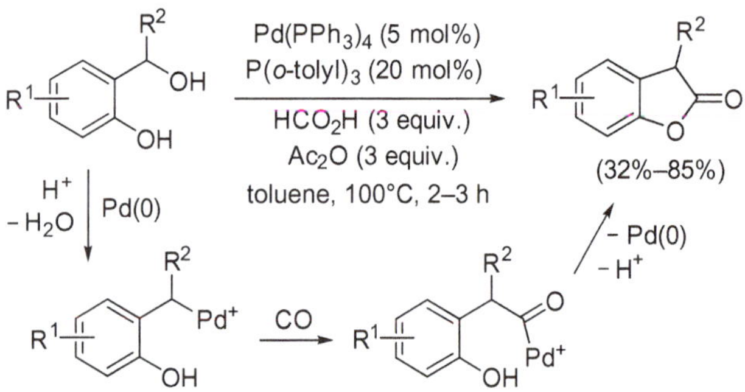

SCHEME 3.59 Synthesis of benzofuran-2(3*H*)-ones by cyclocarbonylation of 2-hydroxybenzyl alcohols (R^1 = H, alkyl, OMe, Br, Cl; R^2 = H, Et).

Our research group reported an innovative synthesis of coumarins under conditions of Pd(II) catalysis, using the PdI_2/KI catalyst already successfully employed for oxidative carbonylations (see Section 3.2.1.1). The method was based on carbonylation of 2-(1-hydroxyprop-2-ynyl)phenols carried out in MeOH as solvent and external nucleophile and under 90 atm of CO.[243] As shown in Scheme 3.60, formation of coumarin derivative could be interpreted as occurring through an ordered sequence of steps, involving: CO insertion into a palladium phenate bond, followed by intramolecular triple bond insertion, alkoxycarbonylation of the ensuing vinylpalladium complex to give an intermediate bearing an allylalcoholic moiety and a palladium hydride species, formation of a π-allyl complex with elimination of water, and protonolysis.[243] As can be seen from Scheme 3.60, Pd(II) was regenerated in the last step, so the process occurred without need for palladium reoxidation, and therefore under nonoxidative conditions. We recently found that the new coumarin derivatives thus obtained possess interesting phytotoxic and herbicidal activity.[244]

The PdI_2/KI catalytic system also proved valuable for the recently reported cyclocarbonylation of 2-(propynylthio)benzimidazoles to give thiadiazafluorenones, via *N*-palladation followed by CO insertion, intramolecular triple bond insertion, protonolysis, and isomerization[245] (Scheme 3.61).

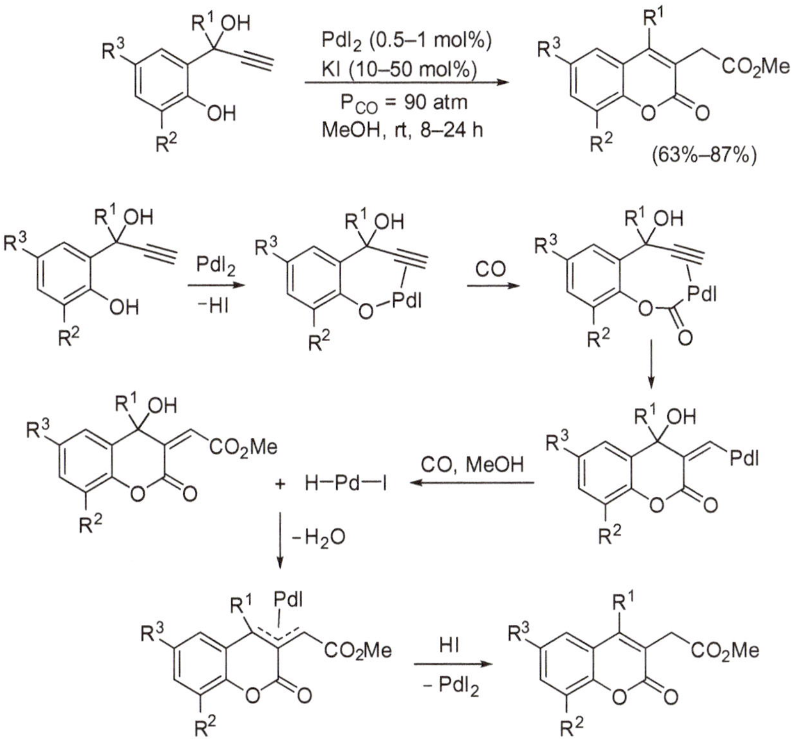

SCHEME 3.60 Synthesis of coumarin derivatives by cyclocarbonylation of 2-(1-hydroxyprop-2-ynyl)phenols (R^1 = H, Me, Ph; R^2 = H, OMe; R^3 = H, OMe, Cl).

Interestingly the reaction was carried out under a 5:2 mixture of $CO-CO_2$; the presence of CO_2 led to better results probably by favoring the final protonolysis step and/or by acting as ligand for palladium.[245]

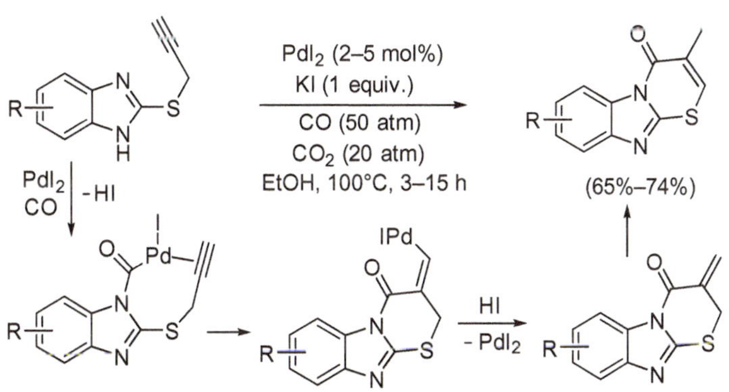

SCHEME 3.61 Synthesis of thiadiazafluorenones by cyclocarbonylation of 2-(propynylthio)benzimidazoles (R = H, Me, Cl, OMe, F).

3.3 PALLADIUM-CATALYZED CARBONYLATION PROCESSES LEADING TO HETEROCYCLES WITHOUT INCORPORATION OF CO INTO THE CYCLE

This section highlights the recent developments in the field of palladium-catalyzed carbonylation processes leading to heterocyclic derivatives, without incorporation of CO into the cycle. This means that, in the final product, CO is incorporated as an exocyclic carbonyl function.

3.3.1 Under Oxidative Conditions

Starting from a suitable precursor, and under appropriate conditions, Pd(II) species may promote carbonylative cyclization processes in which a heterocycle if formed without incorporation of CO into the cycle and with reduction of Pd(II) to Pd(0).[1–3,16–18,90] As explained in Section 3.2.1 a process like this can be made catalytic in palladium in the presence of a suitable oxidant able to reoxidize Pd(0) to Pd(II) (see Scheme 3.1).

The formation of a heterocyclic molecule by Pd(II)-catalyzed oxidative carbonylative annulation of an acyclic precursor without incorporation of CO into the cycle may in principle occur through different mechanistic pathways.[16–18] A first possibility corresponds to carbonylative carbocyclization of suitable substrates bearing a heteroatom into the chain undergoing annulation, as exemplified in Scheme 3.62A for a substrate with two suitably placed unsaturated groups and a heteroatom Y into the main chain. Cyclization through initial insertion of one of the unsaturated bond into a Pd–CONu bond (formed by the reaction between Pd(II), CO, and an external nucleophile NuH, such as an alcohol) followed by intramolecular insertion of the second unsaturated bond will lead to an intermediate that evolves through CO insertion and nucleophilic displacement by NuH to give the final product and Pd(0); the latter is then reoxidated to Pd(II) (Scheme 3.62A).[1–3,16–18,90] By way of example, pyrrole-3,4-diacetic esters[246] and dimethyl thiophene-3,4-diacetate[247] were synthesized in this manner, starting from dipropargylamines and dipropargyl sulfide, respectively, after carbonylation to give the corresponding dihydropyrrole or dihydrothiophene derivatives followed by base-promoted aromatization.

Another possible pathway starts with the intramolecular attack of a heteronucleophilic group –YH to an unsaturated bond activated by coordination to Pd(II), followed by carbon monoxide insertion and nucleophilic displacement by an external nucleophile NuH[1–3,16–18,90] (Scheme 3.62B; only the *exo* cyclization mode is shown). Many important examples were reported in the literature of processes of this kind, leading to important heterocyclic derivatives, from our[95,248–263] and other research groups.[264–284] Interestingly, with 4-ene-1,2-diols, 5-ene-1,2-diols and related substrates, the final

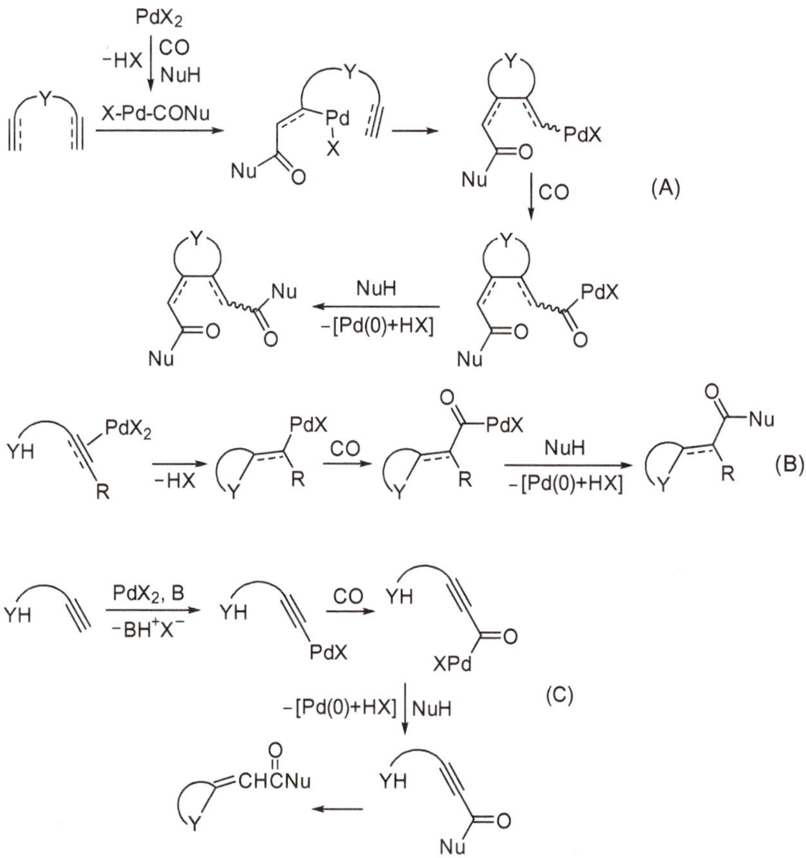

SCHEME 3.62 Different mechanistic pathways leading to carbonylated heterocycles by Pd(II)-catalyzed oxidative carbonylative annulation without incorporation of CO into the cycle (Y = heteroatom; X = counter anion to palladium, e.g., halide; NuH = external nucleophile; B = base). Pd(0) reoxidation occurs as reported in Scheme 3.1.

nucleophilic displacement may occur intramolecularly, with the formation of *bis*-heterocyclic lactones.[1–3,16–18,90,285–289]

Oxidative monocarbonylation of a terminal alkyne with *sp* C–H activation may also occur first (see Section 3.2.1.2), followed by intramolecular conjugate addition to the resulting α,β-unsaturated carbonyl compound (possibly, Pd(II) catalyzed) (Scheme 3.62C). Several examples of this kind of reactivity were reported in the literature from our research group, with PdI_2/KI as the catalytic system and through the formation of 2-ynamide intermediates.[290–293]

In this subsection, we focus on the most recent achievements in the area of oxidative carbonylative heterocyclizations with exocyclic CO incorporation.

3.3.1.1 Through Carbonylative Carbocyclization

A nice example of formation of heterocycles without the incorporation of CO into the cycle through oxidative carbonylative carbocyclization was reported recently. It consisted in the reaction of (3-en-1-yloxy)arenes carried out in the presence of $PdCl_2(MeCN)_2$ as catalyst, $CuCl_2$ as oxidant and an alcohol ROH as external nucleophile under 1 atm of CO at 0°C, to give chromane derivatives (Scheme 3.63).[294] The process probably occurs via electrophilic aromatic substitution by the double bond (activated by the coordination to Pd(II)), followed by alkoxycarbonylation (Scheme 3.63).[294] Interesting, starting from suitable aryl alkenols, intramolecular alkoxycarbonylation took place.

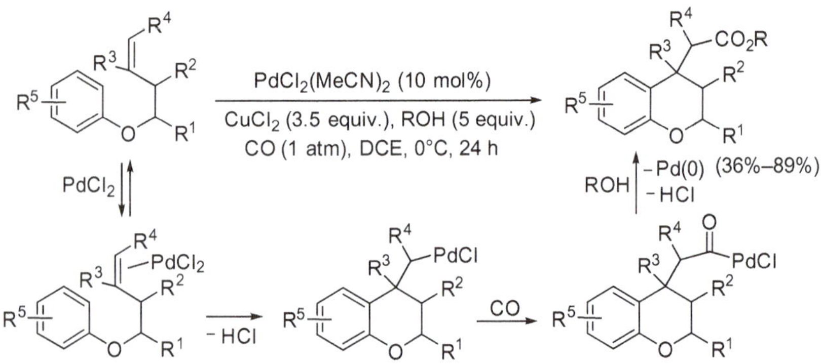

SCHEME 3.63 Synthesis of chromane derivatives from (3-en-1-yloxy)arenes through electrophilic aromatic substitution by the double bond (activated by the coordination to Pd(II)), followed by alkoxycarbonylation (R^1 = H, Me, Ph; R^2 = H, Ph; R^3 = H, Me; R^4=H, Me; R^5 = Me, alkoxy; R = alkyl).

3.3.1.2 Through Heterocyclization–Alkoxycarbonylation

The PdI_2/KI catalytic system, developed by our research group (see Section 3.2.1.1), is an excellent catalyst for realizing the synthesis of carbonylated heterocycles starting from suitably substituted acetylenic derivatives, through intramolecular attack of a heteronucleophilic group –YH to a triple bond activated by coordination to PdI_2, followed by carbon monoxide insertion and nucleophilic displacement by an external alcohol ROH, under oxidative conditions; both *exo* (Scheme 3.64, pathway *a*) and *endo* (Scheme 3.64, pathway *b*) cyclization modes are possible.[95,248–263]

Recently, we exploited this kind of reactivity (with an *endo* cyclization mode) for preparing indole-3-carboxylic esters starting from 2-alkynylanilines bearing an internal triple bond and a secondary amino group (Scheme 3.65).[295] It is worth noting that this process was different and complimentary with the reactivity shown by 2-ethynylanilines bearing a terminal triple bond, which, under similar conditions, afforded oxindole derivatives instead.[94] Interestingly, 2-alkynylanilines bearing an internal triple bond and

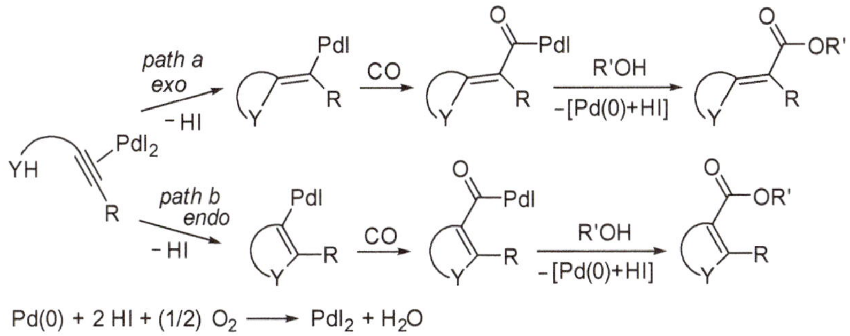

SCHEME 3.64 Formation of heterocycles with exocyclic incorporation of CO by PdI_2-catalyzed oxidative *exo* (pathway *a*) or *endo* (pathway *b*) heterocyclization–alkoxycarbonylation of suitably functionalized acetylenic substrates (Y = O, NR").

a primary amino group could be converted into 1-(dimethoxymethyl)indole-3-carboxylic esters by carrying out the carbonylation reaction in the presence of trimethyl orthoformate, through the formation of *N*-(dimethoxymethyl)-2-alkynylanilines as intermediates.[295]

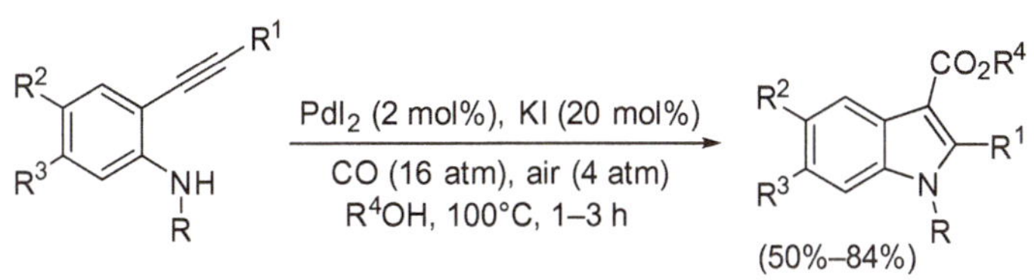

SCHEME 3.65 Synthesis of indole-3-carboxylic esters from 2-alkynylanilines bearing a secondary amino group (R = alkyl, benzyl; R^1 = aryl, alkyl; R^2 = H, Cl, Me; R^3 = H, CF_3, Me; R^4 = alkyl).

A general synthesis of furan-3-carboxylic esters[296] and pyrrole-3-carboxylic esters[297] was developed starting from 3-yne-1,2-diols and *N*-Boc-1-amino-3-yn-2-ols with alcohols ROH as external nucleophiles (Schemes 3.66 and 3.67, respectively). As exemplified in Scheme 3.66 for 3-yne-1,2-diols, the plausible mechanistic pathway involves heterocyclization followed by alkoxycarbonylation and dehydration (or vice versa). Interestingly, under similar conditions, 2-methyl-3-yne-1,2-diols, bearing a tertiary alcoholic group, gave 4-methylene-4,5-dihydrofuran-3-carboxylates.[296] In the case of amine derivatives (Y = NBoc) the carbonylation process led to a mixture of Boc-protected and *N*-unsubstituted pyrrole-3-carboxylic esters, which could be easily and quantitatively converted into the deprotected derivatives by subsequent treatment with NaOR (Scheme 3.67).[297] Moreover, with *N*-Boc-2-alkynyl-1-amino-3-yn-2-ols, bearing two germinal alkynyl groups, *N*-deprotection together with water addition to the second triple bond was observed during the oxidative carbonylation process, with selective formation of highly functionalized pyrroles (Scheme 3.68).[297]

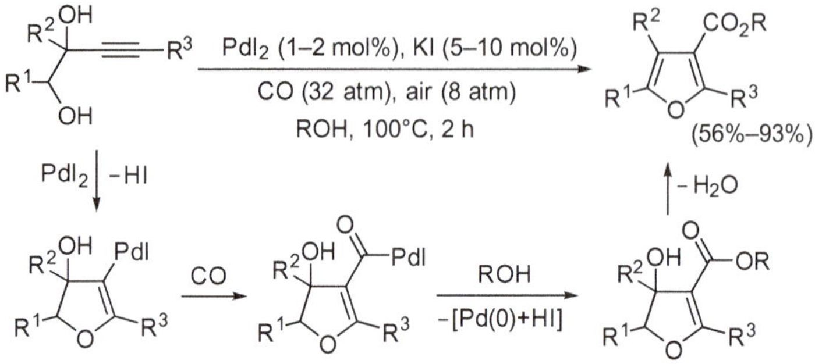

SCHEME 3.66 Synthesis of furan-3-carboxylic esters by PdI_2/KI-catalyzed oxidative carbonylation of 3-yne-1,2-diols (R^1 = H, alkyl; R^2 = H, alkyl, aryl, alkynyl; R^3 = alkyl, aryl, heteroaryl, alkenyl; R = alkyl).

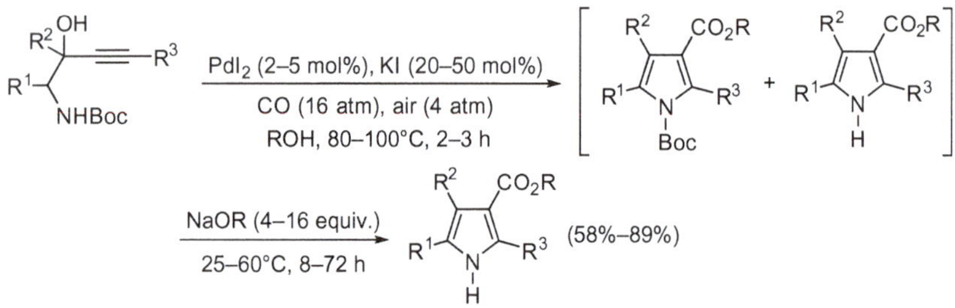

SCHEME 3.67 Synthesis of pyrrole-3-carboxylic esters by PdI_2/KI-catalyzed oxidative carbonylation of *N*-Boc-1-amino-3-yn-2-ols followed by basic treatment of the carbonylation reaction crude (R^1 = H, alkyl, benzyl; R^2 = H, alkyl, Me; R^3 = alkyl, aryl; R = alkyl).

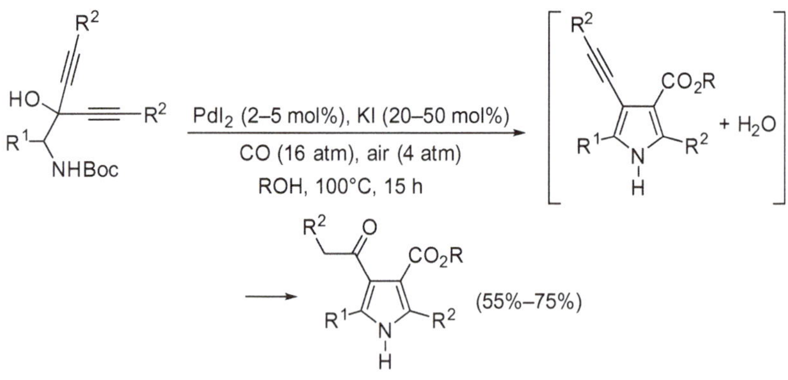

SCHEME 3.68 Synthesis of 4-acylpyrrole-3-carboxylic esters by PdI_2/KI-catalyzed oxidative carbonylation of *N*-Boc-2-alkynyl-1-amino-3-yn-2-ols (R^1 = alkyl, benzyl; R^2 = alkyl, aryl; R = alkyl).

More recently the PdI_2/KI-catalyzed 5-*exo-dig* *O*-heterocyclization–alkoxycarbonylation of 2-alkynylbenzamides bearing a terminal or internal triple bond has been reported to selectively afford 3-[(alkoxycarbonyl)methylene]isobenzofuran-1-(3*H*)imines (Scheme 3.69),[298] which have shown interesting herbicidal activity.[299] The oxidative carbonylation process was carried out in the presence of orthoformate as dehydrating agent, necessary to avoid hydrolysis of the imino group under the reaction conditions.[298] In a similar way, (*Z*)-4-[(methoxycarbonyl)methylene]-2-(4-fluorophenyl)-4*H*-benzo[*d*][1,3]oxazine was synthesized from 4-fluoro-*N*-[2-((trimethylsilyl)ethynyl)phenyl]benzamide (Scheme 3.70).[300] In this latter process, as already reported in the original article leading to nonfluorinated benzoxazines,[257] the catalytically active PdI_4^{2-} species was obtained in situ under the reaction conditions by oxidation of Pd/C with oxygen in the presence of iodide anions.

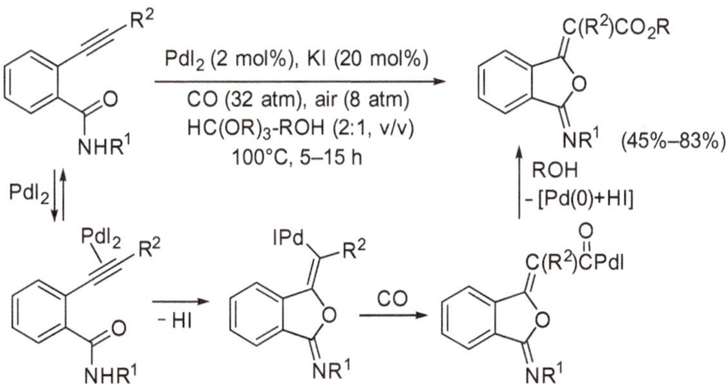

SCHEME 3.69 Synthesis of 3-[(alkoxycarbonyl)methylene]isobenzofuran-1-(3*H*)imines by PdI_2/KI-catalyzed oxidative *O*-heterocyclization–alkoxycarbonylation of 2-alkynylbenzamides (R^1 = alkyl, benzyl, aryl; R^2 = H, alkyl, aryl, alkenyl; R = alkyl).

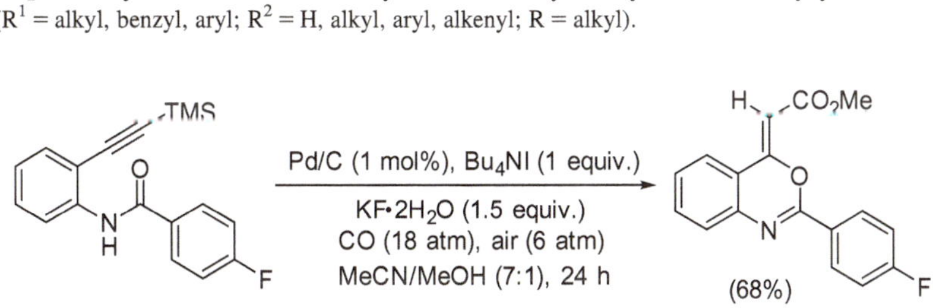

SCHEME 3.70 Synthesis of (*Z*)-4-[(methoxycarbonyl)methylene]-2-(4-fluorophenyl)-4*H*-benzo[*d*][1,3]oxazine by palladium-catalyzed oxidative *O*-heterocyclization–alkoxycarbonylation of 4-fluoro-*N*-[2-((trimethylsilyl)ethynyl)phenyl]benzamide.

Recently the oxidative oxycyclization–alkoxycarbonylation reaction has been used as a key step toward the preparation of natural compounds. Thus a model alkynol was carbonylated through 6-*endo-dig* oxycyclization using $PdCl_2$ as catalyst and benzoquinone as oxidant to give a derivative bearing the dihydropyran substructure of polymaxelonide (a hybrid marine natural product), even though in low yield (Scheme 3.71) (TBSOTf = *tert*-butyldimethylsilyl triflate).[301] On the other hand the synthesis of (+)-gregatins B and E (natural products isolated from *Cephalosporium gregatum*) was accomplished by exploiting the Pd(tfa)$_2$-catalyzed oxidative 5-*exo-dig* oxycyclization–methoxycyclization of (*S*)-2-methylbut-3-yne-1,2-diyl diacetate as a key step (Scheme 3.72).[302] In another recent example a 6-*exo-dig* oxycyclization–methoxycarbonylation reaction was exploited to synthesize a tetrahydropyran subunit useful for the total synthesis of phorborazole A, a potent antitumor agent (Scheme 3.73) (TIPS = triisopropylsilyl).[303]

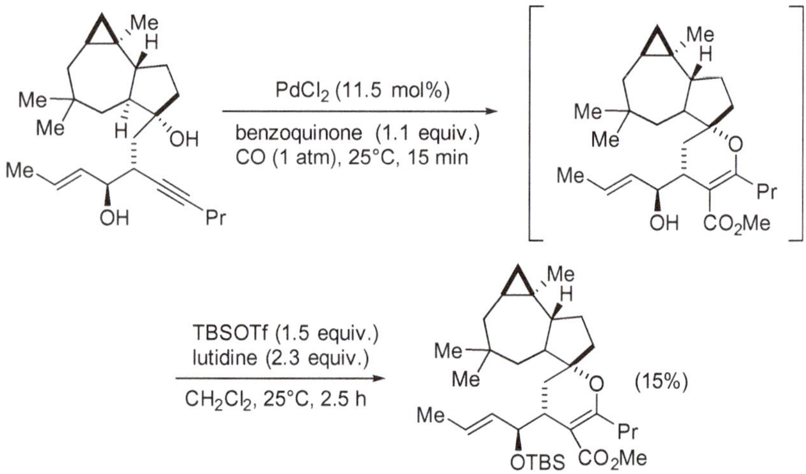

SCHEME 3.71 Synthesis of the dihydropyran substructure of polymaxelonide (a hybrid marine natural product) through PdCl$_2$-catalyzed oxidative 6-*endo-dig* heterocyclization–methoxycarbonylation.

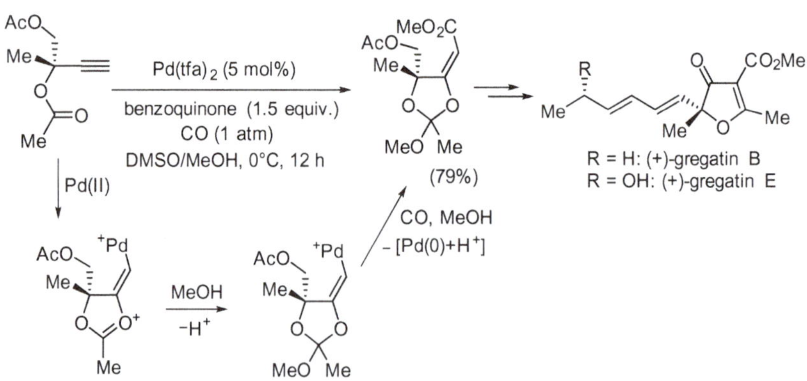

SCHEME 3.72 Carbonylative synthesis of methyl (*E*)-2-[(5*R*)-5-(2-acetoxyethyl)-2-methoxy-2,5-dimethyl-1,3-dioxolan-4-ylidene]acetate as a key step for the total synthesis of (+) gregatins B and E (natural products isolated from *Cephalosporium gregatum*).

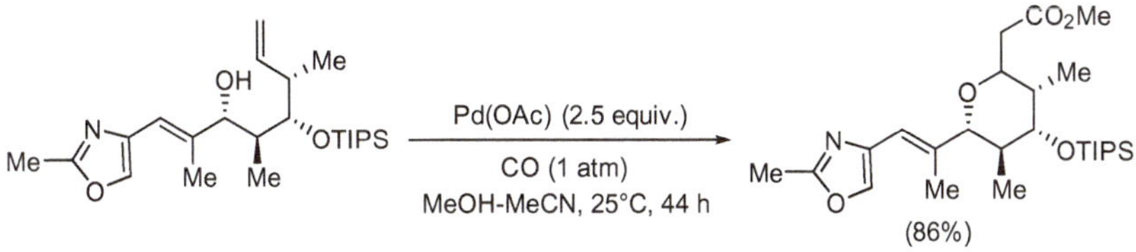

SCHEME 3.73 Carbonylative synthesis of methyl 2-((3*S*,4*S*,5*S*,6*R*)-3,5-dimethyl-6-((*E*)-1-(2-methyloxazol-4-yl)prop-1-en-2-yl)-4-((triisopropylsilyl)oxy)tetrahydro-2*H*-pyran-2-yl)acetate, a subunit for the total synthesis of the potent antitumor agent phorborazole A.

Azacyclization–alkoxycarbonylation was recently reported to occur with 1-aryl-2-(prop-2-yn-1-ylidene)hydrazines in the presence of $Pd(tfa)_2$ as catalyst to afford pyrazole-4-carboxylates (Scheme 3.74A)[304] and with *O*-substituted 1-(pyridin-2-yl)prop-2-yn-1-ol derivatives using $Pd_2(dba)_3$ or Pd/C as catalytic precursors to give indolizine derivative (Scheme 3.74B)[305]; in both cases, benzoquinone was employed as oxidant. The former conditions were also successful for the synthesis of benzothiophene-3-carboxylates starting from methoxymethyl (MOM)-protected 2-alkynylbenzothiols (Scheme 3.74C).[304]

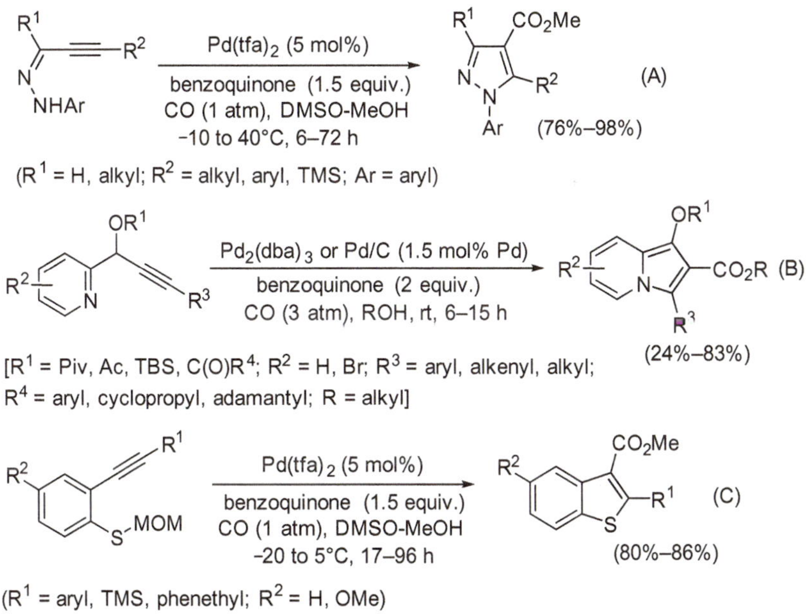

SCHEME 3.74 Synthesis of carbonylated pyrazole, indolizine, and benzothiophene derivatives by oxidative heterocyclization–alkoxycarbonylation of 1-aryl-2-(prop-2-yn-1-ylidene)hydrazines (A), *O*-substituted 1-(pyridin-2-yl)prop-2-yn-1-ol derivatives (B), and MOM-protected 2-alkynylbenzothiols (C), respectively.

3.3.1.3 Through Heterocyclization–Carbonylation–Cyclization Coupling

The group of Kato has disclosed the possibility to realize, under suitable conditions, a carbonylative double heterocyclization process of suitably functionalized alkyne compounds with the formation of diheterocyclic ketones. The authors have called their reaction as "cyclization–carbonylation–cyclization coupling" ("CCC reaction").[306] The use of $Pd(tfa)_2$ as catalyst stabilized by box ligands (box = bisoxazoline) was in most cases essential for the success of the reaction. As shown in Scheme 3.75 the process is believed to occur through an initial Pd(II)-induced heterocyclization followed by CO insertion, coordination of a second molecule of substrate to the ensuing Pd(II) complex, further heterocyclization and reductive elimination (Scheme 3.75).[306]

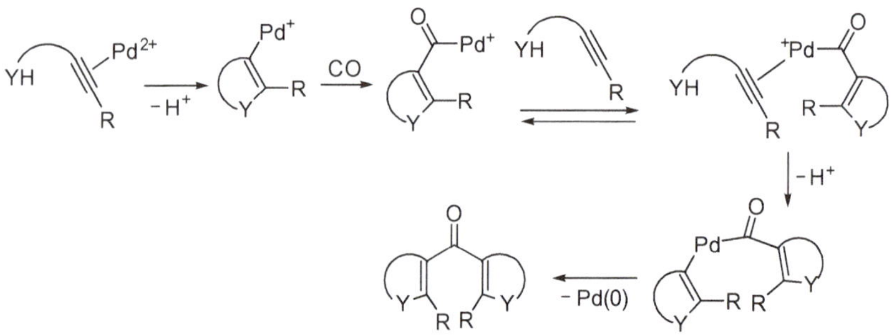

SCHEME 3.75 Heterocyclization–carbonylation–cyclization coupling pathway leading to diheterocyclic ketones.

This kind of reactivity has been recently successfully applied to the synthesis of a variety of diheterocyclic ketone derivatives, including (2*E*,2'*E*)-2,2'-(2-oxopropane-1,3-diylidene)bis(3,3*a*,4,5-tetrahydrobenzofuran-6(2*H*)-ones) (Scheme 3.76A),[307] di(benzo[*b*]thiophen-3-yl)ketones (Scheme 3.76B),[308,309] (1*E*,3*E*)-1,3-bis(2-aminooxazol-5(4*H*)-ylidene)propan-2-ones[310] (Scheme 3.76C), di(pyrazol-4-yl)ketones (Scheme 3.76D),[311] and di (1-benzyl-1*H*-indol-3-yl)ketones (Scheme 3.76E).[312] The external oxidant was usually benzoquinone even though the $CuCl_2/O_2$ system was also employed. It should be noted, however, that in this latter case the $CO-O_2$ ratio (1:1) fell within the explosion limits for a $CO-O_2$ mixture.[110,309] Initial carbocyclization is also possible (see Section 3.3.1.2) as exemplified by the synthesis of di(quinolin-3-yl)ketones (Scheme 3.77).[313]

3.3.1.4 Through Monocarbonylation of Terminal Alkynes Followed by Intramolecular Conjugate Addition

As seen in Section 3.2.1.2, in the presence of a nucleophilic secondary amine the PdI_2/KI catalytic system is able to promote the monoaminocarbonylation

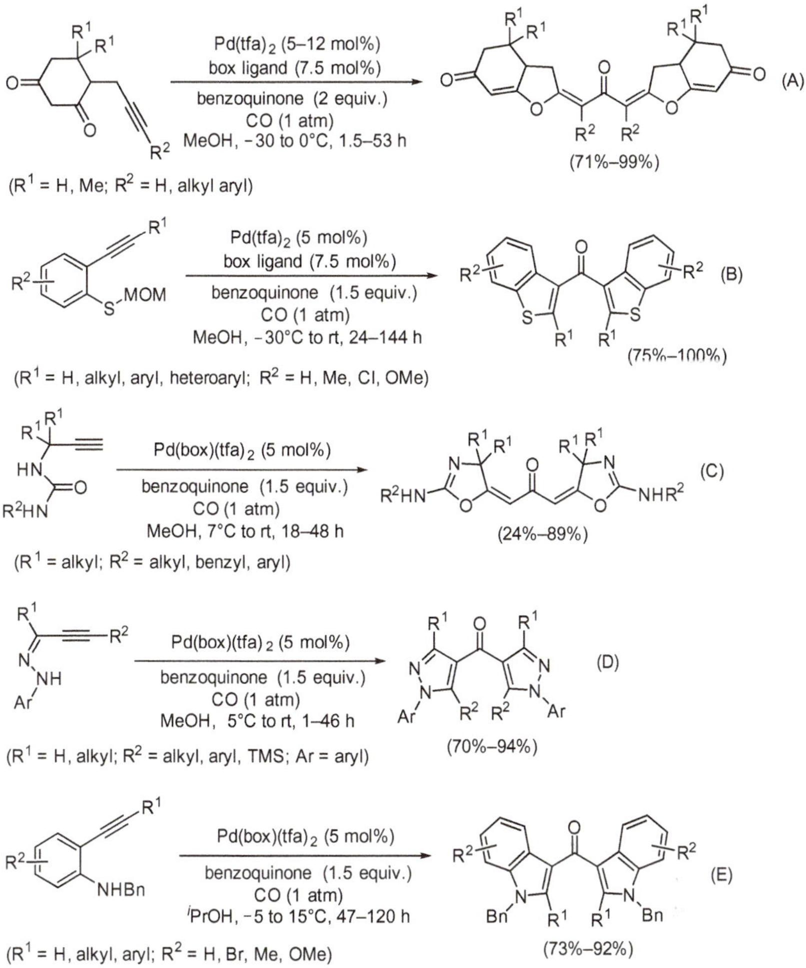

SCHEME 3.76 Synthesis of a variety of diheterocyclic ketones via the heterocyclization–carbonylation–cyclization coupling approach.

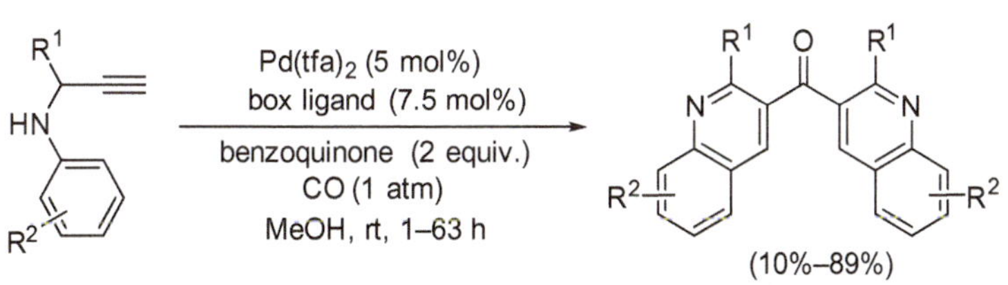

SCHEME 3.77 Synthesis of di(quinolin-3-yl)ketones via carbocyclization–carbonylation–cyclization coupling reaction (R^1 = H, Me; R^2 = H, alkyl, vinyl, $SiEt_3$)

of alkyl- and arylacetylenes to give 2-ynamides, through ≡CH palladation followed by CO insertion and nucleophilic displacement by the amine (Scheme 3.10). This reactivity allows an easy synthesis of heterocyclic derivatives when applied to terminal alkynes functionalized with a heteronucleophilic group in a suitable position for giving an intramolecular conjugate addition to the initially formed 2-ynamide intermediate, according to the general concept shown in Scheme 3.78. Several examples of this kind of reactivity were reported by our research group, leading to the synthesis of a variety of carbonylated heterocycles, including tetrahydrofuran,[290] furan,[291,292] benzodioxine,[293] and benzoxazine[293] derivatives.

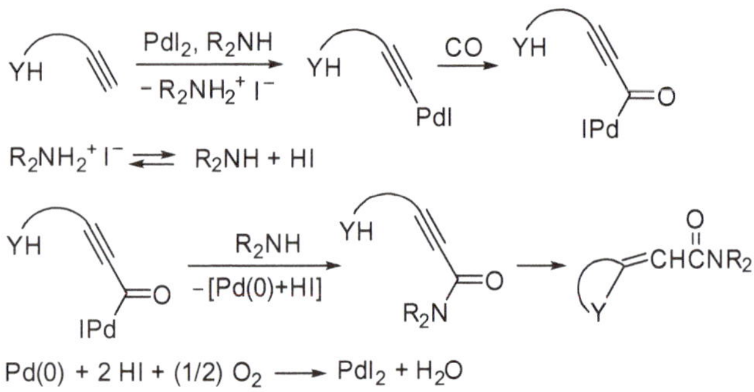

SCHEME 3.78 PdI_2-catalyzed monoaminocarbonylation of terminal alkynes bearing a heteronucleophile on a suitable position followed by intramolecular conjugate addition (Y = heteroatom; R = alkyl).

More recently, we applied this concept to 2-ethynylbenzamides for the synthesis of 3-[(dialkylcarbamoyl)methylene]isoindolinones (Scheme 3.79A),[298,314] which turned out to be useful precursors for the formation of novel spiro[isoindole-1,5′-isoxazolidin]-3(2*H*)-ones with antitumor activity.[315] The possibility to realize the intramolecular conjugate addition by nucleophilic attack of a heterocyclic nitrogen has also been successfully demonstrated, and allowed the direct synthesis of functionalized benzimidazothiazoles starting from 2-prop-2-ynylsulfanyl-3*H*-benzimidazoles (obtained in situ from the corresponding benzimidazolium salts) (Scheme 3.79B).[117] In this latter case the initially formed carbonylated intermediate underwent double bond isomerization to give the final product (Scheme 3.79B).[117]

3.3.2 Under Nonoxidative Conditions

3.3.2.1 *With Halides*

Under certain conditions, suitably functionalized halides may undergo Pd(0)-catalyzed carbonylative cyclization without incorporation of carbon monoxide into the cycle under nonoxidative conditions. For example, 2-(2,2-dihalovinyl) thiophenols were conveniently converted into 2-carbonylbenzothiophene

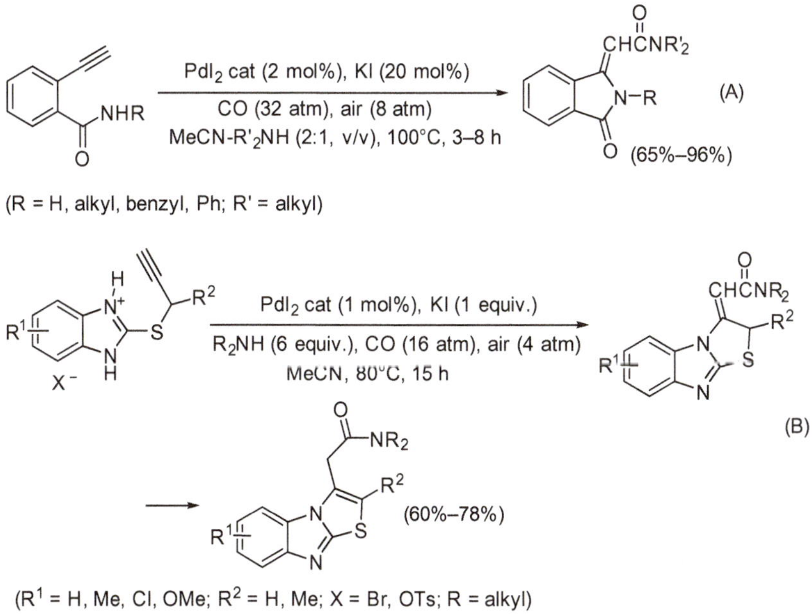

SCHEME 3.79 Synthesis of functionalized isoindolinones (A) and benzimidazothiazoles (B) by PdI_2/KI-catalyzed oxidative monoaminocarbonylation-conjugate addition of 2-ethynylbenzamides and 2-prop-2-ynylsulfanyl-1*H*-benzimidazolium salts, respectively.

derivatives in the presence of a nucleophile (such as alcohol, phenol, or amine), $Pd(OAc)_2$ as catalytic precursor, and Ruphos as ligand (Scheme 3.80) (Ruphos = 2-dicyclohexylphosphino-2',6'-diisopropyloxybiphenyl).[316] The mechanism probably involves a concatenation of two Pd(0)-promoted catalytic cycles, the first cycle leading to a 2-halobenzothiophene intermediate which then undergoes carbonylation (Scheme 3.80).[316]

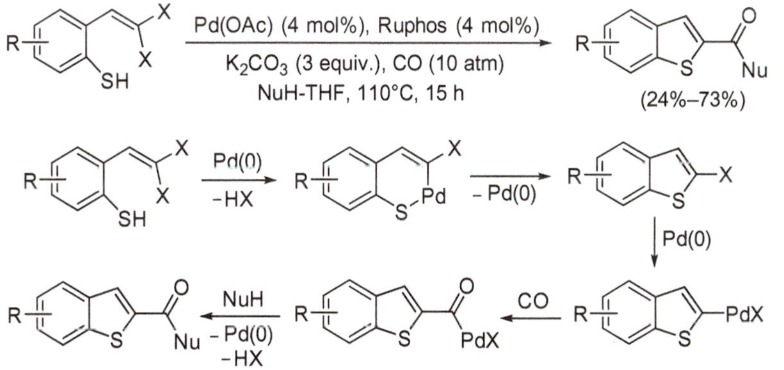

SCHEME 3.80 Synthesis of carbonylated benzothiophenes by Pd(0)-catalyzed carbonylative heterocyclization of 2-(2,2-dihalovinyl)thiophenols (R = H, Cl, F, CF_3, alkoxy; X = Br, Cl; NuH = alcohol, phenol, amine).

More recently a carbonylative coupling of aryl halides with terminal alkynes to give 3-furylketones was reported by Wu, Beller, and coworkers (Scheme 3.81).[317] The mechanism probably involves the formation of an alkynone intermediate followed by isomerization to an allene derivative, insertion of the latter into the Pd-acyl bond, ring closure with palladium shift, and β-H elimination (Scheme 3.81).[317] In a related process, carbonylated thiophene and benzothiophene derivatives were obtained using encapsulated thioglycolate as the sulfur source (Scheme 3.82).[318]

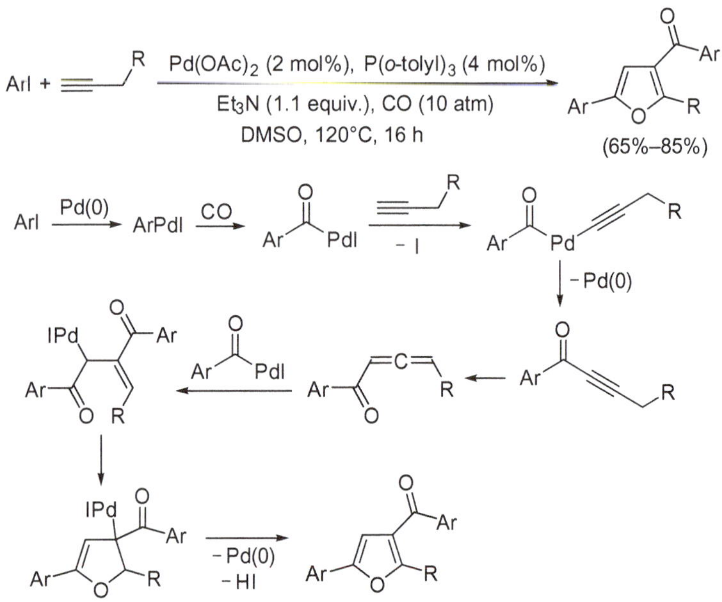

SCHEME 3.81 Synthesis of 3-furyl ketones by Pd(0)-catalyzed carbonylative coupling of aryl halides with terminal alkynes (Ar = aryl; R = alkyl).

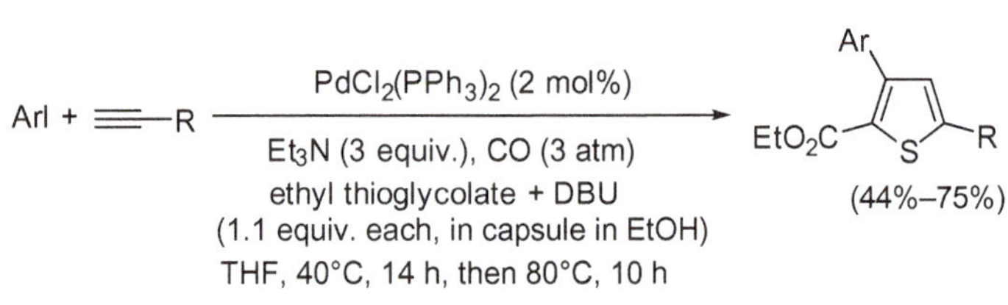

SCHEME 3.82 Synthesis of thiophene-2-carboxylic esters by Pd(0)-catalyzed carbonylative coupling of aryl iodides with terminal alkynes using an encapsulated solution of ethyl thioglycolate and DBU (Ar = aryl; R = alkyl, aryl, heteroaryl).

3.3.2.2 With Imine Derivatives

Imine derivatives can be useful starting materials for the synthesis of heterocycles under nonoxidative conditions without CO incorporation into the cycle.[319] In an interesting recent example, *O*-pentafluorobenzoyl oxime esters were converted into 2-(3,4-dihydro-2*H*-pyrrol-2-yl)-1-arylethan-1-ones through oxidative addition of the N–O bond to Pd(0), followed by 5-*exo-trig* intramolecular double bond insertion, CO insertion, transmetallation by a tetraarylborate, and reductive elimination (Scheme 3.83).[320]

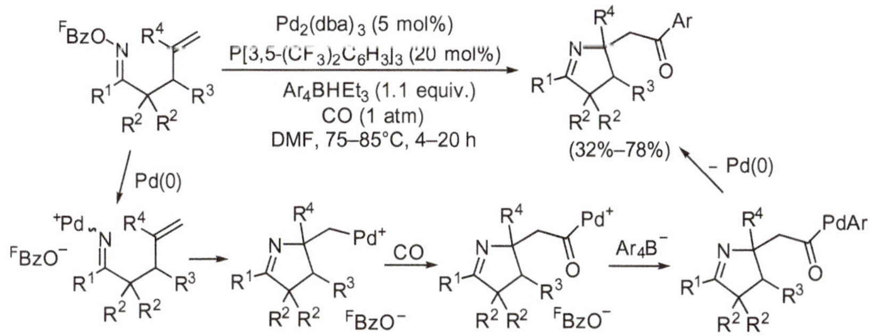

SCHEME 3.83 Synthesis of 2-(3,4-dihydro-2*H*-pyrrol-2-yl)-1-arylethan-1-ones by Pd(0)-catalyzed carbonylative coupling of *O*-pentafluorobenzoyl oxime esters with tetraarylborates (R^1 = Ph, alkyl, cyclopropyl; R^2 = H, alkyl; R^3 = H, alkyl; R^4 = H, alkyl, benzyl; Ar = aryl).

3.3.2.3 With Functionalized Alkynes

Functionalized alkynes are excellent substrates for the nonoxidative heterocyclization with exocyclic CO incorporation. In a recent example, Gevorgyan and coworkers reported the synthesis of 2-aroylindolizines starting from 2-propargylpyridines and aryl iodides, through a proposed mechanism involving oxidative addition of the latter to Pd(0), coordination of the ensuing ArPdI species to the 2-propargylpyridine triple bond, 5-*endo-dig* cyclization (via intramolecular nucleophilic attack by the pyridine nitrogen to the activated triple bond), CO insertion into the Pd–Ar bond, and reductive elimination (Scheme 3.84).[321]

The possibility to convert 2-(1-hydroxyalk-2-yn-1-yl)phenol derivatives into 2-benzofuran-2-ylacetic derivatives under nonoxidative conditions was successfully demonstrated by our research group some years ago, in conventional[322–324] as well as in nonconventional[325] solvents. Owing to the instability of substrates bearing an internal triple bond, 1-(2-allyloxyphenyl)-2-yn-1-ols, with the phenolic hydroxyl protected with an allyl group, were used as the substrate precursors (Scheme 3.85).[322–325]

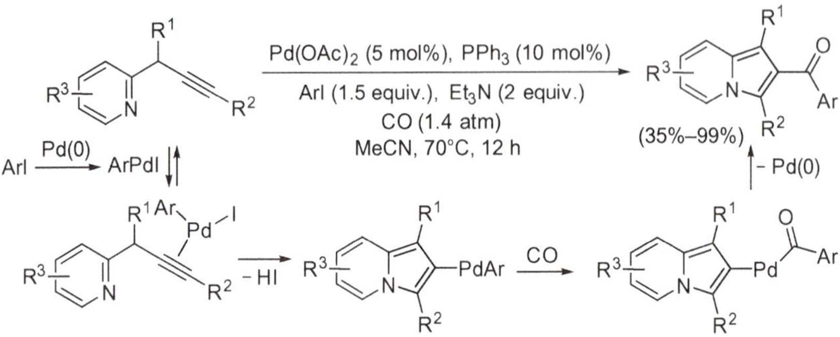

SCHEME 3.84 Synthesis of 2-aroylindolizines by Pd(0)-catalyzed carbonylative coupling of 2-propargylpyridines with aryl iodides (R^1 = OPiv, OTBDMS, 4-morpholino; R^2 = aryl, alkyl, alkenyl; R^3 = H, Me, OMe; Ar = aryl).

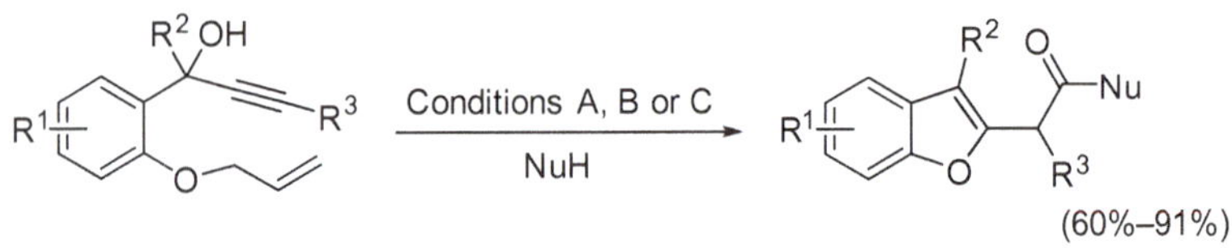

Conditions A: PdI_2 (1 mol%), KI (1–2 equiv.), PPh_3 (4 mol%), H_2O (2 equiv.)
P(CO) = 30–90 atm, MeOH, 100°C, 15–24 h (NuH = MeOH)

Conditions B: PdI_2 (1–2 mol%), KI (0.1–2 equiv.), PPh_3 (4–8 mol%), RR'NH (2 equiv.)
P(CO) = 30 atm, MeOH or MeCN, 100°C, 15–24 h (NuH = RR'NH)

Conditions C: PdI_2 (1 mol%), KI (1 equiv.), PPh_3 (4 mol%), H_2O (2 equiv.), MeOH (28 equiv.)
P(CO) = 30 atm, $BmimBF_4$, 100°C, 24 h (NuH = MeOH)

SCHEME 3.85 Synthesis of 2-benzofuran-2-ylacetic esters and 2-benzofuran-2-ylacetamides by deallylation–heterocyclization–carbonylation of 1-(2-allyloxyphenyl)-2-yn-1-ols (R^1 = H, OMe, Cl; R^2 = H, Me, Ph, alkynyl; R^3 = H, alkyl, aryl; NuH = MeOH or RR′NH, RR′NH = morpholine, Bu_2NH, Et_2NH, PhNHMe, $BnNH_2$, t-$BuNH_2$, $PhNH_2$).

Under the optimized conditions, carbonylation reactions were carried out in the presence of two palladium-based catalytic species: a Pd(0) species stabilized by coordination of triphenylphosphine (either preformed $Pd(PPh_3)_4$ or the complex obtained by reduction in situ of PdI_2 in the presence of PPh_3) and the PdI_2/KI catalytic system already successfully employed in many oxidative carbonylation processes (see Section 3.2.1.1). In fact the Pd(0)-based complex promoted the substrate deallylation, while the Pd(II) species catalyzed the heterocyclization–carbonylation process of the in situ formed free phenolic derivative, eventually leading to benzofurans (Scheme 3.86). For a cascade process like this, involving the concatenation of two catalytic cycles, catalyzed by the same metal but in two different oxidation states, we coined the term "sequential homobimetallic catalysis."[322–325]

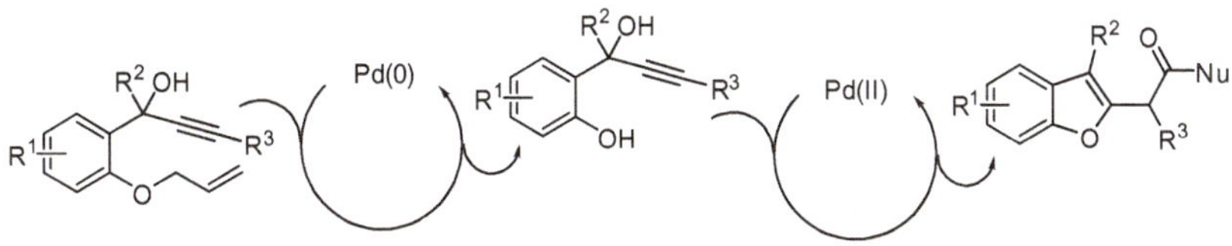

SCHEME 3.86 The sequential homobimetallic catalysis concept (concatenation of two catalytic cycles catalyzed by the same metal but in two different oxidation states) applied to the formation of benzofuran derivatives.

In more detail the deallylation cycle started with the oxidative addition of the allyloxy group to Pd(0), followed by protonolysis (to give an allylpalladium complex) and either alkoxycarbonylation (when the process was carried out in the presence of an alcohol ROH as external nucleophile)[322,323,325] or nucleophilic attack by an amine (when the process was carried out in the presence of an amine RR′NH as external nucleophile).[324] On the other hand the heterocyclization–carbonylation cycle started with the 5-*exo-dig* intramolecular nucleophilic attack by the phenolic hydroxyl to the triple bond activated by coordination to PdI_2, followed by CO insertion and nucleophilic displacement by the external nucleophile NuH (NuH = MeOH or RR′NH). This led to the formation of a heterocyclic intermediate bearing an allylalcoholic moiety together with an iodopalladium hydride species, which reacted with elimination of water to give a π-allylpalladiumiodide complex. Protonolysis of the latter by HI yielded the final benzofuran-2-ylacetic derivative (ester or amide) with regeneration of PdI_2 (Scheme 3.87).[322–325]

We recently reviewed this concept in a highlight in *Chemistry of Heterocyclic Compounds*[326] and showed that some of the benzofuran derivatives synthesized by this approach are promising antitumor agents.[327]

3.4 CONCLUSIONS AND PERSPECTIVES

Carbonylation, the incorporation of CO into an organic substrate, is the most important synthetic method for the direct synthesis of carbonyl compounds from the point of view of atom and step economy. The development of novel, more efficient and selective catalysts together with the fine tuning of the reaction conditions has allowed, during the past years, to realize carbonylation processes under milder reaction conditions and with high selectivity leading to products of industrial interest as well as fine chemicals, heterocyclic derivatives in particular. From an academic point of view the use of carbon monoxide surrogates is attractive since it avoids the handling of gaseous carbon monoxide. However, it should be considered that the direct employment of the simplest C-1 unit (CO) as starting material for the incorporation of the carbonyl group into an organic substrate still remains the most important choice from an economic and ecological point of view.

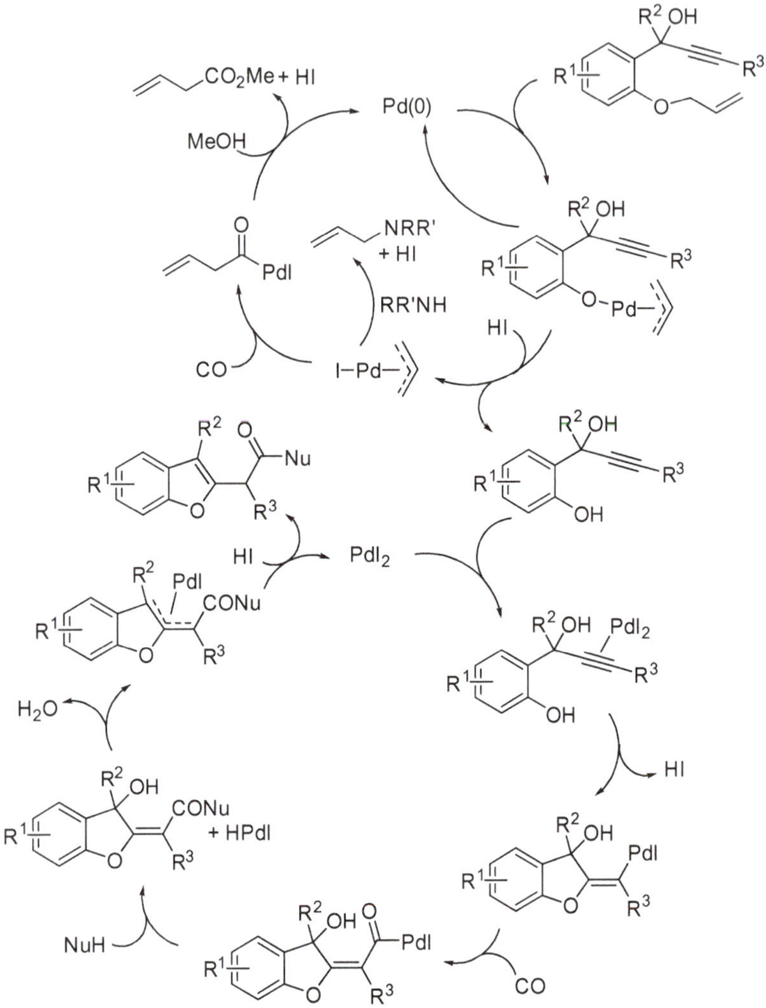

SCHEME 3.87 Formation of 2-benzofuran-2-ylacetic derivatives from 1-(2-allyloxyphenyl)-2-yn-1-ols by the concatenation of two catalytic cycles: Pd(0)-catalyzed deallylation followed by Pd(II)-catalyzed carbonylative heterocyclization (NuH = MeOH or RR′NH).

The future progress in this important field of synthetic chemistry is related, on one hand, to the development of even more active and selective catalytic systems, and, on the other hand, to the possibility of employing "greener" reaction conditions, particularly regarding the solvent. Another important advancement will involve the engineering of novel heterogeneous catalysts characterized by an activity and selectivity similar to those obtained with the homogeneous catalysts known so far, but with the possibility of an much easier recovery from the reaction mixture. This will allow not only their recycling but will also minimize the contamination of the final organic product by the metal species employed for catalyzing the reaction.

REFERENCES

1. Colquhoun, H. M.; Thompson, D. J.; Twigg, M. W. *Carbonylation – Direct Synthesis of Carbonyl Compounds;* Plenum Press: New York, 1991.
2. Beller, M., Ed. *Catalytic Carbonylation Reactions.;* Springer: Berlin, 2006.
3. Kollár, L., Ed. *Modern Carbonylation Methods.;* Wiley-VCH: Weinheim, 2008.
4. Shen, C.; Wu, X.-F. Palladium-Catalyzed Carbonylative Multicomponent Reactions. *Chem. Eur. J.* **2017,** *23*, 2973–2987.
5. Pineiro, M.; Dias, L. D.; Damas, L.; Aquino, G. L. B.; Calvete, M. J. F.; Pereora, M. M. Microwave Irradiation as a Sustainable Tool for Catalytic Carbonylation Reactions. *Inorg. Chim. Acta* **2017,** *455*, 364–377.
6. Gehrtz, P. H.; Hirschbeck, V.; Ciszek, B.; Fleischer, I. Carbonylations of Alkenes in the Total Synthesis of Natural Compounds. *Synthesis* **2016,** *48*, 1573–1596.
7. Wu, X.-F. Palladium-Catalyzed Carbonylative Transformation of Aryl Chlorides and Aryl Tosylates. *RSC Adv.* **2016,** *6*, 83831–83837.
8. Kalck, P.; Urrutigoity, M. Recent Improvements in the Alkoxycarbonylation Reaction Catalyzed by Transition Metal Complexes. *Inorg. Chim. Acta* **2015,** *431*, 110–121.
9. Gadge, S. T.; Bhanage, B. M. Recent Developments in Palladium Catalysed Carbonylation Reactions. *RSC Adv.* **2014,** *4*, 10367–10389.
10. Fang, W.; Zhu, H.; Deng, Q.; Liu, S.; Liu, X.; Shen, Y.; Tu, T. Design and Development of Ligands for Palladium-Catalyzed Carbonylation Reactions. *Synthesis* **2014,** *46*, 1689–1708.
11. Sumino, S.; Fusano, A.; Fukuyama, T.; Ryu, I. Carbonylation Reactions of Alkyl Iodides Through the Interplay of Carbon Radicals and Pd Catalysts. *Acc. Chem. Res.* **2014,** *47*, 1563–1574.
12. Kealey, S.; Gee, A.; Miller, P. W. J. L. Transition Metal Mediated [^{11}C]carbonylation Reactions: Recent Advances and Applications. *J. Label. Compd. Radiopharm.* **2014,** *57*, 195–201.
13. Wu, X.-F.; Fang, X.; Wu, L.; Jackstell, R.; Neumann, H.; Beller, M. Transition-Metal-Catalyzed Carbonylation Reactions of Olefins and Alkynes: A Personal Account. *Acc. Chem. Res.* **2014,** *47*, 1041–1053.
14. Fukuyama, T.; Totoki, T.; Ryu, I. Carbonylation in Microflow: Close Encounters of CO and Reactive Species. *Green Chem.* **2014,** *16*, 2042–2050.
15. Wu, X.-F.; Neumann, H.; Beller, M. Palladium-Catalyzed Oxidative Carbonylation Reactions. *ChemSusChem* **2013,** *6*, 229–241.
16. Wu, X.-F.; Neumann, H.; Beller, M. Synthesis of Heterocycles Via Palladium-Catalyzed Carbonylations. *Chem. Rev.* **2013,** *113*, 1–35.
17. Gabriele, B.; Mancuso, R.; Salerno, G. Oxidative Carbonylation as a Powerful Tool for the Direct Synthesis of Carbonylated Heterocycles. *Eur. J. Org. Chem.* **2012,** 6825–6839.
18. Transition Metal Catalyzed Carbonylative Synthesis of Heterocycles. In *Topics in Heterocyclic Chemistry;* Wu, X.-F.; Beller, M., Eds. *Topics in Heterocyclic Chemistry,* Vol. 42; Springer: Berlin, 2016.
19. Diebolt, O.; van Leeuwen, P. W. N. M.; Kamer, P. C. J. Operando Spectroscopy in Catalytic Carbonylation Reactions. *ACS Catal.* **2012,** *2*, 2357–2370.

20. Wu, X.-F.; Neumann, H. Ruthenium and Rhodium-Catalyzed Carbonylation Reactions. *ChemCatChem* **2012,** *4*, 447–458.
21. Liu, Q.; Zhang, H.; Lei, A. Oxidative Carbonylation Reactions: Organometallic Compounds (RM) or Hydrocarbons (RH) as Nucleophiles. *Angew. Chem. Int. Ed.* **2011,** *50*, 10788–10799.
22. Wu, X.-F.; Neumann, H.; Beller, M. Palladium-Catalyzed Carbonylative Coupling Reactions Between Ar–X and Carbon Nucleophiles. *Chem. Soc. Rev.* **2011,** *40*, 4986–5009.
23. Omae, I. Transition Metal-Catalyzed Cyclocarbonylation in Organic Synthesis. *Coord. Chem. Rev.* **2011,** *255*, 139–160.
24. Liu, J.; Chen, J.; Sun, W.; Xia, C. Recent Progress in Research of Some Carbonylation Reactions. *Chin. J. Catal.* **2010,** *31*, 1–11.
25. Occhiato, E. G.; Scarpi, D.; Prandi, C. Carbonylative Palladium-Catalyzed Reactions of Lactam-, Lactone-, and Thiolactone-Derived Vinyl Triflates and Phosphates for the Synthesis of *N*-, *O*-, and *S*-Heterocycles. *Heterocycles* **2010,** *80*, 697–724.
26. Grigg, R.; Mutton, S. P. Pd-Catalysed Carbonylations: Versatile Technology for Discovery and Process Chemists. *Tetrahedron* **2010,** *66*, 5515–5548.
27. Park, J. H.; Chang, K.-M.; Chung, Y. K. Catalytic Pauson–Khand-Type Reactions and Related Carbonylative Cycloaddition Reactions. *Coord. Chem. Rev.* **2009,** *253*, 2461–2480.
28. Brennfuhrer, A.; Neumann, H.; Beller, M. Palladium-Catalyzed Carbonylation Reactions of Alkenes and Alkynes. *ChemCatChem* **2009,** *1*, 28–41.
29. Brennfuhrer, A.; Neumann, H.; Beller, M. Palladium-Catalyzed Carbonylation Reactions of Aryl Halides and Related Compounds. *Angew. Chem. Int. Ed.* **2009,** *48*, 4114–4133.
30. Ragaini, F. Away From Phosgene: Reductive Carbonylation of Nitroarenes and Oxidative Carbonylation of Amines, Understanding the Mechanism to Improve Performance. *Dalton Trans.* **2009,** 6251–6266.
31. Barnard, C. F. J. Palladium-Catalyzed Carbonylation—A Reaction Come of Age. *Organometallics* **2008,** *27*, 5402–5422.
32. Chaudhari, R. V. Hydroformylation and Carbonylation Processes: New Trends in the Synthesis of Pharmaceuticals. *Curr. Opin. Drug Discov. Dev.* **2008,** *11*, 820–828.
33. Boyarskii, V. P. Catalytic Systems for Carbonylation of Aryl Halides. *Russ. J. Gen. Chem.* **2008,** *78*, 1742–1753.
34. Diaz, D. J.; Darko, A. K.; McElwee-White, L. Transition Metal-Catalyzed Oxidative Carbonylation of Amines to Ureas. *Eur. J. Org. Chem.* **2007,** 4453–4465.
35. Church, T. L.; Getzler, Y. D. Y. L.; Byrne, C. M.; Coates, G. W. Carbonylation of Heterocycles by Homogeneous Catalysts. *Chem. Commun.* **2007,** 657–674.
36. Vasapollo, G.; Mele, G. Synthesis of Heterocycles by Transition Metals-Catalyzed Cyclocarbonylation Reactions. *Curr. Org. Chem.* **2006,** *10*, 1397–1421.
37. Trzeciak, A. M.; Ziółkowski, J. J. Structural and Mechanistic Studies of Pd-Catalyzed C–C Bond Formation: The Case of Carbonylation and Heck Reaction. *Coord. Chem. Rev.* **2005,** *249*, 2308–2322.
38. Muzart, J. Palladium-Catalysed Reactions of Alcohols. Part D: Rearrangements, Carbonylations, Carboxylations and Miscellaneous Reactions. *Tetrahedron* **2005,** *61*, 9423–9463.
39. Gabriele, B.; Salerno, G.; Costa, M.; Chiusoli, G. P. Recent Advances in the Synthesis of Carbonyl Compounds by Palladium-Catalyzed Oxidative Carbonylation Reactions of Unsaturated Substrates. *Curr. Org. Chem.* **2004,** *8*, 919–946.

40. Vizer, S. A.; Yerzhanov, K. B.; Al Quntar, A. A. A.; Dembitsky, V. M. Synthesis of Heterocycles by Carbonylation of Acetylenic Compounds. *Tetrahedron* **2004,** *60*, 5499–5538.
41. Des Abbayes, H.; Salaun, J. Y. Double Carbonylation and Beyond: Systems at Work and Their Organometallic Models. *Dalton Trans.* **2003,** 1041–1052.
42. Gabriele, B.; Salerno, G.; Costa, M.; Chiusoli, G. P. Recent Developments in the Synthesis of Heterocyclic Derivatives by PdI_2-Catalyzed Oxidative Carbonylation Reactions. *J. Organomet. Chem.* **2003,** *687*, 219–228.
43. Skoda-Foldes, R.; Kollár, L. Synthetic Applications of Palladium Catalysed Carbonylation of Organic Halides. *Curr. Org. Chem.* **2002,** *6*, 1097–1119.
44. Peng, J.-B.; Qi, X.; Wu, X.-F. Recent Achievements in Carbonylation Reactions: A Personal Account. *Synlett* **2017,** *28*, 175–194.
45. Cao, J.; Zheng, Z.-J.; Xu, Z.; Xu, L.-W. Transition-Metal-Catalyzed Transfer Carbonylation with HCOOH or HCHO as Non-Gaseous C1 Source. *Coord. Chem. Rev.* **2017,** *336*, 43–53.
46. Friis, S. D.; Lindhardt, A. T.; Srydstrup, T. The Development and Application of Two-Chamber Reactors and Carbon Monoxide Precursors for Safe Carbonylation Reactions. *Acc. Chem. Res.* **2016,** *49*, 594–605.
47. Lia, W.; Wu, X.-F. The Applications of (Para)formaldehyde in Metal-Catalyzed Organic Synthesis. *Adv. Synth. Catal.* **2015,** *357*, 3393–3418.
48. Konishi, H.; Manabe, K. Practical Synthetic Methods Utilizing Formic Acid Derivatives as Carbon Monoxide Sources. *J. Synth. Org. Chem. Jpn.* **2015,** *73*, 911–922.
49. Wu, L.; Liu, Q.; Jackstell, R.; Beller, M. Carbonylations of Alkenes With CO Surrogates. *Angew. Chem. Int. Ed.* **2014,** *53*, 6310–6320.
50. Morimoto, T.; Kakiuchi, K. Evolution of Carbonylation Catalysis: No Need for Carbon Monoxide. *Angew. Chem. Int. Ed.* **2004,** *43*, 5580–5588.
51. Phillips, D.; France, D. J. Palladium-Catalyzed Heterocyclization: A Carbon-Centered Approach. *Asian J. Org. Chem.* **2017,** *6*, 27–40.
52. Ding, G.; Wang, Z.; Yin, Z.; Guizhou, Y. Progress in Palladium-Catalyzed Tandem Reaction of Constructing Benzo-Five-Membered Heterocycles. *Chin. J. Org. Chem.* **2016,** *36*, 43–59.
53. Shaikh, T. M.; Hong, F.-E. Recent Developments in the Preparation of *N*-Heterocycles Using Pd-Catalysed C–H Activation. *J. Organomet. Chem.* **2016,** *801*, 139–156.
54. Kaur, N. Palladium-Catalyzed Approach to the Synthesis of S-Heterocycles. *Catal. Rev.-Sci. Eng.* **2015,** *57*, 478–564.
55. Kaur, N. Palladium Catalysts: Synthesis of Five-Membered *N*-Heterocycles Fused With Other Heterocycles. *Catal. Rev.-Sci. Eng.* **2015,** *57*, 1–78.
56. Herraiz-Cobo, J.; Albericio, F.; Álvarez, M. Chapter One: The Larock Reaction in the Synthesis of Heterocyclic Compounds. *Adv. Heterocycl. Chem.* **2015,** *116*, 1–35.
57. Kaur, N. Palladium-Catalyzed Approach to the Synthesis of Five-Membered O-Heterocycles. *Inorg. Chem. Commun.* **2014,** *49*, 86–119.
58. Mancuso, R.; Gabriele, B. Recent Advances in the Synthesis of Thiophene Derivatives by Cyclization of Functionalized Alkynes. *Molecules* **2014,** *19*, 15687–15719.
59. Yamamoto, Y. Synthesis of Heterocycles Via Transition-Metal-Catalyzed Hydroarylation of Alkynes. *Chem. Soc. Rev.* **2014,** *43*, 1575–1600.
60. Inamoto, K. Synthesis of Heterocyclic Compounds Through Palladium-Catalyzed C–H Cyclization Processes. *Chem. Pharm. Bull.* **2013,** *61*, 987–996.

61. Yoshida, M. Synthesis of Functionalized Cyclic Molecules by Palladium-Catalyzed Cyclization of Propargylic Esters With Bis-Nucleophiles. *Heterocycles* **2013,** *87*, 1835–1864.
62. Gulevich, A. V.; Dudnik, A. S.; Chernyak, N.; Gevorgyan, V. Transition Metal-Mediated Synthesis of Monocyclic Aromatic Heterocycles. *Chem. Rev.* **2013,** *113*, 3084–3213.
63. Ferraccioli, R. Palladium-Catalyzed Synthesis of Carbo- and Heterocycles Through Norbornene-Mediated *Ortho* C–H Functionalization. *Synthesis* **2013,** *45*, 581–591.
64. Ball, C. J.; Willis, M. C. Cascade Palladium- and Copper-Catalysed Aromatic Heterocycle Synthesis: The Emergence of General Precursors. *Eur. J. Org. Chem.* **2013,** 425–441.
65. Mei, T.-S.; Kou, L.; Ma, S.; Engle, K. M.; Yu, J.-Q. Heterocycle Formation Via Palladium-Catalyzed C–H Functionalization. *Synthesis* **2012,** *44*, 1778–1791.
66. Heravi, M. M.; Hashemi, E. Recent Advances in Application of Intramolecular Suzuki Cross-Coupling in Cyclization and Heterocyclization. *Mon. Chem.* **2012,** *143*, 861–880.
67. Majumdar, K. C.; Samanta, S.; Sinha, B. Recent Developments in Palladium-Catalyzed Formation of Five- and Six Membered Fused Heterocycles. *Synthesis* **2012,** *44*, 817–847.
68. Schultz, D. M.; Wolfe, J. P. Recent Developments in Palladium-Catalyzed Alkene Aminoarylation Reactions for the Synthesis of Nitrogen Heterocycles. *Synthesis* **2012,** *44*, 351–361.
69. Platon, M.; Amardeil, R.; Djakovitch, L.; Hierso, J.-C. Progress in Palladium-Based Catalytic Systems for the Sustainable Synthesis of Annulated Heterocycles: A Focus on Indole Backbones. *Chem. Soc. Rev.* **2012,** *41*, 3929–3968.
70. Vlaar, T.; Ruijter, E.; Orru, R. V. A. Recent Advances in Palladium-Catalyzed Cascade Cyclizations. *Adv. Synth. Catal.* **2011,** *353*, 809–841.
71. Guo, L.-N.; Duan, X.-H.; Liang, Y.-M. Palladium-Catalyzed Cyclization of Propargylic Compounds. *Acc. Chem. Res.* **2011,** *44*, 111–122.
72. Sadig, J. E. R.; Willis, M. C. Palladium- and Copper-Catalyzed Aryl Halide Amination, Etherification and Thioetherification Reactions in the Synthesis of Aromatic Heterocycles. *Synthesis* **2011,** (1), 1–22.
73. Majumdar, K. C. Regioselective Formation of Medium-Ring Heterocycles of Biological Relevance by Intramolecular Cyclization. *RSC Adv.* **2011,** *1*, 1152–1170.
74. Herawi, M. M.; Fazeli, A. Recent Advances in the Application of the Heck Reaction in the Synthesis of Heterocyclic Compounds. *Heterocycles* **2010,** *81*, 1979–2026.
75. Majumdar, K. C.; Chattopadhyay, B.; Maji, P. K.; Chattopadhyay, S. K.; Samanta, S. Recent Development in Palladium-Mediated Synthesis of Nitrogen Heterocycles. *Heterocycles* **2010,** *81*, 795–866.
76. Zhang, M. Construction of Heterocycle Scaffolds Via Transition Metal-Catalyzed sp^2 C–H Functionalization. *Adv. Synth. Catal.* **2009,** *351*, 2243–2270.
77. Wolfe, J. P. Stereoselective Synthesis of Saturated Heterocycles Via Palladium-Catalyzed Alkene Carboetherification and Carboamination Reactions. *Synlett* **2008,** *19*, 2913–2937.
78. D'Souza, D. M.; Muller, T. J. J. Multi-Component Syntheses of Heterocycles by Transition-Metal Catalysis. *Chem. Soc. Rev.* **2007,** *36*, 1095–1108.
79. Minatti, A.; Muniz, K. Intramolecular Aminopalladation of Alkenes as a Key Step to Pyrrolidines and Related Heterocycles. *Chem. Soc. Rev.* **2007,** *36*, 1142–1152.
80. Zeni, G.; Larock, R. C. Synthesis of Heterocycles Via Palladium-Catalyzed Oxidative Addition. *Chem. Rev.* **2006,** *106*, 4644–4680.
81. Wolfe, J. P.; Thomas, J. S. Recent Developments in Palladium-Catalyzed Heterocycle Synthesis and Functionalization. *Curr. Org. Chem.* **2005,** *9*, 625–655.

82. Zeni, G.; Larock, R. C. Synthesis of Heterocycles Via Palladium π-Olefin and π-Alkyne Chemistry. *Chem. Rev.* **2004,** *104*, 2285–2309.
83. Gabriele, B.; Salerno, G.; Costa, M. PdI_2-Catalyzed Synthesis of Heterocycles. *Synlett* **2004,** 2468–2483.
84. Rubin, M.; Sromek, A. W.; Gevorgyan, V. New Advances in Selected Transition Metal-Catalyzed Annulations. *Synlett* **2003,** 2265–2291.
85. Balme, G.; Bossharth, E.; Monteiro, N. Pd-Assisted Multicomponent Synthesis of Heterocycles. *Eur. J. Org. Chem.* **2003,** 4101–4111.
86. Balme, G.; Bouyssi, D.; Lomberget, T.; Monteiro, N. Cyclisations Involving Attack of Carbo- and Heteronucleophiles on Carbon–Carbon π-Bonds Activated by Organopalladium Complexes. *Synthesis* **2003,** 2115–2134.
87. Cacchi, S.; Fabrizi, G.; Parisi, L. M. Nitrogen-Containing Heterocycles *Via* Palladium-Catalyzed Reaction of Alkynes With Organic Halides or Triflates. *Heterocycles* **2002,** *58*, 667–682.
88. Occasionally, the carbonylative heterocyclization process may lead to functionalized heterocycles in which carbon monoxide is not incorporated as carbonyl, owing to further transformations (for a recent example, see Kusakabe, T.; Takahashi, T.; Shen, R.; Ikeda, A.; Dhage, Y. D.; Kanno, Y.; Inouye, Y.; Sasai, H.; Mochiada, T.; Kato, K. Carbonylation of Propargyl Carbamates With Palladium(II) Bisoxazoline Catalysts: Efficient Synthesis of 5-Methoxy-3(2H)-Furanones. *Angew. Chem. Int. Ed.* **2013,** *52*, 7845–7849 These processes are beyond the scope of this chapter and will not be treated here.
89. Gabriele, B.; Salerno, G. Cyclocarbonylation. In *Handbook of Organopalladium Chemistry for Organic Synthesis;* Negishi, E., Ed.; , Vol. II; Wiley-Interscience: New York, 2002; pp 2623–2641.
90. Gabriele, B.; Salerno, G.; Costa, M. Oxidative Carbonylations. *Top. Organomet. Chem.* **2006,** *18*, 239–272.
91. See, for example: Gabriele, B.; Costa, M.; Salerno, G.; Chiusoli, G. P. Stereoselective Synthesis of β-Lactones Containing α-Z-Alkoxycarbonylmethylene Chains by Palladium-Catalysed Oxidative Carbonylation of Tertiary α-Hydroxyalkynes. *J. Chem. Soc. Chem. Commun.* **1994,** 1429–1430.
92. See, for example: Gabriele, B.; Salerno, G.; De Pascali, F.; Costa, M.; Chiusoli, G. P. Stereoselective Synthesis of (*Z*)-α-(Alkoxycarbonyl)Methylene β- and γ-Lactones by Palladium-Catalysed Oxidative Carbonylation of Alkynols. *J. Chem. Soc. Perkin Trans.* **1997,** *1*, 147–154.
93. See, for example: Bonardi, A.; Costa, M.; Gabriele, B.; Salerno, G.; Chiusoli, G. P. Versatile Synthesis of Beta-Lactams, Gamma-Lactams or Oxazolines by Palladium-Catalysed Oxidative Carbonylation of 1-Substituted Prop-2-Ynylamines. *Tetrahedron Lett.* **1995,** *36*, 7495–7498.
94. See, for example: Gabriele, B.; Salerno, G.; Veltri, L.; Costa, M. Stereoselective Synthesis of (*E*)-3-(Methoxycarbonyl)Methylene-1,3-Dihydroindol-2-Ones by Palladium-Catalyzed Oxidative Carbonylation of 2-Ethynylanilines. *Eur. J. Org. Chem.* **2001,** 4607–4613.
95. See, for example: Gabriele, B.; Salerno, G.; Fazio, A.; Veltri, L. Versatile Synthesis of Pyrrole-2-Acetic Esters and (Pyridine-2-One)-3-Acetic Amides by Palladium-Catalyzed, Carbon Dioxide-Promoted Oxidative Carbonylation of (Z)-(2-en-4-ynyl)amines. *Adv. Synth. Catal.* **2006,** *348*, 2212–2222.
96. See, for example: Tamaru, Y.; Hojo, M.; Yoshida, Z.-I. Palladium(II)-Catalyzed Carbonylation of 3-Buten-1-ols and 3-Butyn-1-ols: An Efficient Synthesis of γ-Butyrolactones. *J. Org. Chem.* **1991,** *56*, 1099–1105.

97. See, for example: Motodate, S.; Kobayashi, T.; Fuji, M.; Mochida, T.; Kusakabe, T.; Katoh, S.; Kato, K. Synthesis of β-Methoxyacrylate Natural Products Based on Box-Pd^{II}-Catalyzed Intermolecular Methoxycarbonylation of Alkynoles. *Chem. Asian J.* **2010,** *5*, 2221–2230.
98. Gabriele, B.; Salerno, G.; Veltri, L.; Costa, M. Synthesis of 2-Ynamides by Direct Palladium-Catalyzed Oxidative Aminocarbonylation of Alk-1-ynes. *J. Organomet. Chem.* **2001,** *622*, 84–88.
99. Beltrani, M.; Carfagna, C.; Milani, B.; Mancuso, R.; Gabriele, B.; Fini, F. Oxidative Alkoxycarbonylation of Alkynes by Means of Aryl α-Diimine Palladium(II) Complexes as Catalysts. *Adv. Synth. Catal.* **2016,** *358*, 3244–3253, and references therein.
100. Gabriele, B.; Costa, M.; Salerno, G.; Chiusoli, G. P. A New Synthesis of Trimethyl Aconitate by Palladium-Catalysed Triple Carbonylation of Propynyl Alcohol. *J. Chem. Soc. Chem. Commun.* **1992,** 1007–1008.
101. The reactions were carried out under the pressure of a CO–air mixture, air being the source of the O_2 necessary for Pd(0) reoxidation. It should be noted that all reactions were performed under conditions outside the explosion limits for CO–air mixtures. The flammability limits for CO in air are ca. 16% – 70% at 18 – 20°C and atmospheric pressure, and tend to decrease at higher total pressure. See: Bartish, C. M.; Drissel, G. M. In *Kirk – Othmer Encyclopedia of Chemical Technology;* 3rd ed. Grayson, M., Eckroth, D., Bushey, G. J., Campbell, L., Klingsberg, A., van Nes, L., Eds.; , Vol. 4; Wiley-Interscience: New York, 1978; p 7750.
102. Gabriele, B.; Salerno, G.; Costa, M.; Chiusoli, G. P. An Efficient and Selective Palladium-Catalysed Oxidative Dicarbonylation of Alkynes to Alkyl- or Aryl-Maleic Esters. *J. Chem. Soc. Perkin Trans.* **1994,** *1*, 83–87.
103. Gabriele, B.; Veltri, L.; Salerno, G.; Costa, M.; Chiusoli, G. P. Synthesis of Maleic Anhydrides and Maleic Acids by Pd-Catalyzed Oxidative Dicarbonylation of Alk-1-ynes. *Eur. J. Org. Chem.* **2003,** 1722–1728.
104. Gabriele, B.; Salerno, G. PdI_2. In *e-EROS (Electronic Encyclopedia of Reagents for Organic Synthesis);* Crich, D., Ed.; Wiley-Interscience: New York, 2006.
105. Mancuso, R.; Raut, D. S.; Marino, N.; De Luca, G.; Giordano, C.; Catalano, S.; Barone, I.; Andò, S.; Gabriele, B. A Palladium-Catalyzed Carbonylation Approach to Eight-Membered Lactam Derivatives With Antitumor Activity. *Chem. Eur. J.* **2016,** *22*, 3053–3064.
106. Li, W.; Liu, C.; Zhang, H.; Ye, K.; Zhang, G.; Zhang, W.; Duan, Z.; You, S.; Lei, A. Palladium-Catalyzed Oxidative Carbonylation of *N*-Allylamines for the Synthesis of β-Lactams. *Angew. Chem. Int. Ed.* **2014,** *53*, 2443–2446.
107. Ferguson, J.; Zeng, F.; Alper, H. Synthesis of Coumarins Via Pd-Catalyzed Oxidative Cyclocarbonylation of 2-Vinylphenols. *Org. Lett.* **2012,** *14*, 5602–5605.
108. Yang, X.-H.; Li, K.; Song, R.-J.; Li, J.-H. Room-Temperature Palladium-Catalyzed Intramolecular Oxidative Aminocarbonylation of Vinylic $C(sp^2)$–H Bonds With Amines and CO. *Eur. J. Org. Chem.* **2014,** 616–623.
109. Li, J.; Yang, S.; Wu, W.; Jiang, H. Novel Palladium-Catalyzed Cascade Carboxylative Annulation to Construct Functionalized γ-Lactones in Ionic Liquids. *Chem. Commun.* **2014,** *50*, 1381–1383.
110. It should be considered that the CO–O_2 mixtures are potentially explosive over a large range of composition: the flammability range for CO in O_2 is 16.7%–93.5% at room temperature and it becomes even larger at higher temperatures. See: Bartish, C. M.; Drissel, G. M. In *Kirk-Othmer Encyclopedia of Chemical Technology;* 3rd ed. Grayson, M.,

Eckroth, D., Bushey, G. J., Campbell, L., Klingsberg, A., van Nes, L., Eds.; , Vol. 4; John Wiley & Sons: New York, 1978; p 774.

111. Gabriele, B.; Salerno, G.; Plastina, P.; Costa, M.; Crispini, A. Expedient Synthesis of 4-Dialkylamino-5*H*-Furan-2-Ones by One-Pot Sequential Pd-Catalyzed Oxidative Carbonylation of 2-yn-1-ols–Conjugate Addition-Lactonization. *Adv. Synth. Catal.* **2004,** *346*, 351–358.
112. Gabriele, B.; Plastina, P.; Salerno, G.; Costa, M. A New Synthesis of 4-Dialkylamino-1,5-Dihydropyrrol-2-Ones by Pd-Catalyzed Oxidative Aminocarbonylation of 2-Ynylamines. *Synlett* **2005,** 935–938.
113. For the definition of auto-tandem catalysis, See: Shindoh, N.; Takemoto, K.; Takasu, K. Auto-Tandem Catalysis: A Single Catalyst Activating Mechanistically Distinct Reactions in a Single Reactor. *Chem. Eur. J.* **2009,** *15*, 12168–12179.
114. Gabriele, B.; Plastina, P.; Salerno, G.; Mancuso, R.; Costa, M. An Unprecedented Pd Catalyzed, Water-Promoted Sequential Oxidative Aminocarbonylation – Cyclocarbonylation Process Leading to 2-Oxazolidinones. *Org. Lett.* **2007,** *9*, 3319–3322.
115. Mancuso, R.; Maner, A.; Ziccarelli, I.; Pomelli, C.; Chiappe, C.; Della Ca', N.; Veltri, L.; Gabriele, B. Auto-Tandem Catalysis in Ionic Liquids: Synthesis of 2-Oxazolidinones by Palladium-Catalyzed Oxidative Carbonylation of Propargylic Amines in $EmimEtSO_4$. *Molecules* **2016,** *21*, 897.
116. Veltri, L.; Mancuso, R.; Altomare, A.; Gabriele, B. Divergent Multicomponent Tandem Palladium-Catalyzed Aminocarbonylation–Cyclization Approaches to Functionalized Imidazothiazinones and Imidazothiazoles. *ChemCatChem* **2015,** *7*, 2206–2213.
117. Veltri, L.; Grasso, G.; Rizzi, R.; Mancuso, R.; Gabriele, B. Palladium-Catalyzed Carbonylative Multicomponent Synthesis of Functionalized Benzimidazothiazoles. *Asian J. Org. Chem.* **2016,** *5*, 560–567.
118. Chen, M.; Ren, Z.-H.; Wang, Y.-Y.; Guan, Z.-H. Palladium-Catalyzed Oxidative Carbonylation of the Alkenyl C–H Bonds of Enamides: Synthesis of 1,3-Oxazin-6-Ones. *Angew Chem. Int. Ed.* **2013,** *52*, 14196–14199.
119. Liu, K.; Zou, M.; Lei, A. Aerobic Oxidative Carbonylation of Enamides by Merging Palladium With Photoredox Catalysis. *J. Org. Chem.* **2016,** *81*, 7088–7092.
120. Xie, Y.; Chen, T.; Fu, S.; Jiang, H.; Zeng, W. Pd-Catalyzed Carbonylative Cycloamidation of Ketoimines for the Synthesis of Pyrido[1,2-*a*]Pyrimidin-4-Ones. *Chem. Commun.* **2015,** *51*, 9377–9380.
121. Kumazawa, E.; Tokuhashi, T.; Horibata, A.; Koruno, N.; Senboku, H.; Tokuda, M.; Ohkuma, T.; Orito, K. Synthesis of Benzo[*c*]Phenanthridine Alkaloids by $Pd(OAc)_2$-Induced Direct Aromatic Carbonylation. *Eur. J. Org. Chem.* **2012,** 4622–4633.
122. Shun, S.; Jeong, Y.; Jeon, W. H.; Lee, P. H. Phosphaannulation by Palladium-Catalyzed Carbonylation of C–H Bonds of Phosphonic and Phosphinic Acids. *Org. Lett.* **2014,** *16*, 2930–2933.
123. Xu, Y.; Hu, W.; Tang, X.; Zhao, J.; Wu, W.; Jiang, H. Palladium-Catalyzed Csp^2–H Carbonylation of Aromatic Oximes: Selective Access to Benzo[*d*][1,2]Oxazin-1-Ones and 3-Methyleneisoindolin-1-Ones. *Chem. Commun.* **2015,** *51*, 6843–6846.
124. Ji, F.; Li, X.; Wu, W.; Jiang, H. Palladium-Catalyzed Oxidative Carbonylation for the Synthesis of Polycyclic Aromatic Hydrocarbons (PAHs). *J. Org. Chem.* **2014,** *79*, 11246–11253.
125. Wang, P.-L.; Li, Y.; Ma, L.; Luo, C.-G.; Wang, Z.-Y.; Lan, Q.; Wang, X.-S. Palladium-Catalyzed C-7 Selective C–H Carbonylation of Indolines for Expedient Synthesis of Pyrroloquinazolinediones. *Adv. Synth. Catal.* **2016,** *358*, 1048–1053.

126. Chen, J.; Feng, J.-B.; Natte, K.; Wu, X.-F. Palladium-Catalyzed Carbonylative Cyclization of Arenes by C–H Bond Activation With DMF as the Carbonyl Source. *Chem. Eur. J.* **2015,** *21*, 16370–16373.
127. Chen, J.; Natte, K.; Wu, X.-F. Pd/C-Catalyzed Carbonylative C–H Activation with DMF as the CO Source. *Tetrahedron Lett.* **2015,** *56*, 6413–6416.
128. Guan, Z.-H.; Chen, M.; Ren, Z.-H. Palladium-Catalyzed Regioselective Carbonylation of C–H Bonds of *N*-Alkyl Anilines for Synthesis of Isatoic Anhydrides. *J. Am. Chem. Soc.* **2012,** *134*, 17490–17493.
129. Gao, S.; Chen, M.; Zhao, M.-N.; Du, W.; Ren, Z.-H.; Wang, Y.-Y.; Guan, Z.-H. Palladium-Catalyzed Carbonylation of *o*-Iodoanilines for Synthesis of Isatoic Anhydrides. *J. Org. Chem.* **2014,** *79*, 4196–4200.
130. Luo, S.; Luo, F.-X.; Zhang, X.-S.; Shi, Z.-J. Synthesis of Dibenzopyranones Through Palladium-Catalyzed Directed C–H Activation/Carbonylation of 2-Arylphenols. *Angew. Chem. Int. Ed.* **2013,** *52*, 10598–10601.
131. Lee, T.-H.; Jayakumar, J.; Cheng, C.-H.; Chuang, S.-C. One Pot Synthesis of Bioactive Benzopyranones Through Palladium-Catalyzed C–H Activation and CO Insertion into 2-Arylphenols. *Chem. Commun.* **2013,** 11797–11799.
132. Shi, R.; Lu, L.; Xie, H.; Yan, J.; Xu, T.; Zhang, H.; Qi, X.; Lan, Y.; Lei, A. C8–H Bond Activation *vs.* C2–H Bond Activation: From Naphthyl Amines to Lactams. *Chem. Commun.* **2016,** *52*, 13307–13310.
133. Shi, R.; Nius, H.; Lu, L.; Lei, A. Pd/Cu-Catalyzed Aerobic Oxidative Aromatic C–H Bond Activation/N-Dealkylative Carbonylation Towards the Synthesis of Phenanthridinones. *Chem. Commun.* **2017,** *53*, 1908–1911.
134. Li, W.; Duan, Z.; Zhang, X.; Zhang, H.; Wang, M.; Jiang, R.; Zeng, H.; Liu, C.; Lei, A. From Anilines to Isatins: Oxidative Palladium-Catalyzed Double Carbonylation of C–H Bonds. *Angew. Chem. Int. Ed.* **2015,** *54*, 1893–1896.
135. Shi, R.; Lu, L.; Zhang, H.; Chen, B.; Sha, Y.; Liu, C.; Lei, A. Palladium/Copper-Catalyzed Oxidative CH Alkenylation/*N*-Dealkylative Carbonylation of Tertiary Anilines. *Angew. Chem. Int. Ed.* **2013,** *52*, 10582–10585.
136. Chen, J.; Natte, K.; Wu, X.-F. Palladium-Catalyzed Carbonylative C–H Activation of Arenes with Norbornene as the Coupling Partner. *J. Organomet. Chem.* **2016,** *803*, 9–12.
137. Chen, J.; Natte, K.; Spannenberg, A.; Neumann, H.; Beller, M.; Wu, X.-F. Palladium-Catalyzed Carbonylative [3 + 2 + 1] Annulation of *N*-Aryl-Pyridine-2-Amines with Internal Alkynes by C–H Activation: Facile Synthesis of 2-Quinolinones. *Chem. Eur. J.* **2014,** *20*, 14189–14193.
138. Feng, M.; Tang, B.; Xu, H.-X.; Jiang, X. Collective Synthesis of Phenanthridinone Through C–H Activation Involving a Pd-Catalyzed Aryne Multicomponent Reaction. *Org. Lett.* **2016,** *18*, 4352–4355.
139. Zhang, H.; Shi, R.; Gan, P.; Liu, C.; Ding, A.; Wang, A.; Wang, Q.; Lei, A. Palladium-Catalyzed Oxidative Double C–H Functionalization/Carbonylation for the Synthesis of Xanthones. *Angew. Chem. Int. Ed.* **2012,** *51*, 5204–5207.
140. Wen, J.; Tang, S.; Zhang, F.; Shi, R.; Lei, A. Palladium/Copper Co-catalyzed Oxidative C–H/C–H Carbonylation of Diphenylamines: A Way to Access Acridones. *Org. Lett.* **2017,** *19*, 94–97.
141. McNally, A.; Haffemayer, B.; Collins, B. S. L.; Gaunt, M. J.; Palladium-catalysed, C–H. Activation of Aliphatic Amines to Give Strained Nitrogen Heterocycles. *Nature* **2014,** *510*, 129–133.

142. See, for example: Gabriele, B.; Salerno, G.; Brindisi, D.; Costa, M.; Chiusoli, G. P. Synthesis of 2-Oxazolidinones by Direct Palladium-Catalyzed Oxidative Carbonylation of 2-Amino-1-Alkanols. *Org. Lett.* **2000,** *2*, 625–627.
143. See, for example: Gabriele, B.; Mancuso, R.; Salerno, G.; Costa, M. An Improved Procedure for the Palladium-Catalyzed Oxidative Carbonylation of β-Amino Alcohols to Oxazolidin-2-Ones. *J. Org. Chem.* **2003,** *68*, 601–604.
144. See, for example: Gabriele, B.; Salerno, G.; Mancuso, R.; Costa, M. Efficient Synthesis of Ureas by Direct Palladium-Catalyzed Oxidative Carbonylation of Amines. *J. Org. Chem.* **2004,** 4741–4750.
145. See, for example: Gabriele, B.; Mancuso, R.; Salerno, G.; Veltri, L.; Costa, M.; Dibenedetto, A. A General and Expedient Synthesis of 5- and 6-Membered Cyclic Carbonates by Palladium-Catalyzed Oxidative Carbonylation of 1,2- and 1,3-Diols. *ChemSusChem* **2011,** *4*, 1778–1786.
146. See, for example: Hua, J.; Li, J.; Gua, Y.; Guan, Z.; Mo, W.; Ni, Y.; Li, T.; Li, G. Oxidative Carbonylation of Glycerol to Glycerol Carbonate Catalyzed by $PdCl_2$(phen)/KI. *Appl. Catal. A: Gen.* **2010,** *386*, 188–193.
147. See, for example: Hu, J.; Gu, Y.; Guan, Z.; Li, J.; Mo, W.; Li, T.; Li, G. An Efficient Palladium Catalyst System for the Oxidative Carbonylation of Glycerol to Glycerol Carbonate. *ChemSusChem* **2011,** *4*, 1767–1772.
148. See, for example: Troisi, L.; Granito, C.; Perrone, S.; Rosato, F. Synthesis of Benzo-Fused Five- and Six-Membered Heterocycles by Palladium-Catalyzed Cyclocarbonylation. *Tetrahedron Lett.* **2011,** *52*, 4330–4332.
149. Mancuso, R.; Raut, D. S.; Della Ca', N.; Fini, F.; Carfagna, C.; Gabriele, B. Catalytic Oxidative Carbonylation of Amino Moieties to Ureas, Oxamides, 2-Oxazolidinones, and Benzoxazolones. *ChemSusChem* **2015,** *8*, 2204–2211.
150. Wang, Y.; Meng, X.; Yang, Y.; Zhang, L.; Guo, S.; Tang, D.; Li, Y.; Chen, B. Palladium-Catalyzed Oxidative Carbonylation of Hydrazides: Synthesis of 1,3,4-Oxadiazol-2(3*H*)-Ones. *Chem. Commun.* **2015,** *51*, 1905–1907.
151. Ji, F.; Li, X.; Guo, W.; Wu, W.; Jiang, H. Palladium-Catalyzed Oxidative O–H/N–H Carbonylation of Hydrazides: Access to Substituted 1,3,4-Oxadiazole-2(3H)-Ones. *J. Org. Chem.* **2015,** *80*, 5713–5718.
152. See, for example: Yang, Q.; Cao, H.; Robertson, A.; Alper, H. Synthesis of Dibenzo[b,f][1,4]Oxazepin-11(10H)-Ones Via Intramolecular Cyclocarbonylation Reactions Using PdI_2/Cytop 292 as the Catalytic System. *J. Org. Chem.* **2010,** *75*, 6297–6299, and references therein.
153. See, for example: Lu, S.-M.; Alper, H. Synthesis of Large Ring Macrocycles (12–18) by Recyclable Palladium-Complexed Dendrimers on Silica Gel Catalyzed Intramolecular Cyclocarbonylation Reactions. *Chem. Eur. J.* **2007,** *13*, 5908–5916, and references therein.
154. See, for example: Yao, T.; Yue, D.; Larock, R. C. An Efficient Synthesis of Coumestrol and Coumestans by Iodocyclization and Pd-Catalyzed Intramolecular Lactonization. *J. Org. Chem.* **2005,** *70*, 9985–9989.
155. See, for example: Wu, X.; Mahalingam, A. K.; Wan, Y.; Alterman, M. A Novel and Efficient (NHC)CuI (NHC = *N*-Heterocyclic Carbene) Catalyst for the Oxidative Carbonylation of Amino Compounds. *Tetrahedron Lett.* **2007,** *48*, 5883–5886.
156. Bae, Y.; Cho, C. S. Palladium-Catalyzed Carbonylative Cyclization of β-Bromo-α,β-Unsaturated Carboxylic Acids Leading to Maleic Anhydrides. *Synlett* **2013,** *24*, 1848–1850.

157. Khedkar, M. V.; Khan, S. R.; Dhake, K. P.; Bhanage, B. M. Carbonylative Cyclization of *o*-Halobenzoic Acids for Synthesis of *N*-Substituted Phthalimides Using Polymer-Supported Palladium–*N*-Heterocyclic Carbene as an Efficient, Heterogeneous, and Reusable Catalyst. *Synthesis* **2012,** *44*, 2623–2629.
158. Khedkar, M. V.; Shinde, A. R.; Sasaki, T.; Bhanage, B. M. Immobilized Palladium Metal Containing Ionic Liquid Catalyzed One Step Synthesis of Isoindole-1,3-Diones by Carbonylative Cyclization Reaction. *J. Mol. Catal. A: Chem.* **2014,** *385*, 91–97.
159. Barrio, P.; Ibáñez, I.; Herrera, L.; Román, R.; Catalán, S.; Fustero, S. Asymmetric Synthesis of Fluorinated Isoindolinones Through Palladium-Catalyzed Carbonylative Amination of Enantioenriched Benzylic Carbamates. *Chem. Eur. J.* **2015,** *21*, 11579–11584.
160. Harada, R.; Nishida, N.; Uchiito, S.; Onozaki, Y.; Kurono, N.; Sanboku, H.; Masao, T.; Okhuma, T.; Orito, K. Phosphane-Free Pd^0-Catalyzed Cycloamination and Carbonylation with $Pd(OAc)_2$ and $Cu(OAc)_2$ in the Presence of K_2CO_3: Preparation of Benzocyclic Amines and Benzolactams. *Eur. J. Org. Chem.* **2012,** 366–379.
161. Chavan, S.; Bhanage, B. M. Carbonylative Synthesis of Phthalimides and Benzoxazinones by Using Phenyl Formate as a Carbon Monoxide Source. *Eur. J. Org. Chem.* **2015,** 2405–2410.
162. Li, W.; Wu, X.-F. Palladium-Catalyzed Carbonylative Synthesis of Benzoxazinones from *N*-(*o*-Bromoaryl)Amides Using Paraformaldehyde as the Carbonyl Source. *J. Org. Chem.* **2014,** *79*, 10410–10416.
163. Xu, T.; Alper, H. Synthesis of Pyrido[2,1-b]Quinazolin-11-Ones and Dipyrido[1,2-a:2′,3′-d]Pyrimidin-5-Ones by Pd/DIBPP-Catalyzed Dearomatizing Carbonylation. Org. Lett. **2015**, *17*, 1569–1572.
164. Yoo, J. M.; Ho, S. L.; Cho, C. S. Synthesis of 6*H*-Isoindolo[2,1-*a*]Indol-6-Ones Via Palladium-Catalyzed Carbonylative Cyclization of 2-(2-Bromophenyl)-1*H*-Indoles. *Synlett* **2016,** *27*, 1383–1386.
165. Guo, S.; Tao, L.; Wang, F.; Fan, X. Pd-Catalyzed Cyclocarbonylation of 2-(2-Bromoaryl) Indoles with CO as a C1 Source: Selective Access to 6*H*-Isoindolo[2,1-*a*]Indol-6-Ones and Indeno[1,2-*b*]Indol-10(5*H*)-Ones. *Chem. Asian J.* **2016,** *11*, 3090–3096.
166. Frutos-Pedreño, R.; García-López, J.-A. 2-Arylacetamides as Versatile Precursors for 3-Aminoisocoumarin and Homophthalimide Derivatives: Palladium-Catalyzed Cascade Double Carbonylation Reactions. *Adv. Synth. Catal.* **2016,** *358*, 2692–2700.
167. Perrone, S.; Caroli, A.; Cannazza, G.; Granito, C.; Salomone, A.; Troisi, L. A Direct Synthesis of 3-Acyl-4-Hydroxy-2-Pyranone Derivatives Via Palladium-Catalyzed Carbonylation of α-Chloroketones. A Cascade Reaction Involving Acylketenes. *Tetrahedron Lett.* **2015,** *56*, 2773–2776.
168. See, for example: Wu, X.-F.; Schranck, J.; Neumann, H.; Beller, M. A Convenient and General Palladium-Catalyzed Carbonylative Coupling for the Synthesis of 2-Arylbenzoxazinones. *Chem. Eur. J.* **2011,** *17*, 12246–12249.
169. Wu, X.-F.; Neumann, H.; Beller, M. Palladium-Catalyzed Carbonylation Reaction of Aryl Bromides with 2-Hydroxyacetophenones to form Flavones. *Chem. Eur. J.* **2012,** *18*, 12595–12598.
170. Wu, X.-F.; He, L.; Neumann, H.; Beller, M. Palladium-Catalyzed Carbonylative Synthesis of Quinazolinones from 2-Aminobenzamide and Aryl Bromides. *Chem. Eur. J.* **2013,** *19*, 12635–12638.
171. He, H. L.; Neumann, H.; Beller, M.; Wu, X.-F. Cascade Synthesis of Quinazolinones from 2-Aminobenzonitriles and Aryl Bromides *Via* Palladium-Catalyzed Carbonylation Reaction. *Green Chem.* **2014,** *16*, 1336–1343.

172. See, for example: Worlikar, S. A.; Larock, R. C. Palladium-Catalyzed One-Step Synthesis of Isoindole-1,3-Diones by Carbonylative Cyclization of *o*-Halobenzoates and Primary Amines. *J. Org. Chem.* **2008,** *73*, 7175–7180.
173. See, for example: Chouhan, G.; Alper, H. Palladium-Catalyzed Carboxamidation Reaction and Aldol Condensation Reaction Cascade: A Facile Approach to Ring-Fused Isoquinolinones. *Org. Lett.* **2008,** *10*, 4987–4990.
174. See, for example: Chouhan, G.; Alper, H. Synthesis of Ring-Fused Oxazolo- and Pyrazoloisoquinolinones by a One-Pot Pd-Catalyzed Carboxamidation and Aldol-Type Condensation Cascade Process. *J. Org. Chem.* **2009,** *74*, 6181–6189.
175. See, for example: Zeng, F.; Alper, H. One-Step Synthesis of Quinazolino[3,2-*a*] Quinazolinones Via Palladium-Catalyzed Domino Addition/Carboxamidation Reactions. *Org. Lett.* **2010,** *12*, 3642–3644.
176. See, for example: Cao, H.; Vieira, T. O.; Alper, H. Synthesis of Unsaturated Seven-Membered Ring Lactams Through Palladium-Catalyzed Amination and Intramolecular Cyclocarbonylation Reactions of Amines and Baylis – Hillman Acetates. *Org. Lett.* **2011,** *13*, 11–13.
177. See, for example: Cao, H.; Alper, H. Palladium-Catalyzed Double Carbonylation Reactions of *o*-Dihaloarenes with Amines in Phosphonium Salt Ionic Liquids. *Org. Lett.* **2010,** *12*, 4126–4129.
178. See, for example: Khedar, M. V.; Khan, S. R.; Sawant, D. N.; Bagal, D. B.; Bhanage, B. M. Palladium on Carbon: An Efficient, Heterogeneous and Reusable Catalytic System for Carbonylative Synthesis of *N*-Substituted Phthalimides. *Adv. Synth. Catal.* **2011,** *353*, 3415–3422.
179. Chen, J.; Natte, K.; Spannenberg, A.; Neumann, H.; Beller, M.; Wu, X.-F. Efficient Palladium-Catalyzed Double Carbonylation of *o*-Dibromobenzenes: Synthesis of Thalidomide. *Org. Biomol. Chem.* **2014,** *12*, 5578–5581.
180. Chen, J.; Natte, K.; Wu, X.-F. Convenient Palladium-Catalyzed Carbonylative Synthesis of Caprolactam and Butyrolactam Derived Phthalimides and Amides by Using DBU and DBN as the Nitrogen Source. *Tetrahedron Lett.* **2015,** *56*, 342–345.
181. Fuse, S.; Takahashi, R.; Takahashi, T. Facile, One-Step Synthesis of 5-Substituted Thieno [3,4-c]Pyrrole-4,6-Dione by Palladium-Catalyzed Carbonylative Amidation. *Eur. J. Org. Chem.* **2015,** 3430–3434.
182. Chen, J.; Neumann, H.; Beller, M.; Wu, X.-F. Palladium-Catalyzed Synthesis of Isoindoloquinazolinones Via Dicarbonylation of 1,2-Dibromoarenes. *Org. Biomol. Chem.* **2014,** *12*, 5835–5838.
183. Chen, J.; Natte, K.; Spannenberg, A.; Neumann, H.; Langer, P.; Beller, M.; Wu, X.-F. Base-Controlled Selectivity in the Synthesis of Linear and Angular Fused Quinazolinones by a Palladium-Catalyzed Carbonylation/Nucleophilic Aromatic Substitution Sequence. *Angew. Chem. Int. Ed.* **2014,** *53*, 7579–7583.
184. Chen, C.; Neumann, H.; Wu, X.-F. A Highly-Efficient Palladium-Catalyzed Aminocarbonylation/S_NAr Approach to Dibenzoxazepinones. *Green Chem.* **2015,** *17*, 2994–2999.
185. Shen, C.; Wu, X.-F. Base-Regulated Tunable Synthesis of Pyridobenzoxazepinones and Pyridobenzoxazines. *Catal. Sci. Technol.* **2015,** *5*, 4433–4443.
186. Chen, J.; Natte, K.; Neumann, H.; Wu, X.-F. Palladium-Catalyzed Carbonylative Reactions of 1-Bromo-2-Fluorobenzenes with Various Nucleophiles: Effective Combination of Carbonylation and Nucleophilic Substitution. *Chem. Eur. J.* **2014,** *20*, 16107–16110 *Chem. Eur. J.* **2015**, *21*, 17542 (Corrigendum).

187. Shen, C.; Li, W.; Yin, H.; Spannenberg, A.; Skrydstrup, T.; Wu, X.-F. Palladium-Catalyzed Carbonylative Synthesis of 2,3-Disubstituted Chromones. *Adv. Synth. Catal.* **2016,** *358*, 466–479.
188. Sadig, J. E. R.; Foster, R.; Wakenhut, F.; Willis, M. C. Palladium-Catalyzed Synthesis of Benzimidazoles and Quinazolinones from Common Precursors. *J. Org. Chem.* **2012,** *77*, 9473–9486.
189. He, L.; Li, H.; Neumann, H.; Beller, B.; Wu, X.-F. Highly Efficient Four-Component Synthesis of 4(3*H*)-Quinazolinones: Palladium-Catalyzed Carbonylative Coupling Reactions. *Angew. Chem. Int. Ed.* **2014,** *53*, 1420–1424.
190. Natte, K.; Neumann, H.; Wu, X.-F. Pd/C as an Efficient Heterogeneous Catalyst for Carbonylative Four-Component Synthesis of 4(3*H*)-Quinazolinones. *Catal. Sci. Technol.* **2015,** *5*, 4474–4480.
191. He, L.; Sharif, M.; Neumann, H.; Beller, M.; Wu, X.-F. A Convenient Palladium-Catalyzed Carbonylative Synthesis of 4(3*H*)-Quinazolinones from 2-Bromoformanilides and Organo Nitros with $Mo(CO)_6$ as a Multiple Promoter. *Green Chem.* **2014,** *16*, 3763–3767.
192. Shen, C.; Man, N. Y. T.; Stewart, S.; Wu, X.-F. Palladium-Catalyzed Dicarbonylative Synthesis of Tetracycle Quinazolinones. *Org. Biomol. Chem.* **2015,** *13*, 4422–4425.
193. Wu, X.-F.; Neumann, H.; Neumann, S.; Beller, M. Palladium-Catalyzed Synthesis of Phthalazinones: Efficient Carbonylative Coupling of 2-Bromobenzaldehydes and Hydrazines. *Chem. Eur. J.* **2012,** *18*, 8596–8599.
194. Natte, K.; Chen, J.; Li, H.; Neumann, H.; Beller, M.; Wu, X.-F. Palladium-Catalyzed Carbonylation of 2-Bromoanilines with 2-Formylbenzoic Acid and 2-Halobenzaldehydes: Efficient Synthesis of Functionalized Isoindolinones. *Chem. Eur. J.* **2014,** *20*, 14184–14188.
195. Li, H.; Li, W.; Spannenberg, A.; Baumann, W.; Neumann, H.; Beller, M.; Wu, X.-F. A Novel Domino Synthesis of Quinazolinediones by Palladium-Catalyzed Double Carbonylation. *Chem. Eur. J.* **2014,** *20*, 8541–8544.
196. Wu, X.-F.; Wu, L.; Jackstell, R.; Neumann, H.; Beller, M. A General Palladium-Catalyzed Carbonylative Synthesis of Chromenones from Salicylic Aldehydes and Benzyl Chlorides. *Chem. Eur. J.* **2013,** *19*, 12245–12248.
197. Capua, M.; Granito, C.; Perrone, S.; Salomone, A.; Troisi, L. Palladium-Catalyzed Carbonylative Coupling of α-Chloroketones with Hydrazines: A Simple Route to Pyrazolone Derivatives. *Tetrahedron Lett.* **2016,** *57*, 3363–3367.
198. See, for example: Zheng, Z.; Alper, H. Palladium-Catalyzed Carbonylation – Decarboxylation of Diethyl(2-iodoaryl)malonates with Imidoyl Chlorides: An Efficient Route to Substituted Isoquinolin-1(2*H*)-Ones. *Org. Lett.* **2008,** *10*, 4903–4906.
199. See, for example: Chouan, G.; Alper, H. Domino Ring-Opening/Carboxamidation Reactions of *N*-Tosyl Aziridines and 2-Halophenols/Pyridinol: Efficient Synthesis of 1,4-Benzo- and Pyrido-Oxazepinones. *Org. Lett.* **2010,** *12*, 192–195, and references therein.
200. See, for example: Salvadori, J.; Balducci, E.; Zaza, S.; Petricci, E.; Taddei, M. Microwave-Assisted Carbonylation and Cyclocarbonylation of Aryl Iodides Under Ligand Free Heterogeneous Catalysis. *J. Org. Chem* **2010,** *75*, 1841–1847, and references therein.
201. Okuro, K.; Alper, H. Palladium-Catalyzed Intermolecular Cyclocarbonylation of 2-Iodoanilines with the Michael Acceptor, Diethyl Ethoxycarbonylbutendienoate. *J. Org. Chem.* **2012,** *77*, 4420–4424.

202. Wu, X.-F.; Neumann, H.; Beller, M. A General Palladium-Catalyzed Carbonylative Synthesis of 2-Alkylbenzoxazinones from 2-Bromoanilines and Acid Anhydrides. *Chem. Eur. J.* **2012,** *18*, 12599–12602.
203. Wu, X.-F.; Sharif, M.; Shoaib, K.; Neumann, H.; Pews-Davtyan, A.; Langer, P.; Beller, M. A Convenient Palladium-Catalyzed Carbonylative Synthesis of 2-Aminobenzoxazinones from 2-Bromoanilines and Isocyanates. *Chem. Eur. J.* **2013,** *19*, 6230–6233.
204. Li, H.; Spannenberg, A.; Neumann, H.; Beller, M.; Wu, X.-F. Regioselective Synthesis of 2,3-Dihydrobenzodioxepinones from Epoxides and 2-Bromophenols *via* Palladium-Catalyzed Carbonylation. *Chem. Commun.* **2014,** *50*, 2114–2116.
205. See, for example: Ye, F.; Alper, H. Ionic-Liquid-Promoted Palladium-Catalyzed Multicomponent Cyclocarbonylation of *o*-Iodoanilines and Allenes to Form Methylene-2,3-dihydro-1*H*-quinolin-4-ones. *J. Org. Chem.* **2007,** *72*, 3218–3222.
206. See, for example: Umemura, S.; McLaughlin, M.; Micalizio, G. C. Convergent Synthesis of Stereodefined *exo*-Alkylidene-γ-lactams from β-Halo Allylic Alcohols. *Org. Lett.* **2009,** *11*, 5402–5405.
207. Schranck, J.; Wu, X. ,-F.; Tlili, A.; Neumann, H.; Beller, M. A Novel Double Carbonylation Reaction of Aryl Halides: Selective Synthesis of 5-Arylfuranones. *Chem. Eur. J.* **2013,** *19*, 12959–12964.
208. Xie, P.; Qian, B.; Huang, H.; Xia, C. Stereoselective Synthesis of Trans β-Lactams Via Palladium/*N*-Heterocyclic Carbene-Catalyzed Carbonylative [2 + 2] Cycloaddition. *Tetrahedron Lett.* **2012,** *53*, 1613–1616.
209. See, for example: Wu, X.-F.; Sundararaju, B.; Anbarasan, P.; Neumann, H.; Dixneuf, P. H.; Beller, M. A General Cyclocarbonylation of Aryl Bromides and Triflates with Acetylenes: Palladium-Catalyzed Synthesis of 3-Alkylidenefuran-2-ones. *Chem. Eur. J.* **2011,** *17*, 8014–8017.
210. See, for example: Miao, H.; Yang, Z. Regiospecific Carbonylative Annulation of Iodophenol Acetates and Acetylenes to Construct the Flavones by a New Catalyst of Palladium – Thiourea – dppp Complex. *Org. Lett.* **2000,** *2*, 1765–1768.
211. See, for example: Kadnikov, D. V.; Larock, R. C. Synthesis of Coumarins via Palladium-Catalyzed Carbonylative Annulation of Internal Alkynes by *o*-Iodophenols. *Org. Lett.* **2000,** *2*, 3643–3646.
212. See, for example: Kadnikov, D. V.; Larock, R. C. Palladium-Catalyzed Carbonylative Annulation of Internal Alkynes: Synthesis of 3,4-Disubstituted Coumarins. *J. Org. Chem.* **2003,** *68*, 9423–9432.
213. See, for example: Kadnikov, D. V.; Larock, R. C. Synthesis of 2-Quinolones Via Palladium-Catalyzed Carbonylative Annulation of Internal Alkynes by *N*-Substituted *o*-Iodoanilines. *J. Org. Chem.* **2004,** *69*, 6772–6780.
214. See, for example: Awuah, E.; Capretta, A. Access to Flavones Via a Microwave-Assisted, One-Pot Sonogashira – Carbonylation – Annulation Reaction. *Org. Lett.* **2009,** *11*, 3210–3213.
215. See, for example: Yang, Q.; Alper, H. Synthesis of Chromones Via Palladium-Catalyzed Ligand-Free Cyclocarbonylation of *o*-Iodophenols with Terminal Acetylenes in Phosphonium Salt Ionic Liquids. *J. Org. Chem.* **2010,** *75*, 948–950.
216. See, for example: Chen, J.-R.; Liao, J.; Xiao, W.-J. Microwave-Assisted, Palladium-Catalyzed Carbonylative Cyclization—Rapid Synthesis of 2-Quinolones from Unprotected 2-Iodoanilines and Terminal Alkynes. *Can. J. Chem.* **2010,** *88*, 331–337.

217. See, for example: Xue, L.; Shi, L.; Han, Y.; Xia, C.; Huynh, H. V.; Li, F. Pd–Carbene Catalyzed Carbonylation Reactions of Aryl Iodides. *Dalton Trans.* **2011,** *40*, 7632–7638.
218. See, for example: Cao, H.; McNamee, L.; Alper, H. Syntheses of Substituted 3-Methyleneisoindolin-1-Ones by a Palladium-Catalyzed Sonogashira Coupling – Carbonylation – Hydroamination Sequence in Phosphonium Salt-Based Ionic Liquids. *Org. Lett.* **2008,** *10*, 5281–5284.
219. See, for example: Kalinin, V. N.; Shostakovsky, M. V.; Ponomaryov, A. B. A New Route to 2-Aryl-4-quinolones Via Palladium-Catalyzed Carbonylative Coupling of *o*-Iodoanilines with Terminal Arylacetylenes. *Tetrahedron Lett.* **1992,** *33*, 373–376.
220. Zhu, F.; Li, Y.; Wang, Z.; Wu, X.-F. Highly Efficient Synthesis of Flavones via Pd/C-Catalyzed Cyclocarbonylation of 2-Iodophenol with Terminal Acetylenes. *Catal. Sci. Technol.* **2016,** *6*, 2905–2909.
221. Shen, C.; Spannenberg, A.; Wu, X.-F. Palladium-Catalyzed Carbonylative Four-Component Synthesis of Thiochromenones: The Advantages of a Reagent Capsule. *Angew. Chem. Int. Ed.* **2016,** *55*, 5067–5070.
222. Wu, X.-F.; Jiao, H.; Neumann, H.; Beller, M. Palladium-Catalyzed Carbonylative Coupling of Aryl Iodides and Benzyl Acetylenes to 3-Alkylidenefuran-2-Ones Under Mild Conditions and Its Density Functional Theory Modeling. *Chem. Eur. J.* **2012,** *18*, 16177–16185.
223. Feng, M.; Tang, B.; Wang, N.; Xu, H.-X.; Jiang, X. Ligand Controlled Regiodivergent C_1 Insertion on Arynes for Construction of Phenanthridinone and Acridone Alkaloids. *Angew. Chem. Int. Ed.* **2015,** *54*, 14960–14964.
224. See, for example: Gabriele, B.; Salerno, G.; Costa, M. Combined Oxidative and Reductive Carbonylation of Terminal Alkynes with Palladium Iodide-Thiourea Catalysts. *J. Organomet. Chem.* **1995,** *503*, 21–28.
225. See, for example: Gabriele, B.; Salerno, G.; Costa, M.; Chiusoli, G. P. A New Regioselective Synthesis of 3-Substituted Furan-2(5*H*)-ones by Palladium-Catalysed Reductive Carbonylation of Alk-1-ynes. *Tetrahedron Lett.* **1999,** *40*, 989–990.
226. See, for example: Gabriele, B.; Salerno, G.; Costa, M.; Chiusoli, G. P. Palladium-Catalysed Formation of Maleic Anhydrides from CO, CO_2 and Alk-1-ynes. *Chem. Commun.* **1999,** 1381–1382.
227. See, for example: Chiusoli, G. P.; Costa, M.; Cucchia, L.; Gabriele, B.; Salerno, G.; Veltri, L. Carbon Dioxide Effect on Palladium-Catalyzed Sequential Reactions with Carbon Monoxide, Acetylenic Compounds and Water. *J. Mol. Catal. A: Chem.* **2003,** *204*, 133–142.
228. See, for example: Queirolo, M.; Vezzani, A.; Mancuso, R.; Gabriele, B.; Costa, M.; Della Ca', N. Neutral vs Anionic Palladium Iodide-Catalyzed Carbonylation of Terminal Arylacetylenes. *J. Mol. Catal. A: Chem.* **2015,** *398*, 115–126.
229. See, for example: Boeckman, R. K., Jr; Reed, J. E.; Ge, P. A Novel Route to 2,3-Pyrazol-1(5*H*)-Ones Via Palladium-Catalyzed Carbonylation of 1,2-Diaza-1,3-Butadienes. *Org. Lett.* **2001,** *3*, 3651–3653.
230. See, for example: Cao, H.; Xiao, W.-J.; Alper, H. Dramatic Effects of Ionic Liquids on the Palladium-Catalyzed Cyclocarbonylation of Enynols with Thiols. *J. Org. Chem.* **2007,** *72*, 8562–8564.
231. See, for example: Consorti, C. S.; Ebeling, G.; Dupont, J. Carbonylation of Alkynols Catalyzed by Pd(II)/2-$PyPPh_2$ Dissolved in Organic Solvents and in Ionic Liquids: A Facile Entry to α-Methylene γ- and δ-Lactones. *Tetrahedron Lett.* **2002,** *43*, 753–755.

232. See, for example: El Ali, B.; Alper, H. Lactonization of Unsaturated Alcohols Catalyzed by Palladium Complexes Under Neutral Conditions. *J. Org. Chem.* **1991,** *56*, 5357–5360.
233. See, for example: Murray, T. F.; Norton, J. R. The Design and Mechanism of Palladium Catalysts for Synthesis of Methylene Lactones by Cyclocarbonylation of Acetylenic Alcohols. *J. Am. Chem. Soc.* **1979,** *101*, 4107–4111.
234. See, for example: Lu, S.-M.; Alper, H. Sequence of Intramolecular Carbonylation and Asymmetric Hydrogenation Reactions: Highly Regio- and Enantioselective Synthesis of Medium Ring Tricyclic Lactams. *J. Am. Chem. Soc.* **2008,** *130*, 6451–6455.
235. See, for example: El Ali, B.; Okuro, K.; Vasapollo, G.; Alper, H. Regioselective Palladium(II)-Catalyzed Synthesis of Five- or Seven-Membered Ring Lactones and Five-, Six- or Seven-Membered Ring Lactams by Cyclocarbonylation Methodology. *J. Am. Chem. Soc.* **1996,** *118*, 4264–4270.
236. See, for example: Orejon, A.; Alper, H. Cyclocarbonylation of 2-Allylphenols Catalyzed by Palladium-Montmorillonite. *J. Mol. Catal. A: Chem.* **1999,** *143*, 137–142.
237. See, for example: Chanthateyanonth, R.; Alper, H. Recyclable Tridentate Stable Palladium(II) PCP-Type Catalysts Supported on Silica for the Selective Synthesis of Lactones. *Adv. Synth. Catal.* **2004,** *346*, 1375–1384.
238. See, for example: Touzani, R.; Alper, H. PAMAM Dendrimer-Palladium Complex Catalyzed Synthesis of Five-, Six- Or Seven Membered Ring Lactones and Lactams by Cyclocarbonylation Methodology. *J. Mol. Catal. A: Chem.* **2005,** *227*, 197–207.
239. See, for example: Ye, F.; Alper, H. Recyclable Selective Palladium-Catalyzed Synthesis of Five-, Six- Or Seven-Membered Ring Lactones and Lactams by Cyclocarbonylation in Ionic Liquids. *Adv. Synth. Catal.* **2006,** *348*, 1855–1861.
240. See, for example: Amézquita-Valencia, M.; Alper, H. PdI_2-Catalyzed Regioselective Cyclocarbonylation of 2-Allyl Phenols to Dihydrocoumarins. *Org. Lett.* **2014,** *16*, 5827–5829.
241. Li, H.-P.; Ai, H.-J.; Qi, X.; Peng, J.-B.; Wu, X.-F. Palladium-Catalyzed Carbonylative Synthesis of Benzofuran-2(3H)-ones from 2-Hydroxybenzyl Alcohols Using Formic Acid as the CO Source. *Org. Biomol. Chem.* **2017,** *15*, 1343–1345.
242. Qi, X.; Li, H.-P.; Wu, X.-F. A Convenient Palladium-Catalyzed Carbonylative Synthesis of Benzofuran-2(3*H*)-ones with Formic Acid as the CO Source. *Chem. Asian J.* **2016,** *11*, 2453–2457.
243. Gabriele, B.; Mancuso, R.; Salerno, G.; Plastina, P. A Novel Palladium-Catalyzed Dicarbonylation Process Leading to Coumarins. *J. Org. Chem.* **2008,** *73*, 756–759.
244. Araniti, F.; Mancuso, R.; Lupini, A.; Giofrè, S. V.; Sunseri, F.; Gabriele, B.; Abenavoli, M. R. Phytotoxic Potential and Biological Activity of Three Synthetic Coumarin Derivatives as New Natural-Like Herbicides. *Molecules* **2015,** *20*, 17883–17902.
245. Veltri, L.; Paladino, V.; Plastina, P.; Gabriele, B. A Palladium Iodide-Catalyzed Cyclocarbonylation Approach to Thiadiazafluorenones. *J. Org. Chem.* **2016,** *81*, 6106–6111.
246. Chiusoli, G. P.; Costa, M.; Reverberi, S. Synthesis of Pyrrole-3,4-Diacetic Acid and Its Derivatives. *Synthesis* **1989,** 262–265.
247. Fazio, A.; Gabriele, B.; Salerno, G.; Destri, S. Synthesis of 3,4-bis[(Methoxycarbonyl)methyl]thiophene and bis-, ter- and Pentathiophenes with Alternating 3,4-bis[(Methoxycarbonyl)methyl]-Substituted Rings. *Tetrahedron* **1999,** *55*, 485–502.
248. Gabriele, B.; Salerno, G.; De Pascali, F.; Tomasi Scianò, G.; Costa, M.; Chiusoli, G. P. Novel Synthesis of Furan-2-Acetic Esters by Palladium-Catalysed Oxidative Cyclization-Alkoxycarbonylation of (*Z*)-2-en-4-yn-1-ols. *Tetrahedron Lett.* **1997,** *38*, 6877–6880.

249. Bacchi, A.; Chiusoli, G. P.; Costa, M.; Gabriele, B.; Righi, C.; Salerno, G. Palladium-Catalysed Sequential Carboxylation–Alkoxycarbonylation of Acetylenic Amines. *Chem. Commun.* **1997,** 1209–1210.
250. Bacchi, A.; Chiusoli, G. P.; Costa, M.; Sani, C.; Gabriele, B.; Salerno, G. Nitrogen Heterocycles by Palladium-Catalysed Oxidative Cyclization-Alkoxycarbonylation of Acetylenic Ureas. *J. Organomet. Chem.* **1998,** *562*, 35–43.
251. Chiusoli, G. P.; Costa, M.; Gabriele, B.; Salerno, G. Sequential Reaction of Carbon Dioxide and Carbon Monoxide with Acetylenic Amines in the Presence of a Palladium Catalyst. *J. Mol. Catal. A: Chem* **1999,** *143*, 297–310.
252. Gabriele, B.; Salerno, G.; De Pascali, F.; Costa, M.; Chiusoli, G. P. An Efficient and General Synthesis of Furan-2-Acetic Esters by Palladium-Catalyzed Oxidative Carbonylation of (*Z*)-2-en-4-yn-1-ols. *J. Org. Chem.* **1999,** *64*, 7693–7699.
253. Gabriele, B.; Salerno, G.; De Pascali, F.; Costa, M.; Chiusoli, G. P. Palladium-Catalyzed Synthesis of 2*E*-[(Methoxycarbonyl)methylene]tetrahydrofurans: Oxidative Cyclization–Methoxycarbonylation of 4-yn-1-ols Versus Cycloisomerization–Hydromethoxylation. *J. Organomet. Chem.* **2000,** *593*, 409–415.
254. Gabriele, B.; Salerno, G.; Fazio, A.; Campana, F. B. Unprecedented Carbon Dioxide Effect on a Pd-Catalysed Oxidative Carbonylation Reaction: A New Synthesis of Pyrrole-2-Acetic Esters. *Chem. Commun.* **2002,** 1408–1409.
255. Bacchi, A.; Costa, M.; Gabriele, B.; Pelizzi, G.; Salerno, G. Efficient and General Synthesis of 5-(Alkoxycarbonyl)methylene-3-Oxazolines by Palladium-Catalyzed Oxidative Carbonylation of Prop-2-ynylamides. *J. Org. Chem.* **2002,** *67*, 4450–4457.
256. Bacchi, A.; Costa, M.; Della Ca', N.; Fabbricatore, M.; Fazio, A.; Gabriele, B.; Nasi, C.; Salerno, G. Synthesis of 1-(Alkoxycarbonyl)methylene-1,3-Dihydroisobenzofurans and 4-(Alkoxycarbonyl)benzo[*c*]pyrans by Palladium-Catalysed Oxidative Carbonylation of 2-Alkynylbenzyl Alcohols, 2-Alkynylbenzaldehydes and 2-Alkynylphenyl Ketones. *Eur. J. Org. Chem.* **2004,** 574–585.
257. Costa, M.; Della Ca', N.; Gabriele, B.; Massera, C.; Salerno, G.; Soliani, M. Synthesis of 4*H*-3,1-Benzoxazines, Quinazolin-2-Ones, and Quinoline-4-Ones by Palladium-Catalyzed Oxidative Carbonylation of 2-Ethynylaniline Derivatives. *J. Org. Chem.* **2004,** *69*, 2469–2477.
258. Bacchi, A.; Costa, M.; Della Ca', N.; Gabriele, B.; Salerno, G.; Cassoni, S. Heterocyclic Derivative Syntheses by Palladium-Catalyzed Oxidative Cyclization–Alkoxycarbonylation of Substituted γ-Oxoalkynes. *J. Org. Chem.* **2005,** *70*, 4971–4979.
259. Gabriele, B.; Mancuso, R.; Salerno, G.; Lupinacci, E.; Ruffolo, G.; Costa, M. Versatile Synthesis of Quinoline-3-Carboxylic Esters and Indol-2-Acetic Esters by Palladium-Catalyzed Carbonylation of 1-(2-Aminoaryl)-2-yn-1-ols. *J. Org. Chem.* **2008,** *73*, 4971–4977.
260. Della Ca', N.; Campanini, F.; Gabriele, B.; Salerno, G.; Massera, C.; Costa, M. Cascade Reactions: Catalytic Synthesis of Functionalized 1,3-Dihydroisobenzofuran and Tetrahydrofuran Derivatives by Sequential Nucleophilic Ring Opening–Heterocyclization–Oxidative Carbonylation of Alkynyloxiranes. *Adv. Synth. Catal.* **2009,** *351*, 2423–2432.
261. Gabriele, B.; Veltri, L.; Mancuso, R.; Plastina, P.; Salerno, G.; Costa, M. Palladium-Catalyzed Oxidative Heterocyclodehydration-Alkoxycarbonylation of 3-Yne-1,2-Diols: A Novel and Expedient Approach to Furan-3-Carboxylic Esters. *Tetrahedron Lett.* **2010,** *51*, 1663–1665.

262. Gabriele, B.; Veltri, L.; Salerno, G.; Mancuso, R.; Costa, M. Multicomponent Cascade Reactions: A Novel and Expedient Approach to Functionalized Indoles by an Unprecedented Nucleophilic Addition-Heterocyclization-Oxidative Alkoxycarbonylation Sequence. *Adv. Synth. Catal.* **2010,** *352*, 3355–3363.
263. Gabriele, B.; Veltri, L.; Maltese, V.; Spina, R.; Mancuso, R.; Salerno, G. Versatile Synthesis of Isoquinolines and Isochromenes by Pd-Catalyzed Oxidative Carbonylation of (2-Alkynyl)benzylideneamine Derivatives. *Eur. J. Org. Chem.* **2011,** 5626–5635.
264. See, for example: Semmelhack, M. F.; Zask, A. Synthesis of Racemic Frenolicin via Organochromium and Organopalladium Intermediates. *J. Am. Chem. Soc.* **1983,** *105*, 2034–2043.
265. See, for example: Semmelhack, M. F.; Bodurow, C. Intramolecular Alkoxypalladation/Carbonylation of Alkenes. *J. Am. Chem. Soc.* **1984,** *106*, 1496–1498.
266. See, for example: Semmelhack, M. F.; Zhang, N. Stereoselective Formation of Tetrahydrofuran Rings Via Intramolecular Alkoxycarbonylation of Hydroxyalkenes. *J. Org. Chem.* **1989,** *54*, 4483–4485.
267. See, for example: McCormick, M.; Monahan, R., III; Soria, J.; Goldsmith, D.; Liotta, D. Effects of Substitution on Intramolecular Alkoxypalladation Carbonylation Reactions. *J. Org. Chem.* **1989,** *54*, 4485–4487.
268. See, for example: Lathbury, D.; Vernon, P.; Gallagher, T. Palladium(II)-Mediated Routes to Functionalised Heterocycles. *Tetrahedron Lett.* **1986,** *27*, 6009–6012.
269. See, for example: Yadav, J. S.; Rao, E. S.; Rao, C. S.; Choudary, B. M. Synthesis of Spiroacetals by Palladium-Catalyzed Oxidative Carbonylation of Hydroxyenones. *J. Mol. Catal.* **1989,** *49*, L61–L63.
270. See, for example: Tamaru, Y.; Hojo, M.; Higashimura, H.; Yoshida, Z. Urea as the Most Reactive and Versatile Nitrogen Nucleophile for the Palladium(2 +)-Catalyzed Cyclization of Unsaturated Amines. *J. Am. Chem. Soc.* **1988,** *110*, 3994–4002.
271. See, for example: Lütjens, H.; Scammels, P. J. Synthesis of 2-Substituted 3-Acylbenzo[*b*]furans Via the Palladium Catalysed Carbonylative Cyclisation of *ortho*-Hydroxytolans. *Synlett* **1999,** 1079–1081.
272. See, for example: Kato, K.; Teraguchi, R.; Satoki, M.; Uchida, A.; Mochida, T.; Peganova, T. A.; Vologdin, V.; Akita, H. *meso*-Phbox-Pd(II) Catalyzed Tandem Carbonylative Cyclization of 1-Ethynyl-1-Propargyl Acetate. *Chem. Commun.* **2008,** 3687–3689.
273. See, for example: White, J. D.; Kranemann, L.; Kuntiyong, P. Intramolecular Palladium (II)-Mediated Alkoxy Carbonylation as a Route to Functionalized Tetrahydropyrans. Synthesis of the C9 – C32 Segment of Phorboxazole A. *Org. Lett.* **2001,** *3*, 4003–4006.
274. See, for example: Kato, K.; Teraguchi, R.; Yamamura, S.; Mochida, T.; Akita, H.; Peganova, T. A.; Vologdin, N. V.; Gusev, O. V. Ligand-Controlled Intramolecular Carbonylative Cyclization of 1,1-Diethynyl Acetates: New Entry to the Functionalized 4-Cyclopentene-1,3-Diones. *Synlett* **2007,** 638–642.
275. See, for example: Marshall, J. A.; Yanik, M. M. Stereoselective Synthesis of Substituted Ketopyranose Subunits of Polyketide Natural Products by Intramolecular Alkoxycarbonylation of δ-Alkynyl Alcohols. *Tetrahedron Lett.* **2000,** *41*, 4717–4721.
276. See, for example: Kato, K.; Nouchi, H.; Ishikura, K.; Takaishi, S.; Motodate, S.; Tanaka, H.; Okudaira, K.; Mochida, T.; Nishigaki, R.; Shigenobu, K.; Akita, H. A Facile Access to Spiro Furanone Skeleton Based on Pd(II)-Mediated Cyclization–Carbonylation of Propargylic Esters. *Tetrahedron* **2006,** *62*, 2545–2554.

277. See, for example: Kato, K.; Matsuba, C.; Kusakabe, T.; Takayama, H.; Yamamura, S.; Mochida, T.; Akita, H.; Peganova, T. A.; Vologdin, N. V.; Gusev, O. V. 2,2′-Isopropylidenebis[(4*S*,5*R*)-4,5-di(2-naphthyl)-2-oxazoline] Ligand for Asymmetric Cyclization–Carbonylation of Meso-2-Alkyl-2-Propargylcyclohexane-1,3-Diols. *Tetrahedron* **2006,** *62*, 9988–9999.

278. See, for example: Kusakabe, T.; Kato, K.; Takaishi, S.; Yamamura, S.; Mochida, T.; Akita, H.; Peganova, T. A.; Vologdin, N. V.; Gusev, O. V. Asymmetric Cyclization–Carbonylation of 2-Alkyl-2-Propargylcyclohexane-1,3-Diones: Facile Access to Optically Active Hydrindanes. *Tetrahedron* **2008,** *64*, 319–327.

279. See, for example: Kato, K.; Nishimura, A.; Yamamoto, Y.; Akita, H. Improved Method for the Synthesis of (*E*)-Cyclic-β-Alkoxyacrylates Under Mild Conditions. *Tetrahedron Lett.* **2001,** *42*, 4203–4205.

280. See, for example: Kato, K.; Tanaka, M.; Yamamoto, Y.; Akita, H. Asymmetric Cyclization–Carbonylation of Cyclic-2-Methyl-2-Propargyl-1,3-Diols. *Tetrahedron Lett.* **2002,** *43*, 1511–1513.

281. See, for example: Kato, K.; Yamamoto, Y.; Akita, H. Palladium(II)-Mediated Cyclization–Carbonylation of 4-yn-1-Ones: Facile Access to 2-Cyclopentenone Carboxylates. *Tetrahedron Lett.* **2002,** *43*, 4915–4917.

282. See, for example: Kato, K.; Yamamoto, Y.; Akita, H. Unusual Formation of Cyclic-Orthoesters by Pd(II)-Mediated Cyclization–Carbonylation of Propargylic Acetates. *Tetrahedron Lett.* **2002,** *43*, 6587–6590.

283. See, for example: Kato, K.; Tanaka, M.; Yamamura, S.; Yamamoto, Y.; Akita, H. Asymmetric Cyclization–Carbonylation of 2-Propargyl-1,3-Dione. *Tetrahedron Lett.* **2003,** *44*, 3089–3092.

284. See, for example: Kato, K.; Mochida, T.; Takayama, H.; Kimura, M.; Moriyama, H.; Takeshita, A.; Kanno, Y.; Inouye, Y.; Akita, H. Palladium(II) Catalyzed Carbonylative Dimerization of Allenyl Ketones: Efficient Synthesis of Difuranylketones. *Tetrahedron Lett.* **2009,** *50*, 4744–4746.

285. See, for example: Tamaru, Y.; Hojo, M.; Yoshida, Z. Palladium(2 +)-Catalyzed Intramolecular Aminocarbonylation of 3-Hydroxy-4-Pentenylamines and 4-Hydroxy-5-Hexenylamines. *J. Org. Chem.* **1988,** *53*, 5731–5741.

286. See, for example: Li, Z.; Gao, Y.; Tang, Y.; Dai, M.; Wang, G.; Wang, Z.; Yang, Z. Total Synthesis of Crisamicin A. *Org. Lett.* **2008,** *10*, 3017–3020.

287. See, for example: Nesbitt, C. L.; McErlean, C. S. P. Total Synthesis of C_{19} Lipid Diols Containing a 2,5-Disubstituted-3-Oxygenated Tetrahydrofuran. *Org. Biomol. Chem.* **2011,** *9*, 2198–2208.

288. See, for example: Paddon-Jones, G. C.; McErlean, C. S. P.; Hayes, P.; Moore, C. J.; Konig, W. A.; Kitching, W. Synthesis and Stereochemistry of Some Bicyclic γ-Lactones from Parasitic Wasps (Hymenoptera: Braconidae). Utility of Hydrolytic Kinetic Resolution of Epoxides and Palladium(II)-Catalyzed Hydroxycyclization – Carbonylation – Lactonization of Ene-Diols. *J. Org. Chem.* **2001,** *66*, 7487–7495.

289. See, for example: Babjak, M.; Kapitán, P.; Grazca, T. Synthesis of (+)-Goniothalesdiol and (+)-7-Epi-Goniothalesdiol. *Tetrahedron* **2005,** *61*, 2471–2479.

290. Gabriele, B.; Salerno, G.; Plastina, P. A New Synthesis of 2-[(Dialkylcarbamoyl) Methylene]Tetrahydrofuran Derivatives by Pd-Catalyzed Oxidative Aminocarbonylation of 4-yn-1-ols. *Lett. Org. Chem.* **2004,** *1*, 134–136.

291. Gabriele, B.; Plastina, P.; Salerno, P.; Mancuso, R. An Easy and Convenient Synthesis of 2-Furan-2-Ylacetamides by Sequential Palladium-Catalyzed Oxidative Aminocarbonylation of (Z)-2-en-4-yn-1-ols/Conjugate Addition/Aromatization. *Synthesis* **2006,** 4247–4251.
292. Plastina, P.; Gabriele, B.; Salerno, G. Palladium-Catalyzed Oxidative Aminocarbonylation of Alkynols. *Synthesis* **2007,** 3083–3087.
293. Gabriele, B.; Salerno, G.; Veltri, L.; Mancuso, R.; Li, Z.; Crispini, A.; Bellusci, A. A New Synthesis of 2,3-Dihydrobenzo[1,4]Dioxine and 3,4-Dihydro-2*H*-Benzo[1,4]Oxazine Derivatives by Tandem Palladium-Catalyzed Oxidative Aminocarbonylation – Cyclization of 2-Prop-2-Ynyloxyphenols and 2-Prop-2-Ynyloxyanilines. *J. Org. Chem.* **2006,** *71*, 7895–7898.
294. Li, S.; Li, F.; Gong, J.; Yang, Z. Palladium-Catalyzed Carbonylative Cyclization of Aryl Alkenes/Alkenols: A New Reaction Mode for the Synthesis of Electron-Rich Chromanes. *Org. Lett.* **2015,** *17*, 1240–1243.
295. Gabriele, B.; Veltri, L.; Mancuso, R.; Salerno, G.; Costa, M. A General Synthesis of Indole-3-Carboxylic Esters by Palladium-Catalyzed Direct Oxidative Carbonylation of 2-Alkynylaniline Derivatives. *Eur. J. Org. Chem.* **2012,** 2549–2559.
296. Gabriele, B.; Mancuso, R.; Maltese, V.; Veltri, L.; Salerno, G. Synthesis of Furan-3-Carboxylic and 4-Methylene-4,5-Dihydrofuran-3-Carboxylic Esters by Direct Palladium Iodide Catalyzed Oxidative Carbonylation of 3-Yne-1,2-Diol Derivatives. *J. Org. Chem.* **2012,** *77*, 8657–8668.
297. Gabriele, B.; Veltri, L.; Mancuso, R.; Salerno, G.; Maggi, S.; Aresta, B. M. A Palladium Iodide-Catalyzed Carbonylative Approach to Functionalized Pyrrole Derivatives. *J. Org. Chem.* **2012,** *77*, 4005–4016.
298. Mancuso, R.; Ziccarelli, I.; Armentano, D.; Marino, N.; Giofrè, S. V.; Gabriele, B. Divergent Palladium Iodide Catalyzed Multicomponent Carbonylative Approaches to Functionalized Isoindolinone and Isobenzofuranimine Derivatives. *J. Org. Chem.* **2014,** *79*, 3506–3518.
299. Araniti, F.; Mancuso, R.; Ziccarelli, I.; Sunseri, F.; Abenavoli, M. R.; Gabriele, B. 3-(Methoxycarbonylmethylene)Isobenzofuran-1-Imines as a New Class of Potential Herbicides. *Molecules* **2014,** *19*, 8261–8275.
300. Pancrazzi, F.; Motti, E.; Costa, M.; Mancuso, R.; Gabriele, B.; Della Ca', N. (Z)-4-(Carbomethoxymethylene)-2-(4-Fluorophenyl)-4*H*-Benzo[*d*][1,3]Oxazine. *Molbank* **2017,** *2017*, M927.
301. Matsuda, Y.; Koyama, T.; Kato, M.; Kawaguchi, T.; Saikawa, Y.; Nakata, M. Palladium-Catalyzed 6-Endo-Dig Oxycyclization–Methoxycarbonylation Toward the Total Synthesis of Polymaxenolides. *Tetrahedron* **2015,** *71*, 2134–2148.
302. Kusakabe, T.; Kawai, Y.; Kato, K. Total Synthesis of (+)-Gregatin B and E. *Org. Lett.* **2013,** *15*, 5102–5105.
303. Kuntiyong, P.; Lee, T. H.; Kranemann, C. L.; White, J. D. Total Synthesis of the Marine Toxin Phorboxazole A Using Palladium(II)-Mediated Intramolecular Alkoxycarbonylation for Tetrahydropyran Synthesis. *Org. Biomol. Chem.* **2012,** *10*, 7884–7899.
304. Dhage, Y. D.; Daimon, H.; Peng, C.; Kusakabe, T.; Takahashi, K.; Kanno, Y.; Inouye, Y.; Kato, K. Pd(II)-Catalyzed Ligand Controlled Synthesis of Pyrazole-4-Carboxylates and Benzo[b]Thiophene-3-Carboxylates. *Org. Biomol. Chem.* **2014,** *12*, 8619–8626.
305. Xu, T.; Alper, H. Synthesis of Indolizine Derivatives by Pd-Catalyzed Oxidative Carbonylation. *Org. Lett.* **2015,** *17*, 4526–4529.

306. Yasuhara, S.; Sasa, M.; Kusakabe, T.; Takayama, H.; Kimura, M.; Mochida, T.; Kato, K. Cyclization–Carbonylation–Cyclization Coupling Reactions of Propargyl Acetates and Amides with Palladium(II)–Bisoxazoline Catalysts. *Angew. Chem. Int. Ed.* **2011,** *50*, 3912–3915.
307. Kusakabe, T.; Kawai, Y.; Shen, R.; Mochida, T.; Kato, K. Cyclization–Carbonylation–Cyclization Coupling Reaction of γ-Propynyl-1,3-Diketones with Palladium(II)-Bisoxazoline Catalyst. *Org. Biomol. Chem.* **2012,** *10*, 3192–3194.
308. Jiang, Y.; Kusakabe, T.; Takahashi, K.; Kato, K. A Cyclization–Carbonylation–Cyclization Coupling Reaction of (*ortho*-Alkynyl Phenyl) (Methoxymethyl) Sulfides with the Palladium(II)-Bisoxazoline Catalyst. *Org. Biomol. Chem.* **2014,** *12*, 3380–3385.
309. Shen, R.; Kusakabe, T.; Yatsu, T.; Kanno, Y.; Takahashi, K.; Nemoto, K.; Kato, K. Palladium(II) Catalyzed Cyclization-Carbonylation-Cyclization Coupling Reaction of (*ortho*-Alkynyl Phenyl) (Methoxymethyl) Sulfides Using Molecular Oxygen as the Terminal Oxidant. *Molecules* **2016,** *21*, 1177.
310. Kusakabe, T.; Kawaguchi, K.; Kawamura, M.; Niimura, N.; Shen, R.; Takayama, H.; Kato, K. Cyclization-Carbonylation-Cyclization Coupling Reaction of Propargyl Ureas with Palladium(II)-Bisoxazoline Catalyst. *Molecules* **2012,** *17*, 9220–9230.
311. Kusakabe, T.; Sagae, H.; Kato, K. Cyclization–Carbonylation–Cyclization Coupling Reaction of α,β-Alkynic Hydrazones with Palladium(II)-Bisoxazoline Catalyst. *Org. Biomol. Chem.* **2013,** *11*, 4943–4948.
312. Shen, R.; Kusakabe, T.; Takahashi, K.; Kato, K. Pd(II)-Catalyzed Ligand Controlled Synthesis of Methyl 1-Benzyl-1*H*-Indole-3-Carboxylates and Bis(1-Benzyl-1*H*-Indol-3-yl) Methanones. *Org. Biomol. Chem.* **2014,** *12*, 4602–4609.
313. Kusakabe, T.; Sekiyama, E.; Ishino, Y.; Motodate, S.; Kato, S.; Mochida, T.; Kato, K. Cyclization–Carbonylation–Cyclization Coupling Reactions of *N*-Propargylanilines and *o*-Alkynylphenols with Palladium(II)–Bisoxazoline Catalysts. *Synthesis* **2012,** *44*, 1825–1832.
314. Gabriele, B.; Mancuso, R.; Ziccarelli, I.; Salerno, G. A New Approach to Isoindolinone Derivatives by Sequential Palladium Iodide-Catalyzed Oxidative Aminocarbonylation–Heterocyclization of 2-Ethynylbenzamides. *Tetrahedron Lett.* **2012,** *53*, 6694–6696.
315. Giofrè, S. V.; Cirmi, S.; Mancuso, R.; Nicolò, F.; Lanza, G.; Legnani, L.; Campisi, A.; Chiacchio, M. A.; Navarra, M.; Gabriele, B.; Romeo, R. Synthesis of Spiro[Isoindole-1,5'-Isoxazolidin]-3(2*H*)-Ones as Potential Inhibitors of the MDM2-p53 Interaction. *Beilstein J. Org. Chem.* **2016,** *12*, 2793–2807.
316. Zeng, F.; Alper, H. Palladium-Catalyzed Domino C–S Coupling/Carbonylation Reactions: An Efficient Synthesis of 2-Carbonylbenzo[*b*]thiophene Derivatives. *Org. Lett.* **2011,** *13*, 2868–2871.
317. Wu, X.-F.; Zhang, M.; Jiao, H.; Neumann, H.; Beller, M. Palladium-Catalyzed Synthesis of Furans From Double Carbonylative Coupling of Aryl Halides with Terminal Alkynes. *Asian J. Org. Chem.* **2013,** *2*, 135–139.
318. Shen, C.; Spannenberg, A.; Auer, M.; Wu, X.-F. Utilizing an Encapsulated Solution of Reagents to Achieve the Four-Component Synthesis of (benzo)Thiophene Derivatives. *Adv. Synth. Catal.* **2017,** *359*, 941–946.
319. See, for example: Bontemps, S.; Quesnel, J. S.; Worrall, K.; Arndtsen, B. A. Palladium-Catalyzed Aryl Iodide Carbonylation as a Route to Imidazoline Synthesis: Design of a Five-Component Coupling Reaction. *Angew. Chem. Int. Ed.* **2011,** *50*, 8948–8951.

320. Faulkner, A.; Scott, J. S.; Bower, J. F. An Umpolung Approach to Alkene Carboamination: Palladium Catalyzed 1,2-Amino-Acylation, -Carboxylation, -Arylation, -Vinylation, and -alkynylation. *J. Am. Chem. Soc.* **2015,** *137*, 7224–7230.
321. Li, Z.; Chernyak, D.; Gevorgyan, V. Palladium-Catalyzed Carbonylative Cyclization/Arylation Cascade for 2-Aroylindolizine Synthesis. *Org. Lett.* **2012,** *14*, 6056–6059.
322. Gabriele, B.; Mancuso, R.; Salerno, G.; Veltri, L. Sequential Homobimetallic Catalysis: An Unprecedented Tandem Pd(0)-Catalysed Deprotection – Pd(II)-Catalysed Heterocyclisation Reaction Leading to Benzofurans. *Chem. Commun.* **2005,** 271–273.
323. Gabriele, B.; Mancuso, R.; Salerno, G.; Costa, M. Cascade Reactions: Sequential Homobimetallic Catalysis Leading to Benzofurans and β,γ-Unsaturated Esters. *Adv. Synth. Catal.* **2006,** *348*, 1101–1109.
324. Gabriele, B.; Mancuso, R.; Salerno, G.; Costa, M. Cascade Reactions: A New Synthesis of 2-Benzofuran-2-Ylacetamides by Sequential Pd(0)-Catalyzed Deallylation – Pd(II)-Catalyzed Aminocarbonylative Heterocyclization of 1-(2-Allyloxyaryl)-2-yn-1-ols. *J. Org. Chem.* **2007,** *72*, 9278–9282.
325. Gabriele, B.; Mancuso, R.; Lupinacci, E.; Salerno, G.; Veltri, L. Tandem Catalysis in Ionic Liquids: A Recyclable Catalytic Synthesis of Benzofuran Derivatives. *Tetrahedron* **2010,** *66*, 6156–6161.
326. Mancuso, R.; Gabriele, B. The Sequential Homobimetallic Catalysis Concept Applied to the Synthesis of Benzofuran Derivatives (Highlight). *Chem. Heterocycl. Compds.* **2014,** *50*, 160–170.
327. Giordano, C.; Rovito, D.; Barone, I.; Mancuso, R.; Bonofiglio, D.; Giordano, F.; Catalano, S.; Gabriele, B.; Andò, S. Benzofuran-2-Acetic Ester Derivatives Induce Apoptosis in Breast Cancer Cells by Upregulating p21Cip/WAF1 Gene Expression in p53-Independent Manner. *DNA Repair* **2017,** *51*, 20–30.

Chapter 4

Synthesis of Heterocyclic Compounds Based on Transition-Metal-Catalyzed Carbene Coupling Reactions

Fangdong Hu[1] and Jianbo Wang[2]
[1]Linyi University, Linyi City, China, [2]Peking University, Beijing, China

4.1 INTRODUCTION

Heterocyclic compounds, which exhibit various biological activities in many cases, are widely found in natural products, pharmaceuticals, agrochemicals, and organic materials.[1] During the last several decades, great efforts have been devoted toward their synthesis. Therefore a great deal of synthetic approaches for the construction of heterocycles have been developed and reported.[2] However, a fair amount of these synthetic method involves the multistep or pre-functionalization of the substrates. Hence, the development of synthetic strategies for efficient construction of heterocycles from easily available starting materials is of particular interest.

On the other hand, a novel type of cross-coupling reaction based on metal carbene migratory insertion has been developed in the last decades (Scheme 4.1).[3] Diazo compounds including its precursors, *N*-tosylhydrazones, were recognized as new coupling partners in these reactions. In addition to diazo compounds, other carbene precursors such as conjugated enynones[4] and allenyl ketones could also be used as coupling partners in this type of cross-coupling reaction. In contrast to classic cross-coupling reactions, metal

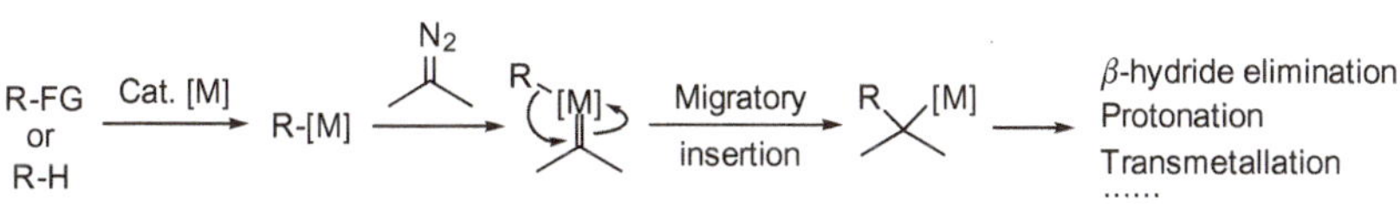

SCHEME 4.1 Cross-coupling reactions involving metal carbene migratory insertion.

Advances in Transition-Metal Mediated Heterocyclic Synthesis.
DOI: https://doi.org/10.1016/B978-0-12-811651-7.00004-2

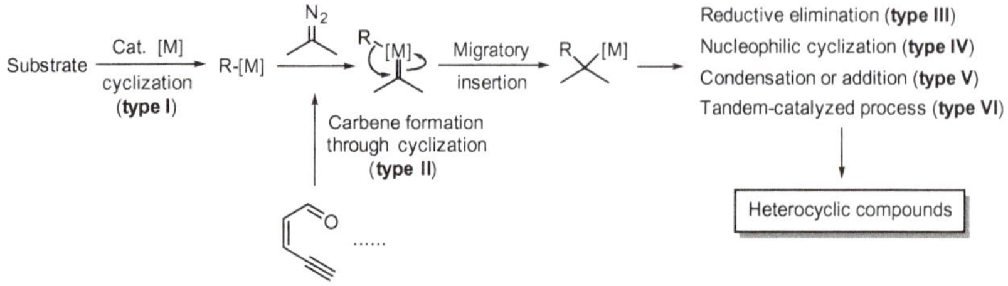

SCHEME 4.2 Synthesis of heterocycles based on metal carbene migratory insertion.

carbene migratory insertion has been proposed as the crucial step in the catalytic cycle. Recent years have witnessed significant progress in this area and the method has evolved into an efficient carbon–carbon bond-forming strategy.

Along with the development of cross-coupling reactions involving metal carbene migratory insertion, such transformations have been introduced for the construction of heterocyclic compounds. In the last five years, a great deal of structurally diverse heterocycles has been constructed based on this strategy. This chapter has summarized the recent publications in this area, and it has been divided into six parts according to the cyclization reaction sequence (Scheme 4.2). In type I cyclization, the heterocyclic fragment was initially formed before the metal carbene formation via transition metal induced cyclization. The furan scaffolds were created in the process of metal carbene formation with conjugated enynones or allenyl ketones as carbene precursors, which was classified as type II cyclization. After metal carbene formation and migratory insertion, a new metal species was generated. Reductive elimination (type III cyclization) and nucleophilic cyclization (type IV cyclization) allowed the formation of corresponding heterocyclic compounds. In addition, the cyclization could also be finished with condensation or addition and this reaction mode was termed as type V cyclization. The last type cyclization involved a tandem-catalyzed process, which could give the heterocycles in an efficient manner.

4.2 CYCLIZATION PRIOR TO METAL CARBENE FORMATION

In the transition-metal-catalyzed cross-coupling reactions involving metal carbene migratory insertion, the metal species were initially generated via several typical processes, such as oxidative addition, transmetallation, C–H bond activation, or C–C bond cleavage. In contrast to these processes, metal-facilitated cyclization of unsaturated substrates is an attractive alternative, which provides a fascinating route to prepare heterocycles. Thus, in this type, the ring systems are firstly created through transition-metal-facilitated cyclization of unsaturated compounds (Scheme 4.3). The cyclized metal species **1** is trapped by diazo compounds to generate the metal carbene

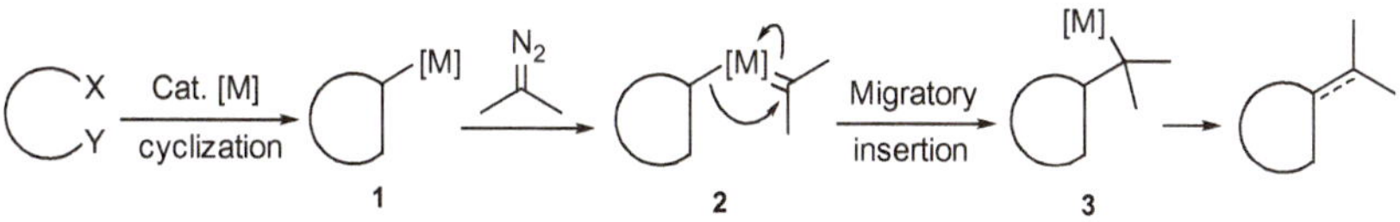

SCHEME 4.3 The general process of type I synthesis of heterocycles.

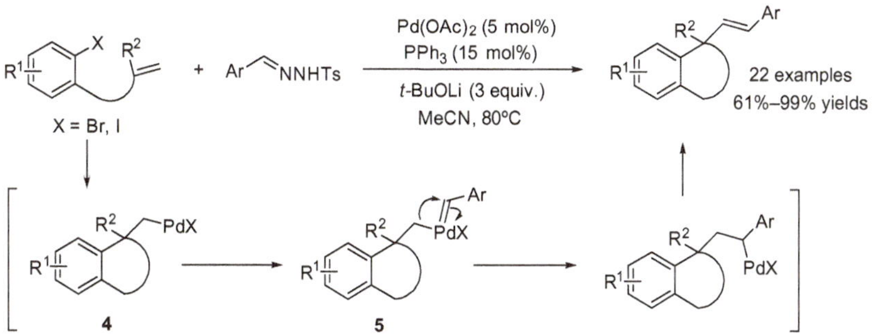

SCHEME 4.4 Pd-catalyzed synthesis of alkene-substituted indolinones.

intermediate **2**. Subsequent migratory insertion leads to the formation of a new metal species **3**, which undergoes β-hydride elimination or protonation to furnish the final heterocycles.

Based on this concept, Gu and coworkers developed a palladium-catalyzed intramolecular Heck-type cascade reaction of organic halides with *N*-tosylhydrazones.[5] The reaction proceeded through alkylpalladium intermediate **4**, which is generated via classical oxidative addition and olefin insertion process. This alkylpalladium species **4** decomposed the in situ generated diazo compounds from *N*-tosylhydrazones giving metal carbene intermediate **5**. Subsequent migratory insertion and β-hydride elimination furnished alkene-substituted indolinone derivatives (Scheme 4.4). Remarkably, spiro compounds could also be accessed by double alkene insertion.

Almost simultaneously, Wang and coworkers demonstrated the synthesis of 3-vinylindoles and 3-vinylbenzofurans through incorporation of Pd-catalyzed cyclization and carbene migratory insertion.[6] The proposed reaction mechanism involves 5-*exo-dig* cyclization, metal carbene formation, migratory insertion, isomerization, and β-hydride elimination sequence (Scheme 4.5). From the synthetic point of view, the protocol provides an efficient method for the construction of 3-vinylindoles and 3-vinylbenzofurans from easily available starting materials. In addition, this transformation shows excellent stereoselectivity.

In addition, Jiang, Wu, and coworkers also reported a mild and efficient methodology for the synthesis of alkene-substituted benzofuran, dihydrobenzofuran, and indoline derivatives using the same strategy (Scheme 4.6).[7] By

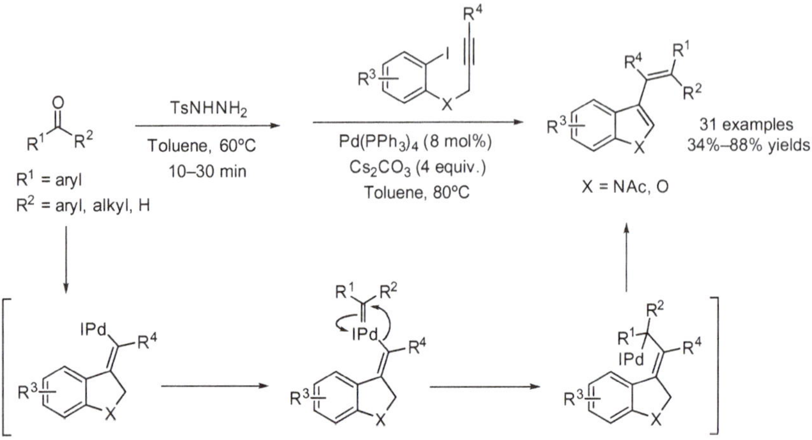

SCHEME 4.5 Pd-catalyzed stereoselective synthesis of 3-vinylindoles and 3-vinylbenzofurans.

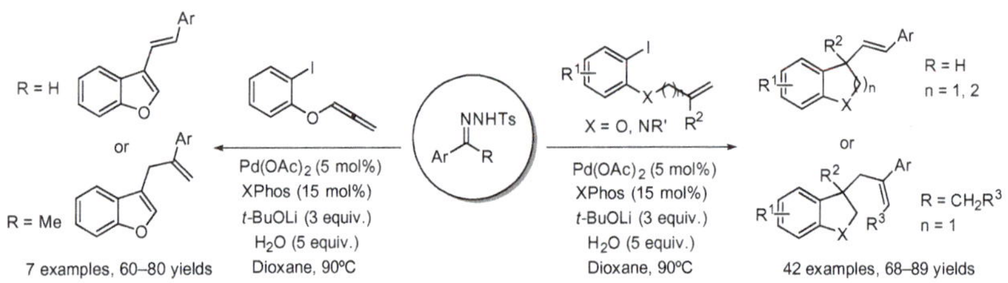

SCHEME 4.6 Synthesis of benzofuran-, dihydrobenzofuran-, and indoline-containing alkenes via Pd-catalyzed Heck-type cascade process.

systemically examining the suitable reaction conditions, the authors achieved this reaction in a high regio- and stereoselectivity manner. Moreover, this reaction exhibits good functional group tolerance and high yields. Compared with Gu's work, this research employed oxygen as a linker and expanded the unsaturated component to allenes in the side chain.

Apart from oxidative addition and unsaturated fragment insertion, the cyclized palladium intermediates could also be generated through intramolecular activation of alkynes. In 2014, Zhu, Luo, and coworkers realized intramolecular aminopalladation of alkynes followed by carbene insertion to access C-3 alkylated indole derivatives.[8] The reaction mechanism was proposed to proceed via palladium-catalyzed intramolecular aminopalladation, Pd-carbene formation, and migratory insertion (Scheme 4.7). After protonation and deprotection during work-up, the final products could be obtained in moderate yields. It should be mentioned that the competitive N–H insertion process was not detected in these reactions. Since indole scaffold demonstrates impressive biological activities and ubiquitous occurrence in

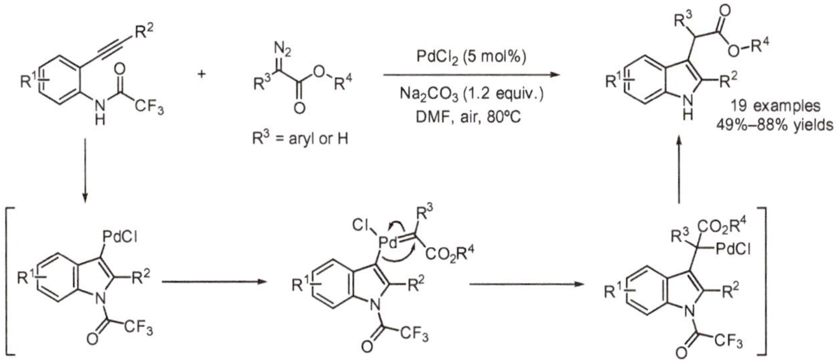

SCHEME 4.7 Palladium-catalyzed intramolecular aminopalladation and carbene insertion for C-3 alkylated indoles.

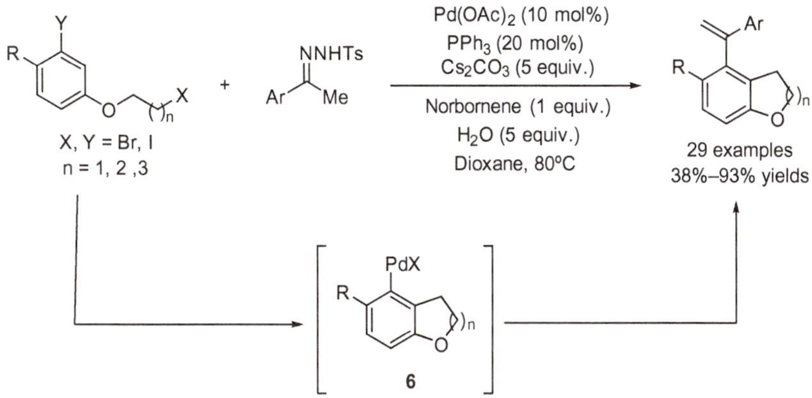

SCHEME 4.8 Palladium-catalyzed synthesis of oxygen or nitrogen containing heterocycles.

natural products, pharmaceuticals, and agrochemicals, this reaction would provide feasible pathway to these privileged structures.

Transition metal carbene migratory insertion could also be merged with Catellani reactions for heterocycle synthesis. In 2014, Liang and coworkers reported palladium-catalyzed synthesis of bicyclic and tricyclic vinylarenes through norbornene-mediated *ortho* C–H alkylation (Scheme 4.8).[9] Mechanistically, the reaction was initiated by oxidative addition and *ortho* C–H activation to form the hetero-fused ring with arylpalladium species **6**. The dediazoniation of the in situ formed diazo compounds by the arylpalladium species **6** led to the formation of the metal carbene intermediate, followed by migratory insertion and β-hydride elimination to produce the final products. Along with numerous organic compounds, such as aryl halides, arylboronic acids, alkenes, or alkynes, used as terminating agents in Catellani reactions, this reaction is featured by selecting *N*-tosylhydrazone as

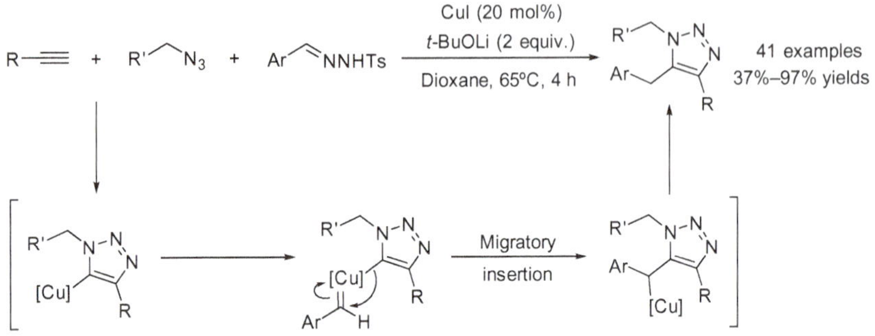

SCHEME 4.9 Copper-catalyzed synthesis of trisubstituted 1,2,3-triazoles.

the reagent to terminate the catalytic cycle through carbene process. Moreover, the oxygen or nitrogen containing heterocyclic skeleton could be rapidly assembled in this transformation.

Apart from the palladium catalyst, Wang and coworkers disclosed a copper(I)-catalyzed protocol for the synthesis of trisubstituted 1,2,3-triazoles from *N*-tosylhydrazones, terminal alkynes, and azides.[10] In this transformation, the click reaction and metal carbene migratory insertion was successfully merged into a single catalytic cycle. A tentative mechanism, involving copper(I) triazolide generation, metal carbene formation, migratory insertion, and protonation sequence, was proposed to rationalize this transformation (Scheme 4.9). From the synthetic point of view, this catalytic routes enabled a convenient method for the synthesis of 1,4,5-trisubstituted 1,2,3-triazoles from simple starting materials. Furthermore, the same group explored similar reactions by selecting trifluoromethyl ketone *N*-tosylhydrazones as carbene precursors. Instead of protonation, this reaction was terminated by β-fluoride elimination. It should be noted that this transformation provided a mild and efficient method for the synthesis of difluoromethylene substituted 1,2,3-triazoles.[11]

Recently, Yao, Lin, and coworkers described the reactions of *o*-ethynylanilines with diazo compounds through Rh(III)-catalyzed cascade annulation and C–H activation.[12] Benzo[*a*]carbazoles were obtained smoothly with high yields and good functional group tolerance. To better understand the reaction mechanism, the authors performed several control and deuterated labeling experiments. They speculated that the reaction was initiated by coordination of alkyne moiety to active Rh(III) species **7**. Subsequent nucleophilic attack of the amine group to activated alkyne generated the benzo heterocyclic Rh(III) species **8**, which then underwent 1,4-rhodium migration to afford aryl-Rh(III) intermediate **9**. Decomposition of the diazo compound furnished a Rh(III) carbene intermediate **10**, followed by migratory insertion and pronation giving intermediate **11**. Final intramolecular condensation constructed another ring system for the formation of benzo[*a*]carbazole scaffold

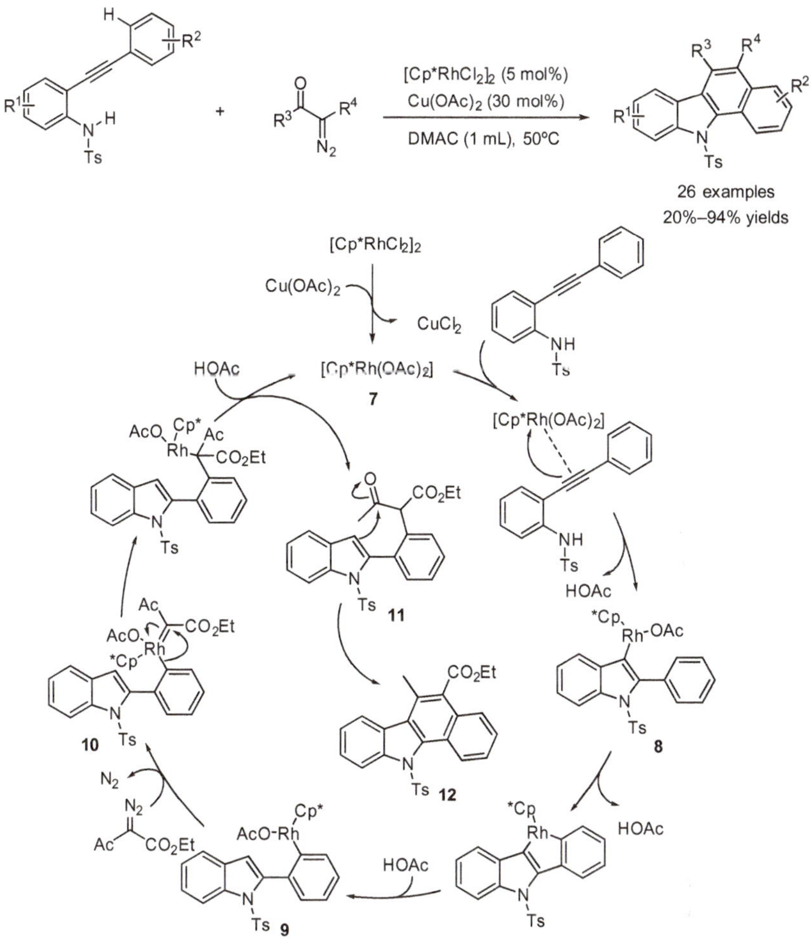

SCHEME 4.10 Rh(III)-catalyzed cascade annulation/C–H activation of *o*-ethynylanilines with diazo compounds

12 (Scheme 4.10). From the step economy point of view, this transformation created one C–N bond and two C–C bond in a one-pot procedure.

4.3 CYCLIZATION THROUGH OTHER CARBENE PRECURSORS

Diazo compounds, either prepared via diazo transfer reactions or generated in situ from *N*-tosylhydrazones, are extensively used as metal carbene precursors. Apart from them, conjugated enynones are attractive alternatives as they are relatively safe in handling and easy in preparation. Alkyne

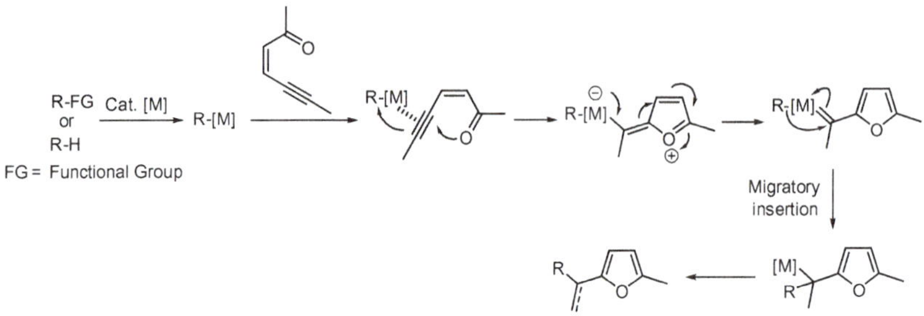

SCHEME 4.11 Synthesis of heterocycles based on carbene migratory insertion using conjugated enynones as the carbene precursors.

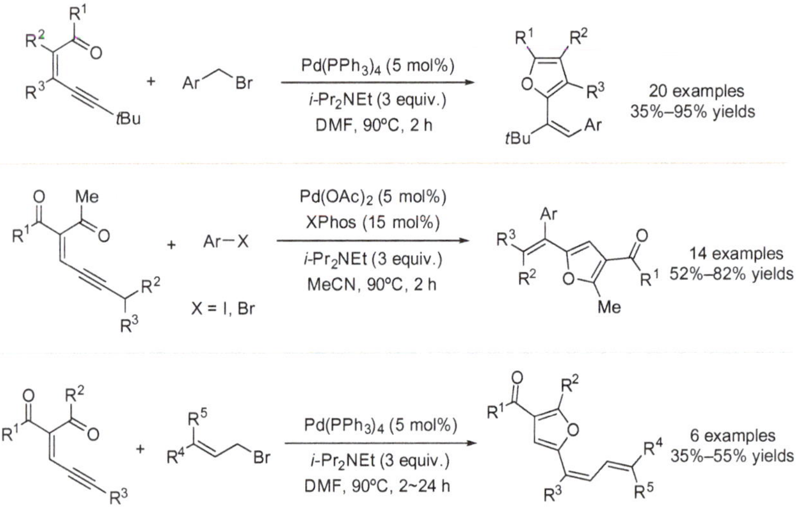

SCHEME 4.12 Palladium-catalyzed cross-coupling reactions of organic halides with conjugated enynones.

activation and cyclization are the accepted procedure to generate furyl-containing carbene species with these compounds (Scheme 4.11). In the past few years, a series of classical reactions including cyclopropanation, insertion, and metathesis have been widely explored.[13] However, the migratory insertion reactions of metal carbene using conjugated enynones as carbene precursors are largely undeveloped.

In 2013, Wang and coworkers conceived a novel palladium-catalyzed carbene migratory insertion reactions, in which conjugated enynones were recognized as carbene precursors (Scheme 4.12).[14] A wide array of 2-alkenyl-substituted furans was produced in moderate to good yields. From the mechanistic point of view, this catalytic reaction was rationalized including oxidative addition, alkyne activation-cyclization, palladium carbene

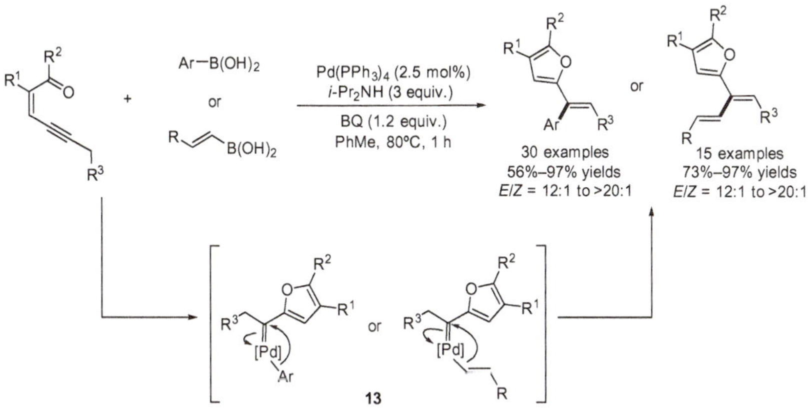

SCHEME 4.13 Palladium-catalyzed cross-coupling of conjugated enynones with organoboronic acids.

migratory insertion, and β-hydride elimination sequence. Density functional theory (DFT) calculations indicated that palladium (2-furyl)carbene was the key intermediate and conjugated enynones cyclization was the rate-limiting step in this transformation. This work expanded carbene migratory insertion reactions from diazo compounds to conjugated enynones, which opened up possibilities to devise new transformations based on carbene migratory insertion using conjugated enynones as carbene precursors.

Under oxidative conditions, conjugated enynones could also be coupled with arylboronic acids. In 2015, Wang and coworkers reported a distinct strategy for the synthesis of 2-alkenylfurans and furan-substituted 1,3-dienes through palladium-catalyzed cross-coupling reaction of conjugated enynones with organoboronic acids.[15] According to the proposed reaction mechanism, the arylpalladium intermediate, which is initially generated through a transmetallation step, activates the alkyne moiety of the enynone to undergo a 5-*exo-dig* cyclization with the carbonyl group as the nucleophile. The furyl Pd-carbene intermediate **13** is thus afforded, followed by migratory insertion and β-hydride elimination to give the final furan derivatives (Scheme 4.13).

Subsequently, the Wang and coworkers explored the reaction between conjugated enynones and arylboronic acids in the presence of Rh(I) catalyst (Scheme 4.14).[16] In contrast to palladium, the rhodium catalyst preferred protonation to terminate the catalytic cycle. As a result, furyl-containing triarylmethanes were synthesized in good to excellent yields with good functional group tolerance. Migratory insertion was also deemed as the key step in proposed mechanism.

Besides, terminal alkynes could also be applied as nucleophiles in transition-metal-catalyzed cross-coupling reactions along with organoboronic acids. The same group developed another oxidative cross-coupling reaction

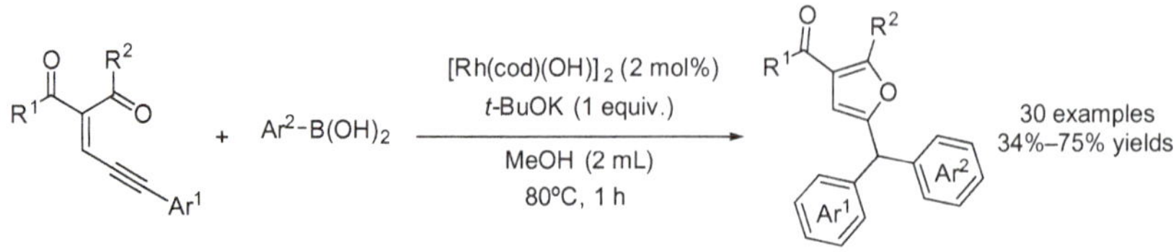

SCHEME 4.14 Rh(I)-catalyzed cross-coupling of conjugated enynones with arylboronic acids.

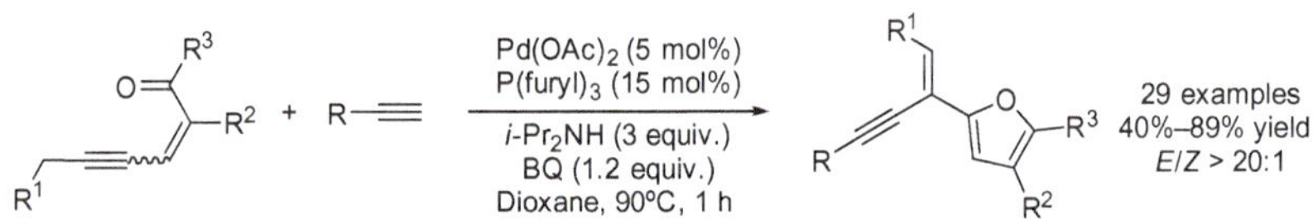

SCHEME 4.15 Pd-catalyzed cross-coupling of conjugated enynones with terminal alkynes.

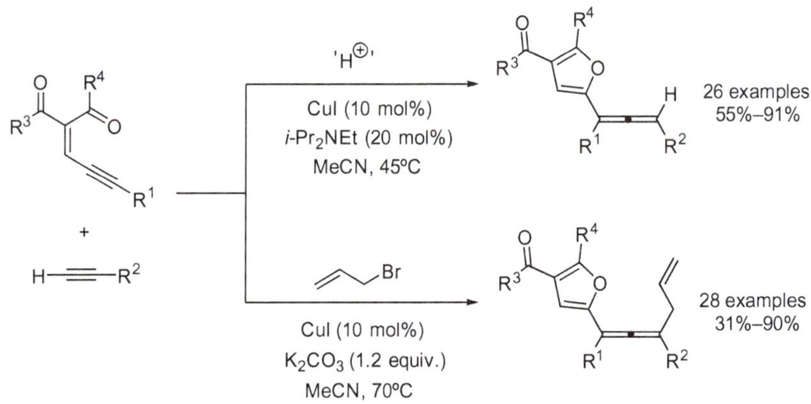

SCHEME 4.16 Copper-catalyzed protocol towards the synthesis of furyl substituted allenes.

of conjugated enynones with terminal alkynes in 2015 (Scheme 4.15).[17] Furan-substituted conjugated enynes were achieved in good yields and with high stereoselectivity. Moreover, a wide range of conjugated enynones and terminal alkynes were compatible in this cross-coupling reaction.

In addition to palladium catalysts that can serve as competent catalysts in conjugated enynone involved cross-coupling reactions, the conjugated enynone cyclization may also be catalyzed by copper complexes. In 2014, Wang and coworkers demonstrated a copper-catalyzed protocol towards the synthesis of furyl substituted allenes (Scheme 4.16).[18,19] Conjugated enynones were employed as carbene precursors in this transformation. The reaction pathway was believed to proceed through copper alkynylation, copper carbene formation, migratory insertion, and protonation. The simple and mild reaction conditions make this protocol more appealing in synthetic

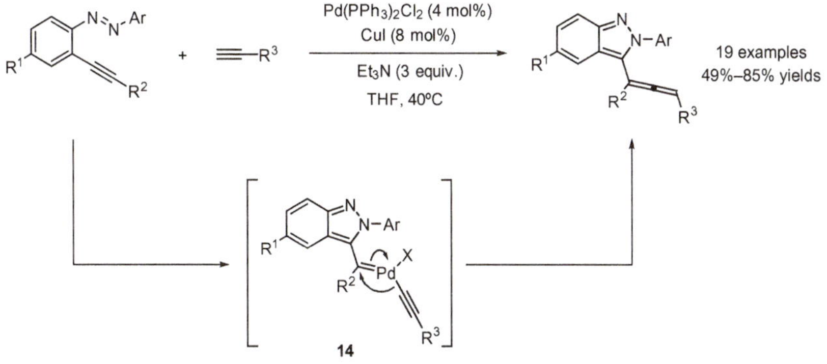

SCHEME 4.17 Cross-coupling of 2-alkynyl azobenzenes and terminal alkynes cooperatively catalyzed by palladium and copper.

community. The authors also examined other electrophiles in this catalytic reaction. As a consequence, they proved that allyl halides could also be incorporated with other coupling partners. Thus diverse furyl substituted allenes were accessed via this tunable approach. Owing to their peculiar structural features of furyl substituted allenes, this novel strategy provide an useful alternative.

Moreover, Yamane and Feng developed a similar reaction using 2-alkynyl azobenzenes as carbene precursor for the synthesis of (3-isoindazolyl)allenes.[20] They found that cooperative $Pd(PPh_3)_2Cl_2$/CuI was the optimized catalytic system, with which Pd catalyst was crucial to prevent the cyclopropene formation. A number of terminal alkynes were effective coupling partners, with a range of electronically different substituents utilized to achieve the corresponding products in good to excellent yield. One of the possible mechanism involves the formation of an alkynyl palladium carbene intermediate **14**, followed by a carbene migratory insertion and protonation to give the desired (3-isoindazolyl)allene derivatives (Scheme 4.17).

Furthermore, conjugated enyones could also be served as carbene precursors in Rh(III)-catalyzed C–H bond activations. Very recently, Chang and coworkers presented Rh(III)-catalyzed [4 + 1] or [4 + 2] cyclization of benzamides **15** with conjugated enyones **16** as coupling partners.[21] The product distribution was largely dependent on the electronic effects of Cp ligand. In that reaction, alkyne insertion gave tricyclic isoquinolinones **17** as products with Cp^*CyRh^{III} catalytic system, while isoindolinone derivatives **18** could be formed with Rh(III)-carbene insertion by employing $[RhCp^ECl_2]_2$ as the rhodium catalyst (Scheme 4.18). It should be mentioned that Rh(III)-carbene migratory insertion was also proposed as crucial step in [4 + 1] annulation. This reaction provided a divergent approach for the construction of isoquinolinones or isoindolinones with high efficiency and good selectivity. DFT

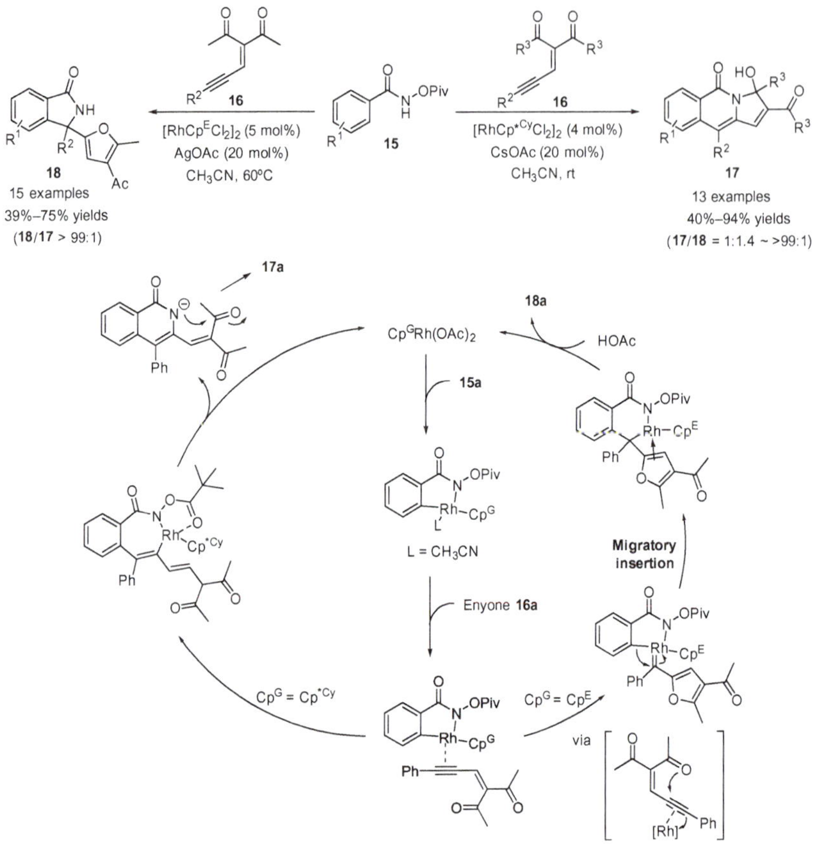

SCHEME 4.18 Rh(III)-catalyzed [4 + 1] and [4 + 2] annulation of benzamides with conjugated enynones.

calculations also indicated that Lewis acidity of the Cp-bound rhodium species affected the reaction pathway.

Otherwise, Wang and coworkers realized another oxidative cross-coupling reaction of allenyl ketones with organoboronic acids through palladium catalysis (Scheme 4.19).[22] Allenyl ketones were served as carbene precursors in this transformation. Oxidation, transmetallation, vinyl carbene formation, migratory insertion, and β-hydride elimination were the essential process in the assumed catalytic cycles. In these processes, carbene migratory insertion was also believed as the key step. This reaction constitutes an efficient methodology for the preparation of highly substituted furan derivatives.

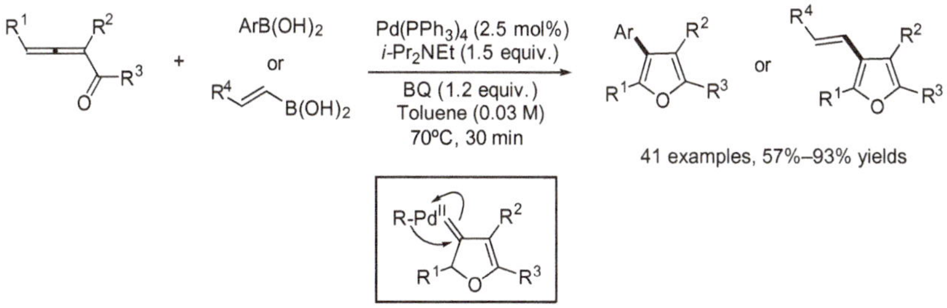

SCHEME 4.19 Oxidative cross-coupling of allenyl ketones and organoboronic acids.

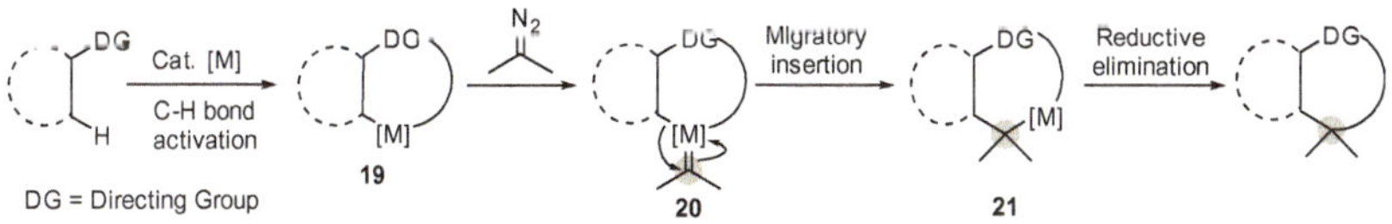

SCHEME 4.20 General process of heterocycle synthesis through reductive elimination.

4.4 CYCLIZATION THROUGH REDUCTIVE ELIMINATION

Transition-metal-catalyzed C–H bond activations have gained tremendous interests and have developed into an attractive strategy for functionalizing unreactive compounds in organic community due to their atom and step economy.[23] Recently, this fascinating strategy has been combined with carbene migratory insertion process.[24] Thus a novel C–H bond activation based on metal carbene migratory insertion has been established and experienced great development for C–C or C-heteroatom bond formation.[3h] In this regard, heterocycles can also be achieved. For general mechanism, the cyclized metal species **19** is generated initially through C–H bond activation, followed by decomposition of diazo substrates to give rise to the formation of metal carbene intermediate **20**. Sequential migratory insertion produces another metal-contained heterocycles **21**. Final reductive elimination provides cyclized heteroatom products (Scheme 4.20).

Based on this concept, Rovis and coworkers achieved the synthesis of isoindolones from *O*-pivaloyl benzhydroxamic acids with donor/acceptor diazo compounds.[25] Both coupling partners including benzhydroxamic acids and diazo compounds were broadly tolerated under optimized reaction conditions. Mechanistically, C–H bond activation, metal carbene formation, migratory insertion, and reductive elimination are hypothesized as indispensable steps in the catalytic cycle (Scheme 4.21). Moreover, the C–H bond activation is found irreversible via deuterium experiment. Electron-deficient diazo compounds are preferred for this transformation.

Moreover, Yu and coworkers also successfully achieved the synthesis of functionalized benzolactams by employing acceptor/acceptor diazo

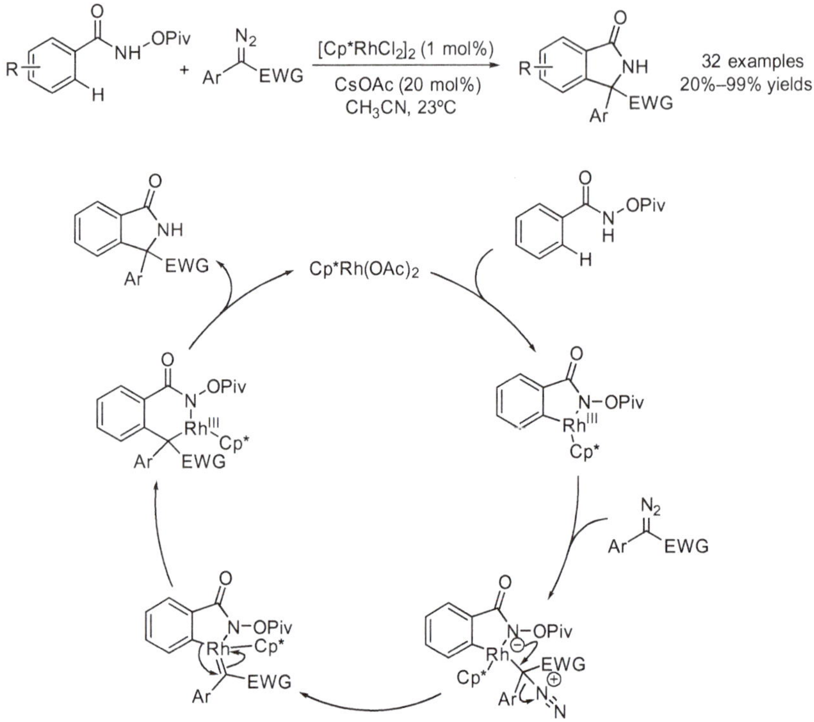

SCHEME 4.21 Isoindolone synthesis through Rh(III)-catalyzed C–H activation.

substrates.[26] In this investigation, a rhodacyclic complex was isolated and its structure was characterized by X-ray crystallography, which would give a direct proof for mechanistic understanding of the reaction.

In 2014, Cramer and coworker devised a novel C_2-symmetric disubstituted Cp ligand with an atropchiral biaryl backbone, which was introduced for symmetric synthesis of isoindolones from aryl hydroxamates and weakly alkyl donor/acceptor diazo derivatives (Scheme 4.22).[27] Remarkably, the relative size of the two diazo substituents has significant effect on the enantioselectivity. This investigation demonstrated new possibilities for asymmetric synthesis of heterocycles based on metal carbene migratory insertion.

In 2013, Cui and coworkers carried out a similar reaction by employing vinyldiazoacetates as carbene precursors.[28] In contrast to the aforementioned work of Rovis group, a suprafacial 1,3-allylic rhodium migration occurred before reductive elimination. Thus a unique azepinone scaffolds were produced (Scheme 4.23). This reaction was characterized by simple starting materials and mild reaction conditions. Furthermore, differentially substituted benzamides and vinyldiazoacetates were smoothly reacted to allow rapid construction of azepinone derivatives. It should be noted that

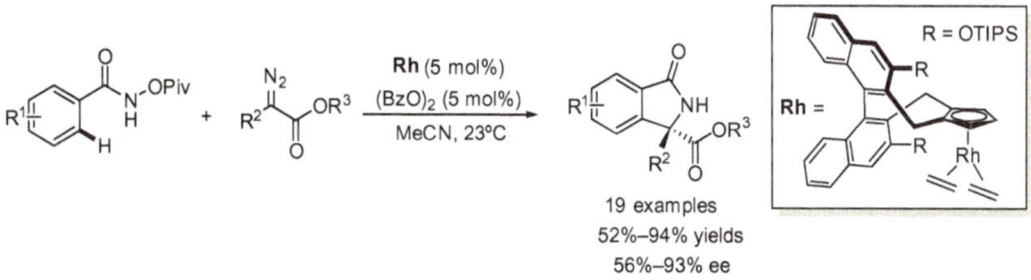

SCHEME 4.22 Asymmetric synthesis of isoindolones by chiral cyclopentadienyl-Rh(III)-catalyzed C–H functionalizations.

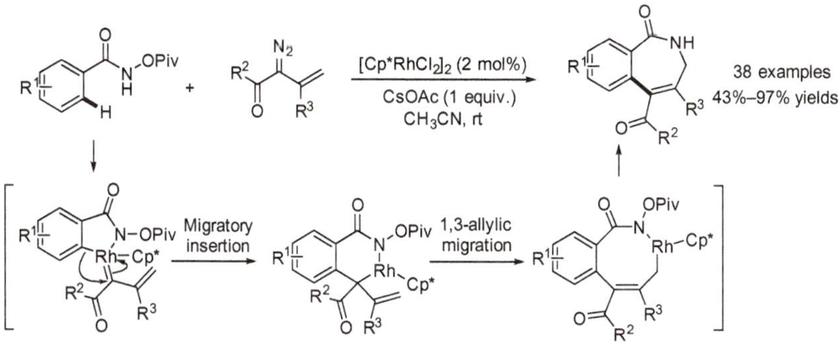

SCHEME 4.23 Rh(III)-catalyzed C–H activation/[4 + 3] cycloaddition for synthesis of azepinone derivatives.

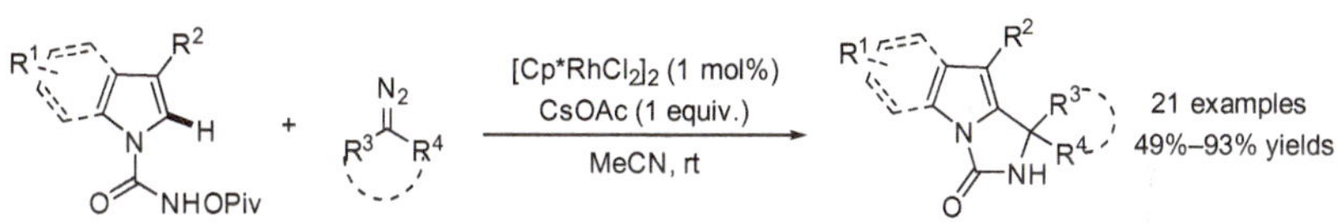

SCHEME 4.24 Synthesis of indole and pyrrole fused heterocycles via Rh(III)-catalyzed C–H activations.

vinyldiazoacetates were not only used as carbene precursors, but also served as three-carbon components.

The same group later on applied the same strategy for the synthesis of indole and pyrrole fused heterocycles (Scheme 4.24).[29] A series of indole and pyrrole derived *N*-carboxamides were compatible in this catalytic route. The migratory insertion and reductive elimination are both vital steps in the final heterocycle formation. In addition to diazo compounds, alkenes and alkynes are also suitable substrates in this transformation, which enables the construction of six-membered fused heterocyclic compounds.

Cui and coworkers subsequently reported a facile synthetic route to isoindolinones.[30] In the precedent literatures, diazo compounds were directly used as the carbene precursors in the preparation of isoindolinones through C–H

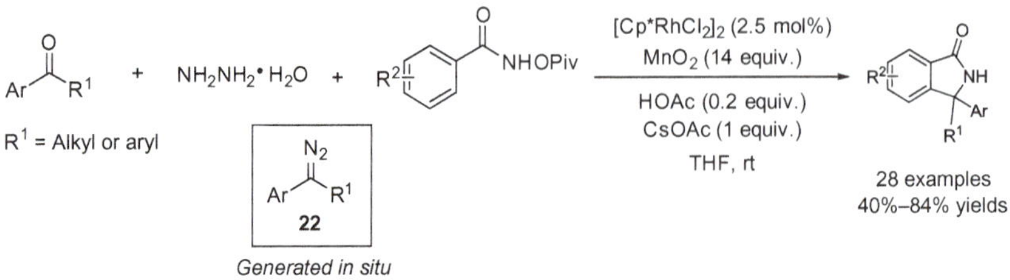

SCHEME 4.25 Rh(III)-catalyzed one-pot reaction of benzamides, ketones, and hydrazines.

activation strategy. While in this transformation, diazo substrates **22** were in situ formed through oxidation of acetophenone hydrazones, which were generated from acetophenones and hydrazine under the reaction conditions (Scheme 4.25). From readily available starting materials, this one-pot three-component reaction constitutes a general and distinct approach toward privileged isoindolinone derivatives. The reaction conditions are rather simple and products are obtained in high efficiency.

More recently, Ma and coworkers presented a tunable strategy for the selective synthesis of α-pyrones and furans via rhodium-catalyzed vinyl C–H bond activation.[31] The hypothesized reaction pathway starts with coordination of the catalyst to the substrate and vinyl C–H activation. The resulted five-membered rhodacycle **23** decomposes diazo compounds to generate rhodium carbene species. Migratory insertion of rhodium carbene give rise to intermediate **24**, which is further converted into rhodacycle **25** via a 1,3-migratory insertion. The catalytic cycle is then completed by protonation, tautomerization, and lactonization to deliver pyrones as the final products. Alternatively, rhodacycle **25** may be transformed into intermediate **26** with the assistance of AgOAc. Reductive elimination leads to fully substituted furans as the corresponding product (Scheme 4.26). This reaction successfully achieved the synthesis of α-pyrones and furans with combination of vinyl C–H activation and metal carbene migratory insertion.

4.5 CYCLIZATION THROUGH NUCLEOPHILIC DISPLACEMENT

Palladium complexes are the privileged catalysts for this type of cyclization for the synthesis of heterocycles. In the general procedure, the metal species **27** is generated through oxidative addition with organic halides. Decomposition of diazo substrates gives metal carbene intermediates **28**. Subsequent migratory insertion affords a new metal species **29**. Instead of pronation, β-hydride elimination, or reductive elimination, nucleophilic cyclization preferentially occurs because a nucleophile is pre-installed in the starting materials. Heterocycles thus produced in this process (Scheme 4.27). In

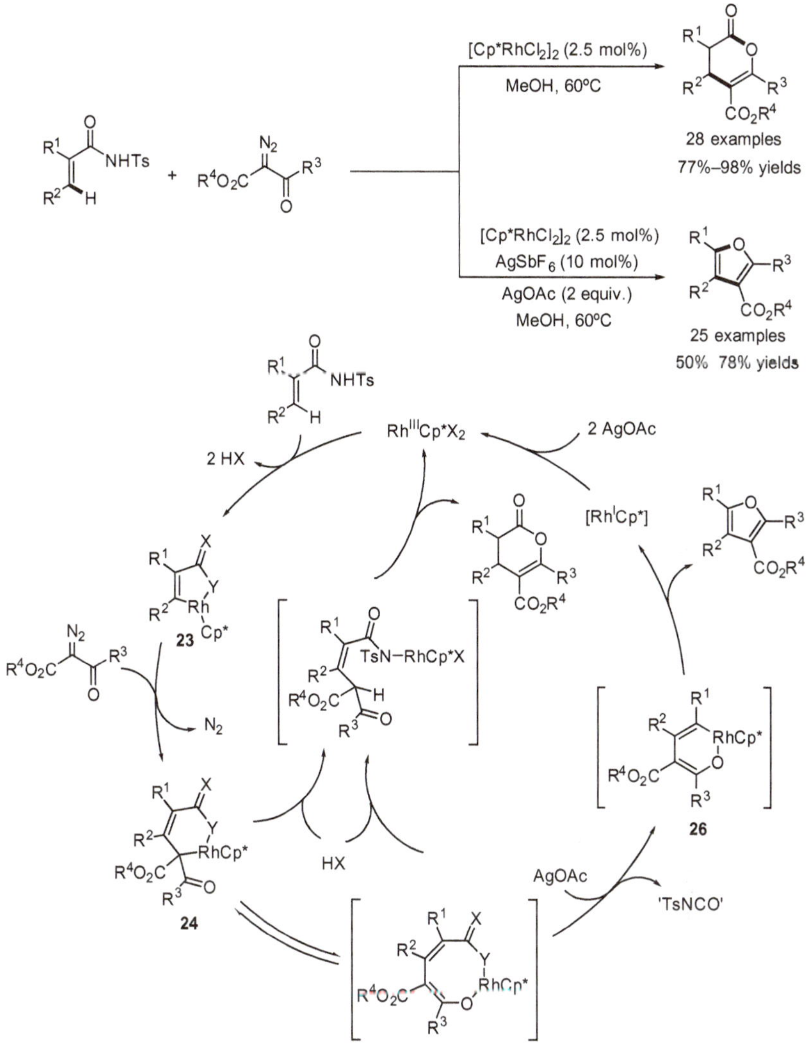

25

SCHEME 4.26 Rhodium-catalyzed selective synthesis of α-pyrones and furans.

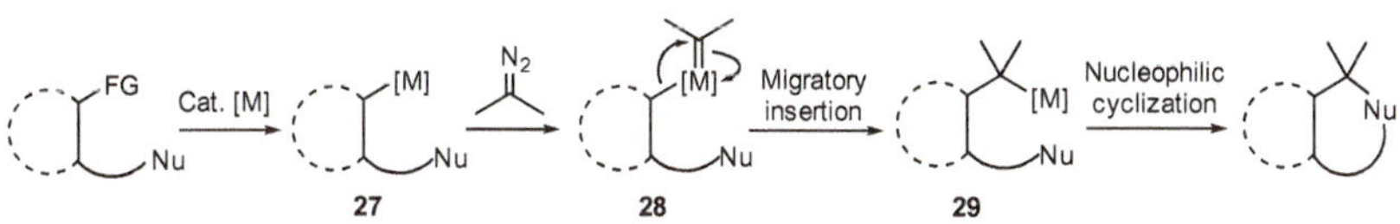

FG = Functional Group
Nu = Nucleophile

SCHEME 4.27 General process of heterocycle synthesis through nucleophilic cyclization.

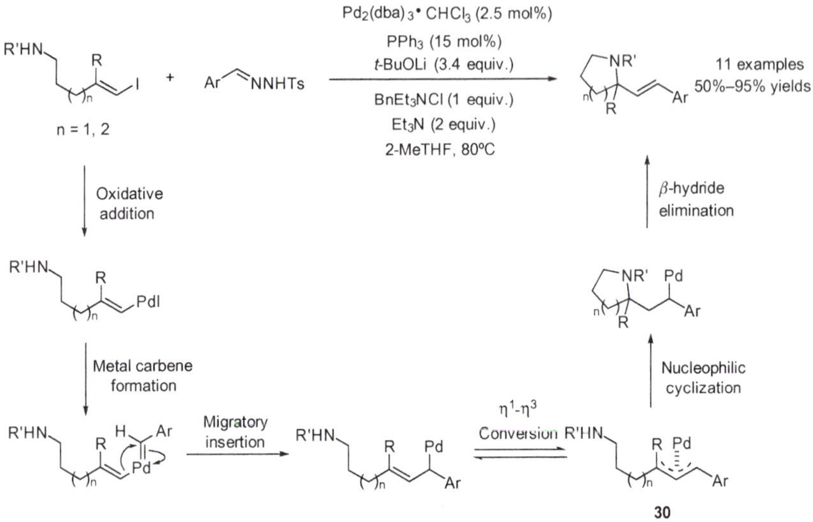

SCHEME 4.28 Pd-catalyzed cross-coupling reaction of vinyl iodides and *N*-tosylhydrazones.

addition, copper could also promote the cyclization by reacting with terminal alkynes.

In 2012, Van Vranken and coworkers demonstrated a palladium-catalyzed carbenylative amination protocol, which enabled the construction of pyrrolidine and piperidine ring systems.[32] Several functional groups on both coupling partners are well tolerated in this reaction. From the mechanistic point of view, the reaction is assumed to proceed with oxidative addition, metal carbene formation, migratory insertion, nucleophilic cyclization, and β-hydride elimination sequence. It should be mentioned that vinyl moiety in the substrates facilitate the generation of η^3-allylpalladium intermediates **30**, which allow for nucleophilic attack of the nitrogen atom (Scheme 4.28). Additionally, alkaloid caulophyllumine B could also be obtained in good yield by this carbenylative amination. Although the yield was not satisfactory, the authors also tried asymmetry carbenylative amination with chiral phosphine ligands.

Liang and coworkers developed a palladium-catalyzed cross-coupling reaction of diazo esters with *N*-substituted-2-iodoanilines for selective synthesis of cyclic α- and γ-amino esters.[33] With regard to the possible reaction mechanism, migratory insertion is the key step for the whole catalytic cycle, while nucleophilic cyclization is also deemed as vital process in the formation of desired heterocycles. It should be aware that cyclic α-amino esters that possessed quaternary carbon center are constructed in the presence of CO with aryldiazoacetates as reactants. On the other hand, arylvinyldiazoacetates gave cyclic α,β-unsaturated γ-amino esters exclusively (Scheme 4.29). Continuously, the same group modified the starting material from

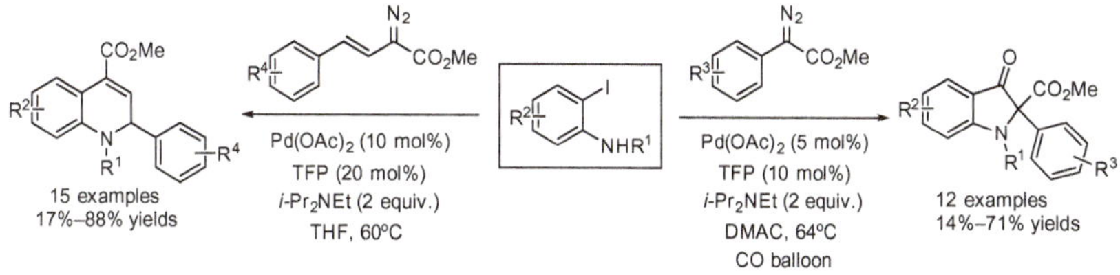

SCHEME 4.29 Palladium-catalyzed cross-coupling reaction of diazo esters with *N*-substituted-2-iodoanilines.

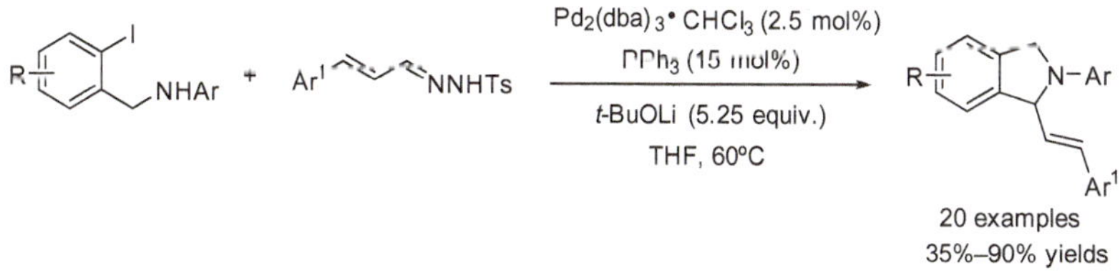

SCHEME 4.30 Palladium-catalyzed cross-coupling reaction of *N*-(2-iodobenzyl) anilines with α,β-unsaturated *N*-tosylhydrazones.

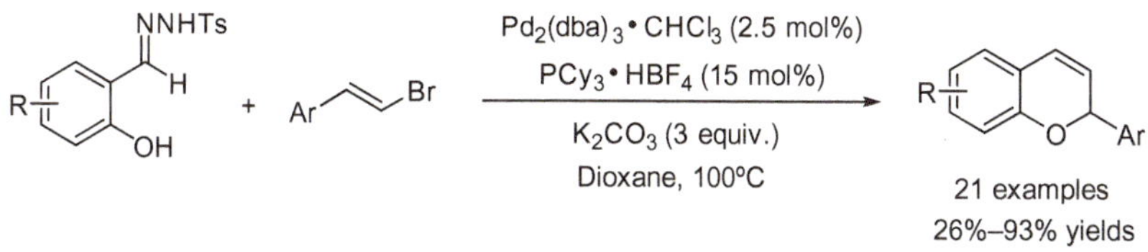

SCHEME 4.31 Palladium-catalyzed cross-coupling reaction of salicyl *N*-tosylhydrazones with β-bromostyrenes.

diazo esters and *N*-substituted-2-iodoanilines to α,β-unsaturated *N*-tosylhydrazones and *N*-(2-iodobenzyl) anilines.[34] As a consequence, they disclosed a new method for synthesis of isoindoline derivatives in the presence of palladium catalyst (Scheme 4.30). This mild protocol exhibited good functional group tolerance, providing an alternative way to access these *N*-containing heterocycles.

In 2014, a new protocol for synthesis of 2*H*-chromene was reported by Wang and coworkers (Scheme 4.31).[35] Salicyl *N*-tosylhydrazones and β-bromostyrenes were selected as the starting materials. In the presence of palladium catalyst, the proposed reaction pathway involves oxidative addition, palladium carbene formation, migratory insertion, nucleophilic cyclization, and β-hydride elimination. Since the substrates are easily prepared, this transformation provides a convenient route toward these *O*-containing heterocycles. Furthermore, Liu and coworkers presented a similar reaction by

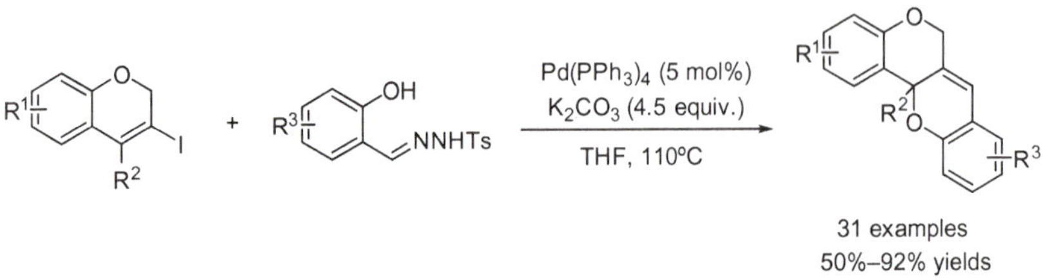

SCHEME 4.32 Palladium-catalyzed cross-coupling reaction of salicyl *N*-tosylhydrazones with cyclic vinyl iodides.

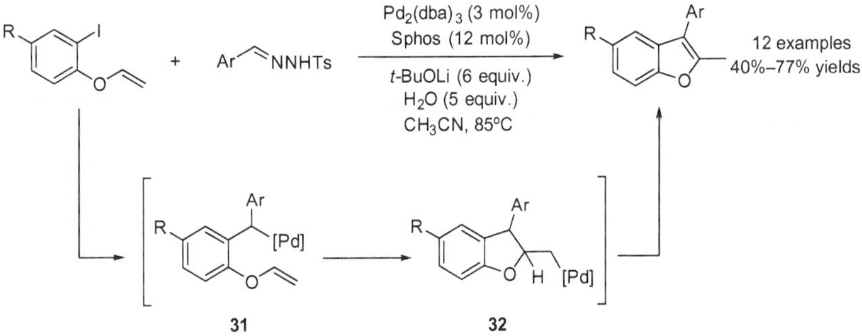

SCHEME 4.33 Palladium-catalyzed synthesis of 2,3-disubstituted benzofurans.

changing the coupling partners from β-bromostyrenes to cyclic vinyl iodides (Scheme 4.32).[36] Instead of formation of 2*H*-chromenes, a series of chromeno[4,3-*b*]chromene derivatives are obtained in moderate to high yields. It is worth mentioning that this reaction tolerates various functional groups, which makes the protocol practically useful.

Alkenes could also be incorporated in this process for the synthesis of heterocycles. In 2015, Valdés and coworkers reported the palladium-catalyzed cross-coupling reaction of iodoenolethers and *N*-tosylhydrazones.[37] This method successfully realized the synthesis of benzofuran derivatives, as well as the construction of two C–C bonds. In this transformation, the benzyl palladium(II) species **31** generated from migratory insertion undergoes intramolecular Heck reaction to give intermediate **32**, which experiences β-hydride elimination and aromatization to afford benzofuran derivatives as the products (Scheme 4.33). Later on, the same group described another Pd-catalyzed cascade reactions between *O*-iodo-*N*-alkenylanilines and *N*-tosylhydrazones (Scheme 4.34).[38] The transformation selectively provided multisubstituted indoles and 1,4-dihydroquinolines in moderate to good yields. It should be mentioned that the substitution on the *N*-alkenyl fragment had significant effect on the outcome. The tosyl substituent featured indole's synthesis through 5-*exo-trig* cyclization, while alkyl substituent preferred

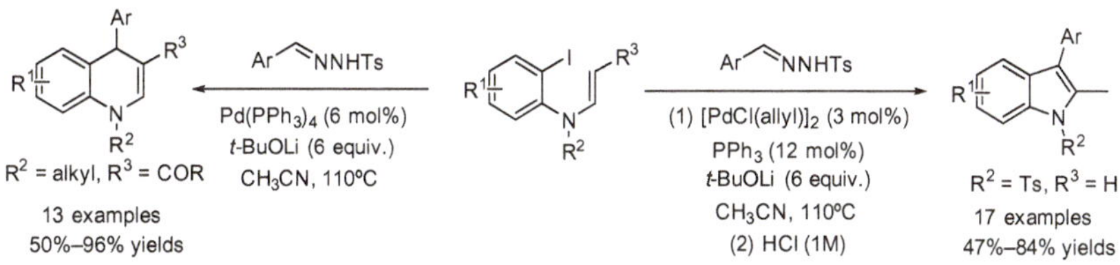

SCHEME 4.34 Pd-catalyzed cascade reactions between *O*-iodo-*N*-alkenylanilines and *N*-tosylhydrazones.

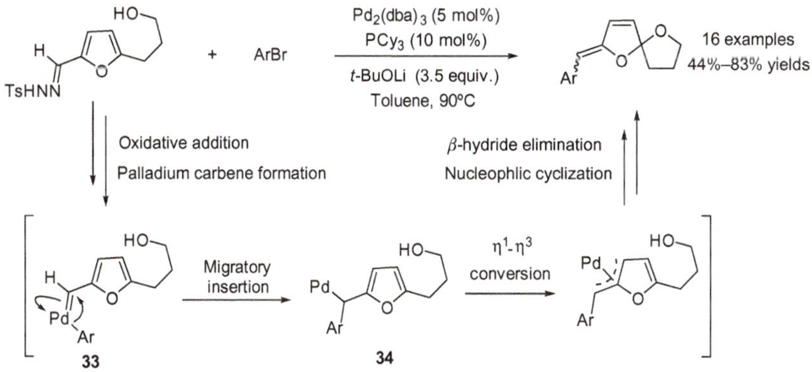

SCHEME 4.35 Pd-catalyzed coupling of furfural tosylhydrazones with aryl halides for spiroacetal enol ethers.

formation of 1,4-dihydroquinolines via 6-*endo-trig* cyclization. Likewise, two C–C bonds were constructed on the hydrazonic carbon atom.

Spiro heterocycles could also be obtained via the same strategy. Yin and colleagues developed a Pd-catalyzed cross-coupling of furfural tosylhydrazones with aryl halides to provide spiroacetal enol ethers via nucleophilic dearomatization.[39] A hydroxypropyl group tethered at the 2-position of the furan ring undergoes intramolecular nucleophilic cyclization as well as dearomatization. In this reaction, 2-furylmethylenepalladium intermediate **34** is generated through migratory insertion from palladium carbene **33**, rather than through oxidative addition from unstable 2-furylmethyl halides (Scheme 4.35).

Subsequently, the same researchers demonstrated palladium-catalyzed transannulation of furfural *N*-tosylhydrazones with 2-iodoanilines or 2-iodothiophenols for the synthesis of poly-substituted indoles or benzothiophenes.[40] According to the proposed reaction mechanism, arylpalladium species **35** is firstly produced through oxidative addition, followed by decomposition of the in situ formed 2-furyl diazoalkane **36** to furnish palladium carbene intermediate **37**. Subsequent migratory insertion affords 2-furylmethylenepalladium iodide **38**, which immediately undergo

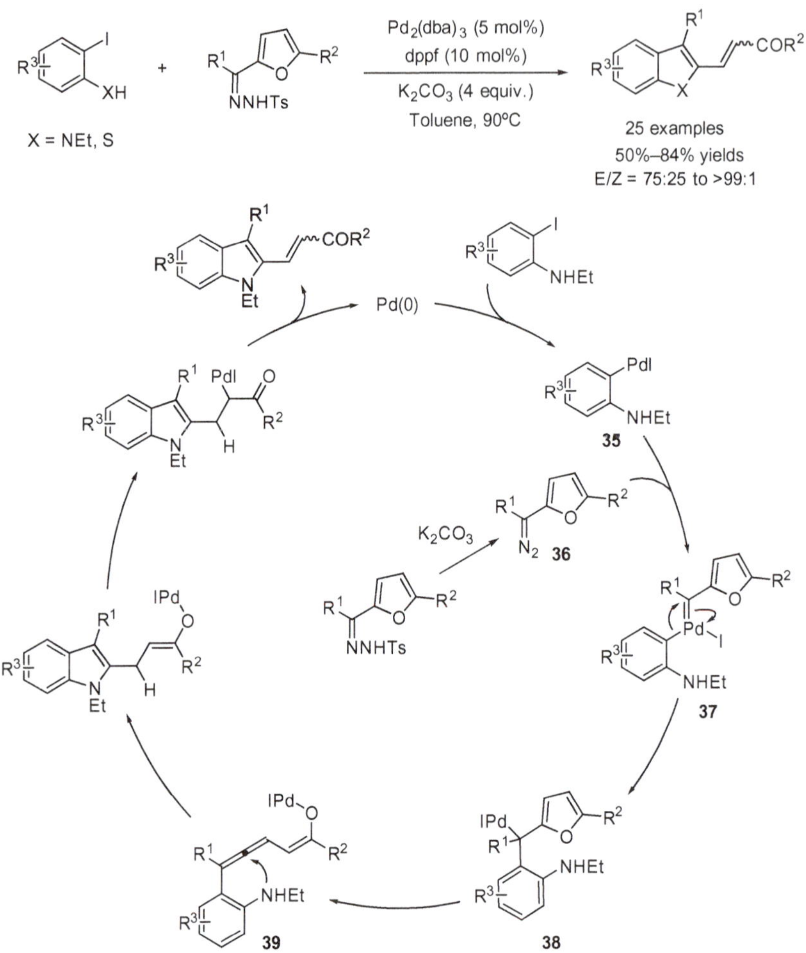

SCHEME 4.36 Palladium-catalyzed cross-coupling of furfural tosylhydrazones with 2-iodoanilines or 2-iodothiophenols.

β-heteroatom elimination and furan ring opening to give allene intermediate **39**. Sequential nucleophilic cyclization and β-hydride elimination complete the final products formation (Scheme 4.36). This palladium-catalyzed route provides a convenient and useful method for the preparation of polysubstituted indoles or benzothiophenes.

Palladium complexes, which incorporated cross-coupling reactions with metal carbene migratory insertion, were extensively employed as catalysts for the synthesis of heterocycles via nucleophilic cyclization. Apart from them, copper complexes are also recognized as qualified catalysts for this purpose. In 2011, Wang and coworkers realized Cu-catalyzed cross-coupling

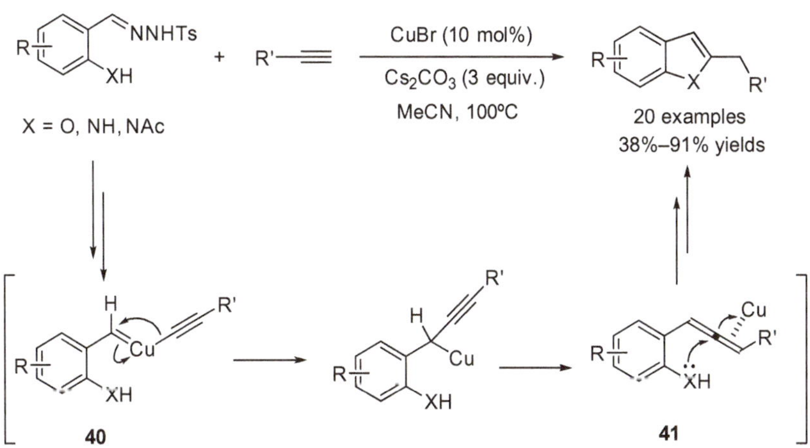

SCHEME 4.37 Cu-catalyzed cross-coupling reaction of terminal alkynes with *N*-tosylhydrazones for the synthesis of indoles or benzofurans.

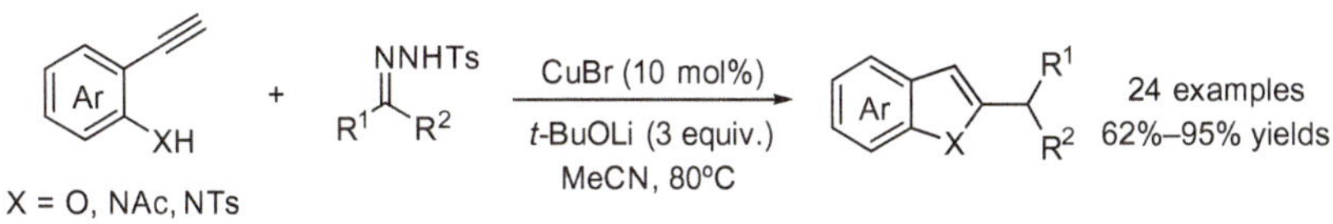

SCHEME 4.38 Cu-catalyzed cross-coupling of *N*-tosylhydrazones with *o*-hydroxy or *o*-amino phenylacetylenes.

reaction of terminal alkynes with *N*-tosylhydrazones derived from *o*-hydroxy- or *o*-aminobenzaldehydes.[41] The reaction conditions tolerate a wide range of functional groups. Thus 2-substituted indoles and benzofurans could be smoothly produced in this transformation. This reaction has been proposed to start with formation of copper acetylide from terminal alkynes and copper catalyst. Then, transient copper carbene species **40** formed through dediazotization of the in situ generated diazo substrate comes into being, which is immediately transformed into allene intermediate **41**. Activated by copper catalyst, the allene moiety is attacked by a hydroxy or an amino group to give the final heterocyclic compounds (Scheme 4.37). Remarkably, this reaction represented a novel and powerful method for accessing 2-substituted indole and benzofuran derivatives from readily available materials. In addition, Zhou and coworkers later on expanded this reaction by employing *N*-tosylhydrazones and *o*-hydroxy or *o*-amino phenylacetylenes as the substrates (Scheme 4.38).[42] Not only primary, but also secondary alkyl substituted benzofurans and indoles can be constructed in this study. In 2014, Sun and coworkers explored synthesis of C_2-functionalized indoles by choosing α-substituted diazoacetates and 2-ethynylanilines as the reactants.[43] Although the substrates are similar, the

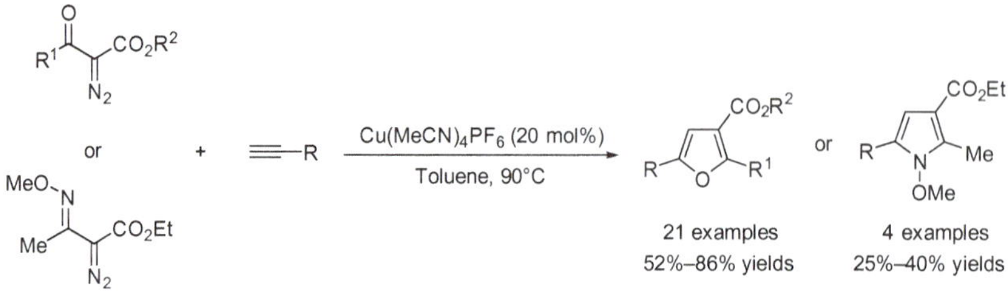

SCHEME 4.39 Cu(I)-catalyzed cross-coupling of terminal alkynes with β-keto-α-diazo esters or β-imine-α-diazo esters.

authors have postulated a different reaction mechanism based on their studies.

Moreover, Wang and coworkers reported a Cu(I)-catalyzed cross-coupling of terminal alkynes with β-keto-α-diazo esters, leading to poly-substituted furan derivatives (Scheme 4.39).[44] This simple and mild reaction proceeded smoothly with various functional groups, thus providing a diverse and versatile method for accessing these important building blocks. Although this strategy can be applied to synthesis of pyrroles, the efficiency was low in some cases. Two possible reaction pathways have been proposed in their study, but the authors are inclined to believe that the reaction proceeds with copper alkynylation, copper carbene formation, migratory insertion, nucleophilic cyclization, and protonation sequence. Later on, the same group reported a Cu(I)-catalyzed cascade reaction of *N*-tosylhydrazones with 3-butyn-1-ol to access tetrahydrofurans.[45]

In 2015, Sun and coworkers developed a stereo-divergent synthesis of five-membered *N*-heterocycles including 2,3-dihydropyrroles, 2-methylene, and 3-methylene pyrrolidines from amino alkynes and diazo compounds (Scheme 4.40).[46] Either copper or rhodium was employed as the optimal catalyst, depending on the diazo compounds tested. In the copper-catalyzed process, the allenoate was initially formed via carbene cross-coupling reaction, which was followed by 5-*exo-dig* or 5-*endo-dig* cycloaddition to give the corresponding 2,3-dihydropyrroles, 2-methylene pyrrolidines. With rhodium as catalyst, the reaction was proposed to proceed with carbenoid insertion and Conia-ene cyclization. In addition to [$Rh_2(esp)_2$] catalyst, $ZnCl_2$ as an additive greatly improved the reaction yield. Afterward, the same group expanded this reaction by choosing other functionalized alkynes. They disclosed a controllable protocol for selective synthesis of dihydropyrans through cascade process (Scheme 4.41).[47] This transformation was shown to proceed in good yields for both coupling partners including terminal alkynes and diazo compounds. Serving as a powerful and versatile protocol, this method has been applied to the synthesis of other poly-functionalized carbo-/heterocycles, such as cyclohexenes, tetrahydropyridines, and dihydropyrans (Scheme 4.41).[48]

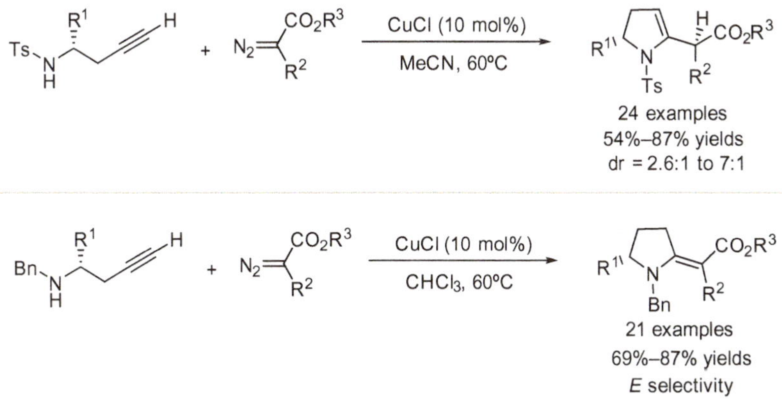

SCHEME 4.40 Copper-catalyzed synthesis of 2,3-dihydropyrroles, 2-methylene pyrrolidines.

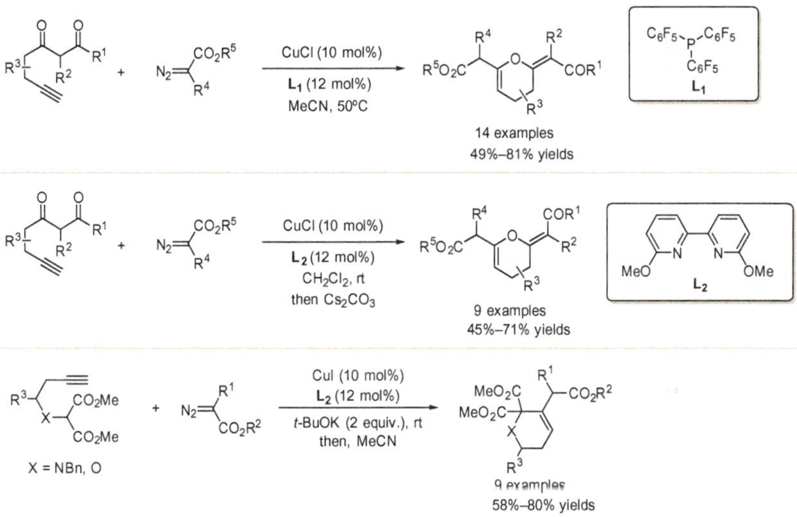

SCHEME 4.41 Copper-catalyzed synthesis of six-membered heterocycles.

4.6 SYNTHESIS OF HETEROCYCLES THROUGH CONDENSATION OR ADDITION

In recent years, C–H bond activation based on metal carbene migratory insertion has been developed into an intriguing strategy for C–H bond functionalization.[3h] Along with type III synthesis of heterocycles, this category also involves the combination of C–H bond activation and metal carbene migratory insertion. The steps of C–H metalation, metal carbene formation, and migratory insertion are similar in these two classifications. Instead of

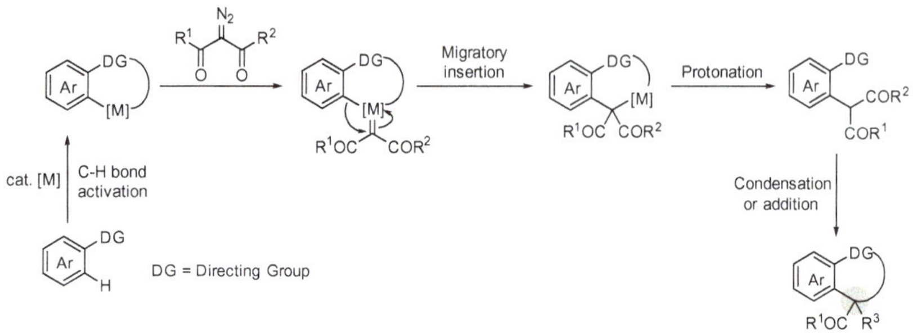

SCHEME 4.42 The general process of heterocycle synthesis through condensation or addition.

reductive elimination, pronation followed by condensation or metal-assist addition terminates the catalytic cycle for final heterocycles formation in this type (Scheme 4.42). In this classification, rhodium has been the most exploited catalyst. In addition, iridium, cobalt, and ruthenium are also explored for this process. It should be mentioned that at least one nitrogen atom is contained in directing group (DG) in most cases, which allows the condensation or addition processes for the generation of final heterocyclic compounds.

In 2012, Yu and coworkers made a pioneering advance, in which they employed diazo compounds in rhodium-catalyzed directed C–H bond activations.[24] A broad array of arenes with several DGs including oximes, carboxylic acids, amines, and pyridines are well coupled with diazomalonates. This novel C–H activation allows the preparation of diverse products bearing different functional groups with excellent regioselectivities. In this report, unprotected benzylamines are also tolerated, which provide possibilities for further annulation through nucleophilic addition. Consequently, isoquinolones are afforded in good to excellent yields through tandem process (Scheme 4.43).

Valdés and colleagues discovered a Pd-catalyzed cross-coupling reaction of α-alkoxytosylhydrazones with sulfonates derived from salicylaldehydes (Scheme 4.44).[49] This protocol furnishes protected 1,5-dicarbonyl compounds, which then reacts with ammonium hydroxide to provide 3,4-disubstituted isoquinolines as the final products. On the other hand, a similar reaction by employing *o*-cyanononaflates and α-alkoxytosylhydrazones as starting materials were also carried out. This cross-coupling reaction gave *o*-cyanoalkenes, which could further be transformed into 1-substituted isoquinolines through the treatment with organolithium. The reported protocol in this study constituted a versatile and efficient method for construction of poly-substituted isoquinoline derivatives.

Isoquinolines could also be achieved by rhodium-catalyzed process. More recently, the Cheng and coworkers developed a novel protocol for constructing isoquinoline framework via rhodium-catalyzed annulation of primary

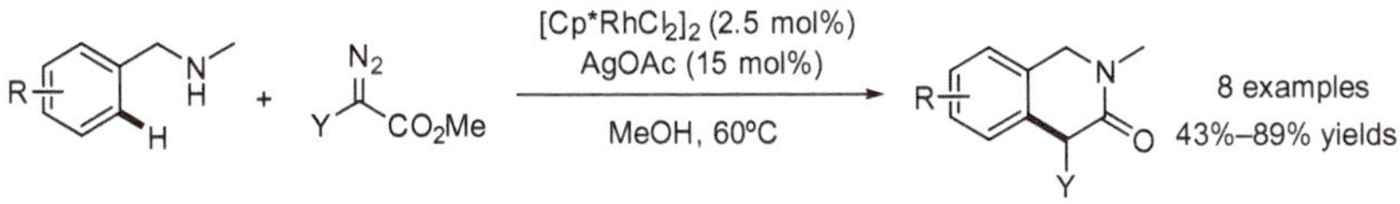

SCHEME 4.43 Isoquinolones synthesis through Rh-catalyzed C–H bond activation.

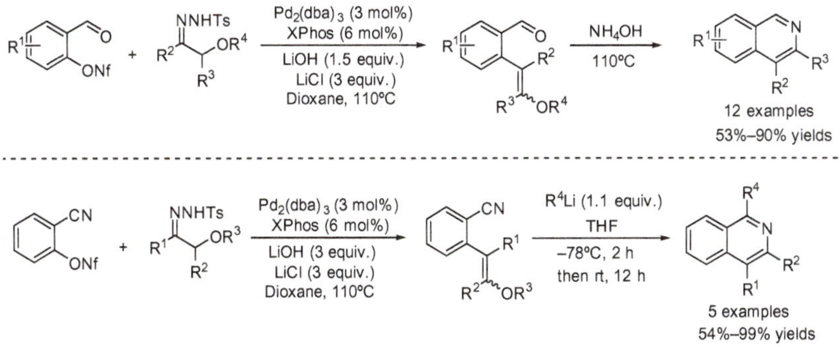

SCHEME 4.44 Synthesis of poly-substituted isoquinolines through cross-coupling reactions with α-alkoxytosylhydrazones.

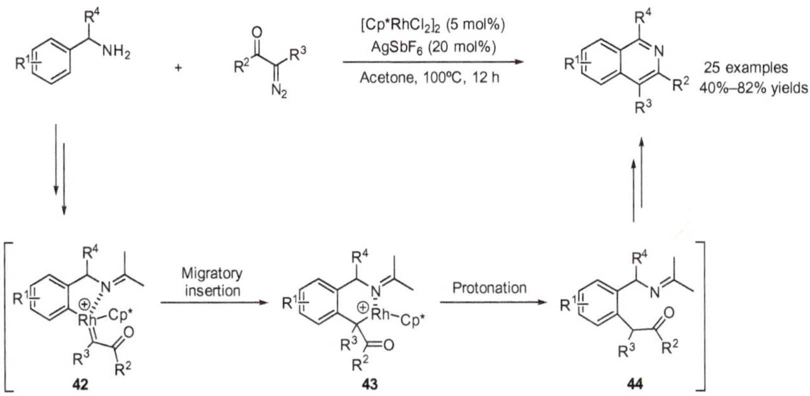

SCHEME 4.45 Rhodium-catalyzed annulation of primary benzylamines with α-diazo ketones.

benzylamines with α-diazo ketones.[50] The reaction pathway started with C–H activations by employing the in situ formed imine as DG. In the presence of α-diazo ketones, rhodium carbene **42** is generated with the loss of N_2. Migratory insertion leads to the formation of six-membered rhodacycle intermediate **43**, followed by protonation to give *ortho*-alkylated intermediate **44**. Proton-catalyzed equilibration, intramolecular condensation, and oxidation successively occur to furnish the final products (Scheme 4.45).

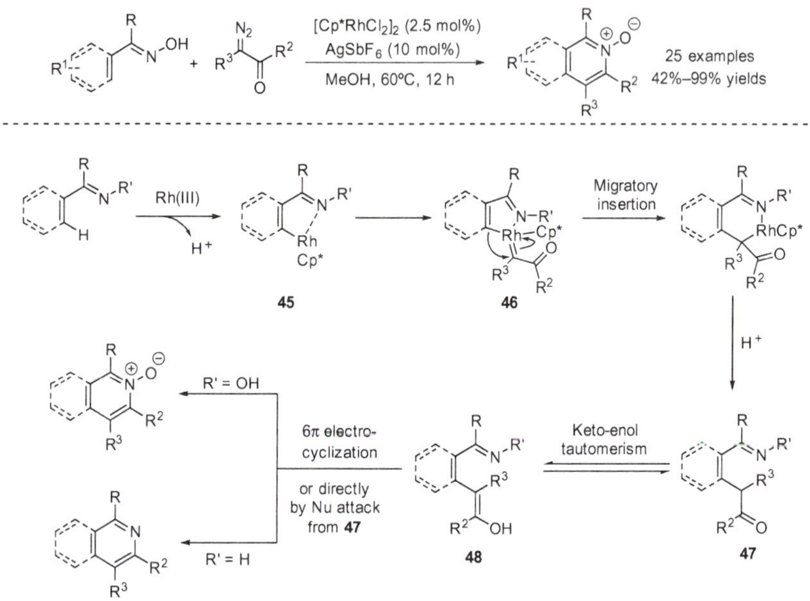

SCHEME 4.46 Rh(III)-catalyzed synthesis of multisubstituted isoquinoline or pyridine *N*-oxides from oximes and diazo compounds.

Typically, this reaction afforded diverse 3,4-disubstituted isoquinolines in moderate to good yields with high regioselectivities.

Since isoquinoline and pyridine *N*-oxides are important and useful synthetic intermediates in organic community, Glorius and coworkers developed synthetic methods to access these scaffolds by Rh(III)-catalyzed cross-coupling of oximes and diazo compounds via aryl and vinylic C–H activation.[51] In this tandem reaction, the rhodacycle **45** is initially formed through regioselective C–H bond cleavage. Coordination of diazo compounds gives Rh-carbene intermediate **46**, followed by migratory insertion and pronation to furnish alkylated intermediate **47**. Then, isomer **48** is generated through keto-enol tautomerization from intermediate **47**. Subsequent 6π electrocyclization and dehydration give the corresponding products. Alternatively, direct nucleophilic addition and sequential dehydration are also possible processes for final products formation (Scheme 4.46).

Recently, Yao, Lin, and coworkers expanded this strategy to achieve the synthesis of heterocycle-fused pyridine *N*-oxides and their derivatives.[52] A wide range of substrates with different kinds of substituents were smoothly coupled with carbonyl-containing diazo compounds. Furthermore, the heterocycle-fused pyridine *N*-oxides could be transformed to heterocycle-fused pyridine derivatives in high efficiency. In addition, Patel and coworkers also accomplished the synthesis of isoquinoline *N*-oxides with iridium

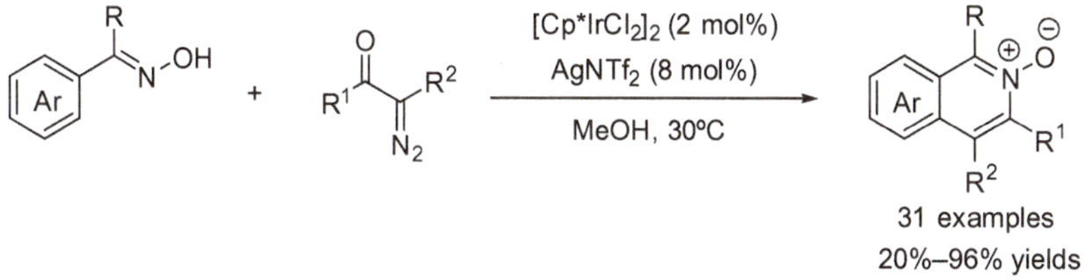

SCHEME 4.47 Ir(III)-catalyzed synthesis of isoquinoline *N*-oxides from aryloximes and diazocarbonyl compounds.

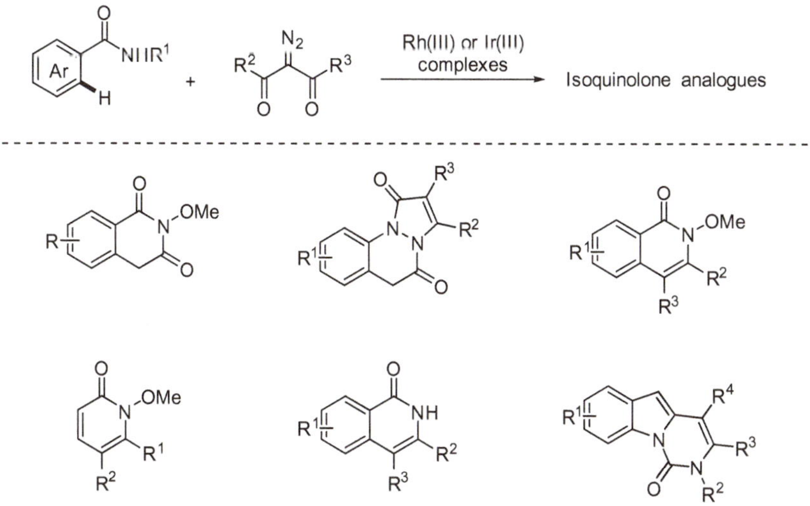

SCHEME 4.48 Isoquinolone analogs synthesis through Rh-catalyzed C–H bond activations.

complexes from similar starting materials (Scheme 4.47).[53] A dimeric iridacyclic complex, which was linked with a dimeric silver bridge, was isolated and characterized by X-ray crystallography in this report. This dimeric iridacyclic complex could catalyze the standard reaction in efficient manner, providing an important evidence for the formation of five-membered iridacycle in the proposed reaction mechanism.

Furthermore, with amides as privileged DGs in rhodium-catalyzed C–H bond activations, a wide range of isoquinolone analogs including *N*-methoxyisoquinolinediones,[54–56] 1*H*-pyrazolo[1,2-*a*]cinnoline-1,5(6*H*)-diones,[54,55] *N*-methoxyisoquinolones,[55,57] *N*-methoxypyridones,[57] 2*H*-pyrimido[1,6-*a*]indol-1-ones,[58] and *N*-unprotected isoquinolinones[59] had been prepared in good to excellent yields (Scheme 4.48). The general process for accessing these scaffolds involved C–H activation, metal carbene formation, migratory insertion, protonation, and condensation or addition. From easily available starting materials, these Rh- or Ir-catalyzed carbene-based C–H

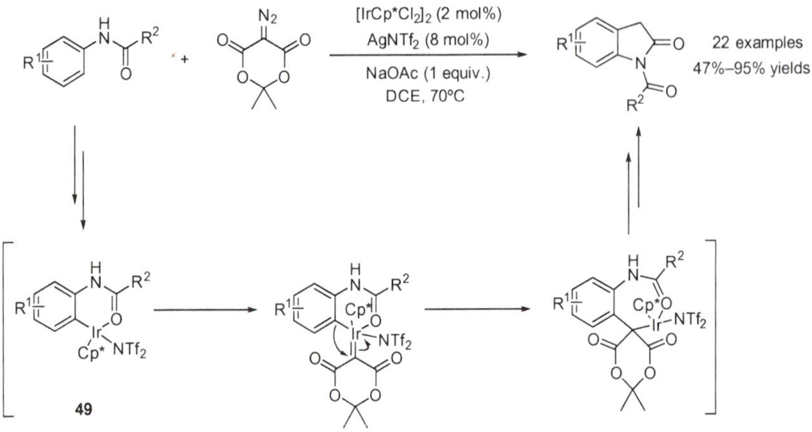

SCHEME 4.49 Ir(III)-catalyzed C–H activation of acetanilides with diazotized Meldrum's acid.

functionalizations provided convenient routes to acquire different kinds of isoquinolone derivatives.

Another application of amide as DG was reported by Patel and coworker for the effective synthesis of oxindole derivatives.[60] Employing diazotized Meldrum's acid as the carbene precursors, a range of electron-rich or -poor anilides were found as efficient coupling partners. The carbonyl chelating groups such as ester and acetyl groups could not hinder the reaction and the corresponding products were also obtained in good yields. However, the commonly *N*-protected anilines such as *N*-Ts, *N*-Boc, and *N*-Cbz were proved sluggish in this transformation. This reaction was proposed to proceed with the formation of cyclo-metalated Ir(III) complex **49**, followed by metal carbene formation, migratory insertion, and protonation sequence. Final condensation afforded the oxindoles with the release of acetone and carbon dioxide (Scheme 4.49). The authors also demonstrated the synthetic utility of oxindole derivatives by further derivatization.

Furthermore, imine was recognized as competent DG in this strategy. More recently, the groups of Li,[61] Liu,[62] and Zhu[63] almost simultaneously reported the Rh(III)-catalyzed C–H activations of arylimidates, followed by intramolecular C–N bond formation, affording isoquinoline derivatives under mild conditions. A wide range of diazo compounds including α-diazo ketoesters, α-diazomalonates, α-diazo imidamide, and diazotized Meldrum's acid could be applied as suitable reactants in these cascade transformations. Thus structurally diversified isoquinoline analogs were conveniently synthesized in efficient manner (Scheme 4.50). Specifically, the Li group also demonstrated rhodium/copper-catalyzed annulation of benzylamines with diazo compounds for isoquinoline's synthesis.[64] In this reaction, aldimines were generated in situ from benzylamines via Cu-catalyzed oxidation.

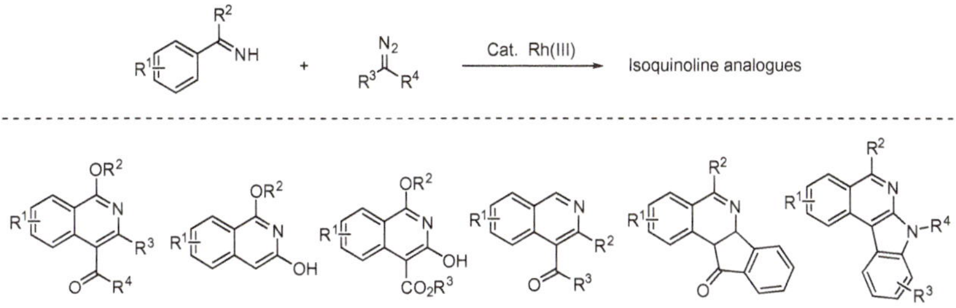

SCHEME 4.50 Isoquinoline analogs synthesis via Rh-catalyzed C–H bond activations.

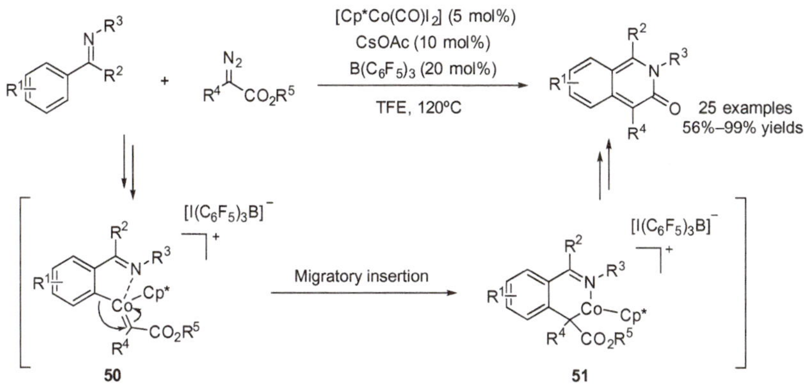

SCHEME 4.51 Cooperative $B(C_6F_5)_3$- and Cp^*Co^{III}-catalyzed C–H bond activation of imines with diazo compounds.

At the same time, Glorius and coworkers developed a Co(III)-catalyzed imine-directed C–H activations for rapid access to highly substituted isoquinolin-3-ones with broad substrate scope and good functional group tolerance.[65] $B(C_6F_5)_3$ was found to be the most suitable Lewis acid in this reaction, which co-catalyzed this reaction in efficient manner. The C–H activation was found reversible and not the rate-limiting step through deuterium labeling experiments and kinetic studies, respectively. In the proposed reaction mechanism, the catalytic cycle is started with the formation of chelated cobaltacycle, followed by decomposition of diazo compounds to generate metal carbene species **50**. Subsequent migratory insertion then occurs to provide six-membered cobaltacycle **51**, followed by nucleophilic addition, elimination, and isomerization to produce isoquinolin-3-ones as the final products (Scheme 4.51).

Quinolinone derivatives could be obtained by multicomponent coupling reactions. In 2016, Cheng and coworkers presented a palladium-catalyzed three-component coupling reaction of *N*-tosylhydrazones, 2-iodoanilines, and

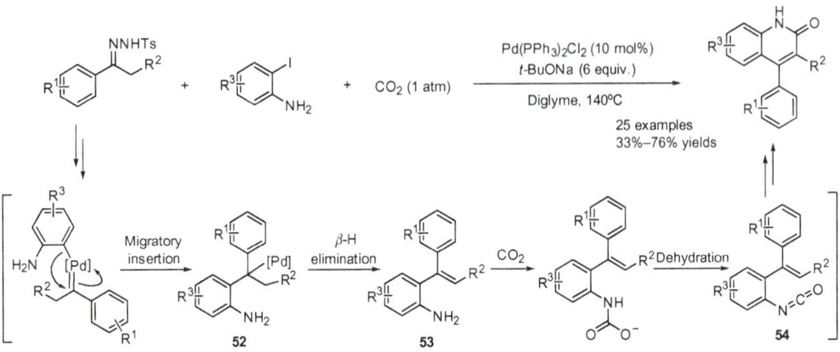

SCHEME 4.52 Palladium-catalyzed three-component coupling reaction of N-tosylhydrazones, 2-iodoanilines, and CO_2.

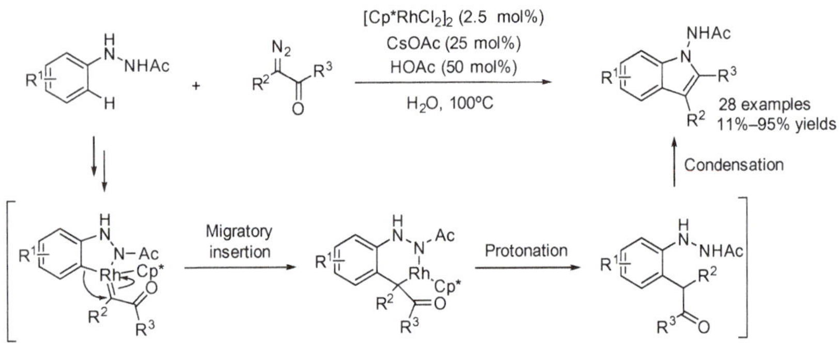

SCHEME 4.53 Rh(III)-catalyzed C−H bond activation of 2-acetyl-1-arylhydrazines with diazocarbonyl compounds.

CO_2, affording 4-aryl-2-quinolinones in moderate to good yields.[66] In this cascade reaction, the benzyl palladium species **52** generated from migratory insertion undergoes β-hydride elimination to give olefin intermediate **53**. Incorporation of carbon dioxide and subsequent dehydration deliver isocyanate intermediate **54**. 6π-Electrocyclization and isomerization furnish the desired products (Scheme 4.52). Alternatively, the amine group can be firstly transformed into isocyanate intermediate before olefin formation. Notably, carbon dioxide is a novel coupling partner which is employed in carbene-based cross-coupling reactions.

In 2014, Wang and coworkers established a new methodology for the synthesis of 1-aminoindole derivatives via Rh(III)-catalyzed C−H bond activation of 2-acetyl-1-arylhydrazines with diazocarbonyl compounds.[67] They proposed a mechanism involving tandem C−H activation, migratory insertion, and condensation steps (Scheme 4.53). A series of 2-acetyl-1-arylhydrazines and diazocarbonyl compounds with different substituent patterns worked well under the reaction conditions. It should be noted that

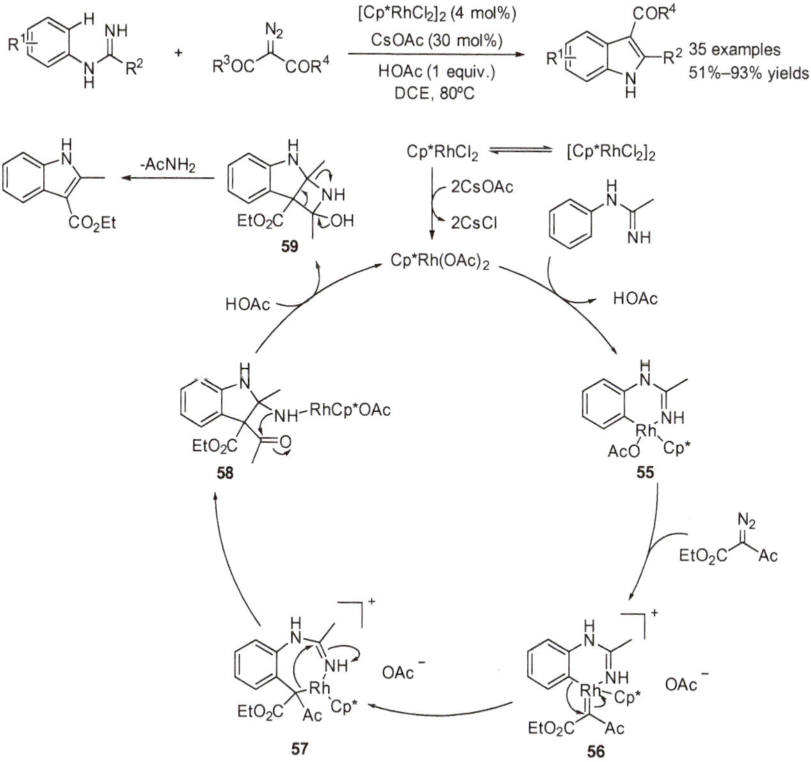

SCHEME 4.54 Rh(III)-catalyzed synthesis of *N*-unprotected indoles from imidamides and diazo ketoesters.

water, which served as the solvent in this reaction, significantly simplified the separation procedure. From the synthetic and separation point of view, this protocol offered a convenient method for accessing these important building blocks.

Subsequently, Li and coworkers reported Rh(III)-catalyzed C−H activation/annulation for the synthesis of free indoles by using imidamides and α-diazo-β-ketoesters as the substrates.[68] In the proposed reaction mechanism, rhodium carbene **56** undergoes migratory insertion to give seven-membered rhodacycle intermediate **57**. Instead of protonation, the Rh−C bond then undergoes nucleophilic addition to the C = N bond, affording an amide species **58**. Subsequent protonation and intramolecular nucleophilic addition generate intermediate **59**. Finally, elimination of acetamide affords the final product (Scheme 4.54). Remarkably, a six-membered rhodacyclic complex **55** was isolated and fully characterized, which was then applied in model reaction and the corresponding product was obtained in a comparable yield. This result provided a direct evidence for the proposed reaction mechanism.

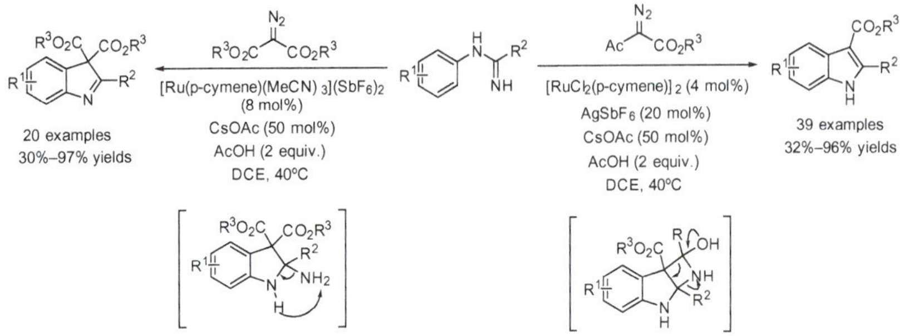

SCHEME 4.55 Ru(II)-catalyzed synthesis of *NH*-indoles and 3*H*-indoles from imidamides and α-diazo ketoesters or α-diazomalonates.

Furthermore, the same group reported Ru(II)-catalyzed C–H activation of imidamides with α-diazo ketoesters and α-diazomalonates, affording *NH*-indoles and 3*H*-indoles, respectively.[69] A sequence of Ru-C(alkyl) migratory insertion, protonation, intramolecular nucleophilic addition, and final elimination of one molecule of amide afforded *NH*-indoles as the final products, whereas protonation and subsequent elimination of one molecule of ammonia occur to give 3*H*-indoles exclusively (Scheme 4.55). Both gram-scale synthesis and further transformation of 3*H*-indoles demonstrated the synthetic utility of this reaction. In this transformation, ruthenium was also proved as qualified transition metal catalyst for the synthesis of heterocycles based on metal carbene migratory insertion. In addition, this catalytic reaction is compatible with a wide range of functional groups on both coupling partners.

Very recently, Zhu and coworkers disclosed a distinct Rh(III)-catalyzed C–H activation strategy for the construction of highly substituted indoles by employing a traceless DG.[70] With $[Cp^*RhCl_2]_2/AgSbF_6$ as a suitable catalytic system, a variety of *N*-nitrosoanilines and α-diazo-β-ketoesters reacted smoothly to furnish the corresponding indole derivatives in good to excellent yields. This reaction has been proposed to occur via C–H activation, metal carbene formation, migratory insertion, and protonation to generate alkylated intermediate **60**, followed by the elimination of nitrous acid, which is the major side product of the reaction (Scheme 4.56).

The pyridyl and pyrimidyl moieties, which act as a popular and efficient DG, have been widely employed in C–H activations for indoles synthesis. Recently, several groups[71–75] merged Rh(III)-catalyzed carbene-involved C–H activation with intramolecular cascade reaction together, assembling various highly substituted indole derivatives in high efficiency. They selected readily available *N*-phenylpyridin-2-amines or *N*-phenylpyrimidin-2-amines and diazo acetoacetate as the substrates. The common features of these reactions are tandem C–H activation, migratory insertion, and condensation steps (Scheme 4.57). From the environmental benign as well as atom economy point of view, these reactions all released nitrogen and water as side

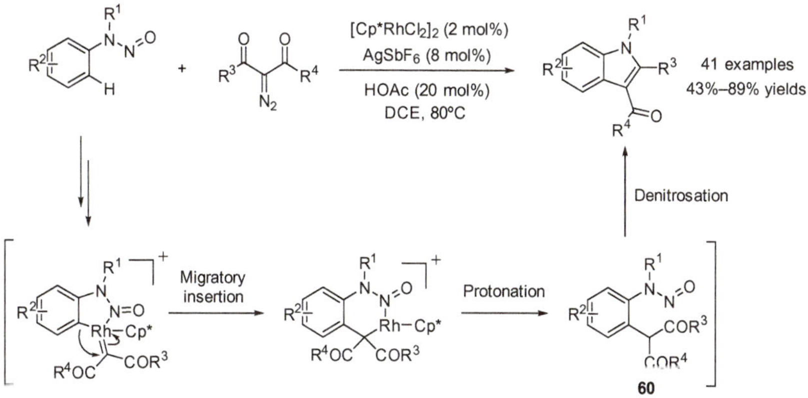

SCHEME 4.56 Rhodium(III)-catalyzed C–H activation of *N*-nitrosoanilines with α-diazo-β-keto compounds.

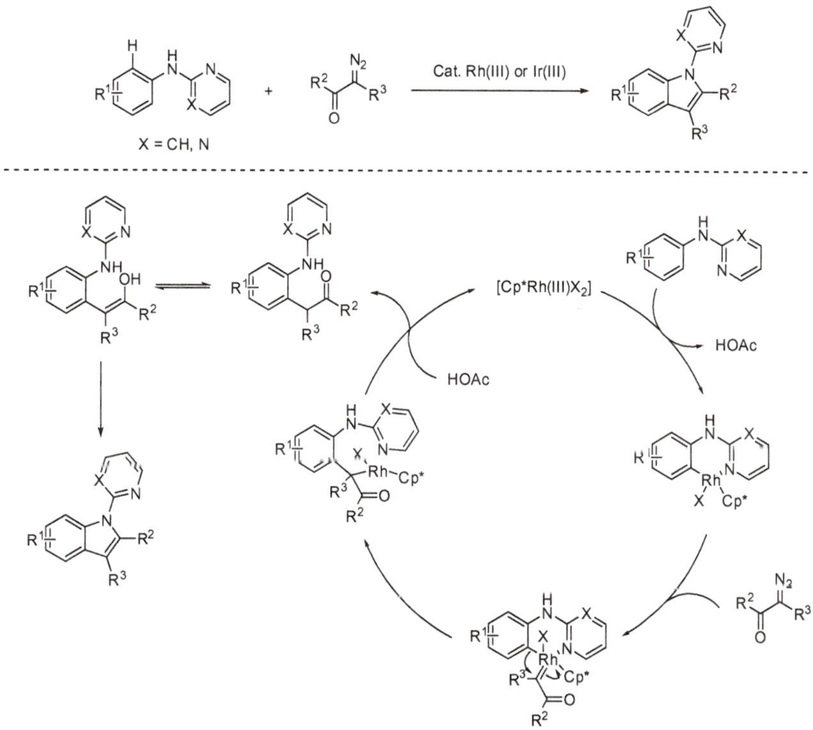

SCHEME 4.57 Rh(III)- or Ir(III)-catalyzed pyridyl- or pyrimidyl-directed C–H activations for regioselective synthesis of indoles.

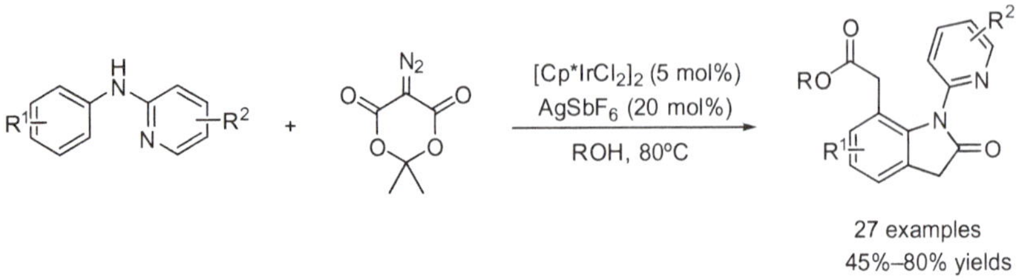

SCHEME 4.58 Ir(III)-catalyzed C–H activation to access 1,3-dihydroindol-2-ones.

products. Along with traditional method for indoles synthesis, this C–H activation strategy provided a straightforward method toward these privileged heterocycles.

More recently, Zeng and coworkers expanded the scope of carbene precursors to diazotized Meldrum's acid by employing the same DG, affording multifunctionalized 1,3-dihydroindol-2-ones in good yields (Scheme 4.58).[76] A wide range of *N*-aryl-2-pyridinamines with electronically different substituents could smoothly attend this Ir(III)-catalyzed C–H annulation. In addition to diazotized Meldrum's acid, other types of diazo compounds, including 1,3-diketo-2-diazo compounds and 1,3-ketoester-2-diazo compounds, were all suitable coupling partners in this transformation.

Pyridine and pyrimidine groups can also assist direct C(sp^3)–H bond activations for carbene-induced heterocycle's synthesis. In 2015, the same group described the Rh(III)-catalyzed chelation-assisted sequential aliphatic C–H activation and intramolecular condensation for efficient synthesis of poly-substituted *N*-(2-pyridyl)pyrroles from ketimines and α-acyl diazo compounds.[77] This reaction involves carbene insertion into C(sp^3)–H bond followed by protonation and intramolecular cyclization (Scheme 4.59). From the synthetic point of view, this newly designed methodology enables efficient construction of poly-substituted *N*-(2-pyridyl)pyrroles bearing numerous functional groups.

Subsequently, the same group disclosed Rh(III)-catalyzed double C–H bond activation of *N*-phenyl-ketoimines with α-acyl diazocarbonyl compounds, providing efficient access to π-conjugated 1-azaphenalenes.[78] In this reaction, imine-directed C–H activation initially occurred to give five-membered rhodacycle **61**. Coordination of the diazoacetate substrate to the rhodacycle affords rhodium carbene species **62**, followed by migratory insertion and protonation to generate the mono-functionalized ketimine intermediate. The second imine-directed C–H bond carbene functionalization occurs similarly. Final cyclization gives π-conjugated 1-azaphenalenes as the products (Scheme 4.60). Remarkably, three C–C bonds and one C–N bond is formed in this cascade reaction, which offers a step-economic method for these π-conjugated fused heteroarenes.

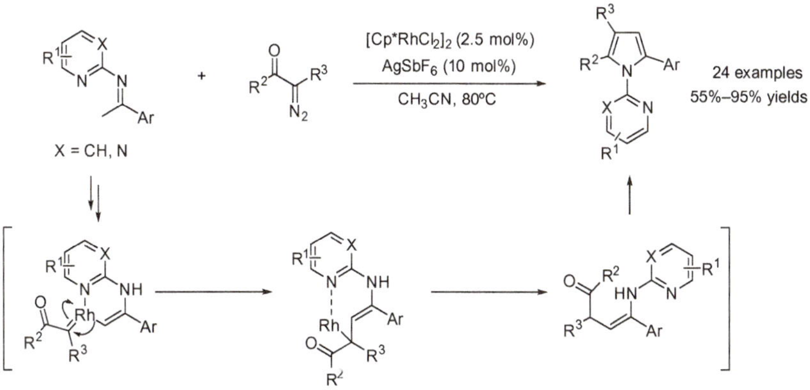

SCHEME 4.59 Rh(III)-catalyzed C(sp^3)–H activation for synthesis of poly-substituted *N*-(2-pyridyl)pyrroles.

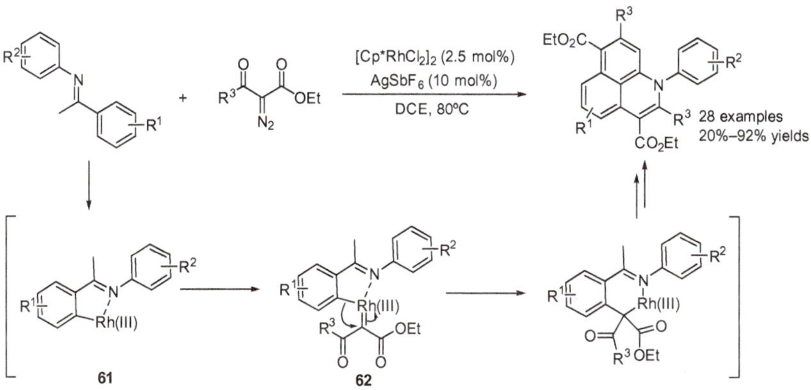

SCHEME 4.60 Rh(III)-catalyzed double C–H bond activation of *N*-phenyl-ketoimines to access π-conjugated 1-azaphenalenes.

Carbazoles could also be synthesized by using the same DG. Recently, Wang, Huang, and coworkers developed a tandem rhodium(III)/Brønsted acid-catalyzed benzannulation of indoles with enaldiazo compounds, providing an efficient avenue for substituted carbazoles under mild conditions (Scheme 4.61).[79] In this reaction, enal or aldehyde intermediate generated from protonation undergoes intramolecular cyclization and dehydration to give the final products. A wide range of pyrimidyl-substituted indoles and enaldiazo compounds with series functional groups are well tolerated in this transformation. Moreover, this protocol had been extended to the synthesis of indole derivatives by selecting pyrroles as substrates.

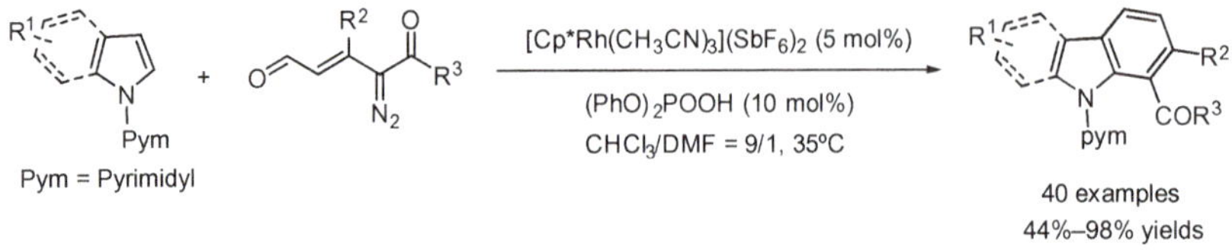

SCHEME 4.61 Tandem rhodium(III)/Brønsted acid-catalyzed benzannulation of indoles with enaldiazo compounds.

In 2015, Liu and coworkers developed a Rh(III)-catalyzed C–H annulation of benzamides and diazocarbonyl compounds.[80] Instead of giving isoquinolone derivatives, isocoumarins and α-pyrones were obtained in moderate to good yields. In this reaction, the *ortho* C–H bond of benzamides is firstly cleaved with concomitant formation of rhodacycle **63**. The Rh(III)-carbene **64** is then formed with extrusion of N_2. Next, aryl group migrates to Rh(III)-carbene center, giving six-membered rhodacycle **65**, which can either tautomerize to intermediate **66** or be protonated by acetic acid. From intermediate **66**, metal-assisted nucleophilic addition and elimination occurs to finish the catalytic cycle and provides isocoumarins or α-pyrones as the final products. In another process, intermediate **68** can be generated from **67** via enolization, followed by esterification to afford the desired product (Scheme 4.62).

Later, Yao, Lin, and coworkers described a controllable Rh(III)-catalyzed annulation between salicylaldehydes and diazo compounds, providing a divergent synthetic route to chromone and benzofuran derivatives.[81] The product distribution in this transformation is largely dependent on the additives. $AgNTf_2$ preferred benzofuran derivatives, while AcOH afforded chromones as the products. According to the reaction mechanism, the activation of the C–H bond of aldehyde firstly occurred to produce five-membered rhodacycle **69**, which then undergoes decarbonylation in the presence of a weaker coordinating counterion to generate the four-membered oxa-rhodacycle **70**. Subsequent rhodium carbene formation, migratory insertion, protonation, and dehydration successively occur to yield 2,3-disubstituted benzofurans. On the other hand, the five-membered rhodacycle **69** with a stronger coordinating counterion can directly react with diazo substrate to produce rhodium carbene intermediate **71**. The similar process subsequently occurs to produce chromones as the products (Scheme 4.63).

Recently, Bolm and coworker reported a novel protocol for the regioselective synthesis of 1,2-benzothiazines through rhodium-catalyzed C–H activation of NH-sulfoximines with diazo compounds.[82] This reaction is initiated by the C–H activations and decomposition of diazo compounds to generate rhodium carbene species **72**, which undergoes migratory insertion, protonation, and condensation to afford the products with good to excellent

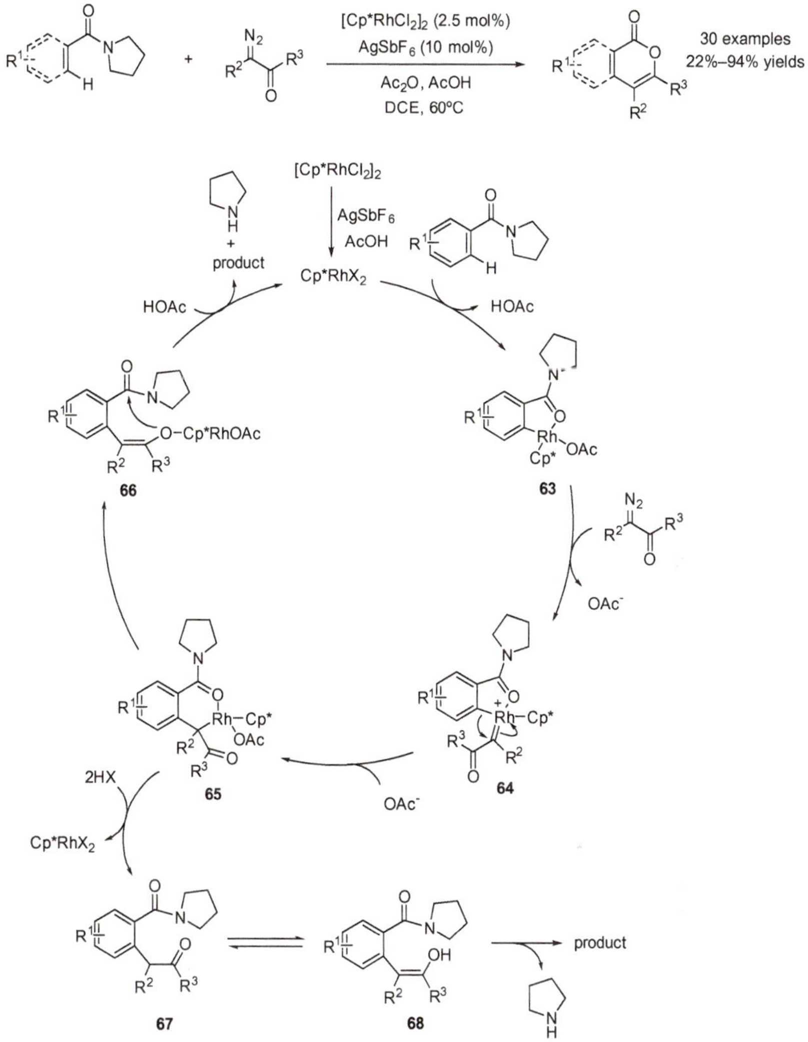

SCHEME 4.62 Rh(III)-catalyzed C–H annulation of benzamides and diazocarbonyl compounds.

yields (Scheme 4.64). In view of synthetic utility of 1,2-benzothiazines, these scaffold can be easily converted into several important building blocks under mild reaction conditions.

In addition, sulfoximines can also couple with pyridotriazoles. In 2016, Lee and coworkers developed an efficient catalytic strategy for regioselectively synthesis of 1,2-benzothiazines through rhodium-catalyzed C–H activation of *S*-aryl sulfoximines and pyridotriazoles (Scheme 4.65).[83] In this

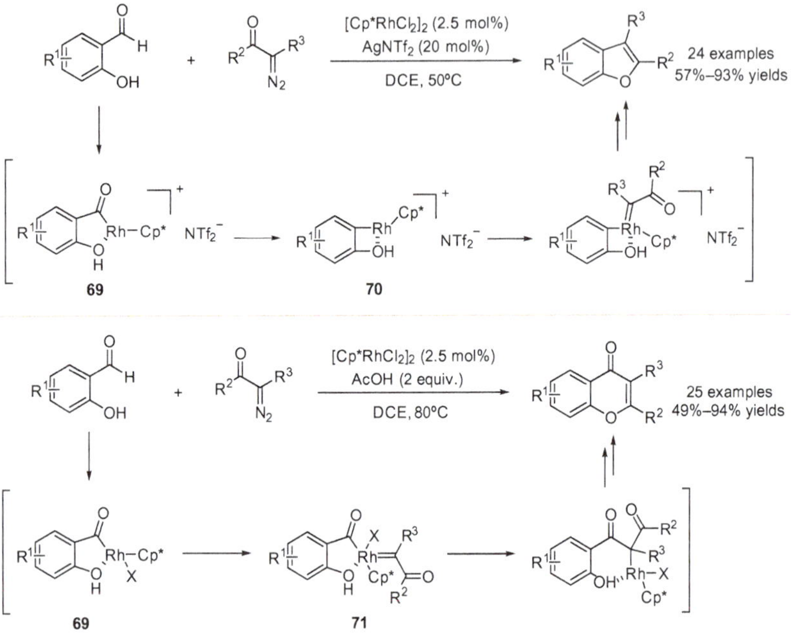

SCHEME 4.63 Rh(III)-catalyzed annulation between salicylaldehydes and diazo compounds for divergent synthesis of chromones and benzofurans.

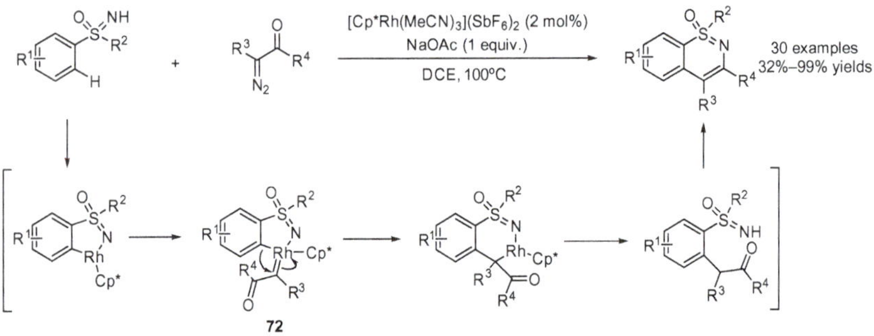

SCHEME 4.64 Regioselective synthesis of 1,2-benzothiazines through rhodium-catalyzed C–H activation of NH-sulfoximines with diazo compounds.

reaction, pyridotriazoles serve as carbene precursors, which can equilibrate to the diazoimine form. Similar to the Bolm's work, the proposed mechanism proceeded via C–H activation, metal carbene formation, migratory insertion, protonation, nucleophilic addition, and elimination pathway. The method enables the synthesis of a series of 1,2-benzothiazines bearing pyridyl as well as carbonyl groups with good to excellent yields.

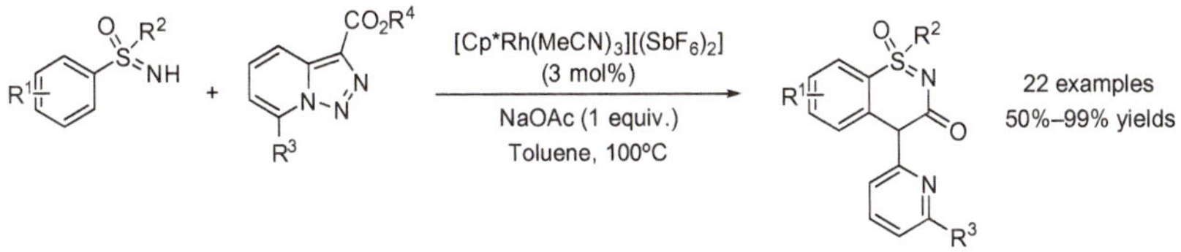

SCHEME 4.65 Rhodium-catalyzed C–H activation of *S*-aryl sulfoximines with pyridotriazoles.

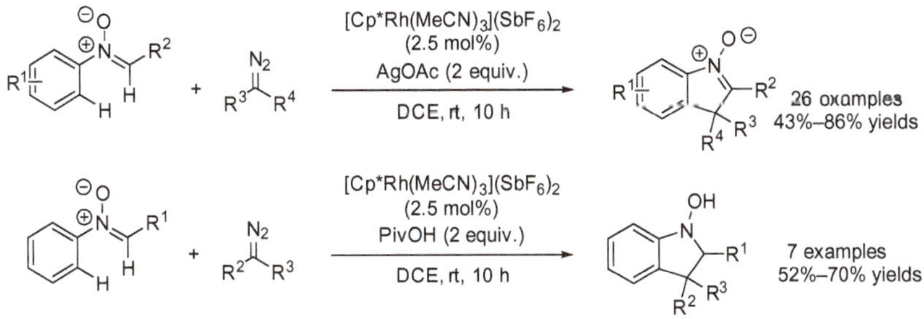

SCHEME 4.66 Rhodium(III)-catalyzed C–H oxidative cyclization of nitrones with diazo compounds.

In 2015, Zhou and coworkers explored this catalytic C–H bond activation by employing nitrone as DG.[84] They developed rhodium(III)-catalyzed C–H oxidative cyclization of nitrones with diazo compounds as one-carbon components, affording various 3*H*-indole-*N*-oxides. Under slightly modified reaction conditions, *N*-hydroxyindoline can be selectively achieved in good yields (Scheme 4.66). In addition, the present methodology enables the synthesis of *N*-hydroxyindoles by decarbonylation in one-pot fashion. Mechanistically, the authors have assumed that the reaction proceeds through a directed C–H activation, metal carbene formation, migratory insertion, Grignard-type addition, and in situ oxidation sequence. Remarkably, this transformation can be conducted under mild reaction conditions, which makes this protocol quite appealing in organic synthesis.

Later, Chang and coworker have demonstrated another Rh(III)-catalyzed C–H activation for synthesis of *N*-hydroxyindolines by using same DG.[85] A great deal of arylnitrones and diazo compounds were tolerated in this cross-coupling reaction. In contrast to Zhou's study, this reaction exclusively produced *N*-hydroxyindoline derivatives as the final products. The reaction conditions for this approach are mild and the cyclization can be completed at room temperature. Based on mechanistic studies and literature precedent, the authors have proposed a mechanism that involves C–H bond activation, rhodium carbene formation, migratory insertion, rhodium-assisted addition, and protonation as vital steps in catalytic cycle (Scheme 4.67). Another possible

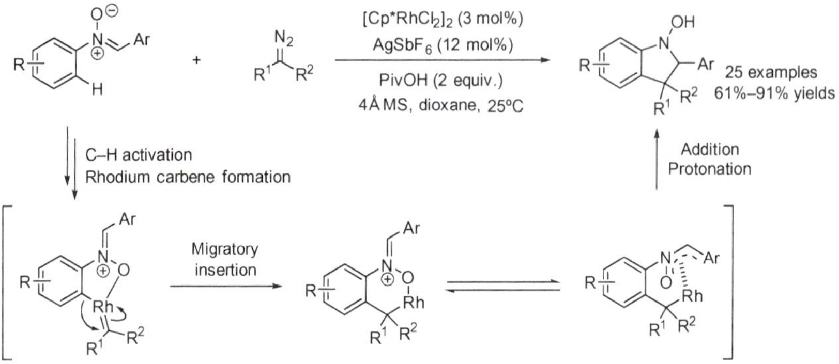

SCHEME 4.67 Rh(III)-catalyzed C–H activation of arylnitrones with diazocarbonyl compounds.

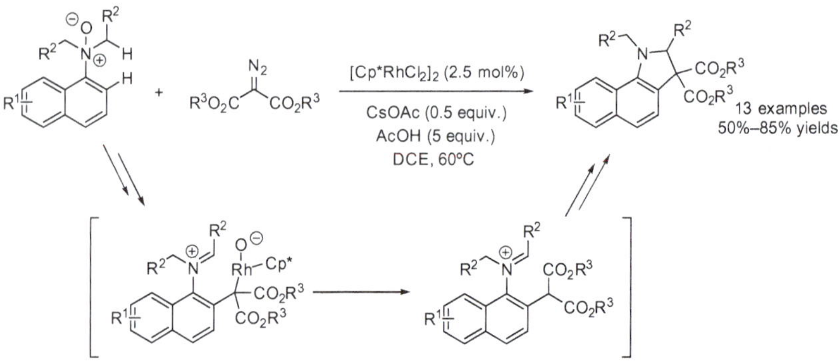

SCHEME 4.68 Rh-catalyzed regioselective synthesis of 1*H*-benzo[*g*]indolines.

process involves 1,2-aryl shift rather than rhodium carbene migratory insertion.

Furthermore, Rh(III)-catalyzed regioselective synthesis of 1*H*-benzo[*g*]indolines have been developed by Zhu, Yang, Zhou, and coworkers.[86] This catalytic reaction used 1-naphthylamine *N*-oxides and diazo compounds as the substrates, through dual functionalization of unactivated primary C(sp^3)–H bond and C(sp^2)–H bond, affording 1*H*-benzo[*g*]indoline derivatives in high efficiency (Scheme 4.68). In addition to 1-naphthylamine *N*-oxides, aniline *N*-oxides could also serve as suitable coupling partners, which reacted with α-diazomalonates under external oxidant-free conditions, giving aminomandelic acid derivatives as the final products. Specifically, this mild reaction can be carried out under redox-neutral conditions in the absence of external oxidants. Although three possible pathways have been proposed, DFT calculations indicate that an intermediate iminium is most likely involved in the catalytic cycle.

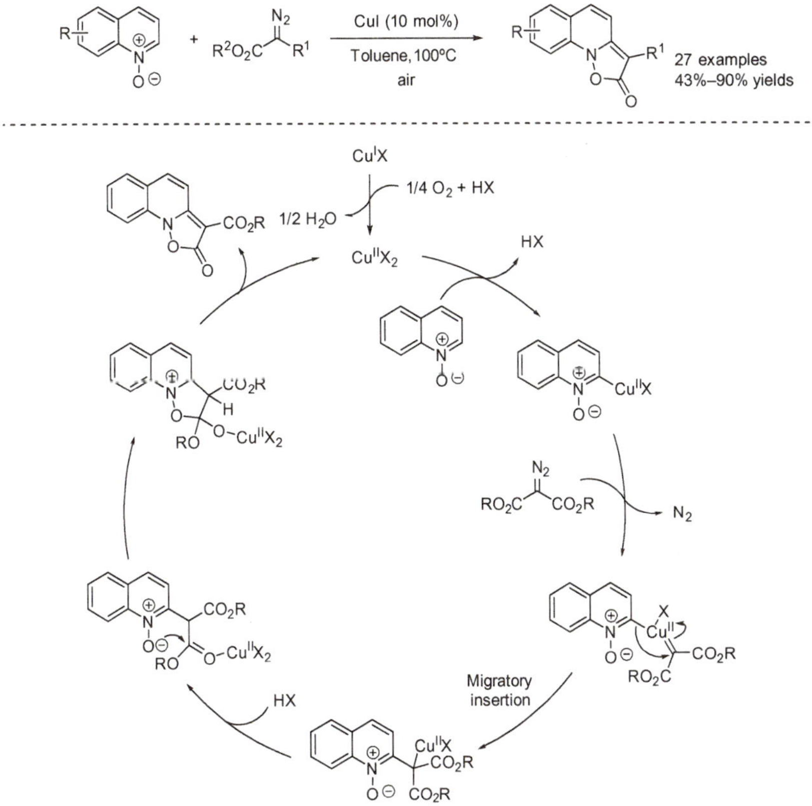

SCHEME 4.69 Copper-catalyzed regioselective cascade alkylation and cyclo-condensation of quinoline *N*-oxides with diazo esters.

In 2016, Samanta and coworkers reported a copper-catalyzed regioselective cascade alkylation and cyclo-condensation of quinoline *N*-oxides with diazo esters for accessing a new class of heteroarene ring-fused conjugated π-systems.[87] In this reaction, the copper not only serves as the catalyst for C–H activation, but also as Lewis acid to promote the following cyclization. The proposed mechanism starts with concerted metalation deprotonation, followed by metal carbene formation and migratory insertion. Protodemetalation gives the *ortho*-alkylated intermediate, followed by final cyclo-condensation to provide the desired products (Scheme 4.69).

With azo as the DG, cinnoline analogs can be constructed by Rh-catalyzed C–H activation strategy. In recent years, the groups of Lee,[88] Kim,[89] Lin and Yao[90] reported the Rh(III)-catalyzed cross-coupling of azo derivatives with diazo compounds (Scheme 4.70). These reactions allowed the efficient preparation of cinnoline derivatives from different diazo compounds including α-diazo esters and diazotized Meldrum's acid. In the

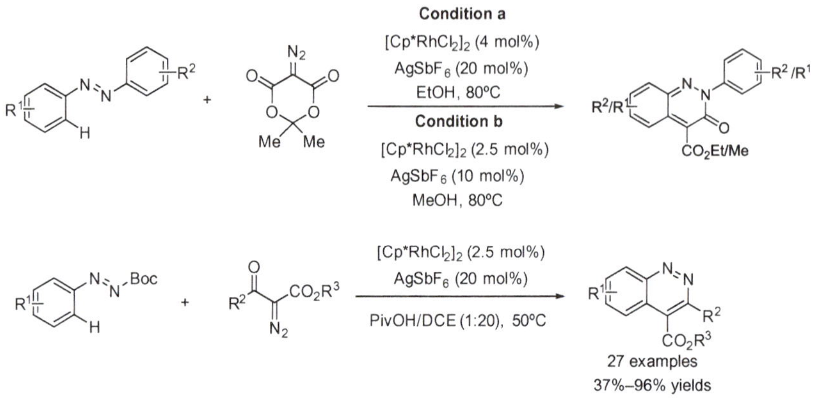

SCHEME 4.70 Synthesis of cinnoline derivatives via Rh-catalyzed C−H bond activation.

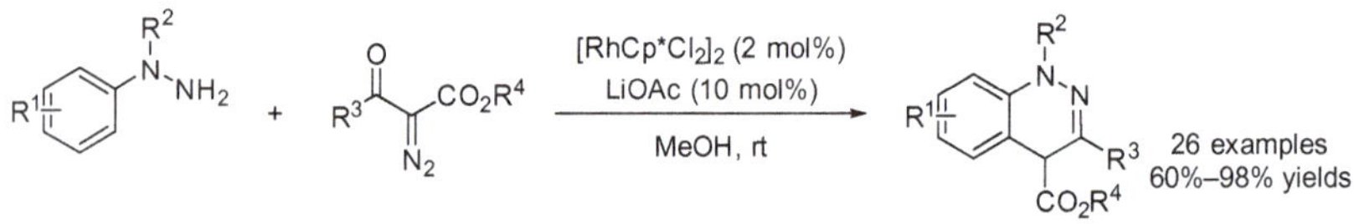

SCHEME 4.71 Rh(III)-catalyzed C−H activation of 1-alkyl-1-phenylhydrazines with α-diazo-β-ketoesters.

proposed reaction mechanism, the C−H bond activation is the initial step. Then rhodium carbene formation, migratory insertion, protonation, and condensation are successively followed, providing the final products. In another alternative process, the intramolecular 1,2-migratory insertion of the aryl group can also afford six-membered rhodacycles instead of metal carbene migratory insertion.

Apart from azo group, hydrazine can also act as DG for the synthesis of cinnoline scaffold. More recently, a unique Rh(III)-catalyzed C−H activation of 1-alkyl-1-phenylhydrazines with α-diazo-β-ketoester has been reported by Zhu and coworkers (Scheme 4.71).[91] In this transformation, C−H bond activation, metal carbene formation, migratory insertion, and intramolecular dehydration successively occurred under the optimized reaction conditions. This coupling reaction proceeds with mild and switchable reaction conditions as well as wide substituents tolerance. Moreover, only water and nitrogen gas are released as the byproducts, which makes this reaction an environmentally benign method for the construction of these important building blocks.

The construction of seven-membered heterocycles can also be achieved by this strategy. In 2016, Huang and coworkers developed a free-amine directed C−H activation of 2-aminobiaryls with diazo compounds, affording a series of azepinone derivatives in moderate to excellent yields (Scheme

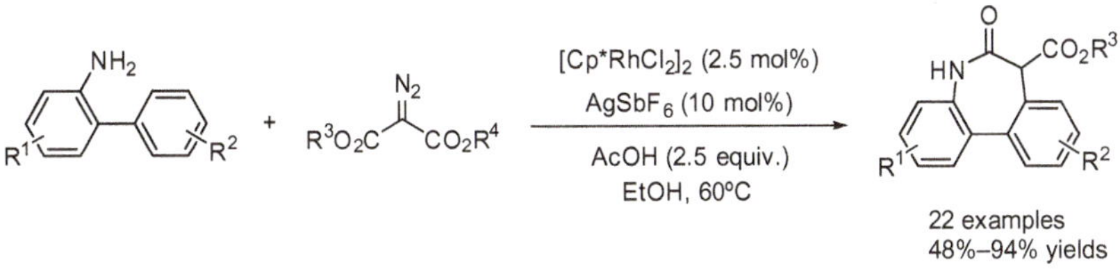

SCHEME 4.72 Cascade C–H functionalization/amidation reaction for synthesis of azepinone derivatives.

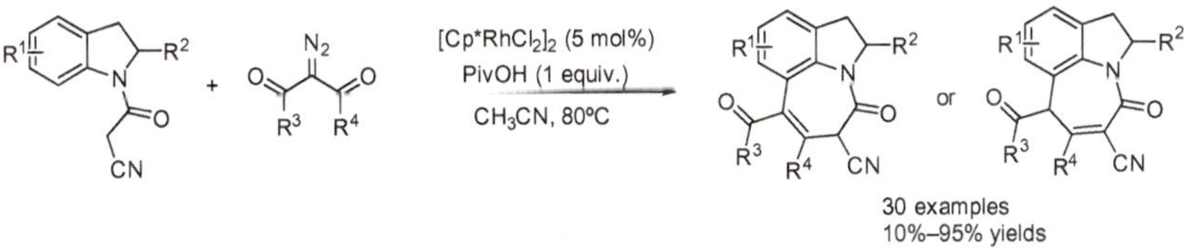

SCHEME 4.73 Rh(III)-catalyzed C–H activation of 3-(indolin-1-yl)-3-oxopropanenitriles with diazo compounds.

4.72).[92] This cascade transformation involves directed C–H activation, metal carbene formation, migratory insertion, protonation, and intramolecular amidation sequence. Since azepinones are important structural motifs in alkaloids and pharmaceuticals, this cascade reaction provides a straightforward and efficient method for accessing these distinctive heterocycles. The mild reaction conditions as well as broad substrate scope make the method practically useful in organic synthesis.

Subsequently, Wang and coworkers developed a synthetic route for the synthesis of azepino[3,2,1-*hi*]indole skeletons via Rh(III)-catalyzed C–H activations (Scheme 4.73).[93] They used 3-(indolin-1-yl)-3-oxopropanenitriles and diazo compounds as the substrates. This tandem reaction have combined aldol condensation with Rh(III)-carbene induced C–H alkylations. It should be mentioned that a fused tricyclic or tetracyclic structure containing five-, six-, and seven-membered rings can be constructed in this tandem reaction.

In addition to diazo substrates, pyridotriazoles are another entrance to carbene species. In 2015, Glorius, Strassert, and coworkers reported Rh(III)-catalyzed C–H bond functionalization of 2-arylpyridines by employing pyridotriazoles as carbene precursors, providing a direct access to π-conjugated polycyclic ring systems which can be used as fluorescent scaffolds for detecting metal ions.[94] Instead of being directly dediazonized by transition metals, pyridotriazoles initially transformed into their equilibrium form pyridyl-substituted diazo compounds (Scheme 4.74). After similar sequence including C–H metalation, metal carbene formation, migratory insertion, and protonation, *ortho*-alkylated products are generated. In contrast to direct condensation for cyclized products, {Cp*Rh(III)} activates ester carbonyl

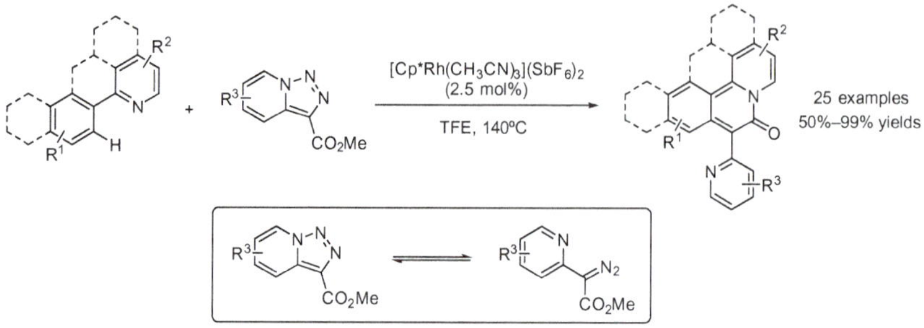

SCHEME 4.74 Rh(III)-catalyzed C–H bond functionalization of 2-arylpyridines with pyridotriazoles for direct access to π-conjugated polycyclic ring systems.

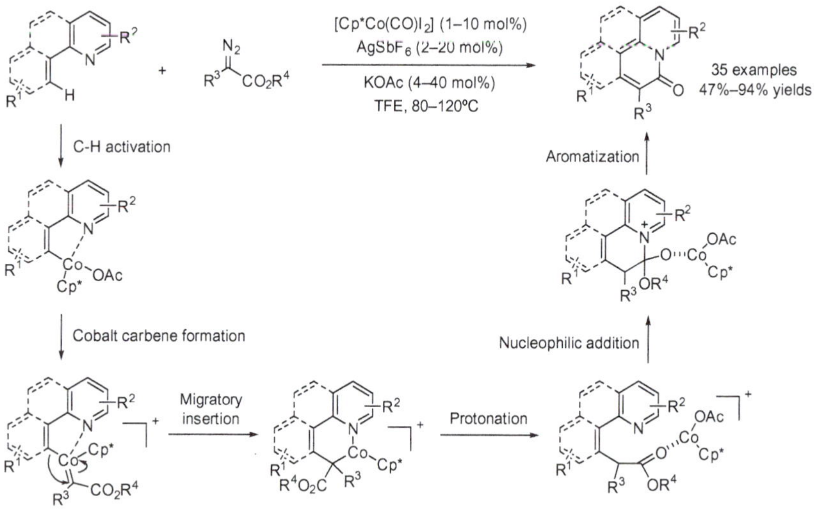

SCHEME 4.75 Co(III)-catalyzed C–H functionalization of *N*-heteroarylarenes with diazo compounds towards extended π-systems.

group as a Lewis acid for nucleophilic addition by pyridyl nitrogen atom. Notably, the authors also investigate the fluorescence properties of their products by taking advantage of the dual chelation ability.

Continuously, Glorius, Zhao, and coworkers demonstrated the Co(III)-catalyzed C–H functionalization of *N*-heteroarylarenes with diazo compounds, affording a new class of conjugated polycyclic hydrocarbons with tunable emission wavelengths (Scheme 4.75).[95] In this transformation, cobalt was not only used as a transition metal catalyst but also served as Lewis acid catalyst, which catalyzed C–H bond activations and promoted nucleophilic addition, respectively. The fluorescent properties of these conjugated polycyclic fused heterocycles were investigated. As extended π-systems, these

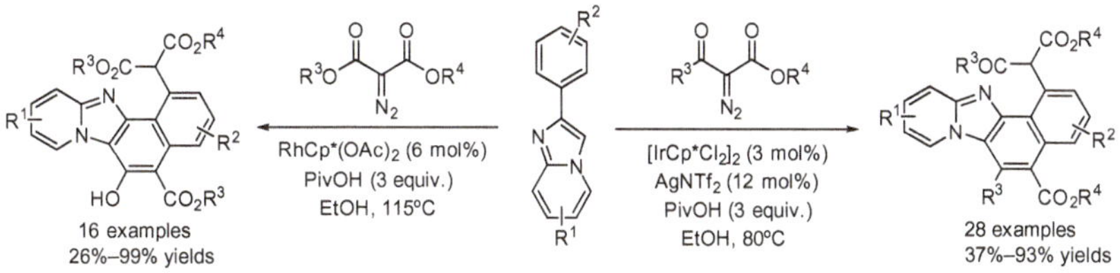

SCHEME 4.76 Rh(III)- and Ir(III)-catalyzed C–H activations of imidazo[1,2-*a*]pyridine with α-diazo esters.

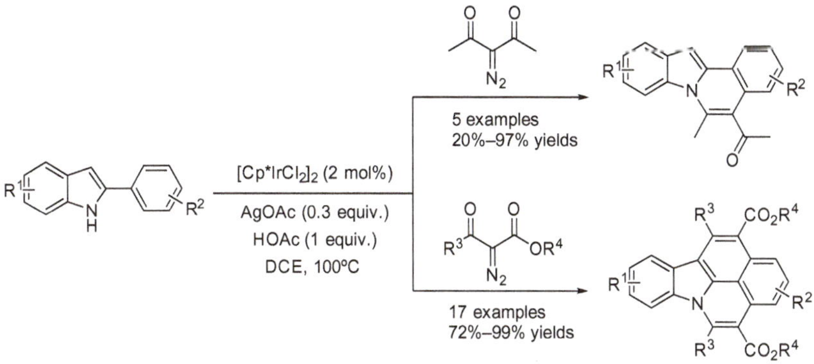

SCHEME 4.77 Ir(III)-catalyzed C–H activation of 2-phenyl indoles with α-diazo carbonyl compounds.

compounds showed broad and intense absorption and emission bands, and exhibited high extinction coefficients and large Stokes shifts.

Li and coworkers recently reported Rh(III)- and Ir(III)-catalyzed C–H activations of imidazo[1,2-α]pyridine with α-diazo esters, affording annulated products in moderate to good yields (Scheme 4.76).[96] Mechanistically, the reaction is initiated by C–H activation, followed by dialkylation and termination by intramolecular annulation. Rhodium and iridium complexes are both suitable catalysts in this transformation, but they played complementary role for different diazo substrates. To substantiate the reaction mechanism, an iridacycle was isolated and characterized by X-ray crystallography, which was found capable of catalyzing the reaction in efficient manner.

Almost at the same time, Dong and coworkers described a similar procedure by selecting 2-aryl indoles as the substrates (Scheme 4.77).[97] In contrast to Li's report, this reaction creates two new cyclic moieties with three C–C bonds and one C–N bond formation. Therefore the unique pentacyclic-fused carbazole structures are constructed in this catalytic cascade cyclization. Similarly, metal carbene formation and migratory insertion are involved in the catalytic cycle. More recently, Fan, Zhang, and coworkers introduced an efficient method for the synthesis of benzo[*a*]carbazoles or 6-aminobenzo[*a*]

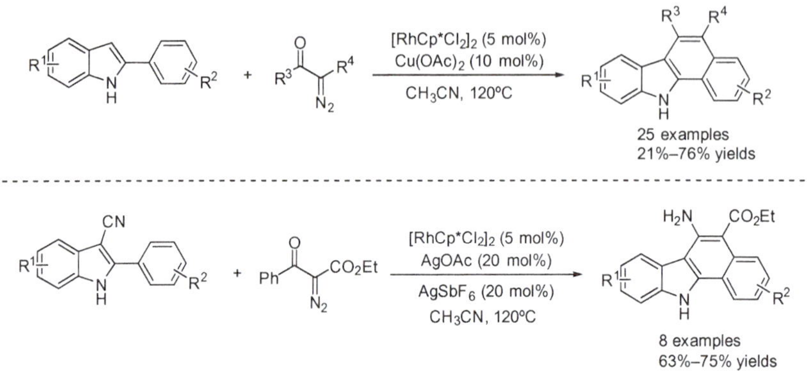

SCHEME 4.78 Rh(III)-catalyzed C–H annulation of 2-arylindoles or 2-arylindole-3-carbonitriles with α-diazo carbonyl compounds.

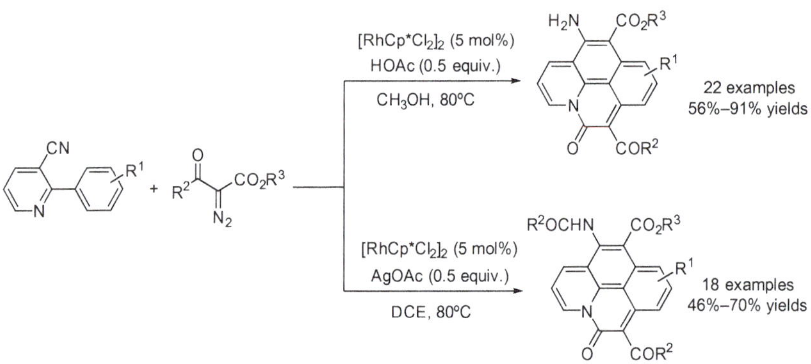

SCHEME 4.79 Rh(III)-catalyzed double C(sp^2)–H bond activation and annulation of 2-aryl-3-cyanopyridines with α-diazo carbonyl compounds.

carbazoles with 2-arylindoles or 2-arylindole-3-carbonitriles and α-diazo carbonyl compounds as substrates (Scheme 4.78).[98] In view of the easily obtainable substrates and good functional group tolerance, this reaction is a good approach for the regioselective synthesis of benzo[*a*]carbazoles or 6-aminobenzo[*a*]carbazoles. The authors also demonstrated that this C–H annulation could be successfully applied for the synthesis of key precursor toward antitumor agents.

In addition, the same group developed a Rh(III)-catalyzed double C(sp^2)–H bond activation and annulation of 2-aryl-3-cyanopyridines with α-diazo carbonyl compounds, affording naphthoquinolizinone derivatives in efficient manner (Scheme 4.79).[99] Migratory insertion was also postulated as the indispensable step for the formation of final product. This reaction is featured by good selectivity, which gives different naphthoquinolizinones by changing the solvent and additive. As a large π-system, azapyrenes are not

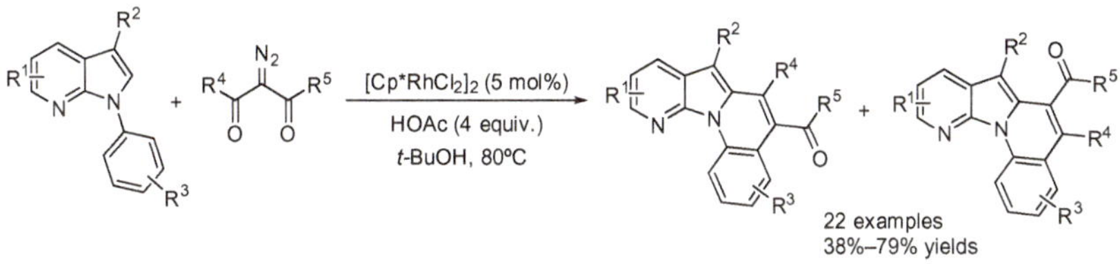

SCHEME 4.80 Rh(III)-catalyzed double C–H activation/cyclization of azaindoles with α-diazo carbonyl compounds.

easily obtained in one-pot with high efficiency. However, this new method provided an efficient and convenient protocol for these products.

Furthermore, rhodium(III)-catalyzed double C–H activation/cyclization by using azaindoles and α-diazo carbonyl compounds was realized, affording π-conjugated azaindole derivatives with high reaction selectivity (Scheme 4.80).[100] Similarly, the formation of Rh(III) carbene intermediate is involved in this reaction and migratory insertion is also regarded as indispensable step in catalytic cycle.

In 2017, Shang and coworkers used carboxylic acid as the DG in rhodium-catalyzed C–H activation with cyclic 2-diazo-1,3-diketones, which afforded a facile and efficient method for the preparation of isocoumarins from benzoic acid and diazo compound.[101] This reaction tolerates various functional groups and affords the corresponding isocoumarins in moderate to excellent yields. Similar to the previous discussed mechanism, the reaction pathway is proposed via C–H activation, metal carbene formation, migratory insertion, protonation, and condensation sequence (Scheme 4.81). This reaction provides a unique strategy for construction of isocoumarin scaffolds from readily available starting materials.

In 2014, Wang and coworkers developed a Rh(III)-catalyzed transannulation of cyclopropenes with *N*-phenoxyacetamides through C–H activation (Scheme 4.82).[102] With cyclopropene as a three-carbon unit, this transformation enabled the efficient synthesis of 2*H*-chromene under mild conditions. From the reaction mechanism, rhodium carbene migratory insertion is proposed as the essential step in one of the two possible catalytic cycles. Cyclopropene insertion and subsequent β-carbon elimination are also considered as alternative steps in hypothesized reaction mechanism. In addition, this mild protocol demonstrates broad substrates scope, providing an alternative route to access these useful building blocks. Recently, Xia[103] and Li[104] groups further investigated this transformation by DFT calculations. The results have indicated that cyclic Rh(V)-nitrenoid is likely the key intermediate.

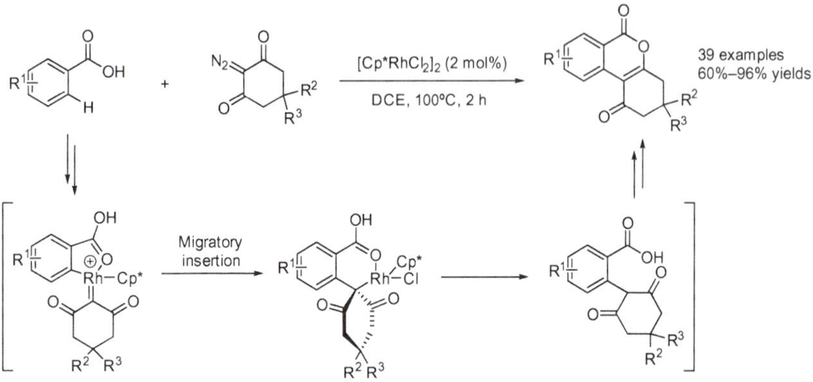

SCHEME 4.81 Rh(III)-catalyzed C–H activation of benzoic acids with cyclic 2-diazo-1,3-diketones.

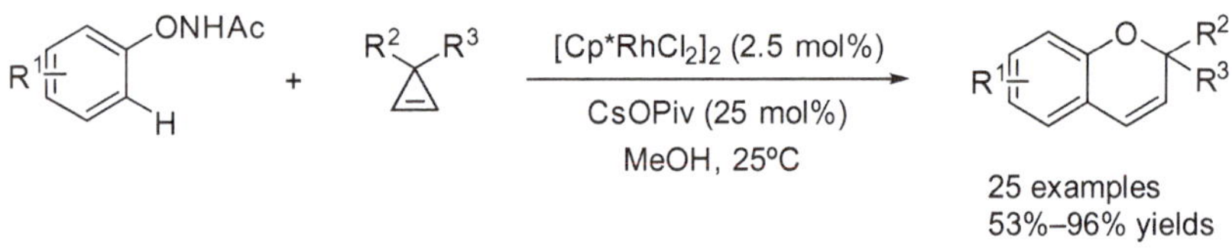

SCHEME 4.82 Rh(III)-catalyzed transannulation of cyclopropenes with *N*-phenoxyacetamides through C–H activation.

4.7 TANDEM CATALYSIS PROCESS

Palladium complexes are the most exploited catalysts in carbene-based cross-coupling reactions. In this regard, a series of transformations have been devised and the reactions experienced fruitful developments in the past decade.[3a,3b,3d–g] In order to achieve step economy and high efficiency, palladium complexes have been introduced for the synthesis of heterocycles based on metal carbene migratory insertion, which allows for the construction of multiple C–C or C-heteroatom bonds in one-pot fashion. For general process, oxidative addition, metal carbene formation, migratory insertion, and β-hydride elimination are similar to the aforementioned reactions and migratory insertion is also considered as the key step in this type of reactions. In contrast, auto-tandem or tandem-catalyzed processes involving carbene process has been devised in one-pot reaction, leading to the cyclized products (Scheme 4.83). A wide array of heterocycles can be synthesized in high efficiency.

In 2011, Barluenga, Valdés, and coworkers invoked a palladium-catalyzed cross-coupling reaction of *N*-tosylhydrazones derived from β-aminoketones with *o*-dihalobenzenes (Scheme 4.84).[105] In this reaction, C = C double bond was initially formed through a sequence involving

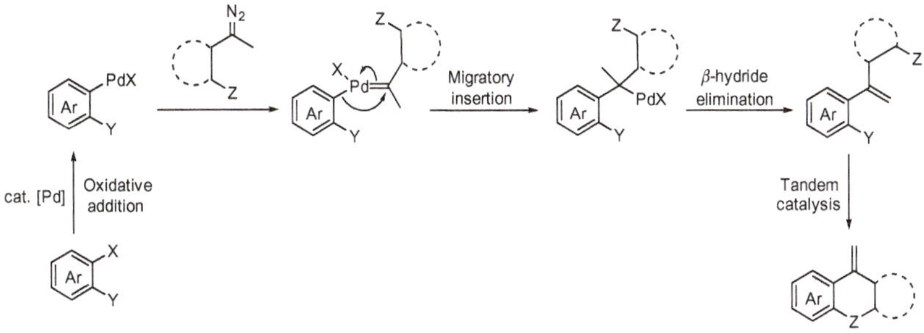

SCHEME 4.83 General process for the synthesis of heterocycles through tandem catalysis.

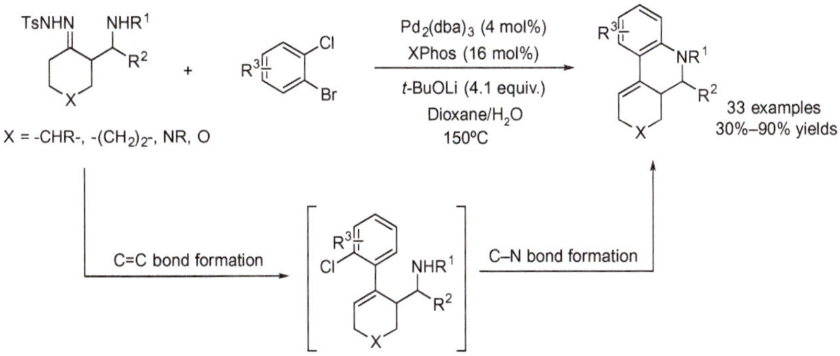

SCHEME 4.84 Auto-tandem Pd-catalyzed cross-coupling reaction of *N*-tosylhydrazones derived from β-aminoketones with *o*-dihalobenzenes.

oxidative addition, metal carbene formation, migratory insertion, and β-hydride elimination. Palladium-catalyzed intramolecular C–N bond formation successively occurred, affording substituted tetrahydroquinolines in good yields. Notably, this tandem reaction generated one C = C double bond and one C–N single bond, resulting in hetero-cyclization in the final products. In addition, enantiomerically enriched *N*-tosylhydrazones, which derived from asymmetric organocatalyzed Mannich reactions, were also successfully employed as suitable substrates in the synthesis of corresponding products with high enantiomeric excess.

Afterward, another auto-tandem reaction for the synthesis of acridines was developed by Wang and coworkers (Scheme 4.85).[106] Similarly, the C = C double bond was constructed through Pd-catalyzed cross-coupling reaction of *N*-tosylhydrazones with aryl halides, while the C–N single bond was formed by Buchwald and Hartwig amination. This approach provides a new method for the synthesis of acridine derivatives with wide substrates scope. The one-pot reaction also expanded the synthetic utility of this interesting transformation.

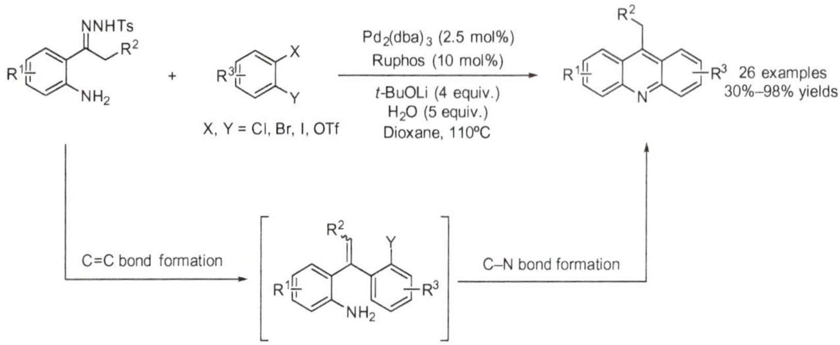

SCHEME 4.85 Auto-tandem Pd-catalyzed synthesis of acridine derivatives.

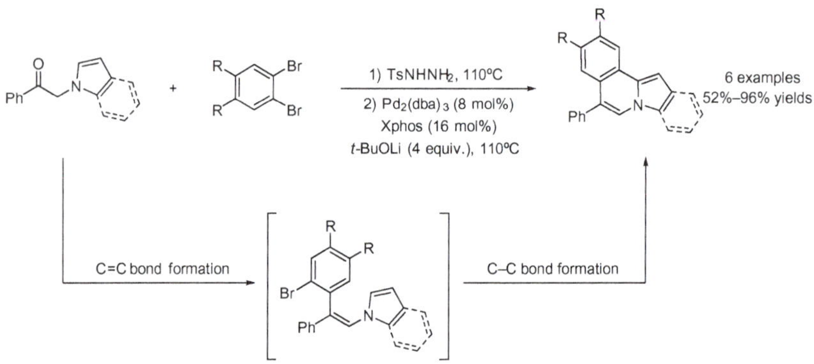

SCHEME 4.86 Palladium-catalyzed auto-tandem reaction to pyrrolo[2,1-*a*]isoquinoline.

Subsequently, Valdés and coworker reported an auto-tandem palladium-catalyzed synthesis of pyrrolo[2,1-*a*]isoquinoline derivatives by employing *N*-tosylhydrazones and 1,2-dibromobenzenes as the substrates (Scheme 4.86).[107] In contrast to C = C/C−N bond formation, this auto-tandem reaction is featured by the construction of C = C and C−C bonds in a simple manner. The *ortho* halogen substituent provides a platform for intramolecular palladium-catalyzed C−H arylation. Since pyrrolo[2,1-*a*]isoquinoline skeletons are not easily obtained or usually require stepwise processes, this auto-tandem reaction provided a modular approach to construct these structurally complex compounds.

In 2017, the same group reported another palladium-catalyzed auto-tandem reaction of *N*-tosylhydrazones with dibromides which contained a benzylic bromide and an aryl bromide in different phenyls (Scheme 4.87).[108] The reaction was proposed to occur via two consecutive processes including carbene cross-coupling and intramolecular Heck reaction. A mechanistic study by exchanging the reactive functional groups of two substrates

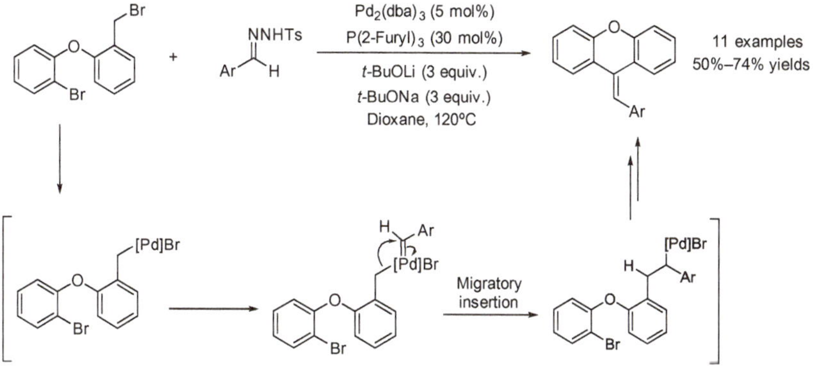

SCHEME 4.87 Palladium-catalyzed auto-tandem reaction of *N*-tosylhydrazones with dibromides.

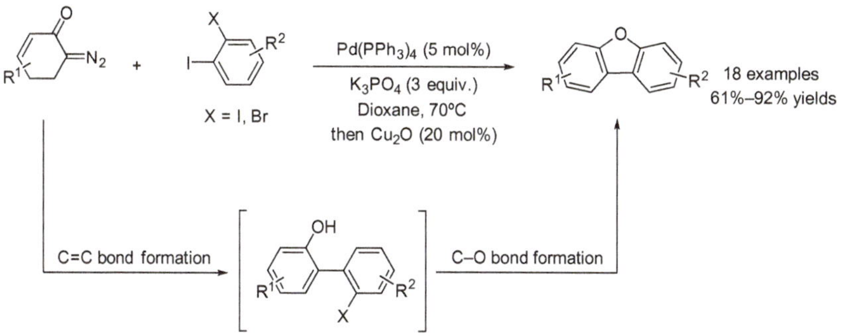

SCHEME 4.88 Tandem Pd/Cu-catalyzed protocol for the synthesis of dibenzofuran motifs.

supported that carbene cross-coupling reaction proceeded before intramolecular Heck reaction. This palladium-catalyzed auto-tandem reaction has a broad scope for both coupling partners. Thus 9-methylene-9*H*-fluorenes, 9-methylene-9*H*-xanthenes, 9-methylene-9,10-dihydroacridines could all be obtained in good to excellent yields. In addition, the heterocyclic *N*-tosylhydrazones can be transformed to pyrroloisoquinoline and indoloisoquinoline derivatives in 52% and 54% yields, respectively. The variability of substrates demonstrates that this reaction is a suitable synthetic method for these carbo- and heterocyclic structures. Moreover, the 9-benzylidene-9,10-dihydroacridines could also be prepared with *N*,*N*-diarylaniline and benzyl bromides as starting material.

Recently, Zhai and coworkers have reported a unique tandem Pd/Cu-catalyzed protocol for convenient synthesis of dibenzofuran motifs from 6-diazo-2-cyclohexenone and *ortho*-haloiodobenzene (Scheme 4.88).[109] Mechanistically, the reaction has been assumed to proceed through Pd-catalyzed cross-coupling/aromatization, followed by a Cu-catalyzed Ullmann coupling. This tandem reaction creates one C = C double bond and one

C–O single bond with high efficiency in one-pot fashion. In addition, this mild reaction exhibits broad substrate scope and high yields.

4.8 CONCLUSION AND OUTLOOK

The transition-metal-catalyzed carbene-involved cross-coupling reactions have been growing rapidly in the last decades. Diazo compounds including α-diazo esters and *N*-tosylhydrazones are recognized as unique cross-coupling partners in these novel cross-coupling reactions. The various examples based on these strategies have indicated that migratory insertion is the key step in well-established catalytic process. Although being a newly emerging field, remarkable applications especially in the synthesis of heterocycles have been actively pursued. As summarized in this chapter, six types based on cyclization sequence have been independently discussed. A diverse set of *N*- or *O*-containing heterocycles have been ingeniously constructed in efficient manner. Several transition metals such as Pd, Cu, Rh, Ir, and Co are surveyed in the synthesis of heterocycles based on metal carbene migratory insertion. Especially, rhodium complexes are the most used catalysts, which incorporated C–H bond activations with carbene migratory insertion for the synthesis of heterocycles. In addition to diazo compounds, conjugated enynones are a type of alternative carbene precursors, which enable the construction of functionalized furan derivatives.

Despite the myriad recent advances, it is expected that the scope of both coupling partners including functionalized substrates and carbene precursors can still be expanded to reach further diversified structures and useful heterocyclic compounds. Besides, in spite of a great deal of examples, the development of other cyclization mode is highly desirable and can further enhance the application of this novel methodology. In addition, in many cases the reaction mechanism is not suitably supported by experiments. Detailed mechanistic studies will provide useful insights into the reaction and thus open up possibilities to design new cascade cyclization reactions. The exploration of enantioselective synthesis of heterocycles based on metal carbene migratory insertion would be highly demanded, but this direction has not been actively pursued.

Finally, it is foreseeable that in the near future one will witness the applications of these methods in total synthesis of natural products and other functional molecules. Thus these newly emerging methods will evolve into useful and powerful tools for the construction of both simple heterocycles and complicated natural products.

REFERENCES

1. (a) Eicher, T.; Hauptmann, S.; Speicher, A. *The Chemistry of Heterocycles.*, 2nd ed; Wiley-VCH Verlag GmbH & Co, Weinheim, 2003. (b) Khan, M. T. H., Ed. *Top. Heterocycl.*

Chem. **2007,** *vol,* 7–11. (c) Katritzky, A. R.; Ramsden, C. A.; Scriven, E. F. V.; Taylor, R. J. K. *Comprehensive Heterocyclic Chemistry III*, Pergamon: Oxford, U.K, vol. 2008 1–13.

2. (a) Zhang, B.; Studer, A. Recent Advances in the Synthesis of Nitrogen Heterocycles via Radical Cascade Reactions using Isonitriles as Radical Acceptors. *Chem. Soc. Rev.* **2015,** *44*, 3505–3521. (b) Guo, X.-X.; Gu, D.-W.; Wu, Z.; Zhang, W. Copper-Catalyzed C-H Functionalization Reactions: Efficient Synthesis of Heterocycles. *Chem. Rev.* **2015,** *115*, 1622–1651. (c) Kurahashi, T.; Matsubara, S. Nickel-Catalyzed Reactions Directed toward the Formation of Heterocycles. *Acc. Chem. Res.* **2015,** *48*, 1703–1716. (d) Properzi, R.; Marcantoni, E. Construction of Heterocyclic Structures by Trivalent Cerium Salts Promoted Bond Forming Reactions. *Chem. Soc. Rev.* **2014,** *43*, 779–791. (e) Wu, X.-F.; Neumann, H.; Beller, M. Synthesis of Heterocycles via Palladium-Catalyzed Carbonylations. *Chem. Rev.* **2013,** *113*, 1–35.
3. (a) Barluenga, J.; Valdés, C. Tosylhydrazones: New Uses for Classic Reagents in Palladium-Catalyzed Cross-Coupling and Metal-Free Reactions. *Angew. Chem. Int. Ed.* **2011,** *50*, 7486–7500. (b) Zhang, Y.; Wang, J. Recent Developments in Pd-Catalyzed Reactions of Diazo Compounds. *Eur. J. Org. Chem.* **2011,** 1015–1026. (c) Zhao, X.; Zhang, Y.; Wang, J. Recent Developments in Copper-Catalyzed Reactions of Diazo Compounds. *Chem. Commun.* **2012,** *48*, 10162–10173. (d) Shao, Z.; Zhang, H. *N*-Tosylhydrazones: Versatile Reagents for Metal-Catalyzed and Metal-Free Cross-Coupling Reactions. *Chem. Soc. Rev.* **2012,** *41*, 560–572. (e) Xiao, Q.; Zhang, Y.; Wang, J. Diazo Compounds and *N*-Tosylhydrazones: Novel Cross-Coupling Partners in Transition-Metal-Catalyzed Reactions. *Acc. Chem. Res.* **2013,** *46*, 236–247. (f) Xia, Y.; Zhang, Y.; Wang, J. Catalytic Cascade Reactions Involving Metal Carbene Migratory Insertion. *ACS Catal.* **2013,** *3*, 2586–2598. (g) Liu, Z.; Wang, J. Cross-Coupling Reactions Involving Metal Carbene: From C = C/C – C Bond Formation to C-H Bond Functionalization. *J. Org. Chem.* **2013,** *78*, 10024–10030. (h) Hu, F.; Xia, Y.; Ma, C.; Zhang, Y.; Wang, J. C-H Bond Functionalization based on Metal Carbene Migratory Insertion. *Chem. Commun.* **2015,** *51*, 7986–7995. (i) Xia, Y.; Wang, J. *N*-Tosylhydrazones: Versatile Synthons in the Construction of Cyclic Compounds. *Chem. Soc. Rev.* **2017,** *46*, 2306–2362. (j) Xia, Y.; Qiu, D.; Wang, J. Transition-Metal-Catalyzed Cross-Couplings Through Carbene Migratory Insertion. *Chemical Reviews* **2017,** *117*, 13810–13889.
4. (a) Miki, K.; Uemura, S.; Ohe, K. Transition Metal-Catalyzed Reactions using Alkynes as Precursors of Carbene and Vinylidene Complexes. *Chem. Lett.* **2005,** *34*, 1068–1073. (b) Kusama, H.; Iwasawa, N. Cycloaddition Reactions of Transition Metal-Containing Benzopyrylium and Related Zwitterionic Intermediates. *Chem. Lett.* **2006,** *35*, 1082–1087. (c) Ohe, K.; Miki, K. Development of Catalytic Carbene Transfer Reactions using Alkynes as a Source of Carbenes. *Yuki Gosei Kagaku Kyokaishi* **2009,** *67*, 1161–1171. (d) Fürstner, A.; Davies, P. W. Catalytic Carbophilic Activation: Catalysis by Platinum and Gold π acids. *Angew. Chem. Int. Ed.* **2007,** *46*, 3410–3449. (e) Jia, M.; Ma, S. New Approaches to the Synthesis of Metal Carbenes. *Angew. Chem. Int. Ed.* **2016,** *55*, 9134–9166.
5. Liu, X.; Ma, X.; Huang, Y.; Gu, Z. Pd-Catalyzed Heck-Type Cascade Reactions with N-Tosylhydrazones: An Efficient Way to Alkenes via in Situ Generated Alkylpalladium. *Org. Lett.* **2013,** *15*, 4814–4817.
6. Liu, Z.; Xia, Y.; Zhou, S.; Wang, L.; Zhang, Y.; Wang, J. Pd-Catalyzed Cyclization and Carbene Migratory Insertion: New Approach to 3-Vinylindoles and 3-Vinylbenzofurans. *Org. Lett.* **2013,** *15*, 5032–5035.

7. Gao, Y.; Xiong, W.; Chen, H.; Wu, W.; Peng, J.; Gao, Y.; Jiang, H. Pd-Catalyzed Highly Regio- and Stereoselective Formation of C-C Double Bonds: An Efficient Method for the Synthesis of Benzofuran-, Dihydrobenzofuran-, and Indoline-Containing Alkenes. *J. Org. Chem.* **2015,** *80*, 7456–7467.
8. Hu, Z.; Luo, S.; Zhu, Q. Aminopalladation-Triggered Carbene Insertion Reaction: Synthesis of 2-(1*H*-indol-3-yl)Acetates. *Adv. Synth. Catal.* **2015,** *357*, 1060–1064.
9. Wu, X.-X.; Zhou, P.-X.; Wang, L.-J.; Xu, P.-F.; Liang, Y.-M. Synthesis of Polycyclic Substituted Vinylarenes *via* a One-Pot Intramolecular Aryl Alkylation-*N*-tosylhydrazone Insertion Reaction. *Chem. Commun.* **2014,** *50*, 3882–3884.
10. Zhang, Z.; Zhou, Q.; Ye, F.; Xia, Y.; Wu, G.; Hossain, M. L.; Zhang, Y.; Wang, J. Copper (I)-Catalyzed Three-Component Coupling of *N*-Tosylhydrazones, Alkynes and Azides: Synthesis of Trisubstituted 1,2,3-Triazoles. *Adv. Synth. Catal.* **2015,** *357*, 2277–2286.
11. Zhang, Z.; Zhou, Q.; Yu, W.; Li, T.; Zhang, Y.; Wang, J. Cu(I)-Catalyzed Three-Component Coupling of Trifluoromethyl Ketone *N*-Tosylhydrazones, Alkynes and Azides: Synthesis of Difluoromethylene Substituted 1,2,3-Triazoles. *Chin. J. Chem.* **2017,** *35*, 387–391.
12. Guo, S.; Yuan, K.; Gu, M.; Lin, A.; Yao, H. Rh(III)-Catalyzed Cascade Annulation/C-H Activation of *o*-Ethynylanilines with Diazo Compounds: One-Pot Synthesis of Benzo[a]carbazoles *via* 1,4-Rhodium Migration. *Org. Lett.* **2016,** *18*, 5236–5239.
13. (a) Miki, K.; Nishino, F.; Ohe, K.; Uemura, S. Novel Approach for Catalytic Cyclopropanation of Alkenes *via* (2-Furyl)carbene Complexes from 1-Benzoyl-*cis*-1-buten-3-yne. *J. Am. Chem. Soc.* **2002,** *124*, 5260–5261. (b) Miki, K.; Yokoi, T.; Nishino, F.; Kato, Y.; Washitake, Y.; Ohe, K.; Uemura, S. Catalytic Cyclopropanation of Alkenes *via* (2-Furyl)carbene Complexes from 1-Benzoyl-cis-1-buten-3-yne with Transition Metal Compounds. *J. Org. Chem.* **2004,** *69*, 1557–1564. (c) Vicente, R.; González, J.; Riesgo, L.; González, J.; López, L. A. Catalytic Generation of Zinc Carbenes from Alkynes: Zinc-Catalyzed Cyclopropanation and Si-H Bond Insertion Reactions. *Angew. Chem. Int. Ed.* **2012,** *51*, 8063–8067. (d) González, J.; González, J.; Pérez-Calleja, C.; López, L. A.; Vicente, R. Zinc-Catalyzed Synthesis of Functionalized Furans and Triarylmethanes from Enynones and Alcohols or Azoles: Dual X-H Bond Activation by Zinc. *Angew. Chem. Int. Ed.* **2013,** *52*, 5853–5857.
14. Xia, Y.; Qu, S.; Xiao, Q.; Wang, Z.-X.; Qu, P.; Chen, L.; Liu, Z.; Tian, L.; Huang, Z.; Zhang, Y.; Wang, J. Palladium-Catalyzed Carbene Migratory Insertion using Conjugated Ene-yne-ketones as Carbene Precursors. *J. Am. Chem. Soc.* **2013,** *135*, 13502–13511.
15. Xia, Y.; Ge, R.; Chen, L.; Liu, Z.; Xiao, Q.; Zhang, Y.; Wang, J. Palladium-Catalyzed Oxidative Cross-Coupling of Conjugated Enynones with Organoboronic Acids. *J. Org. Chem.* **2015,** *80*, 7856–7864.
16. Xia, Y.; Chen, L.; Qu, P.; Ji, G.; Feng, S.; Xiao, Q.; Zhang, Y.; Wang, J. Rh(I)-Catalyzed Coupling of Conjugated Enynones with Arylboronic Acids: Synthesis of Furyl-Containing Triarylmethanes. *J. Org. Chem.* **2016,** *81*, 10484–10490.
17. Xia, Y.; Liu, Z.; Ge, R.; Xiao, Q.; Zhang, Y.; Wang, J. Pd-Catalyzed Cross-Coupling of Terminal Alkynes with Ene-yne-ketones: Access to Conjugated Enynes via Metal Carbene Migratory Insertion. *Chem. Commun.* **2015,** *51*, 11233–11235.
18. Hu, F.; Xia, Y.; Ma, C.; Zhang, Y.; Wang, J. Cu(I)-Catalyzed Cross-Coupling of Conjugated Ene-yne-ketones and Terminal Alkynes: Synthesis of Furan-substituted Allenes. *Org. Lett.* **2014,** *16*, 4082–4085.

19. Hu, F.; Xia, Y.; Ma, C.; Zhang, Y.; Wang, J. Cu(I)-Catalyzed Synthesis of Furan-Substituted Allenes by Use of Conjugated Ene-yne Ketones as Carbene Precursors. *J. Org. Chem.* **2016,** *81*, 3275–3285.
20. Zhu, C.; Feng, C.; Yamane, M. Pd/Cu Cooperative Catalysis: An Efficient Synthesis of (3-Isoindazolyl)allenes via Cross-Coupling of 2-Alkynyl Azobenzenes and Terminal Alkynes. *Chem. Commun.* **2017,** *53*, 2606–2609.
21. Hong, S. Y.; Jeong, J.; Chang, S. [4 + 2] or [4 + 1] Annulation: Changing the Reaction Pathway of a Rhodium-Catalyzed Process by Tuning the Cp Ligand. *Angew. Chem. Int. Ed.* **2017,** *56*, 2408–2412.
22. Xia, Y.; Xia, Y.; Ge, R.; Liu, Z.; Xiao, Q.; Zhang, Y.; Wang, J. Oxidative Cross-Coupling of Allenyl Ketones and Organoboronic Acids: Expeditious Synthesis of Highly Substituted Furans. *Angew. Chem. Int. Ed.* **2014,** *53*, 3917–3921.
23. (a) Shang, R.; Ilies, L.; Nakamura, E. Iron-Catalyzed C-H Bond Activation. *Chem. Rev.* **2017,** *117*, 9086–9139. (b) Wei, Y.; Hu, P.; Zhang, M.; Su, W. Metal-Catalyzed Decarboxylative C–H Functionalization. *Chem. Rev.* **2017,** *117*, 8864–8907. (c) Newton, C. G.; Wang, S.-G.; Oliveira, C. C.; Cramer, N. Catalytic Enantioselective Transformations Involving C-H bond Cleavage by Transition-Metal Complexes. *Chem. Rev.* **2017,** *117*, 8908–8976. (d) Dong, Z.; Ren, Z.; Thompson, S. J.; Xu, Y.; Dong, G. Transition-Metal-Catalyzed C-H Alkylation using Alkenes. *Chem. Rev.* **2017,** *117*, 9333–9403. (e) Yang, Y.; Lan, J.; You, J. Oxidative C-H/C-H Coupling Reactions between Two (Hetero)arenes. *Chem. Rev.* **2017,** *117*, 8787–8863. (f) Kim, D.-S.; Park, W.-J.; Jun, C.-H. Metal-Organic Cooperative Catalysis in C-H and C-C Bond Activation. *Chem. Rev.* **2017,** *117*, 8977–9015. (g) Park, Y.; Kim, Y.; Chang, S. Transition Metal-Catalyzed C-H Amination: Scope, Mechanism, and Applications. *Chem. Rev.* **2017,** *117*, 9247–9301. (h) Hummel, J. R.; Boerth, J. A.; Ellman, J. A. Transition-Metal-Catalyzed C-H Bond Addition to Carbonyls, Imines, and Related Polarized π Bonds. *Chem. Rev.* **2017,** *117*, 9163–9227. (i) He, J.; Wasa, M.; Chan, K. S. L.; Shao, Q.; Yu, J.-Q. Palladium-Catalyzed Transformations of Alkyl C-H Bonds. *Chem. Rev.* **2017,** *117*, 8754–8786. (j) Labinger, J. A. Platinum-Catalyzed C-H Functionalization. *Chem. Rev.* **2017,** *117*, 8483–8496. (k) Gensch, T.; Hopkinson, M. N.; Glorius, F.; Wencel-Delord, J. Mild Metal-Catalyzed C-H Activation: Examples and Concepts. *Chem. Soc. Rev.* **2016,** *45*, 2900–2936. (l) C-H Bond Activation and Catalytic Functionalization I. In *;* Dixneuf, P. H.; Doucet, H., Eds. , vol. 55; Springer International Publishing, Cham, 2016. (m) C-H Bond Activation and Catalytic Functionalization II. In *;* Dixneuf, P. H.; Doucet, H., Eds. , vol. 56; Springer International Publishing, Cham, 2016. (n) Gulías, M.; Mascareñas, J. L. Metal-Catalyzed Annulations through Activation and Cleavage of C-H Bonds. *Angew. Chem. Int. Ed.* **2016,** *55*, 11000–11019.
24. For pioneering report Chan, W.-W.; Lo, S.-F.; Zhou, Z.; Yu, W.-Y. Rh-Catalyzed Intermolecular Carbenoid Functionalization of Aromatic C-H Bonds by α-Diazomalonates. *J. Am. Chem. Soc.* **2012,** *134*, 13565–13568.
25. Hyster, T. K.; Ruhl, K. E.; Rovis, T. A Coupling of Benzamides and Donor/Acceptor Diazo Compounds to form γ-Lactams *via* Rh(III)-catalyzed C-H Activation. *J. Am. Chem. Soc.* **2013,** *135*, 5364–5367.
26. Lam, H.-W.; Man, K.-Y.; Chan, W.-W.; Zhou, Z.; Yu, W.-Y. Rhodium(III)-Catalyzed Formal Oxidative [4 + 1] Cycloaddition of Benzohydroxamic Acids and α-Diazoesters. A Facile Synthesis of Functionalized Benzolactams. *Org. Biomol. Chem.* **2014,** *12*, 4112–4116.

27. Ye, B.; Cramer, N. Asymmetric Synthesis of Isoindolones by Chiral Cyclopentadienyl-rhodium(III)-catalyzed C-H Functionalizations. *Angew. Chem. Int. Ed.* **2014,** *53*, 7896–7899.
28. Cui, S.; Zhang, Y.; Wang, D.; Wu, Q. Rh(III)-Catalyzed C-H Activation/[4 + 3] Cycloaddition of Benzamides and Vinylcarbenoids: Facile Synthesis of Azepinones. *Chem. Sci* **2013,** *4*, 3912–3916.
29. Zhang, Y.; Zheng, J.; Cui, S. Rh(III)-Catalyzed C-H Activation/Cyclization of Indoles and Pyrroles: Divergent Synthesis of Heterocycles. *J. Org. Chem.* **2014,** *79*, 6490–6500.
30. Zhang, Y.; Wang, D.; Cui, S. Facile Synthesis of Isoindolinones *via* Rh(III)-Catalyzed One-Pot Reaction of Benzamides, Ketones, and Hydrazines. *Org. Lett.* **2015,** *17*, 2494–2497.
31. Wu, J.; Wang, D.; Wan, Y.; Ma, C. Rhodium-Catalyzed Tunable Oxidative Cyclization toward the Selective Synthesis of α-Pyrones and Furans. *Chem. Commun.* **2016,** *52*, 1661–1664.
32. Khanna, A.; Maung, C.; Johnson, K. R.; Luong, T. T.; Van Vranken, D. L. Carbenylative Amination with *N*-Tosylhydrazones. *Org. Lett.* **2012,** *14*, 3233–3235.
33. Zhou, P.-X.; Zhou, Z.-Z.; Chen, Z.-S.; Ye, Y.-Y.; Zhao, L.-B.; Yang, Y.-F.; Xia, X.-F.; Luo, J.-Y.; Liang, Y.-M. Palladium-Catalyzed Insertion of α-Diazocarbonyl Compounds for the Synthesis of Cyclic Amino Esters. *Chem. Commun.* **2013,** *49*, 561–563.
34. Zhou, P.-X.; Luo, J.-Y.; Zhao, L.-B.; Ye, Y.-Y.; Liang, Y.-M. Palladium-Catalyzed Insertion of *N*-Tosylhydrazones for the Synthesis of Isoindolines. *Chem. Commun.* **2013,** *49*, 3254–3256.
35. Xia, Y.; Xia, Y.; Zhang, Y.; Wang, J. Palladium-Catalyzed Coupling of *N*-Tosylhydrazones and β-Bromostyrene Derivatives: New Approach to 2*H*-Chromenes. *Org. Biomol. Chem.* **2014,** *12*, 9333–9336.
36. Shang, X. S.; Li, N. T.; Siyang, H. X.; Liu, P. N. Palladium-Catalyzed Tandem Carbene Migratory Insertion and Intramolecular Cyclization: Synthesis of Chromeno[4,3-*b*]chromene Compounds. *J. Org. Chem.* **2015,** *80*, 4808–4815.
37. Paraja, M.; Pérez-Aguilar, M. C.; Valdés, C. The Pd-Catalyzed Synthesis of Benzofused Carbo- and Heterocycles through Carbene Migratory Insertion/Carbopalladation Cascades with Tosylhydrazones. *Chem. Commun.* **2015,** *51*, 16241–16243.
38. Paraja, M.; Valdés, C. Pd-Catalyzed Cascade Reactions between *o*-Iodo-*N*-alkenylanilines and Tosylhydrazones: Novel Approaches to the Synthesis of Polysubstituted Indoles and 1,4-Dihydroquinolines. *Chem. Commun.* **2016,** *52*, 6312–6315.
39. Yin, B.; Zhang, X.; Liu, J.; Li, X.; Jiang, H. Practical Access to Spiroacetal Enol Ethers *via* Nucleophilic Dearomatization of 2-Furylmethylenepalladium Halides Generated by Pd-Catalyzed Coupling of Furfural Tosylhydrazones with Aryl Halides. *Chem. Commun.* **2014,** *50*, 8113–8116.
40. Yin, B.; Zhang, X.; Zhang, X.; Peng, H.; Zhou, W.; Liu, B.; Jiang, H. Access to Polysubstituted Indoles or Benzothiophenes *via* Palladium-Catalyzed Cross-Coupling of Furfural Tosylhydrazones with 2-Iodoanilines or 2-Iodothiophenols. *Chem. Commun.* **2015,** *51*, 6126–6129.
41. Zhou, L.; Shi, Y.; Xiao, Q.; Liu, Y.; Ye, F.; Zhang, Y.; Wang, J. CuBr-Catalyzed Coupling of *N*-Tosylhydrazones and Terminal Alkynes: Synthesis of Benzofurans and Indoles. *Org. Lett.* **2011,** *13*, 968–971.
42. Xiao, T.; Dong, X.; Zhou, L. Benzofuran and Indole Synthesis *via* Cu(I)-Catalyzed Coupling of *N*-Tosylhydrazone and *o*-Hydroxy or *o*-Amino Phenylacetylene. *Org. Biomol. Chem.* **2013,** *11*, 1490–1497.

43. Liu, G.; Xu, G.; Li, J.; Ding, D.; Sun, J. Copper-Catalyzed Annulation of α-Substituted Diazoacetates with 2-Ethynylanilines: The Direct Synthesis of C2-functionalized Indoles. *Org. Biomol. Chem.* **2014,** *12*, 1387–1390.
44. Hossain, M. L.; Ye, F.; Zhang, Y.; Wang, J. Cu(I)-Catalyzed Reaction of Diazo Compounds with Terminal Alkynes: A Direct Synthesis of Trisubstituted Furans. *Tetrahedron* **2014,** *70*, 6957–6962.
45. Hossain, M. L.; Wang, K.; Ye, F.; Zhang, Y.; Wang, J. Cu(I)-Catalyzed Cascade Reaction of *N*-Tosylhydrazones with 3-Butyn-1-ol: A New Synthesis of Tetrahydrofurans. *Chin. J. Catal.* **2017,** *38*, 115–122.
46. Liu, K.; Zhu, C.; Min, J.; Peng, S.; Xu, G.; Sun, J. Stereodivergent Synthesis of N-heterocycles by Catalyst-Controlled, Activity-Directed Tandem Annulation of Diazo Compounds with Amino Alkynes. *Angew. Chem. Int. Ed.* **2015,** *54*, 12962–12967.
47. Min, J.; Xu, G.; Sun, J. A controlled selective synthesis of dihydropyrans through tandem reaction of alkynes with diazo compounds. *Chem. Commun.* **2017,** *53*, 4350–4353.
48. Min, J.; Xu, G.; Sun, J. Synthesis of Six-Membered Carbo-/Heterocycles *via* Cascade Reaction of Alkynes and Diazo Compounds. *J. Org. Chem.* **2017,** *82*, 5492–5498.
49. Florentino, L.; Aznar, F.; Valdés, C. Synthesis of Polysubstituted Isoquinolines through Cross-Coupling Reactions with α-Alkoxytosylhydrazones. *Org. Lett.* **2012,** *14*, 2323–2325.
50. Chu, H.; Xue, P.; Yu, J.-T.; Cheng, J. Rhodium-Catalyzed Annulation of Primary Benzylamine with α-Diazo Ketone toward Isoquinoline. *J. Org. Chem.* **2016,** *81*, 8009–8013.
51. Shi, Z.; Koester, D. C.; Boultadakis-Arapinis, M.; Glorius, F. Rh(III)-Catalyzed Synthesis of Multi-substituted Isoquinoline and Pyridine *N*-Oxides from Oximes and Diazo Compounds. *J. Am. Chem. Soc.* **2013,** *135*, 12204–12207.
52. Sun, P.; Wu, Y.; Yang, T.; Wu, X.; Xu, J.; Lin, A.; Yao, H. Synthesis of Heterocycle-Fused Pyridine *N*-Oxides from Oximes and Diazo Compounds *via* Rh^{III}-catalyzed C-H Activation and Annulation. *Adv. Synth. Catal.* **2015,** *357*, 2469–2473.
53. Phatake, R. S.; Patel, P.; Ramana, C. V. Ir(III)-Catalyzed Synthesis of Isoquinoline *N*-Oxides from Aryloxime and α-Diazocarbonyl Compounds. *Org. Lett.* **2016,** *18*, 292–295.
54. Shi, J.; Zhou, J.; Yan, Y.; Jia, J.; Liu, X.; Song, H.; Xu, H. E.; Yi, W. One-Pot Cascade Synthesis of *N*-Methoxyisoquinolinediones *via* Rh(III)-Catalyzed Carbenoid Insertion C-H Activation/Cyclization. *Chem. Commun.* **2015,** *51*, 668–671.
55. Lv, H.; Xu, W. L.; Lin, K.; Shi, J.; Yi, W. Iridium(III)-Catalyzed Regioselective Carbenoid Insertion C-H Alkylation by α-Diazotized Meldrum's Acid. *Eur. J. Org. Chem.* **2016,** 5637–5641.
56. Phatake, R. S.; Patel, P.; Ramana, C. V. Ir(III)-Catalyzed Carbenoid Functionalization of Benzamides: Synthesis of *N*-Methoxyisoquinolinediones and *N*-Methoxyisoquinolinones. *Org. Lett.* **2016,** *18*, 2828–2831.
57. Shi, L.; Yu, K.; Wang, B. Regioselective Synthesis of Multi-Substituted Isoquinolones and Pyridones *via* Rh(III)-Catalyzed Annulation Reactions. *Chem. Commun.* **2015,** *51*, 17277–17280.
58. Chen, X.; Hu, X.; Bai, S.; Deng, Y.; Jiang, H.; Zeng, W. Rh(III)-Catalyzed [4 + 2] Annulation of Indoles with Diazo Compounds: Access to Pyrimido[1,6-a]indole-1(2H)-ones. *Org. Lett.* **2016,** *18*, 192–195.
59. Wu, Y.; Sun, P.; Zhang, K.; Yang, T.; Yao, H.; Lin, A. Rh(III)-Catalyzed Redox-Neutral Annulation of Primary Benzamides with Diazo Compounds: Approach to Isoquinolinones. *J. Org. Chem.* **2016,** *81*, 2166–2173.

60. Patel, P.; Borah, G. Synthesis of Oxindole from Acetanilide *via* Ir(III)-Catalyzed C-H Carbenoid Functionalization. *Chem. Commun.* **2017,** *53*, 443–446.
61. Wang, H.; Li, L.; Yu, S.; Li, Y.; Li, X. Rh(III)-Catalyzed C-C/C-N Coupling of Imidates with α-Diazo Imidamide: Synthesis of Isoquinoline-Fused Indoles. *Org. Lett.* **2016,** *18*, 2914–2917.
62. Li, X. G.; Sun, M.; Jin, Q.; Liu, K.; Liu, P. N. Access to Isoquinolines and Isoquinolin-3-ols *via* Rh(III)-Catalyzed Coupling/Cyclization Cascade Reaction of Arylimidates and Diazo Compounds. *J. Org. Chem.* **2016,** *81*, 3901–3910.
63. Wang, J.; Zha, S.; Chen, K.; Zhang, F.; Zhu, J. Synthesis of Isoquinolines *via* Rh-Catalyzed C-H Activation/C-N Cyclization with Diazodiesters or Diazoketoesters as a C_2 Source. *Org. Biomol. Chem.* **2016,** *14*, 4848–4852.
64. Wang, Q.; Li, X. Rhodium/Copper Co-Catalyzed Annulation of Benzylamines with Diazo Compounds: Access to Fused Isoquinolines. *Org. Chem. Front.* **2016,** *3*, 1159–1162.
65. Kim, J. H.; Greßies, S.; Glorius, F. Cooperative Lewis Acid/Cp*Co^{III} Catalyzed C-H Bond Activation for the Synthesis of Isoquinolin-3-ones. *Angew. Chem. Int. Ed.* **2016,** *55*, 5577–5581.
66. Sun, S.; Hu, W.-M.; Gu, N.; Cheng, J. Palladium-Catalyzed Multi-Component Reactions of *N*-Tosylhydrazones, 2-Iodoanilines and CO_2 towards 4-Aryl-2-quinolinones. *Chem. Eur. J.* **2016,** *22*, 18729–18732.
67. Liang, Y.; Yu, K.; Li, B.; Xu, S.; Song, H.; Wang, B. Rh(III)-Catalyzed Synthesis of 1-Aminoindole Derivatives from 2-Acetyl-1-arylhydrazines and Diazo Compounds in Water. *Chem. Commun.* **2014,** *50*, 6130–6133.
68. Qi, Z.; Yu, S.; Li, X. Rh(III)-Catalyzed Synthesis of *N*-Unprotected Indoles from Imidamides and Diazo Ketoesters *via* C-H Activation and C-C/C-N Bond Cleavage. *Org. Lett.* **2016,** *18*, 700–703.
69. Li, Y.; Qi, Z.; Wang, H.; Yang, X.; Li, X. Ruthenium(II)-Catalyzed C-H Activation of Imidamides and Divergent Couplings with Diazo Compounds: Substrate-Controlled Synthesis of Indoles and 3*H*-Indoles. *Angew. Chem. Int. Ed.* **2016,** *55*, 11877–11881.
70. Wang, J.; Wang, M.; Chen, K.; Zha, S.; Song, C.; Zhu, J. C-H Activation-Based Traceless Synthesis *via* Electrophilic Removal of a Directing Group. Rhodium(III)-Catalyzed Entry into Indoles from *N*-Nitroso and α-Diazo-β-Keto Compounds. *Org. Lett.* **2016,** *18*, 1178–1181.
71. Mishra, N. K.; Choi, M.; Jo, H.; Oh, Y.; Sharma, S.; Han, S. H.; Jeong, T.; Han, S.; Lee, S.-Y.; Kim, I. S. Direct C-H Alkylation and Indole Formation of Anilines with Diazo Compounds under Rhodium Catalysis. *Chem. Commun.* **2015,** *51*, 17229–17232.
72. Tang, G.-D.; Pan, C.-L.; Li, X. Iridium(III)- and Rhodium(III)-Catalyzed Coupling of Anilines with α-Diazoesters *via* Chelation-Assisted C-H Activation. *Org. Chem. Front* **2016,** *3*, 87–90.
73. Jiang, H.; Gao, S.; Xu, J.; Wu, X.; Lin, A.; Yao, H. Multiple Roles of the Pyrimidyl Group in the Rhodium-Catalyzed Regioselective Synthesis and Functionalization of Indole-3-carboxylic Acid Esters. *Adv. Synth. Catal.* **2016,** *358*, 188–194.
74. Yu, K.; Liang, Y.; Li, B.; Wang, B. Rhodium(III)-Catalyzed Synthesis of Indole Derivatives from Pyrimidyl-Substituted Anilines and Diazo Compounds. *Adv. Synth. Catal.* **2016,** *358*, 661–666.
75. Allu, S.; Ravi, M.; Swamy, K. C. K. Rhodium(III)-Catalysed Carbenoid $C(sp^2)$-H Functionalisation of Aniline Substrates with α-Diazo Esters: Formation of Oxindoles and Characterisation/Utility of an Intermediate-Like Rhodacycle. *Eur. J. Org. Chem.* **2016,** 5697–5705.

76. Bai, S.; Chen, X.; Hu, X.; Deng, Y.; Jiang, H.; Zeng, W. An Ir(III)-Catalyzed Aryl C-H Bond Carbenoid Functionalization Cascade: Access to 1,3-Dihydroindol-2-ones. *Org. Biomol. Chem.* **2017,** *15*, 3638–3647.
77. Chen, X.; Xie, Y.; Xiao, X.; Li, G.; Deng, Y.; Jiang, H.; Zeng, W. Rh(III)-Catalyzed Chelation-Assisted Intermolecular Carbenoid Functionalization of α-Imino Csp^3-H bonds. *Chem. Commun.* **2015,** *51*, 15328–15331.
78. Xie, Y.; Chen, X.; Liu, X.; Su, S.-J.; Li, J.; Zeng, W. Rh(III)-Catalyzed Relay Carbenoid Functionalization of Aromatic C-H Bonds: Access to π-Conjugated Fused Heteroarenes. *Chem. Commun.* **2016,** *52*, 5856–5859.
79. Wu, J.-Q.; Yang, Z.; Zhang, S.-S.; Jiang, C.-Y.; Li, Q.; Huang, Z.-S.; Wang, H. From Indoles to Carbazoles: Tandem Cp*Rh(III)-Catalyzed C-H Activation/Brønsted Acid-Catalyzed Cyclization Reactions. *ACS Catal* **2015,** *5*, 6453–6457.
80. Li, X. G.; Sun, M.; Liu, K.; Jin, Q.; Liu, P. N. Rh(III)-Catalyzed C-H Activation/Cyclization of Benzamides and Diazo Compounds to Form Isocoumarins and α-Pyrones. *Chem. Commun.* **2015,** *51*, 2380–2383.
81. Sun, P.; Gao, S.; Yang, C.; Guo, S.; Lin, A.; Yao, H. Controllable Rh(III)-Catalyzed Annulation between Salicylaldehydes and Diazo Compounds: Divergent Synthesis of Chromones and Benzofurans. *Org. Lett.* **2016,** *18*, 6464–6467.
82. Cheng, Y.; Bolm, C. Regioselective Syntheses of 1,2-Benzothiazines by Rhodium-Catalyzed Annulation Reactions. *Angew. Chem. Int. Ed.* **2015,** *54*, 12349–12352.
83. Jeon, W. H.; Son, J.-Y.; Kim, J. E.; Lee, P. H. Synthesis of 1,2-Benzothiazines by a Rhodium-Catalyzed Domino C-H Activation/Cyclization/Elimination Process from *S*-Aryl Sulfoximines and Pyridotriazoles. *Org. Lett.* **2016,** *18*, 3498–3501.
84. Yang, Y.; Wang, X.; Li, Y.; Zhou, B. A [4 + 1] Cyclative Capture Approach to 3*H*-Indole-*N*-oxides at Room Temperature by Rhodium(III)-Catalyzed C-H Activation. *Angew. Chem. Int. Ed.* **2015,** *54*, 15400–15404.
85. Dateer, R. B.; Chang, S. Rh(III)-Catalyzed C-H Cyclization of Arylnitrones with Diazo Compounds: Access to N-hydroxyindolines. *Org. Lett.* **2016,** *18*, 68–71.
86. Zhou, B.; Chen, Z.; Yang, Y.; Ai, W.; Tang, H.; Wu, Y.; Zhu, W.; Li, Y. Redox-Neutral Rhodium-Catalyzed C-H Functionalization of Arylamine N-oxides with Diazo Compounds: Primary $C(sp^3)$-H/$C(sp^2)$-H Activation and Oxygen-Atom Transfer. *Angew. Chem. Int. Ed.* **2015,** *54*, 12121–12126.
87. Biswas, A.; Karmakar, U.; Pal, A.; Samanta, R. Copper-Catalyzed Regioselective Cascade Alkylation and Cyclocondensation of Quinoline *N*-oxides with Diazo Esters: Direct Access to Conjugated π-Systems. *Chem. Eur. J* **2016,** *22*, 13826–13830.
88. Son, J.-Y.; Kim, S.; Jeon, W. H.; Lee, P. H. Synthesis of Cinnolin-3(2*H*)-one Derivatives from Rh-Catalyzed Reaction of Azobenzenes with Diazotized Meldrum's Acid. *Org. Lett.* **2015,** *17*, 2518–2521.
89. Sharma, S.; Han, S. H.; Han, S.; Ji, W.; Oh, J.; Lee, S.-Y.; Oh, J. S.; Jung, Y. H.; Kim, I. S. Rh(III)-catalyzed direct coupling of azobenzenes with α-diazo esters: facile synthesis of cinnolin-3(2*H*)-ones. *Org. Lett.* **2015,** *17*, 2852–2855.
90. Sun, P.; Wu, Y.; Huang, Y.; Wu, X.; Xu, J.; Yao, H.; Lin, A. Rh(III)-Catalyzed Redox-Neutral Annulation of Azo and Diazo Compounds: One-Step Access to Cinnolines. *Org. Chem. Front.* **2016,** *3*, 91–95.
91. Song, C.; Yang, C.; Zhang, F.; Wang, J.; Zhu, J. Access to the cinnoline scaffold *via* rhodium-catalyzed intermolecular cyclization under mild conditions. *Org. Lett.* **2016,** *18*, 4510–4513.

92. Bai, P.; Huang, X.-F.; Xu, G.-D.; Huang, Z.-Z. Cascade C-H Functionalization/Amidation Reaction for Synthesis of Azepinone Derivatives. *Org. Lett.* **2016,** *18*, 3058–3061.
93. Zhou, T.; Li, B.; Wang, B. Rhodium-Catalyzed C-H Activation of 3-(Indolin-1-yl)-3-oxo-propanenitriles with Diazo Compounds and Tandem Cyclization Leading to Hydrogenated Azepino[3,2,1-hi]indoles. *Chem. Commun.* **2016,** *52*, 14117–14120.
94. Kim, J. H.; Gensch, T.; Zhao, D.; Stegemann, L.; Strassert, C. A.; Glorius, F. Rh^{III}-catalyzed C-H activation with pyridotriazoles: direct access to fluorophores for metal-ion detection. *Angew. Chem. Int. Ed.* **2015,** *54*, 10975–10979.
95. Zhao, D.; Kim, J. H.; Stegemann, L.; Strassert, C. A.; Glorius, F. Cobalt(III)-Catalyzed Directed C-H Coupling with Diazo Compounds: Straightforward Access towards Extended π-Systems. *Angew. Chem. Int. Ed.* **2015,** *54*, 4508–4511.
96. Li, Y.; Wang, F.; Yu, S.; Li, X. Iridium- and Rhodium-Catalyzed Carbocyclization between 2-Phenylimidazo[1,2-*a*]pyridine and α-Diazo Esters. *Adv. Synth. Catal.* **2016,** *358*, 880–886.
97. Li, S.-S.; Xia, Y.-Q.; Hu, F.-Z.; Liu, C.-F.; Su, F.; Dong, L. Ir^{III}-Catalyzed One-Pot Cascade Synthesis of Pentacyclic-Fused Carbazoles from Indoles and Diazoes. *Chem. Asian J* **2016,** *11*, 3165–3168.
98. Li, B.; Zhang, B.; Zhang, X.; Fan, X. Regio-Selective Synthesis of Diversely Substituted Benzo[*a*]carbazoles through Rh(III)-Catalyzed Annulation of 2-Arylindoles with α-Diazo Carbonyl Compounds. *Chem. Commun.* **2017,** *53*, 1297–1300.
99. Zhang, B.; Li, B.; Zhang, X.; Fan, X. Synthesis of Naphthoquinolizinones through Rh (III)-Catalyzed Double C(sp^2)-H Bond Carbenoid Insertion and Annulation of 2-Aryl-3-cyanopyridines with α-Diazo Carbonyl Compounds. *Org. Lett.* **2017,** *19*, 2294–2297.
100. Liu, C.-F.; Zhang, G.-T.; Sun, J.-S.; Dong, L. Access to π-Conjugated Azaindole Derivatives *via* Rhodium(III)-Catalyzed Cascade Reaction of Azaindoles and Diazo Compounds. *Org. Biomol. Chem.* **2017,** *15*, 2902–2905.
101. Yang, C.; He, X.; Zhang, L.; Han, G.; Zuo, Y.; Shang, Y. Synthesis of Isocoumarins from Cyclic 2-Diazo-1,3-diketones and Benzoic Acids *via* Rh(III)-Catalyzed C-H Activation and Esterification. *J. Org. Chem.* **2017,** *82*, 2081–2088.
102. Zhang, H.; Wang, K.; Wang, B.; Yi, H.; Hu, F.; Li, C.; Zhang, Y.; Wang, J. Rhodium (III)-Catalyzed Transannulation of Cyclopropenes with *N*-Phenoxyacetamides through C-H Activation. *Angew. Chem. Int. Ed.* **2014,** *53*, 13234–13238.
103. Guo, W.; Xia, Y. Mechanistic Understanding of the Divergent Reactivity of Cyclopropenes in Rh(III)-Catalyzed C-H Activation/Cycloaddition Reactions of *N*-Phenoxyacetamide and *N*-Pivaloxybenzamide. *J. Org. Chem.* **2015,** *80*, 8113–8121.
104. Li, J.; Qiu, Z. DFT Studies on the Mechanism of the Rhodium(III)-Catalyzed C-H Activation of *N*-Phenoxyacetamide. *J. Org. Chem.* **2015,** *80*, 10686–10693.
105. Barluenga, J.; Quiñones, N.; Cabal, M.-P.; Aznar, F.; Valdés, C. Tosylhydrazide-Promoted Palladium-Catalyzed Reaction of β-Aminoketones with *o*-Dihaloarenes: Combining Organocatalysis and Transition-Metal Catalysis. *Angew. Chem. Int. Ed.* **2011,** *50*, 2350–2353.
106. Huang, Z.; Yang, Y.; Xiao, Q.; Zhang, Y.; Wang, J. Auto-Tandem Catalysis: Synthesis of Acridines by Pd-Catalyzed C = C Bond Formation and C(sp^2)-N Cross-Coupling. *Eur. J. Org. Chem.* **2012,** 6586–6593.
107. Florentino, L.; Aznar, F.; Valdés, C. Synthesis of (Z)-*N*-alkenylazoles and Pyrroloisoquinolines from α-*N*-Azoleketones through Pd-Catalyzed Tosylhydrazone Cross-Couplings. *Chem. Eur. J.* **2013,** *19*, 10506–10510.

108. Paraja, M.; Valdés, C. Pd-Catalyzed Autotandem Reactions with *N*-Tosylhydrazones. Synthesis of Condensed Carbo- and Heterocycles by Formation of a C-C Single Bond and a C = C Double Bond on the Same Carbon Atom. *Org. Lett.* **2017,** *19*, 2034–2037.
109. Zhao, H.; Yang, K.; Zheng, H.; Ding, R.; Yin, F.; Wang, N.; Li, Y.; Cheng, B.; Wang, H.; Zhai, H. A One-Pot Synthesis of Dibenzofurans from 6-Diazo-2-cyclohexenones. *Org. Lett.* **2015,** *17*, 5744–5747.

Chapter 5

Synthesis of Heterocycles With Iron Salts as Sustainable Metal Catalysts

Victoria Sinka[1,2], Daniel A. Cruz[1,2], Juan M. López-Soria[1,2], Víctor S. Martín[1,2], Pedro O. Miranda[1,2] and Juan I. Padrón[1,2]

[1]*Instituto de Productos Naturales y Agrobiología-CSIC, La Laguna, Santa Cruz de Tenerife, Spain,* [2]*Instituto Universitario de Bio-Orgánica Antonio González (CIBICAN), Departamento de Química Orgánica, Universidad de La Laguna, Avda. Astrofísico Fco., La Laguna, Tenerife, Spain*

5.1 INTRODUCTION

According to the Lambda-Cold Dark Matter model the current measurement of the age of the universe is estimated to be $13{,}799 \pm 0.021$ billion years when an event called "Big Bang" took place.[1] The matter (which was so compacted and with such high temperatures that could not even form neutral atoms) started cooling, allowing the formation of light elements such as hydrogen and helium via fusion reactions. At that time the universe was formed entirely by 75% hydrogen and 25% helium with trace amounts of lithium and beryllium.[2] Much later the formation of early stars allowed the fusion of hydrogen to helium, while massive stars bigger than the sun allowed further reactions that convert helium into carbon and oxygen. In the very massive stars (supernovae) the reaction chain continued to produce elements like silicon up to iron. Elements higher than iron cannot be formed through fusion as one has to supply energy for the reaction to take place, making it the most stable nucleus in the universe. Iron is an abundant element on Earth and is a biologically essential component of every living organism. However, iron is usually present on earth in the form of oxides, which are highly insoluble and are not readily available for uptake by organisms.[3–5] Still, several cellular mechanisms have evolved to capture iron from the environment in biologically useful forms.[6,7] Many of the mechanisms found in lower organisms have analogous counterparts in humans. In fact, iron is the most abundant metal in humans with healthy

Advances in Transition-Metal Mediated Heterocyclic Synthesis.
DOI: https://doi.org/10.1016/B978-0-12-811651-7.00005-4

adults possessing up to 3–4 g of iron in complex forms bound to protein (hemoprotein) as heme compounds (hemoglobin or myoglobin), heme enzymes, or nonheme compounds (Flavin-iron enzymes, transferrin, ferritin).[8] The human body requires iron for the synthesis of the oxygen transport proteins and, therefore, is recycled and highly conserved and not really lost from the body.[9] Although living organisms have taken advantage of the iron stability, organic chemistry has overlooked the possibilities of iron when developing catalysts. In fact, chemists have used catalysts (transition metal catalysts like Pd, Ni, etc.) to achieve reactions which were almost impossible by traditional methods. Catalysts are an important part in many chemical processes since they speed up reactions allowing to work at a lower temperature and consequently economizing the outcome of the process. However, organic chemistry has focused mainly on the use of transition metal catalysts which are toxic, expensive, and scarce. Nowadays the development of catalysts that avoid these shortcomings has contributed enormously to the bloom of green catalysis,[10] which basically is formed by three main research lines: (1) biocatalysis, which involves the use of natural catalysts such as enzymes to perform chemical reactions[11]; (2) organocatalysis, based on the use of organic molecules such as proline and oxazolidines that show synergistic effects[12]; (3) sustainable metal catalysis, residing in the use of nonexpensive, nontoxic, and widely available metals.[13]

Due to its abundance on earth and far less toxicity than common metal catalysts currently used, iron fits with the element strategy initiative, meaning the systematic exploration of widespread elements instead of rare and precious elements in the interest of sustainability and economy.[14] Centered in the middle of the d-block and able to support formal oxidation states ranging from − II to + VI, iron catalysts hold the promise of being able to circumscribe a wide array of chemical processes. Its ability to take a variety of spin rates and to expand its oxidation state makes it attractive in reductive and oxidative processes. Also, its Lewis acidity varies from modest to high, requiring to be tuned via the introduction of ligands, which opens a wide range of possible conditions to perform organic reactions.[15] Although iron-mediated chemistry has been underrated until recent years, the first report on iron-mediated reactions was reported in early 1941 by Kharasch and Fields, when they reported a cross-coupling reactivity catalyzed by iron salts.[16] Ten years later, in 1951 the discovery of ferrocene was a step forward in organometallic chemistry after its structure was confirmed by X-ray crystallography.[17] Ferrocene molecule features an iron atom *sandwiched* between two cyclopentadienyl rings. The discovery of this sandwich-type compound was a milestone in the study of organometallic compounds and developed the field of metallocenes, which acquired high importance in the 80s in the design of transition metal catalysts. Despite the implications of these two discoveries, it took another 20 years until the field of iron chemistry really got started. In 1971 Kochi reported the coupling of alkenyl halides with

Grignard reagents catalyzed by iron in which the active species was Fe(I) formed by reduction of the Fe(III) precatalyst with the Grignard reagent to an active Fe(0) species which is ligated to a source of nucleophile.[18] Following Kochi's report the iron-catalyzed couplings with alkenyls electrophiles developed rapidly with various alkyl and aryl Grignard reagents to achieve sp^2-sp^2 or sp^3-sp^3 bonds.[19] In the early years of the 21st century, Fürstner reported the first coupling of arylelectrophiles with Grignard reagents giving rise to a renaissance of iron-mediated catalysis with a vast amount of works.[20,21] It seemed that organic chemistry was definitely entering in the iron age when a controversy began in 2009 with the publication by Buchwald and Bolm where they reported that impurities of copper presented in iron catalysts were responsible for the catalysis.[22] This discovery originated an intense debate between organic chemists,[23,24] which led to a number of articles on the topic, concluding that for processes catalyzed by copper or by iron/copper caution is advised.[25–27] Since then, some iron-catalyzed processes have been revisited and there are, indeed, examples where impure iron salts produce better results than highly pure ones. However, a clear trend cannot be given and it is always advisable to test impurities in the iron source. These problems have depreciated iron-mediated catalysis toward more established strategies based on other transition metals, helped by its somehow capricious reactivity. Still, there is a *modus operandi* in iron catalysis in which reviews in iron catalysis has not rigorously delve into. Based on the substrate reactivity, iron catalysis takes place via three different possibilities: (1) σ-bond activation, (2) π-bond activation, and (3) radical processes. The first two approaches usually involve iron(III) species while the radical process generally involves iron(II) species. In this chapter, our goal is not a comprehensive inventory of each novel reaction reported in the literature describing the formation of heterocycles, but the rationalization of the behavior of the iron based on these three reactivities to provide a systematic approach that is usually avoided in reviews on the topic.

5.2 σ-BOND ACTIVATION

σ-Bonds are the strongest type of covalent bonds due to the direct overlap of orbitals. Nevertheless, the ability of transition metals to activate them has been widely used in organic synthesis.[28,29] Although the details of the interactions of these metals with the corresponding organic molecules are not completely understood due to the inherent complexity of the metal core, they can be classified into three main categories:

1. σ-bond activation through oxidative addition of the metal,[30,31] resulting in an increase of the oxidation state by +2.

2. σ-activation via insertion of the molecule into the coordination sphere of the metal using metal pincer complexes.[32–34]
3. σ-bond activation via complexation with functional groups containing nitrogen or oxygen (amines, carbonyls, hydroxyls, etc.).[35,36] This complexation induces a change of polarization in the molecule, enhancing its reactivity by increasing its electrophilicity. The oxidative addition of iron requires the use of phosphine ligands to improve the reactivity of the catalytic species. It is, therefore, a route that demands a rational design of catalyst. Since it has not been applied to the synthesis of heterocycles in recent years, this approach is out of the scope of the present chapter, which will be focused on the other two methods.

5.2.1 σ-Activation Via Insertion of the Molecule into the Coordination Sphere of the Metal

Pincer ligands, in general, are tridentate, 6-electron donor ligands that enforce a meridional coordination geometry on the metal center.[37] Generally a pincer ligand traditionally consists of a central benzene ring, which is 1,3-disubstituted with two chelating side arms. Both the central atom and the donors at the two arms bind to the metal center, resulting in strong chelation. Replacement of the central aryl ring by a pyridinyl or diarylamine group leads to two major classes of variants of the original pincer ligands. The advantages of using pincer ligands are showcased by their strong chelating ability and its structural rigidity.

Fe-MACHO-BH **1** is an iron pincer complex that is typically used in dehydrogenative processes due to its ability to release hydrogen with no alterations in the oxidation state of the iron core.[38] The synthesis of lactones and lactams **3** (with size rings varying from four up to seven) using Fe-MACHO-BH complex was achieved. As it is shown in Scheme 5.1, a dehydrogenative process followed by intramolecular cyclization and final oxidation take place. Fe-MACHO complexes allowed the dehydrogenative process in a rather efficient manner.

5.2.2 σ-Bond Activation Via Complexation With Functional Groups

Different types of activations have been reported depending on the functional group that is activated.

5.2.2.1 Hydroxy Activation

Alcohols are one of the most important functional groups in organic chemistry and a good source of reagents for synthetic reactions. Besides its versatility that allows its transformation into other functional groups, alcohols

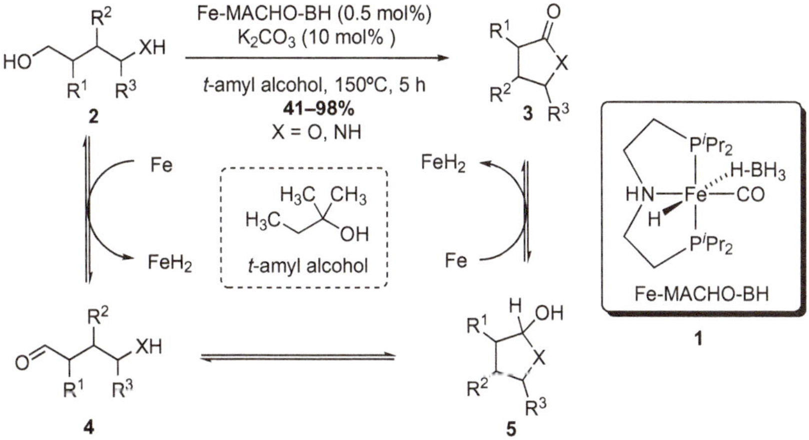

SCHEME 5.1 Dehydrogenative process using Fe-MACHO complexes.

interact with Lewis acid in a rather strong manner. Hydroxyl groups have two unshared electrons on the oxygen that occupy the highest occupied molecular orbital (HOMO). When a Lewis acid is added, the empty valence orbital, the lowest unoccupied molecular orbital (LUMO), can accept the electrons from HOMO. Thus an interaction takes place stabilizing the unshared pair of the oxygen and increasing the HOMO–LUMO interaction.

Many synthetic strategies have taken advantage of this behavior to achieve a wide diversity of chemical reactions ranging from cyclization reactions or coupling reactions that often take place via this interaction.

Herein, we describe the synthesis of a diversity of heterocycles following this strategy:

1. Substituted thiazoles

   Iron-mediated formation of propargylic carbocation **8** is firstly carried out in a similar manner to the Nicholas reaction and with no need of protecting the alkyne moiety with cobalt octacarbonyl complex. Trapping of intermediate **8** via nucleophilic attack through the nitrogen atom of the amide **7** gives alkynyl-amide **9**, which is easily converted to alkynyl-thioamide **10** in the presence of Lawesson's reagent. Subsequent cycloisomerization gives final thiazole **11**[39] (Scheme 5.2).

2. Trisubstituted *oxane* rings can be produced via intramolecular Michael addition activated with iron hydrated salts. The process is stereoselective at 2,6-position. The stereochemistry at 4-position is determined prior to the cyclization process, that is, in the starting material **12** (Scheme 5.3).[40,41]

   Prins cyclization catalyzed by iron is also a valuable tool to synthesize 2,6-disubstituted oxanes. The reaction takes place via the in situ formation of an oxocarbenium intermediate **III**. As it can be seen in

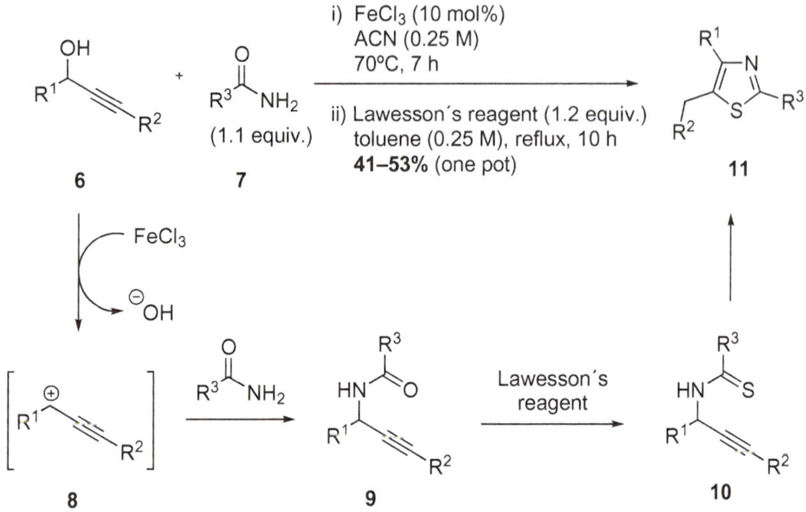

SCHEME 5.2 Synthesis of thiazoles through propargylic carbocation formation.

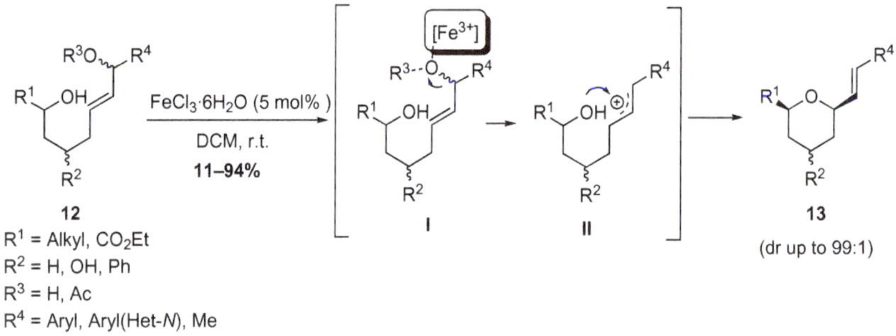

SCHEME 5.3 Synthesis of trisubstituted oxane rings via intramolecular Michael addition.

Scheme 5.4, homoallylic alcohol embedded in a carbohydrate derivative from α-methyl-D-glucopyranoside gives rise to enantioenriched bicyclic oxanes **15** with good yields and using substoichiometric amounts of iron salts.

Under these conditions and using this type of substrate, the competitive oxonia-Cope rearrangement over intermediate **III** is avoided.[42] Still the outcome of the reaction is controversial since it has been recently claimed that a 2 + 2 process may be taking place instead of the classical Prins cyclization approach.[43] A systematic study made by Padrón, Fernández, and Martín and coworkers has manifested that the outcome of the reaction is highly dependent on the alcohol moiety as well as the grade of substitution of the olefin used as starting material.[44] Thus when using homoallylic alcohols **16** with no substitution pattern at the olefin

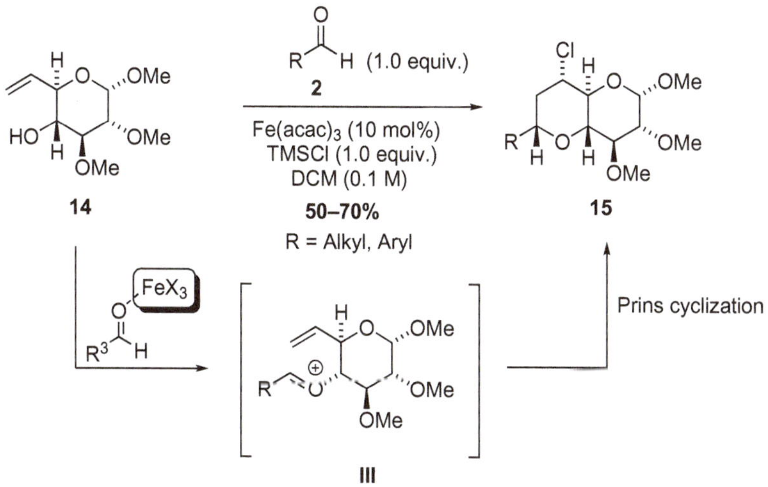

SCHEME 5.4 Synthesis of 2,6-disubstituted bicyclic oxanes through Prins cyclization.

moiety, the reaction follows the classical oxocarbenium intermediate **IV** and subsequent cyclization (typical Prins cyclization behavior). In contrast, when a substituted olefin alongside with an alcohol bearing an electron withdrawing group in α position was used, a 2 + 2 cycloaddition was taking place through intermediate **V** (Scheme 5.5).

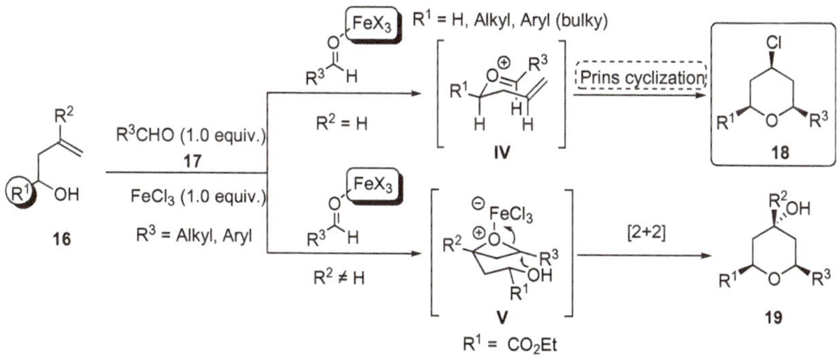

SCHEME 5.5 Prins cyclization versus [2 + 2] cycloaddition process.

Moreover, the lack of aromatic substituents at R^1 of the homoallylic alcohol proved to be essential to eliminate the competitive sigmatropic rearrangement that usually takes place.[45] Thus bulky substituents at the R^1 position are preferred since they do not rearrange.

Another useful variation in Prins cyclization to synthesize 5-membered ring heterocycles **21**, **22**, and **23** is depicted in Scheme 5.6. In

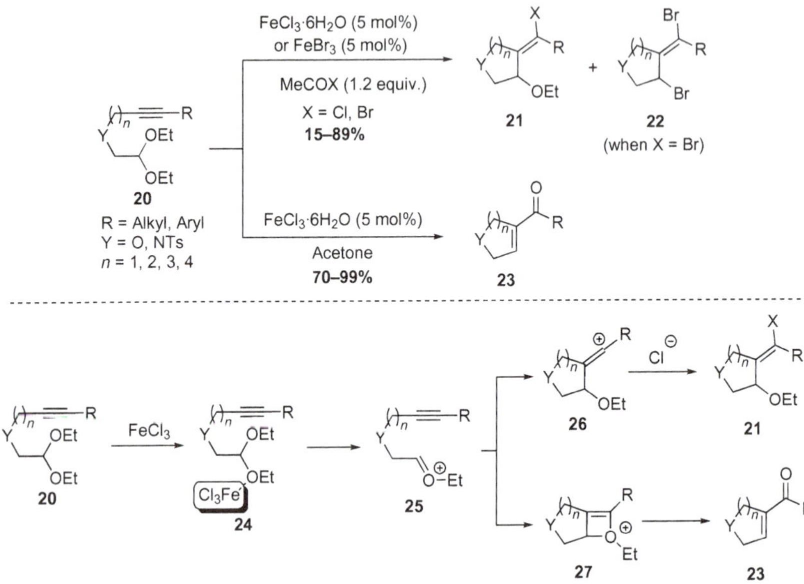

SCHEME 5.6 Prins cyclization in the synthesis of 5-membered ring heterocycles.

this case an intramolecular reaction of alkynes with the acetal present in **20** to form enone **23** takes place catalyzed by iron(III) hexahydrate.[46] As a halogen source, acyl chlorides or bromides are used. When no halogen source is present, ketone **23** is obtained exclusively. Hence, after initial formation of the oxocarbenium **25** target products are formed via vinylogous carbenium cation **26** and oxete **27** intermediates according to DFT calculations which are also supported by works of Padrón and coworkers[47] (Scheme 5.6).

3. *2H-chromenes* have been synthesized using iron(III)-catalyzed cyclization of 2-(1-hydroxyallyl)phenols **28**.[48] To modulate the Lewis acidity of iron salts, 2-2′-bipyridine ligand was added with good results. Consequently the complexation of iron with the hydroxyl group leads to the carbocationic intermediates **VI** or **VII** depending on the stereochemistry of the olefin. Nucleophilic attack of the alcohol moiety affords the cyclized *2H*-chromene **31** (Scheme 5.7).
4. *Furan* and *pyrrolidine* rings via activation of the hydroxyl moiety with an iron diselenide complex **33**.[49] In this case the process is triggered by the coordination of the iron and diselenide species with the oxygen atom, leading to the formation of cation **IX**, which is a resonance structure of allene **X**. The nucleophilic attack of the selenolate anion at the carbon center of allene **X** affords allenic selenide **XI**. The allene coordination to the iron-selenide provides the seleniranium intermediate **XII**. The subsequent nucleophilic *anti*-attack at the activated double bond gives finally

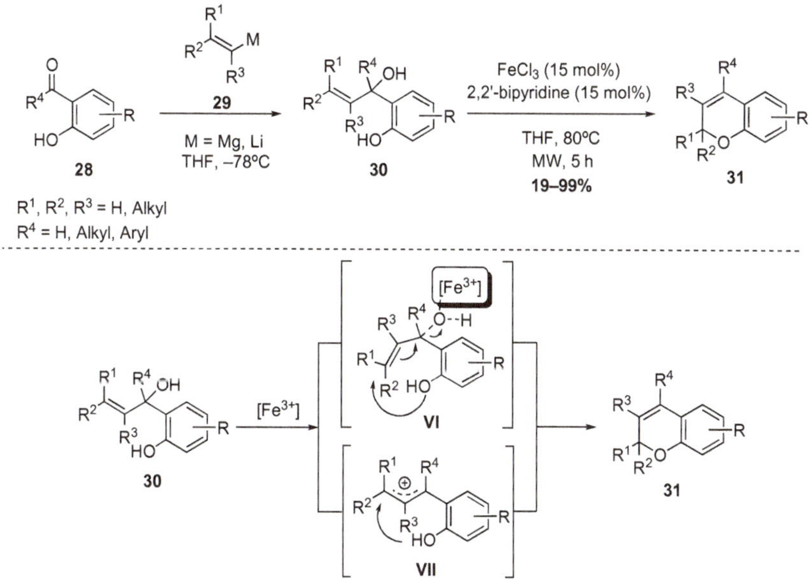

SCHEME 5.7 The role of iron(III) in the synthesis of 2*H*-chromenes.

furan/pyrrolidine **34**. Once more, iron salts promoted the formation of a propargylic cation **IX** as an alternative route to the Nicholas reaction (Scheme 5.8).

5. Different size ring *lactams* using Beckmann rearrangement of oximes **35**. In this case, Lewis acidity needs to be increased by adding silver (I) salts. The superior Lewis acidity of the catalytic system allows the successful formation of the corresponding lactam **36** with good yields[50] (Scheme 5.9).

5.2.2.2 Amine Activation

Parallel strategies have been developed using amines based on the similarities in reactivity between some nitrogen derivatives and oxygen atoms. For instance the reactivity of NTs is quite similar to OH[51].

1. *Benzothiazoles* via a tandem process of halobenzenamines **37** and isothiocyanates **38**. The Lewis acidity of the iron(III) salt was tuned using different ligands. 1,10-Phenanthroline resulted in the best reaction conditions to achieve the cyclization reaction[52] (Scheme 5.10).
2. *1,2,4-Benzothiazidine 1,1-dioxides* using a similar strategy. No tuning of the Lewis acidity of the iron(III) salt was required. Excess of cesium carbonate was used to neutralize the hydrochloric salt of the amidine **41** (Scheme 5.11).[53]

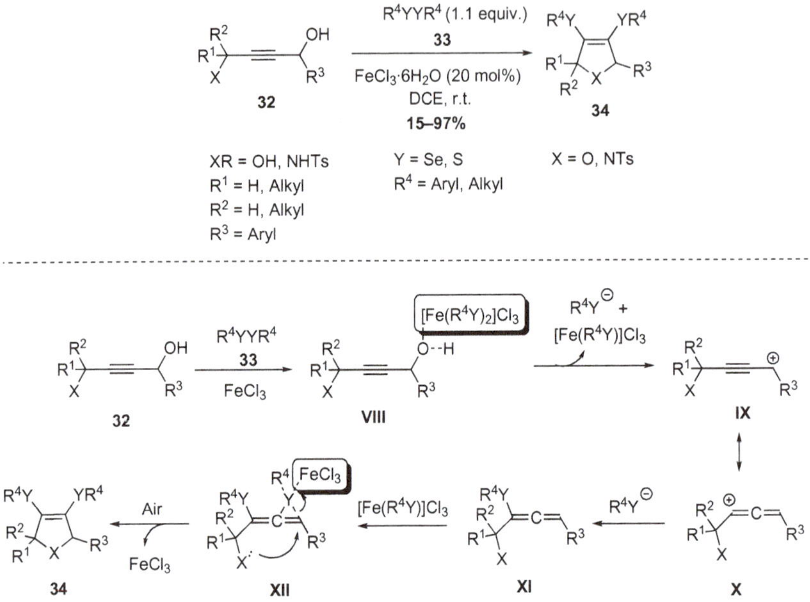

SCHEME 5.8 The iron diselenide complex activates the propargyl hydroxy group.

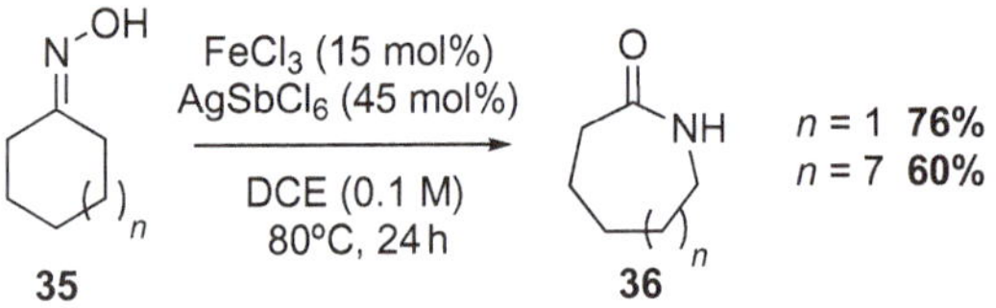

SCHEME 5.9 Lactam formation via iron(III)-catalyzed Beckmann rearrangement.

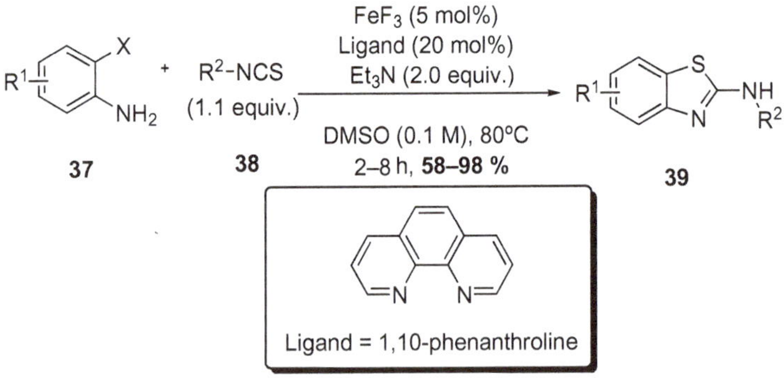

SCHEME 5.10 Iron(III)-1,10-phenanthroline complex in the synthesis of benzothiazoles.

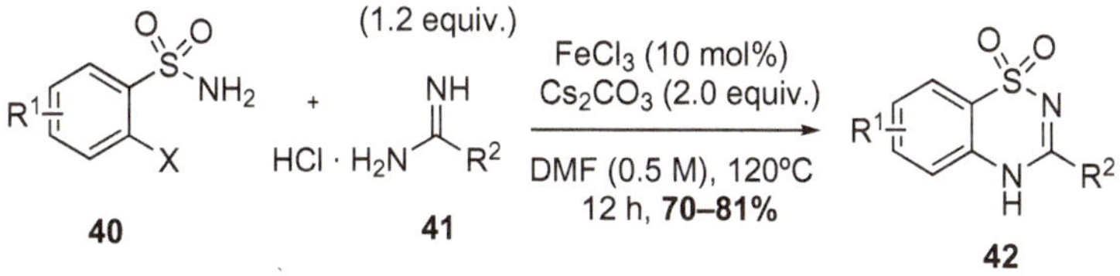

SCHEME 5.11 Iron(III) chloride in the synthesis of 1,2,4-Benzothiazidine 1,1-dioxides.

3. *Pyrrolidines trans*-2,5-disubstituted pyrrolidines **44** have been synthesized using iron catalysis.[54] Attempts to synthesize them using aza-Prins cyclization were fruitless since 5-*endo*-trig cyclization is disfavored according to Baldwin's rules. Thus an intramolecular hydroamination reaction leading to these heterocycles and *trans*-5-substituted proline derivatives has been developed. The role of the iron metal is essential to exclusively obtain the *trans*-2,5-disubstituted pyrrolidine through the specific interaction between the oxygen atom of the sulfonyl group and the iron core. Other methods always produce a mixture of the *trans* and *cis* isomers with no further explanation of the outcome of the reaction (Scheme 5.12).

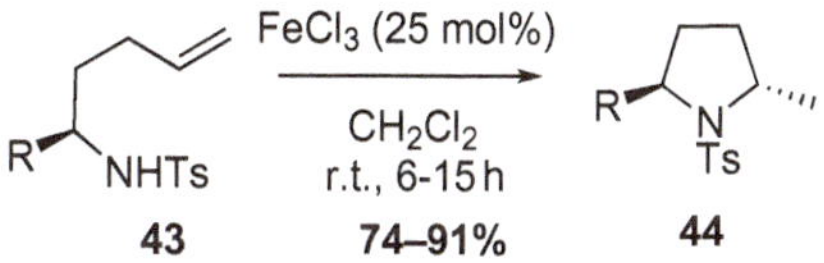

SCHEME 5.12 Intramolecular hydroamination catalized by $FeCl_3$.

The scope of the methodology is highlighted by the enantiodivergent approach to the synthesis of both (+) and (−)-pyrrolidine 197B alkaloids (**51** and **57**, respectively) from L-glutamic acid as a unique starting material, as it is shown in Schemes 5.13 and 5.14.

The (+)-pyrrolidine 197B pathway is based on the formation of the tosylaziridine **47**, which was accomplished in three steps from the ester **45**. *N*-Boc cleavage using iron(III) salts generated the corresponding *N*-tosyl-α-aminoester, which was subsequently reduced to the corresponding *N*-tosyl-β-amino alcohol. The synthesis of *N*-tosylaziridine **47** was carried out in a single step involving a sequential *O*-tosylation followed by intramolecular cyclization. Regioselective ring opening by treatment with allylmagnesium bromide afforded *N*-tosyl-*bis*-homoallyl amine **48**. Then the second aliphatic chain was installed by an alkene-metathesis reaction which was followed by iron(III)-mediated cyclization to obtain *N*-protected (+)-pyrrolidine 197B **50** (Scheme 5.13).

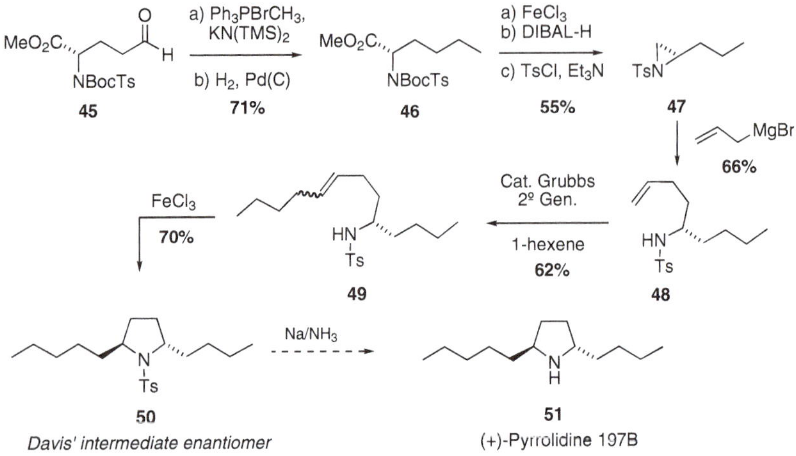

SCHEME 5.13 Enantiodivergent approach. Synthesis (+)-pyrrolidine 197B synthesis using iron(III)-mediated cyclization as a key step.

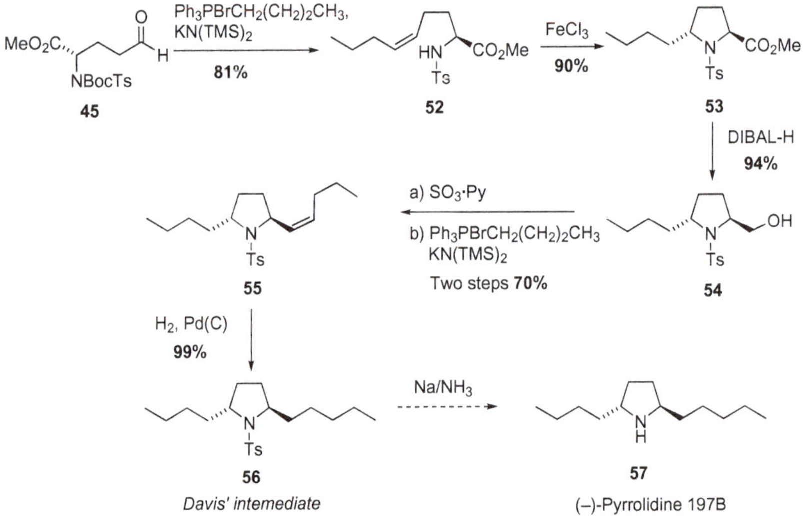

SCHEME 5.14 Enantiodivergent approach. Synthesis of (-)-pyrrolidine 197B with an intramolecular hydroamination step.

(−)-Pyrrolidine 197B was synthesized using again L-glutamic acid as the enantiopure starting material. The synthesis was conducted via the *trans*-5-substituted proline pathway, in which the α-substituent ester group was modified after the intramolecular hydroamination/cyclization reaction. Wittig olefination of aldehyde **45** and *N*-Boc deprotection followed by iron-mediated hydroamination/cyclization led to the *trans*-proline derivative **53**. Reduction of the ester group with Diisobutylaluminium hydride (DIBAL-H) provided the primary alcohol **54**, in which the alkyl chain was installed via Parikh–Doering oxidation, Wittig olefination, and final hydrogenation. The resulting pyrrolidine **56** constitutes the Davis' intermediate which can be submitted to *N*-tosyl cleavage using standard conditions to afford final product **57** (Scheme 5.14).

The same authors describe the synthesis of 3-formylpyrrolidines via aza Cope–Mannich between 2-hydroxy homoallyl (or propargyl) tosylamine **58** and aldehydes **59**. This method is of special relevance since the preparation of 3-formylpyrrolidines **60** is particularly challenging due to polymerization byproducts. Therefore oxazolidine **XIII** formation via reaction of tosylamine **58** and aldehyde **59** evolves to the iminium ion **XIV** which then undergoes a 2-oxonia-[3,3]-sigmatropic rearrangement to give iminium **XV**. This species reacts through an irreversible intramolecular Mannich reaction to give the corresponding pyrrolidine **60**[55] (Schemes 5.15 and 5.16).

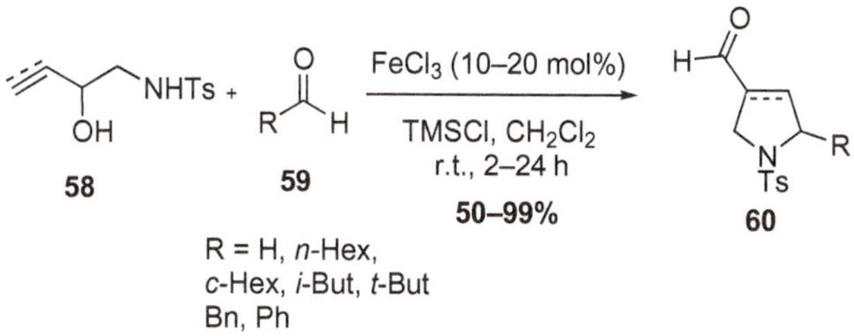

SCHEME 5.15 3-Formylpyrrolidines iron(III)-mediated aza Cope-Mannich process.

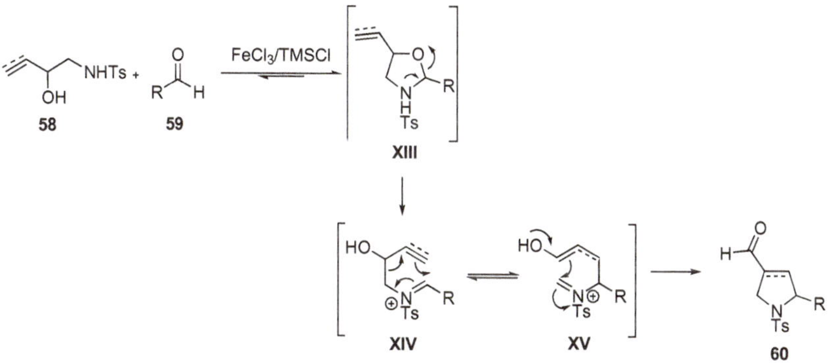

SCHEME 5.16 Mechanistic proposal of 3-formylpyrrolidines synthesis.

This methodology has been used by Cao and coworkers for the synthesis of *daphniphyllum* alkaloids,[56,57] which are a family of complex natural products isolated from trees of the genus *Daphniphyllum*, widely used in

Chinese traditional herbal medicine. The first of these alkaloids synthesized by using this strategy is (±)-cycloclavine, firstly isolated in 1969.[58] The first racemic synthesis was not reported until 2008 by Szántay,[59] followed by Wipf's synthesis in 2011.[60] The strategy followed by Cao and coworkers consists of using the known aldehyde **62** that is subjected to aza Cope–Mannich conditions with tosylamine **61** to generate unsaturated iminium **XVI** that rearranges via 2-azonia-[3,3]-sigmatropic rearrangement to **XVII**. Subsequent intramolecular Mannich reaction affords aldehyde **63**, which is submitted to intramolecular Heck reaction conditions to afford Szántay's intermediate **64** (Scheme 5.17).

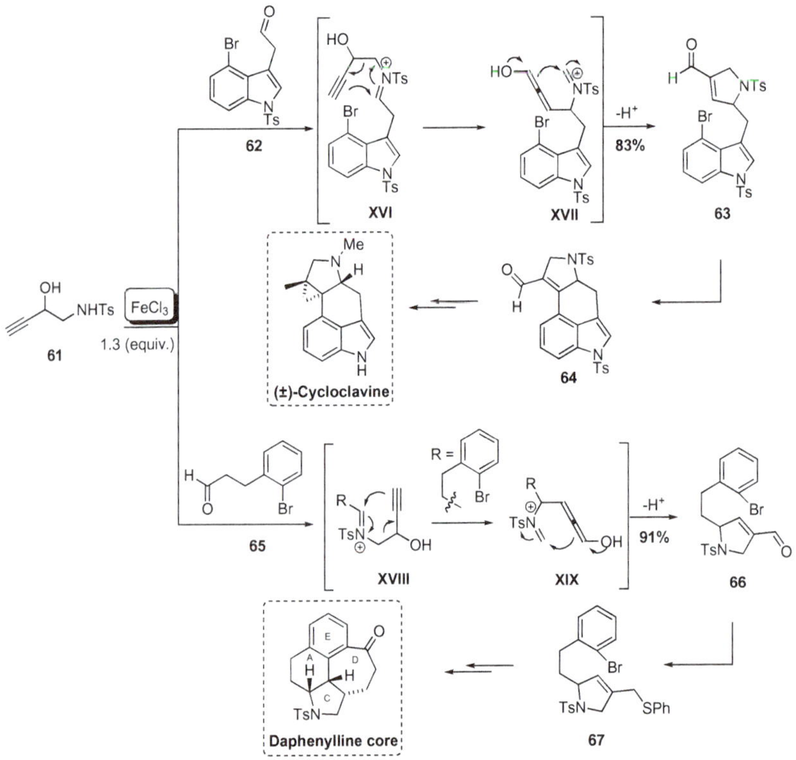

SCHEME 5.17 Applications of 3-formylpyrrolidines formation in total synthesis.

Daphenylline is another alkaloid from the same family discovered in 2009. The main difference from other substrates of the family is the presence of a fused arene motif bridging six-membered ring systems. Following the same strategy, tosylamine **61** was reacted with aldehyde **65** (synthesized from commercially available starting materials). The generation of γ-unsaturated iminium ion **XVIII**, 2-azonia-[3,3]-sigmatropic rearrangement intermediate **XIX**, and further intramolecular Mannich reaction afforded aldehyde **66**, which was transformed into sulfide **67** prior final cyclization (Scheme 5.17).

5.2.2.3 Diselenide Activation

Diselenides are the selenium equivalents of peroxides and disulfides. They are useful precursors for more reactive selenium species and they are more nucleophilic than sulfur compounds. Selenium–selenium bonds are considered to be dynamic covalent bonds (DCBs), a covalent bond that can form or cleave under certain conditions.[61] A comparison between selenium–selenium bond energy (172 kJ mol^{-1}) and sulfur–sulfur bond energy (240 kJ mol^{-1}), points out that the energies of selenium–selenium bonds are low enough to be used as DCBs.[62] The strength of the selenium–selenium bond is therefore weaker than that of the sulfur–sulfur bond and tends to break easily. The interaction of selenium–selenium bond with other transition metals can break the DCBs and release reactive species of organoselenium reagents.

5.2.2.3.1 Benzo[*b*]furan-Fused Selenophenes

For instance it is reported in the literature that the activation of diselenides by iron(III) salts leads to a tandem cyclization of 1,3-diynyl chalcogen derivatives **68** affording benzo[*b*]furan-fused selenophenes **70**.[63]

A cooperative action between diphenyl diselenide and the iron(III) chloride is essential to provide good yields of the cyclized product. The reaction takes place via the pathway outlined in Scheme 5.18. The iron(III) coordination with one selenium atom from the diselenide activates the other selenium toward nucleophilic attack by the alkyne to give the seleniranium ion **XXI**. According to mechanistic studies, it is proposed that the iron salts react with the diorganyl diselenide promoting heterogeneous cleavage of the Se–Se bond to give an organoselenyl cation and an organoselenyl anion. Nucleophilic antiattack of the selenium atom on the activated seleniranium **XXI** leads to the cyclized species **XXII** and selenolate anion. The removal of the alkyl group via a S_N2 displacement by the selenolate anion present in solution gives rise to benzoselenophene **XXIII**. Iteration of the process involving the other alkyne functionality of **XXIII** affords the final product **70**.

5.2.2.4 Ester Activation

Esters constitute general motifs for metal coordination due to the presence of lone pairs in both oxygen atoms. This structural feature makes them ideal targets for metal reactivity and triggering effects in a wide type of reactions. From the vast amount of examples found in the literature, only a few of

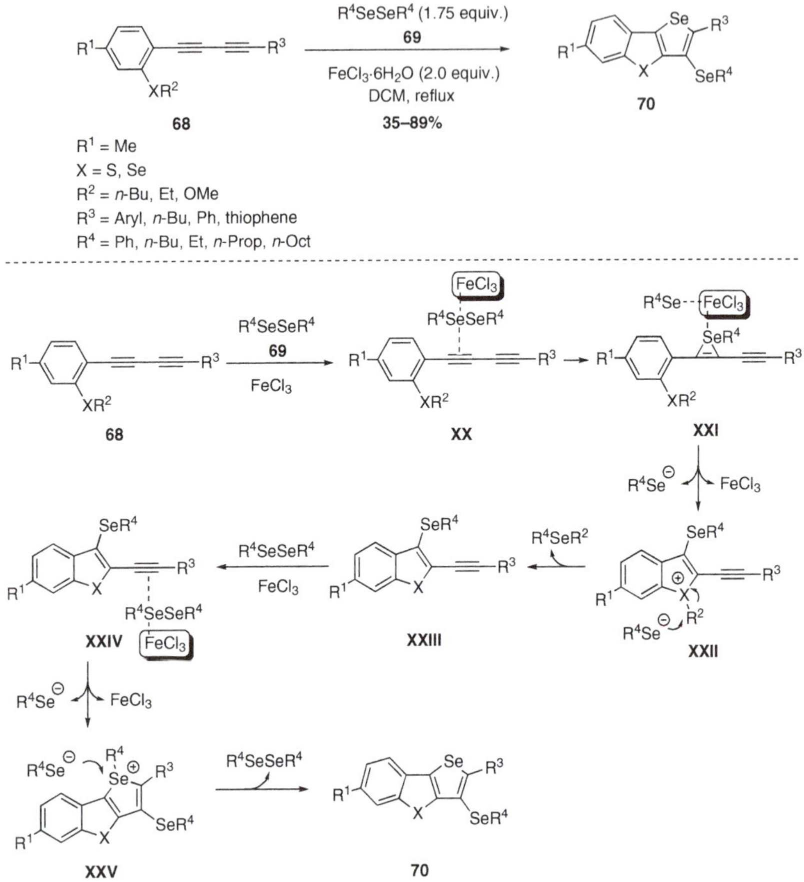

SCHEME 5.18 Iron(III)-diselenide cooperative multiple cyclization, conditions and mechanism.

them have been used to perform cyclization reactions to heterocyclic structures.

1. *Lactones* via intramolecular cyclization of alkenyl ethenetricarboxylates **71**. The process is triggered by the activation of the ester moiety via iron (III) salt catalyzed C–O scission to achieve intermediate **XXVI**. Subsequent cycloaddition reaction of the 2-tertiary alkyl ester of ethenetricarboxylate produces lactone **73**[64] (Scheme 5.19).
2. *Pyrrolo[1,2-a]quinoxaline* derivatives via intramolecular C(sp^2)-N cyclization of ethanones **74**. The reaction most likely begins with the coordination of ethanone **74** to iron(III) to allow the activation of the acetoxy group and the formation of intermediate **XXVII**, which undergoes an intramolecular cyclization to give intermediate **XXVIII** with the help of the lone pair of electrons on the nitrogen atom of the pyrrole moiety. **75** is released by the deprotonation of **XXVIII**[65] (Scheme 5.20).

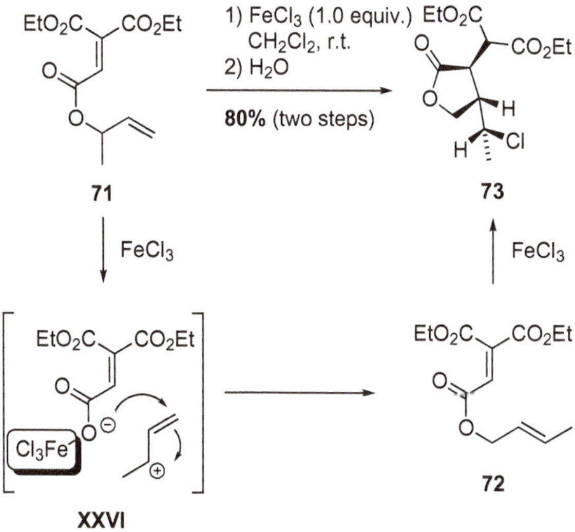

SCHEME 5.19 Synthesis of lactones via intramolecular cyclization.

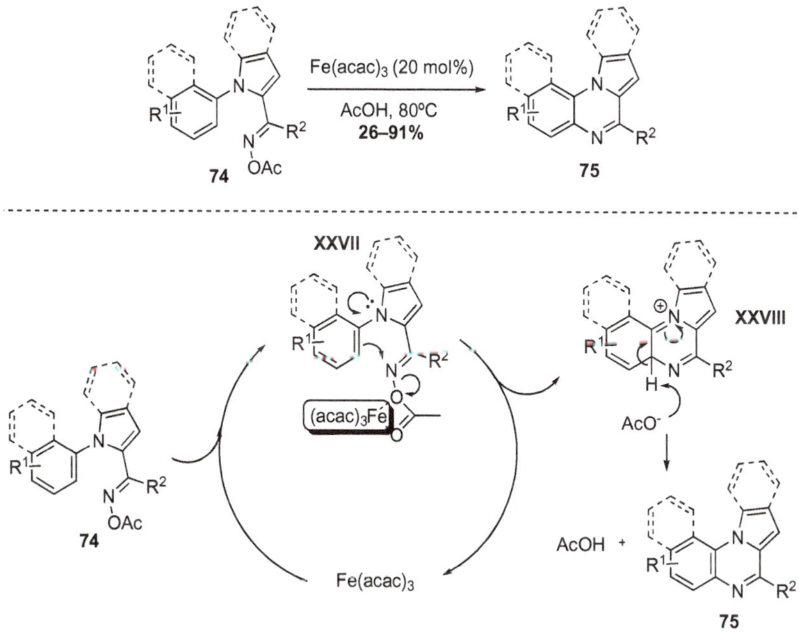

SCHEME 5.20 Synthesis of pyrrolo[1,2-a]quinoxaline derivatives via intramolecular C(sp^2)-N cyclization.

5.3 π-BOND ACTIVATION

π-Bonds are more versatile than σ-bonds since they are prone to activation by transition metals.[66] For instance the π-bond of an alkene can donate electrons into an empty orbital of a transition metal. In turn an occupied metal *d* orbital has the correct symmetry to overlap with a π* orbital at the olefin allowing the formation of a metallacycle. These organometallic species are highly reactive and have a great potential as synthetic intermediates in organic synthesis. The following examples report the π-bond activation with iron catalysts in different ways but always with the common feature of an iron metallacycle formation in its mechanistic proposal (Fig. 5.1).

1. *Benzo[b]furans* have been synthesized from 2-alkynylanisoles **76** via a dual system consisting of iron(III) chloride and diselenyl dichalcogenides **77**.[67] A more complicated outcome has been speculated by the authors with the following steps: (1) Fe(III) could be first reduced to Fe(II); (2) Fe(III) tetracoordinated square-planar selenoate is formed from diphenyl diselenide **77** and Fe(II) via an oxidative addition; (3) alkyne coordination to the metal core to afford the cationic organoiron complex **XXIX**; (4) attack of the oxygen atom from the methoxy group on the activated alkynyl moiety to afford intermediate **XXX**; (5) a coupling reaction to produce the salt **XXXI** and; (6) nucleophilic attack of selenoate anion on the methyl group to give rise to the final benzo-[*b*]-furan **78** (Scheme 5.21).
2. *Furan* derivatives via domino process involving an intramolecular rearrangement followed by cyclization and oxidation as a final step.[68] In this particular case the reaction pathway depicted in Scheme 5.22 consists of the activation of enyne adduct **81** to form an iron(III)−alkyne complex **XXXII** that undergoes a 6-endo-dig addition of the enol ether resulting in the formation of a six-membered intermediate **XXXIII**. This intermediate **XXXIII** collapses into the β-allenic ketone **XXXIV**, which undergoes sequential cyclization, carbine oxidation, and dehydration-oxidation to form furan **82** (Scheme 5.22).
3. *Pyrrolidines* and *tetrahydrofuran* derivatives have been also synthesized via reductive cyclization of 1,6-enynes **83**.[69] The process is thought to

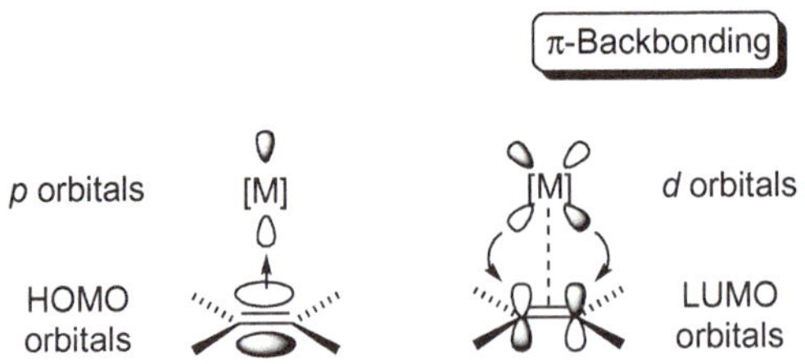

FIGURE 5.1 Interaction of π-bond of an alkene and empty orbital of a transition metal.

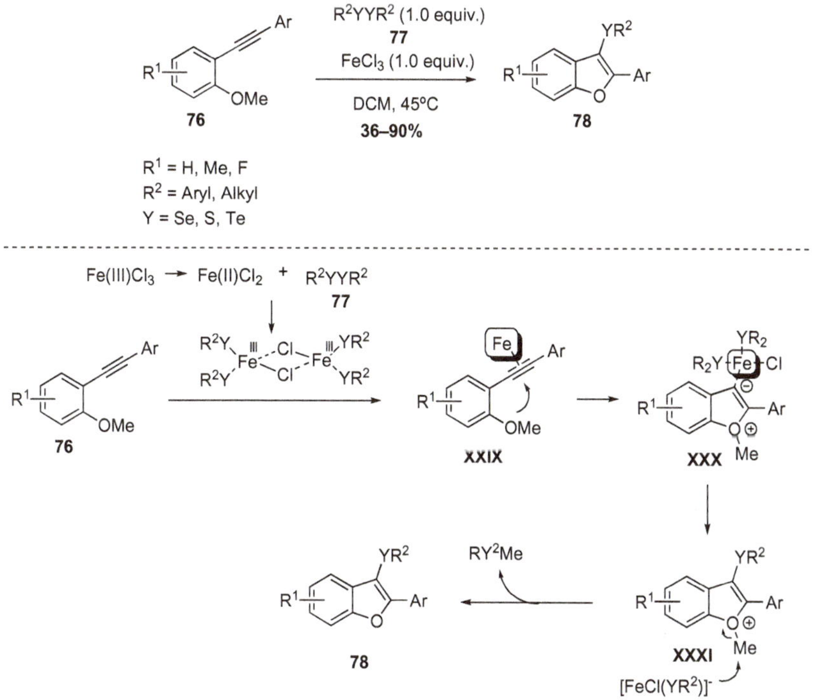

SCHEME 5.21 Synthesis of benzo[b]furans through iron(III) and diselenyl dichalcogenides.

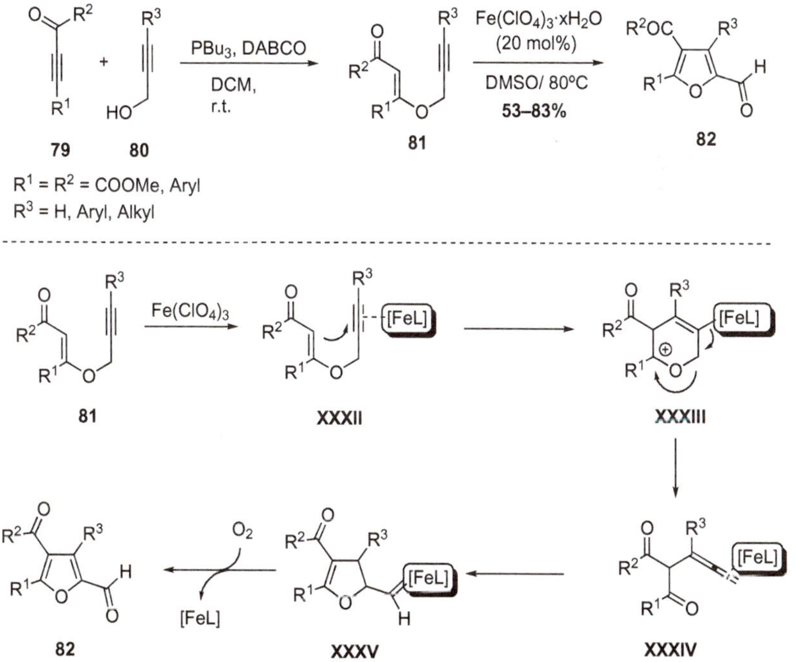

SCHEME 5.22 Sequential intramolecular rearrangement, cyclization and oxidation in the synthesis of furan derivatives.

take place via an initial preactivation of the iron catalyst with diethy zinc and magnesium bromide to form a highly reactive Fe(0) species. The formation of this reactive species forms the five-membered metallacycle **XXXVI**. Then, transmetallation with diethylzinc cleaves the metallacycle **XXXVI** giving intermediate **XXXVII**, which can undergo β-hydride elimination followed by reductive elimination of the iron hydride intermediate **XXXVIII** to afford pyrrolidine **84** (Scheme 5.23).

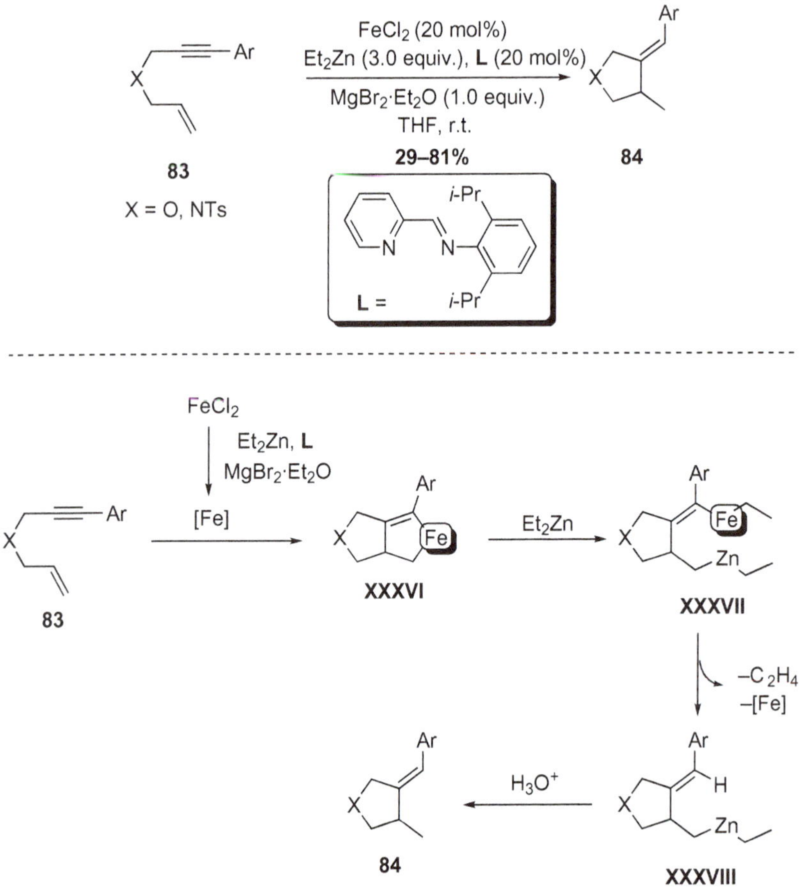

SCHEME 5.23 Reductive cyclization in the synthesis of pyrrolidines and tetrahydrofurans.

4. *Isochromenones* **86** and **87** and *2H-coumarins* **89**.[70,71] The use of selenium reagents has been reported for the synthesis of organoselenyl isochromenones and coumarins. Since the selenium atom can act as a hydrogen-bond acceptor or electron donor, it alters the biological properties of such molecules, opening a wide field of research in medicinal

chemistry, that is, the selenyl derivatives of biologically active products. Organoselenyl isochromenones **86** and **87** have been synthesized from the corresponding alkynylaryl esters **85**. Using a similar procedure the synthesis of 2*H*-coumarins **89** have been carried out starting from arylpropiolates or arylpropiolamides **88**. In this case the proposed mechanism is based on the assumption that the reaction proceeds through initial activation of the alkyne moiety by the complex $[Fe(SeR)_2]L_n$ to form intermediate **XXXIX**. Nucleophilic attack from the aromatic ring to **XXXIX** produces the cyclized species **XL**, which is followed by reductive elimination to afford the stabilized carbocation **XLI**. Final aromatization step leads to final product **89** (Scheme 5.24).

5. *Organoselenylisoxazoles* **91** starting from alkynone *O*-methyloximes **90**.[72] In this reaction mechanism the authors proposed the formation of a

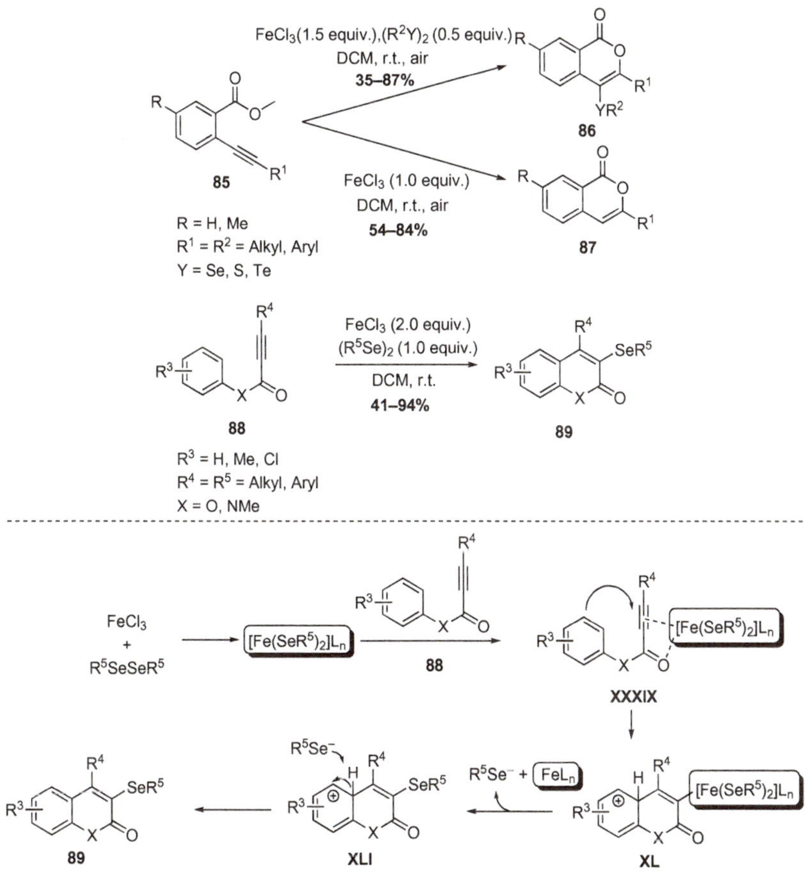

SCHEME 5.24 Synthesis of isochromenones and 2*H*-coumarins using iron(III) and selenium reagents.

reactive species by the association of iron(III) chloride with the diorganyl diselenide to furnish the reactive species R^3SeCl or/and $R^3SeFeCl_2$. At this stage, two plausible pathways can be envisaged: (1) the reaction of the alkynyl moiety with R^3SeCl leads to selenonium intermediate **XLIV**, which is followed by nucleophilic *anti*-attack from the oxygen atom into the activated C-sp atom affording the cationic species **XLV**, which gives the desired product **91** by eliminating methyl chloride; (2) the coordination of the alkynyl moiety with $R^3SeFeCl_2$ affords intermediate **XLII** followed by a nucleophilic *anti*-attack of the oxygen lone pair into the activated triple bond furnishing the isoxazolyliron species **XLIII** which then suffers an elimination process to give rise to **91** (Scheme 5.25).

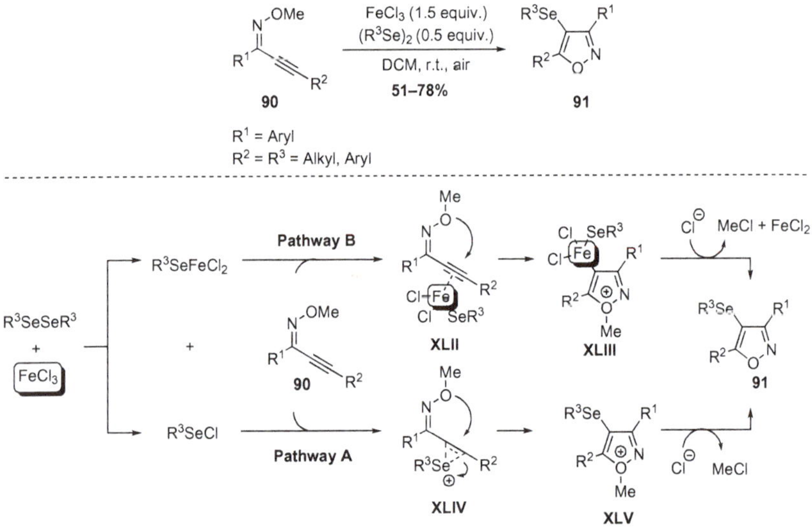

SCHEME 5.25 Synthesis of organoselenylisoxazoles by the association of iron(III) chloride with the diorganyl diselenide.

5.4 RADICAL REACTIONS

Radical intermediates and transition metal intermediates are ideally made for each other.[73] To perform radical reactions the concentration of radical has to be kept low to promote the reaction and to avoid competitive homocoupling processes. The combination of both radical intermediates and transition metal process is thus ideal to keep the radical concentration low since their maximum concentration cannot exceed that of the metal catalyst. On the other hand, radicals are highly reactive species, and the involvement of radicals in transition metal catalysis often leads to an acceleration of the reaction compared to the analogous process occurring in the absence of radicals[74] (Scheme 5.27).

The ability of iron to adopt many oxidation states as well as transferring one electron to substrates makes it a powerful tool for the development of radical reactions.[75,76] Depending on the oxidation state of iron the ability to sustain and perform radical processes varies. Thus while Fe(0) and Fe(I) are valid oxidation states suitable for radical involvement, Fe(II) is the ideal oxidation state to promote and perform radical reactions. Electronic configuration of Fe(0) is $1s^2\ 2s^2\ 2p^6\ 3s^2\ 3p^6\ 4s^2\ 3d^6$. When Fe gets oxidized to Fe(II), it donates the two valance electrons at the highest molecular energy, resulting in $1s^2\ 2s^2\ 2p^6\ 3s^2\ 3p^6\ 3d^6$. Removal of electrons from the fourth shell results in octahedral stereochemistry with two possible spin states: high and low spin. High spin states are kinetically labile while low spin states are inert. The reorganization of the electron distribution associated with the change from high to low spin Fe(II) is particularly favorable, which contributes to its significant stability relative to Fe(III). However, the unique suitability of iron comes from the extreme variability of Fe^{2+}/Fe^{3+} redox potentials which can be tuned by judicious choice of ligands, so that iron sites can encompass almost the entire significant range of redox potentials, from about -0.5 V to +0.6 V.[77,78] Iron-promoted radical reactions are often sensitive to the oxidation state of the iron salt or complex. In general, while Fe(III) salts display a better performance compared with Fe(II) salts, Fe(II) salts contribute to maintaining catalytic radical reactions.

In the examples reported later a variety of radical processes using iron(II) salts are described.

1. *(-)-Agelastatin A*, a cytotoxic alkaloid isolated from marine sponges *Agelas dendromorpha*, has been synthesized using an iron(II)-mediated radical cyclization strategy.[79] *N*-Tosyl carbamate **92** when treated with iron(II) salts gives the *N*-iron complex **XLVI**, which has free-radical character leading to cyclization to obtain **XLVII** followed by halogen transfer to give **XLVIII** and a regeneration of iron(II) species in the presence of tetrabutylammonium halide (Scheme 5.26).
2. *3,4-Dihydro-2H-pyrans* by reaction of 1,6-enones **95** with aldehydes **94** via a 6-endo-trig radical cyclization onto the keto oxygen as the key step. Hydrogen abstraction of aldehyde **94** by *tert*-butoxyl radical leads to the corresponding acyl radical **XLIX**. The addition of the acyl radical **XLIX** to the olefin moiety of 1,6-enone **95** gives the carbo-intermediate **L**, which undergoes a 6-endo-trig cyclization to afford intermediate **LI**. **LI** is then irreversibly oxidized by the iron-oxo species to give carbocation **LII** which is later deprotonated by *tert*-butanolate to afford dihydro-2H-pyran **96**[80] (Scheme 5.27).
3. *Oxazolidinones*

   Two approaches have been reported in the literature: an intramolecular and an intermolecular process, both using a diastereoselective aminohydroxylation of olefins **97** and **99** with a functionalized hydroxylamine.

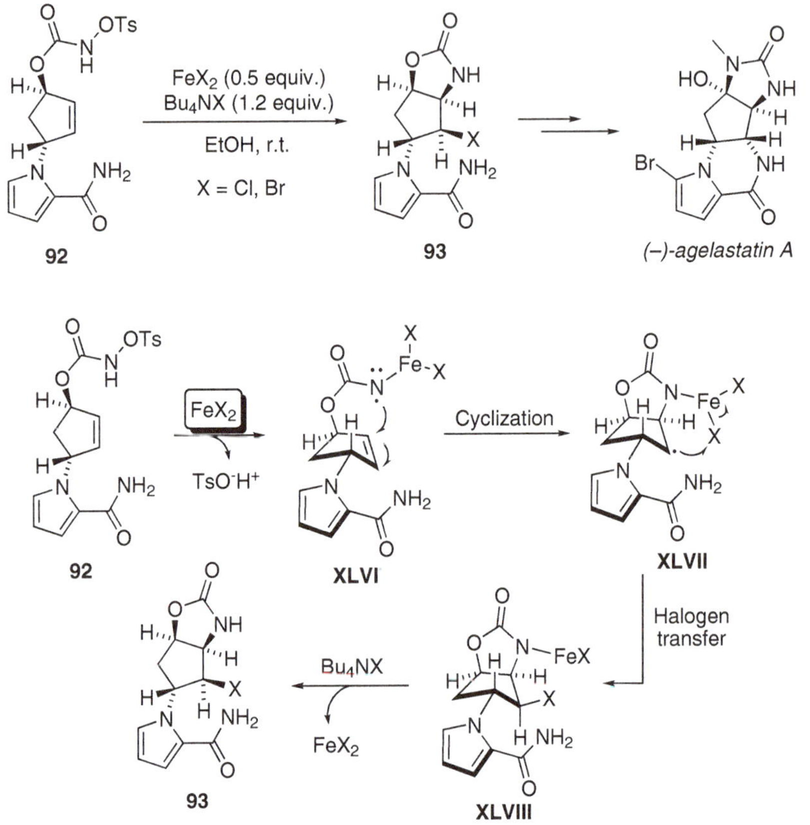

SCHEME 5.26 Synthesis of (-)-Agelastatin A using an iron(II)-mediated radical cyclization strategy.

The corresponding heterocycles **98** and **101** are obtained by using a Fe (II)-chiral bisoxazoline (BOX) catalyst with good stereoselectivity[81,82] (Scheme 5.28).

The outcome of the intramolecular process has been studied and the mechanistic proposal is as follows: The Fe(II)–ligand complex can reductively cleave the N–O σ-bond and possibly convert the styrenyl substrate **97** into iron-nitrenoid **LIII**. Then a stepwise cycloamination occurs affording two carbo-radical species **LIV** and **LV** that are in fast equilibrium. The aminohydroxylation occurs predominately from intermediate **LV** leading to the oxazolidone **98**.

Although the intermolecular process works in a similar way, it proceeds with slight differences. Thus again the iron(II)–ligand complex reductively cleaves the N–O bond converting the olefin **99** to an iron-nitrenoid intermediate **LVI**. This nitrenoid then initiates the radical

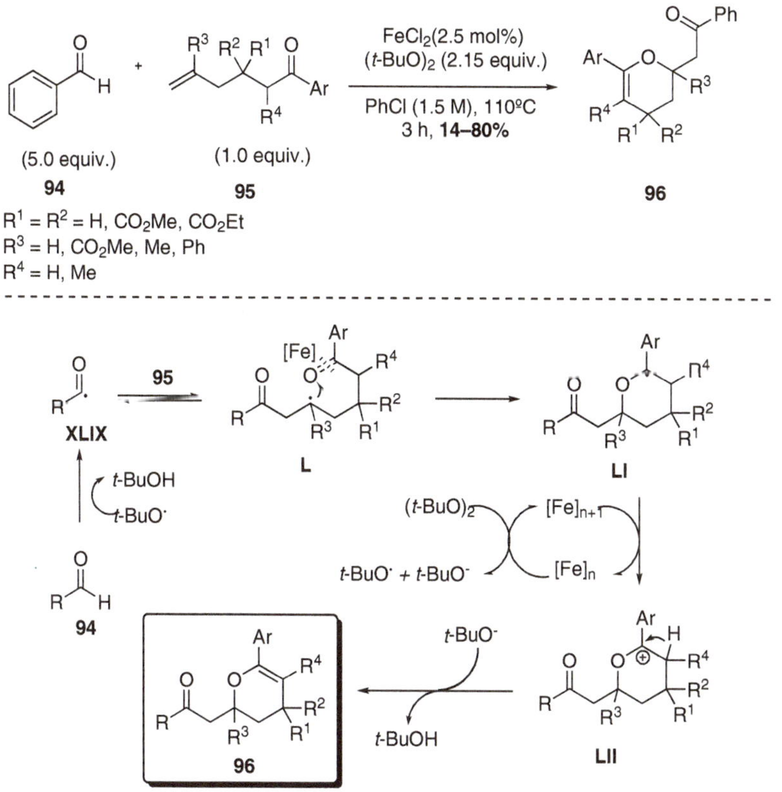

SCHEME 5.27 6-endo-trig radical cyclization at the synthesis of 3,4-dihydro-2*H*-pyrans.

amination to afford the radical species **LVII**. Then, **LVII** is presumably oxidized by the iron center to a carbocation **LVIII**, which can be rapidly captured by the neighboring carbamate group affording the corresponding oxazolidinone **101** (Scheme 5.28).

4. *Pyrrolidines and furans*

Arylmethyl-substituted pyrrolidines and tetrahydrofurans **103** are produced by iron(II)-catalyzed tandem cyclization and cross-coupling reaction of alkyl iodides **102** and aryl Grignard reagents.[83] The substituents of the alkene greatly influence the progress of the tandem reaction, with di- and tri-substituted alkenes affording cyclized pyrrolidines. The catalytic cycle proposed is as follows: The iron salt ($FeCl_2$) reacts with the Grignard reagent to form an aryliron(I) complex. The aryl iron(I) complex undergoes a single electron transfer reaction to yield the radical anion of the substrate **LIX**, which promotes 5-exo-dig cyclization rapidly to attain **LX**. The radical **LX** is added to the iron center to form an iron (III) intermediate **LXI**, which undergoes reductive elimination affording

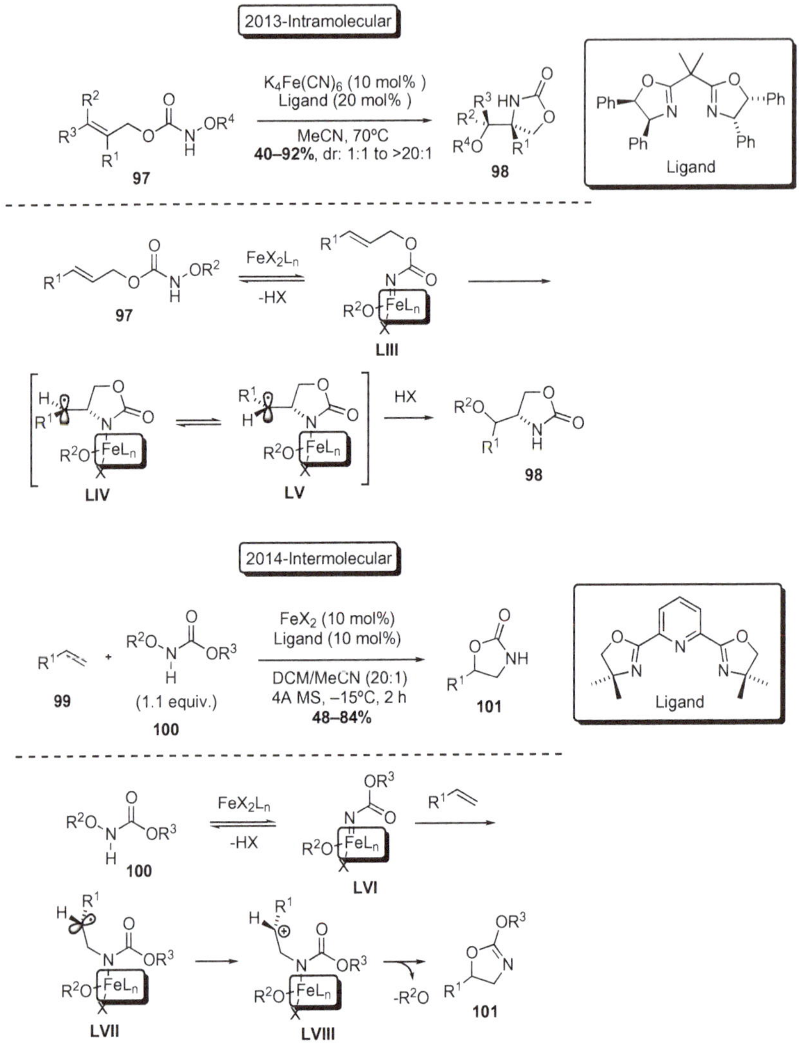

SCHEME 5.28 Intra- and intermolecular strategies for the synthesis of oxazolidinones.

the final product **103**. Transmetallation with Grignard reagent regenerates the reactive iron species to complete the catalytic cycle (Scheme 5.29).

5. *Quinoline fused lactones*

A dual-oxidative dehydrogenative (DOD) tandem annulation of glycine derivatives **104** with tetrahydrofurans **105** has been reported for the synthesis of quinoline-fused lactones **106** (Scheme 5.30).[84]

The proposed mechanism for this iron-catalyzed process is based on the conversion of the glycine derivative **104** and tetrahydrofuran **105** to

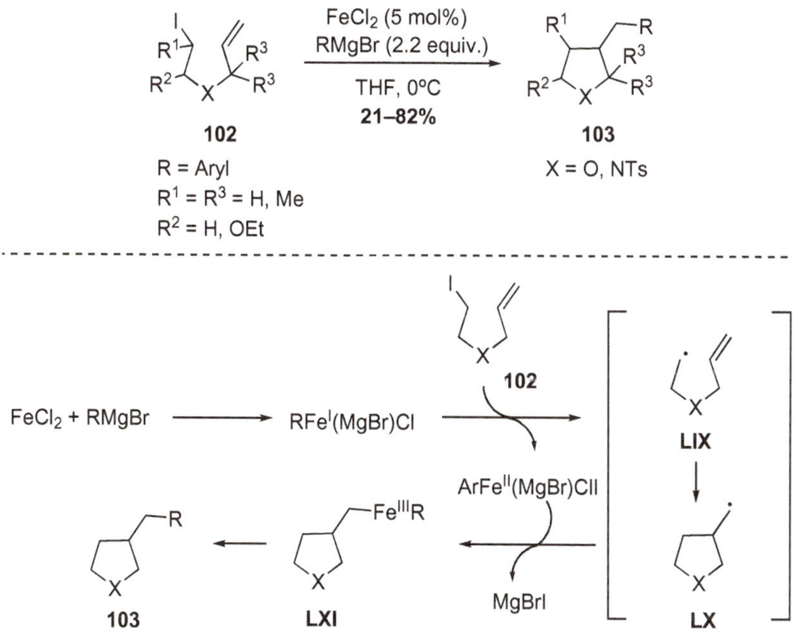

SCHEME 5.29 Iron(II) catalyzed the synthesis of pyrrolidines and furans.

aryl imine **LXVII** and dihydrofuran **LXVI** intermediates using *tert*-butylhydroperoxide as oxidant and the iron(II) salt as the catalyst. The radical process involves the formation of iron(III) species intermediates. Then an imino-Diels-Alder reaction between **LXVII** and **LXVI** occurs to generate the tetrahydroquinazoline intermediate **LXVIII**. Subsequently, **LXVIII** is transformed to 2,3-disubstituted quinolone intermediate **LXIX** under acidic conditions. Finally, aromatization of this intermediate followed by intramolecular ester exchange ring closure takes place to afford the corresponding heterocycle **106** (Scheme 5.30).

6. *Furo[2,3-b]oxane*

5-Exo-trig radical reductive cyclizations of bromoacetals or iodocetals **107** catalyzed by iron(II) in the presence of a reducing agent effectively mediates the synthesis of furo[2,3-*b*]oxane cores **108**.[85,86] The reaction is carried out using $Fe^{II}Cl_2(dppe)_2$ as precatalyst since it reacts with sodium borohydride to afford $HFe^{II}Cl(dppe)_2$. The one-electron reduction of $HFe^{II}Cl(dppe)_2$ gives the anionic hydride $[HFe^{I}Cl(dppe)_2]^-$ which transfers one electron to **107** to generate the radical anion intermediate **LXXIII**. Halogen atom transfer followed by the nucleophilic substitution of the resulting primary alkyl halide **LXXIV** with the borohydride leads to the final product **108**. An alternative route is proposed via the formation of intermediate **LXXV** which leads to the final product **107** with no substantial changes in the yield.

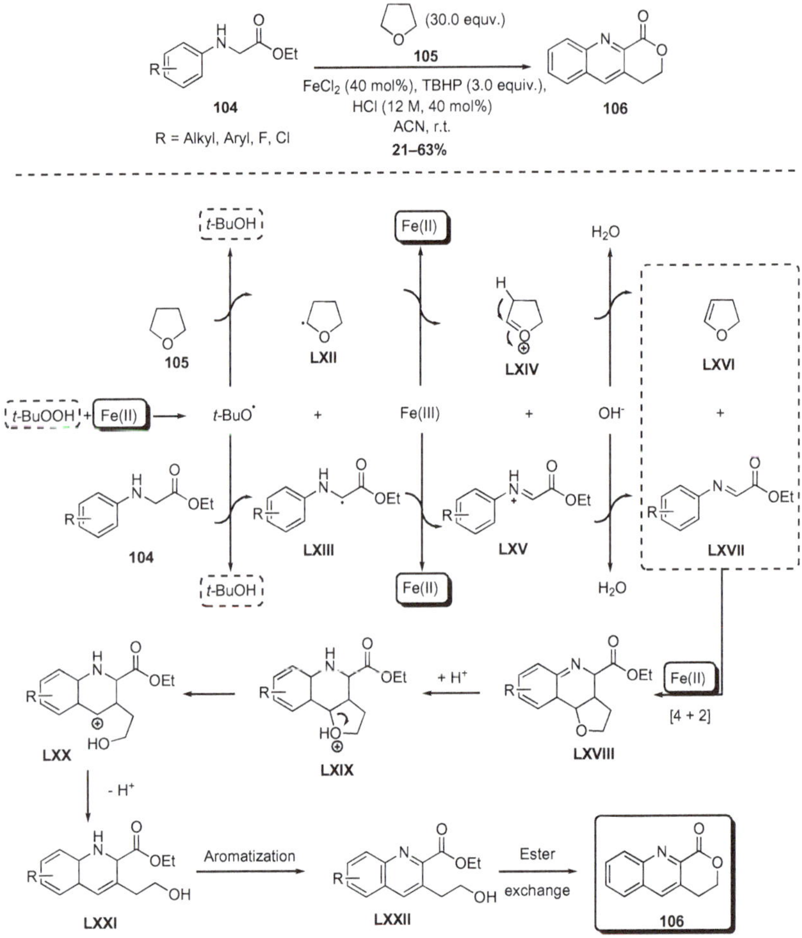

SCHEME 5.30 Proposed mechanism of a dual-oxidative dehydrogenative (DOD) tandem annulation in the synthesis of quinoline fused lactones.

Initially, this process was developed using iron(II) chloride salts, but the catalytic iron species formed was highly reactive to control the formation of by-products. Thus the reactivity was modulated by introducing *bis*-phosphine ligands. The catalytic activity of $Fe^{II}Cl_2(dppe)_2$ proved to be more versatile than that of iron(II) chloride (Scheme 5.31).

7. *Pyrrolidines*

Exposure of organic azides **109** to iron(II) dipyrrinato catalyst **110** furnishes pyrrolidines **111** bearing complex substitution patterns. A direct amination of C–H bond takes place, which showcases the advantage of the iron-mediated catalysis in activating C–H bonds[87] (Scheme 5.32).

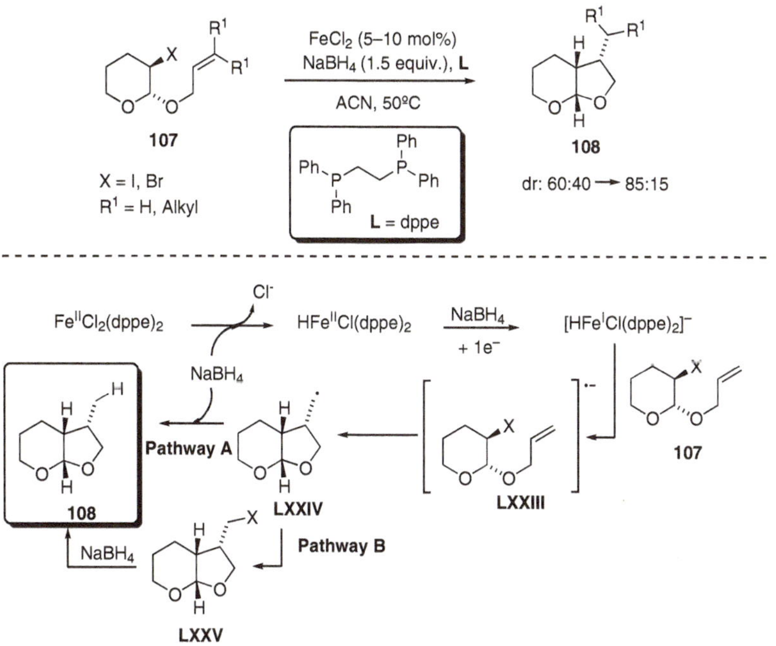

SCHEME 5.31 5-Exo-trig radical reductive cyclizations catalyzed by iron(II) in the presence of a reducing agent mediates the synthesis of furo[2,3-*b*]oxane cores.

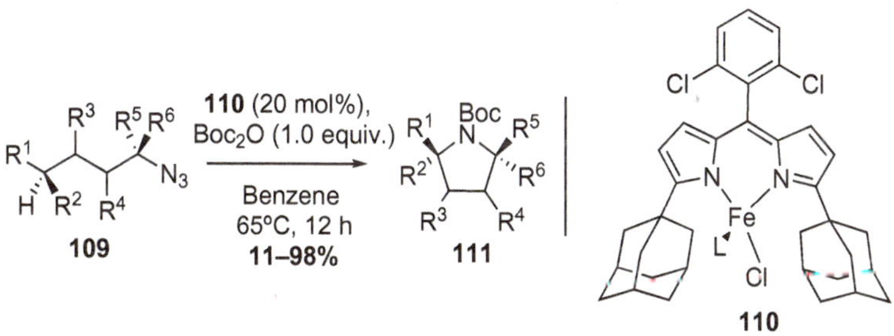

SCHEME 5.32 Iron-mediated catalysis, activating C-H bond, in the synthesis of pyrrolidines.

The process is hypothesized to occur in three steps: (1) oxidation of the Fe(II) catalyst to an Fe(III) imido radical **LXXVI** by the alkyl azide **109** substrate, (2) intramolecular hydrogen atom abstraction to generate an alkyl radical and an Fe(III) amide **LXXVII**, and (3) a radical recombination to form the observed pyrrolidine product **111**. Alternatively a direct C–H bond insertion by Fe(III) to afford the imido radical intermediate **LXXVIII** can take place. Both reaction pathways require the substrate C–H bond to be in close proximity to the reactive iron-imido radical. Therefore the imido radical resides in the plane defined by the iron and the dipyrrin ligand, surrounded by large pyrrolide adamantly substituents. This conformation requires that the C–H bond substrate approaches the imido radical opposite the chloride ligand. Once this orientation is obtained, C–H functionalization is fast, resulting in retention of configuration (Scheme 5.33).

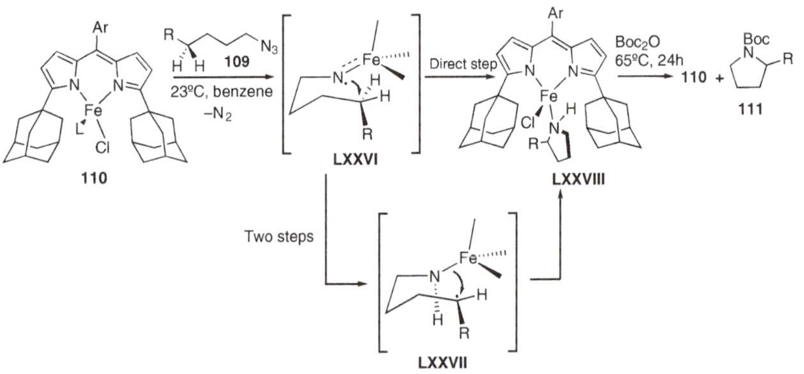

SCHEME 5.33 Mechanistic proposal of pyrrolidines through activation of C-H bond.

8. *Benzothiazole*

   Direct C–H functionalization followed by C–S bond formation was achieved to attain the synthesis of benzothiazoles **113**. The main difference of this procedure with respect to the examples described previously is the use of iron(III) salts to trigger the radical process. Sodium persulfate was used as the oxidant according to the mechanism proposed in Scheme 5.34. *N*-Phenyl benzothiamide **112** is oxidized by iron(III) losing one electron and a proton to form the radical intermediate **LXXIX**. In the meantime, iron(III) species are reduced to iron(II) species which are then reoxidized by sodium persulfate to regenerate iron(III). Then the cyclization of the thioyl radical **LXXIX** intermediate followed by the oxidation in the presence of sodium persulfate gave the final benzothiazole **113**. By-product **114** is also observed albeit in low yields.[88]
9. A wide variety of quinazolines and quinazolinones have been synthesized by using aerobic functionalization of sp^3 C–H bonds from carboxylic acid derivatives **114** and *O*-substituted anilines **115** and **116**.[89] A mechanism proposed for this transformation resides in the oxidation of the C–H bond with the iron/oxygen system to form an α-hydroxycarboxylic acid **LXXXII** via a radical pathway. Dehydrogenation affords oxocarboxylic acid **LXXXIII**. Subsequently the resulting intermediate **LXXXIII** undergoes decarboxylation, generating an aldehyde **LXXXIV** which is then captured by *O*-substituted anilines to produce the quinazoline **118** and quinazolinone **117** compounds under oxygen atmosphere (Scheme 5.35).

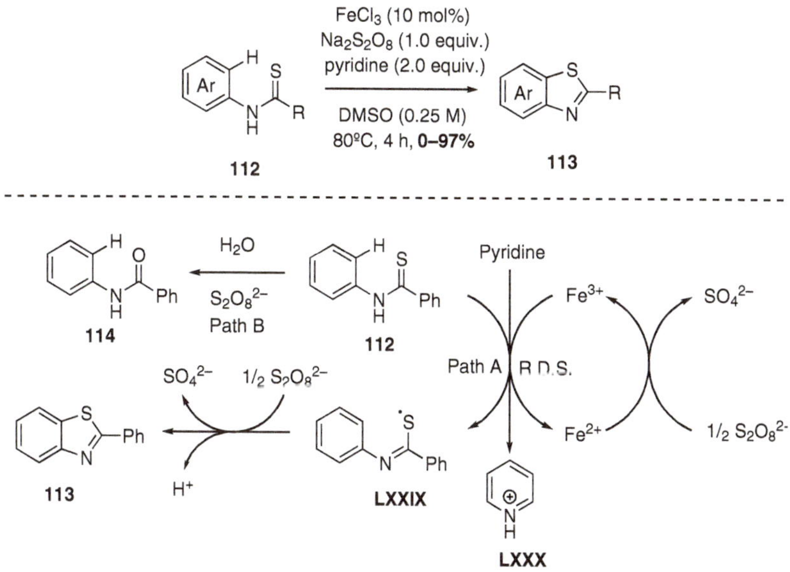

SCHEME 5.34 Iron trichloride catalyzed direct C-H functionalization followed by C-S bond formation. Synthesis of benzothiazoles.

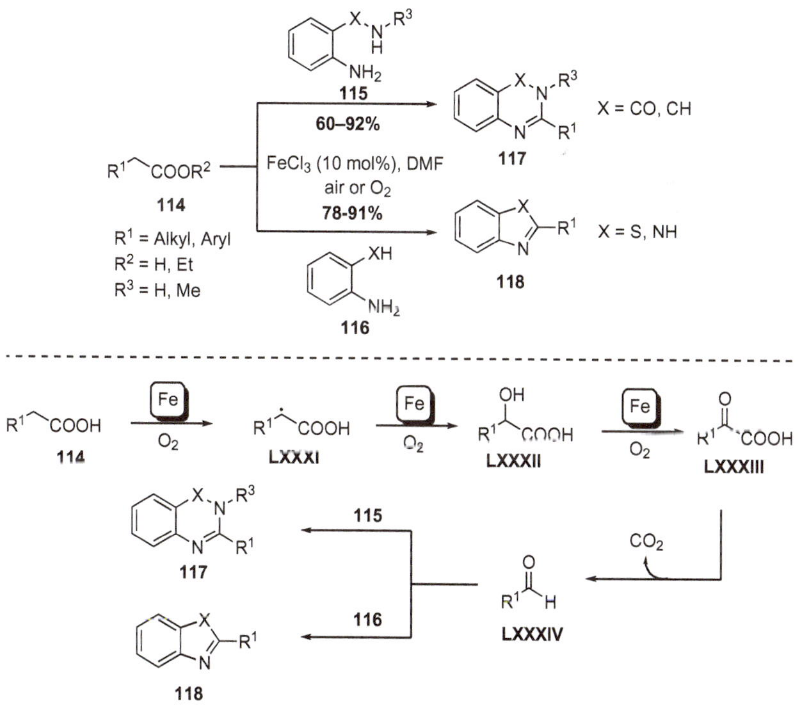

SCHEME 5.35 Oxidation of the C-H bond with an iron/oxygen system. Synthesis of quinazoline and quinazolinone compounds.

5.5 CONCLUSIONS AND OUTLOOK

As it has been reported throughout the chapter, iron is involved in a wide number of chemical reactions to synthesize different types of heterocyclic structures. Although initially its behavior may appear to be merely chaotic, there is a trend in its reactivity which can be modulated by judicious choice of different ligands. Synthetic methods spanning from radical processes to σ- and π-bond activations have spotlighted the versatility of iron as a suitable eco-friendly metal for performing a substantial number of cyclization processes. We strongly believe that iron is a reliable and viable choice for catalysis with a high potential for industrial applications. Several fruitless attempts to dismiss the potential of iron as a catalyst has been chased, but there will come a time when the use of iron(III) salts and derivatives will be used for enantioselective cyclizations not for a minority of organic chemists but for a majority of them. Likewise, its participation in the synthesis of medium-sized ring heterocycles has yet to be explored in detail. We expect that the rationalization attempt that we have pursued in this chapter may be helpful for the future development of novel catalysts for different types of reactions.

ACKNOWLEDGMENTS

This research was supported by the Spanish MINECO and Canary Island goverment (ACIISI), cofinanced by the European Regional Development Fund (ERDF) (CTQ2014-56362-C2-1-P and ProID2017010118). D.A.C. thanks the Spanish MINECO for an F.P.I fellowship. J.M.L.-S. thanks the Canary Island Government (ACIISI) for a predoctoral contract. P.O.M. was supported by a Marie Curie IOF from the European Union's Seventh Framework Program FP7/2007–2013 under REA Grant Agreement No. 623155.

REFERENCES

1. Spergel, D. N.; Verde, L.; Peiris, H. V.; Komatsu, E.; Nolta, M.; Bennett, C.; Halpern, M.; Hinshaw, G.; Jarosik, N.; Kogut, A. First-Year Wilkinson Microwave Anisotropy Probe (WMAP)* Observations: Determination of Cosmological Parameters. *Astrophys. J. Suppl. Ser.* **2003,** *148*, 175–194.
2. Pospelov, M.; Pradler, J. Big Bang Nucleosynthesis as a Probe of New Physics. *Annu. Rev. Nucl. Part. Sci.* **2010,** *60*, 539–568.
3. Quintero-Gutiérrez, A. G.; González-Rosendo, G.; Sánchez-Muñoz, J.; Polo-Pozo, J.; Rodríguez-Jerez, J. J. Bioavailability of Heme Iron in Biscuit Filling Using Piglets as an Animal Model for Humans. *Int. J. Biol. Sci.* **2008,** *4*, 58–62.
4. Aisen, P.; Enns, C.; Wessling-Resnick, M. Chemistry and Biology of Eukaryotic Iron Metabolism. *Int. J. Biochem. Cell Biol.* **2001,** *33*, 940–959.
5. Shils, M. E.; Shike, M.; Ross, A. C.; Caballero, B.; Cousins, R. J. *Modern Nutrition in Health and Disease;* Lippincott Williams & Wilkins: Philadelphia, PA, 2006248–270.
6. Guerinot, M. L. Microbial Iron Transport. *Annu. Rev. Microbiol.* **1994,** *48*, 743–772.
7. Askwith, C.; Kaplan, J. Iron and Copper Transport in Yeast and Its Relevance to Human Disease. *Trends Biochem. Sci.* **1998,** *23*, 135–138.

8. McDowell, L. R. *Minerals in Animal and Human Nutrition;* Academic Press Inc: Cambridge, MA, 1992.
9. Abbaspour, N.; Hurrell, R.; Kelishadi, R. Review on Iron and Its Importance for Human Health. *J. Res. Med. Sci.* **2014,** *19*, 164–174.
10. Kaneda, K.; Mizugaki, T.; Mitsudome, T. *Green Catalysis. Encyclopedia of Catalysis;* John Wiley & Sons, Inc.: Hoboken, NJ, 2010.
11. Bommarius, A. S. Biocatalysis: A Status Report. *Annu. Rev. Chem. Biomol. Eng.* **2015,** *6*, 319–345.
12. List, B. Introduction: Organocatalysis. *Chem. Rev.* **2007,** *107*, 5413–5415.
13. Holzwarth, M. S.; Plietker, B. Biorelevant Metals in Sustainable MetalCatalysis—A Survey. *Chem. Cat. Chem.* **2013,** *5*, 1650–1679.
14. Plietker, B. Sustainability in Catalysis-Concept or Contradiction? *Synlett* **2010,** 2049–2058.
15. Plietker, B. *Iron Catalysis in Organic Chemistry: Reactions and Applications;* John Wiley & Sons: Hoboken, NJ, 2008.
16. Kharasch, M. S.; Fields, E. K. Factors Determining the Course and Mechanisms of Grignard Reactions. IV. The Effect of Metallic Halides on the Reaction of Aryl Grignard Reagents and Organic Halides. *J. Am. Chem. Soc.* **1941,** *63*, 2316–2320.
17. Kealy, T.; Pauson, P. A New Type of Organo-Iron Compound. *Nature* **1951,** *168*, 1039–1040.
18. Tamura, M.; Kochi, J. K. Vinylation of Grignard Reagents. Catalysis by Iron. *J. Am. Chem. Soc.* **1971,** *93*, 1487–1489.
19. Dyker, G. Transition Metal Catalyzed Coupling Reactions Under C–H Activation. *Angew. Chem. Int. Ed.* **1999,** *38*, 1698–1712.
20. Fürstner, A.; Leitner, A. Iron-Catalyzed Cross-Coupling Reactions of Alkyl-Grignard Reagents with Aryl Chlorides, Tosylates, and Triflates. *Angew. Chem. Int. Ed.* **2002,** *41*, 609–612.
21. Fürstner, A.; Leitner, A.; Méndez, M.; Krause, H. Iron-Catalyzed Cross-Coupling Reactions. *J. Am. Chem. Soc.* **2002,** *124*, 13856–13863.
22. Buchwald, S. L.; Bolm, C. On the Role of Metal Contaminants in Catalyses with $FeCl_3$. *Angew. Chem. Int. Ed.* **2009,** *48*, 5586–5587.
23. Kylmälä, T.; Valkonen, A.; Rissanen, K.; Xu, Y.; Franzén, R. *trans*-Tetrakis(pyridine) Dichloroiron(II) as Catalyst for Suzuki Cross-Coupling in Ethanol and Water. *Tetrahedron Lett.* **2008,** *49*, 6679–6681.
24. Bedford, R. B.; Nakamura, M.; Gower, N. J.; Haddow, M. F.; Hall, M. A.; Huwe, M.; Hashimoto, T.; Okopie, R. A. Iron-Catalysed Suzuki Coupling? A Cautionary Tale. *Tetrahedron Lett.* **2009,** *50*, 6110–6111.
25. Volla, C. M. R.; Vogel, P. Iron/Copper-Catalyzed C–C Cross-Coupling of Aryl Iodides with Terminal Alkynes. *Tetrahedron Lett.* **2008,** *49*, 5961–5964.
26. Taillefer, M., Xia, N.; Ouali, A. Efficient Iron/Copper Co-Catalyzed Arylation of Nitrogen Nucleophiles. *Angew. Chem. Int. Ed.* **2007,** *46*, 934–936.
27. Mao, J.; Xie, G.; Wu, M.; Guo, J.; Ji, S. Ligand-Free Iron/Copper Cocatalyzed Alkynylation Coupling Reactions. *Adv. Synth. Catal.* **2008,** *350*, 2477–2482.
28. Collman, J. P. *Principles and Applications of Organotransition Metal Chemistry.; University Science Books*: California, 1987324.
29. Guan, W.; Sayyed, F. B.; Zeng, G.; Sakaki, S. σ-Bond Activation of Small Molecules and Reactions Catalyzed by Transition-Metal Complexes: Theoretical Understanding of Electronic Processes. *Inorg. Chem.* **2014,** *53*, 6444–6457.

30. Arockiam, P. B.; Bruneau, C.; Dixneuf, P. H. Ruthenium(II)-Catalyzed C–H Bond Activation and Functionalization. *Chem. Rev.* **2012,** *112*, 5879–5918.
31. Lyons, T. W.; Sanford, M. S. Palladium-Catalyzed Ligand-Directed C–H Functionalization Reactions. *Chem. Rev.* **2010,** *110*, 1147–1169.
32. Ateşin, T. A.; Li, T.; Lachaize, S.; Brennessel, W. W.; García, J. J.; Jones, W. D. Experimental and Theoretical Examination of C–CN and C–H Bond Activations of Acetonitrile Using Zerovalent Nickel. *J. Am. Chem. Soc.* **2007,** *129*, 7562–7569.
33. Senn, H. M.; Ziegler, T. Oxidative Addition of Aryl Halides to Palladium(0) Complexes: A Density-Functional Study Including Solvation. *Organometallics* **2004,** *23*, 2980–2988.
34. Amgoune, A.; Bourissou, D. σ-Acceptor, Z-Type Ligands for Transition Metals. *Chem. Commun.* **2011,** *47*, 859–871.
35. Qiu, R.; Chen, Y.; Yin, S.-F.; Xu, X.; Au, C.-T. A Mini-Review on Air-Stable Organometallic Lewis Acids: Synthesis, Characterization, and Catalytic Application in Organic Synthesis. *RSC Adv.* **2012,** *2*, 10774–10793.
36. Guan, W.; Sakaki, S.; Kurahashi, T.; Matsubara, S. Theoretical Mechanistic Study of Novel Ni(0)-Catalyzed [6–2 + 2] Cycloaddition Reactions of Isatoic Anhydrides with Alkynes: Origin of Facile Decarboxylation. *Organometallics* **2013,** *32*, 7564–7574.
37. Morales-Morales, D.; Jensen, C. G. *The Chemistry of Pincer Compounds;* Elsevier: Amsterdam, 2011.
38. Peña-López, M.; Neumann, H.; Beller, M. Iron(II) Pincer-Catalyzed Synthesis of Lactones and Lactams Through a Versatile Dehydrogenative Domino Sequence. *Chem. Cat. Chem.* **2015,** *7*, 865–871.
39. Gao, X.; Pan, Y.-M.; Lin, M.; Chen, L.; Zhan, Z.-P. Facile One-Pot Synthesis of Three Different Substituted Thiazoles From Propargylic Alcohols. *Org. Biomol. Chem.* **2010,** *8*, 3259–3266.
40. Bosset, C.; Angibaud, P.; Stanfield, I.; Meerpoel, L.; Berthelot, D.; Guérinot, A.; Cossy, J. Iron-Catalyzed Synthesis of C2 Aryl- and N-Heteroaryl-Substituted Tetrahydropyrans. *J. Org. Chem.* **2015,** *80*, 12509–12525.
41. Cornil, J.; Gonnard, L.; Bensoussan, C.; Serra-Muns, A.; Gnamm, C.; Commandeur, C.; Commandeur, M.; Reymond, S.; Guérinot, A.; Cossy, J. Iron- and Indium-Catalyzed Reactions Toward Nitrogen- and Oxygen-Containing Saturated Heterocycles. *Acc. Chem. Res.* **2015,** *48*, 761–773.
42. Pérez, S. J.; Miranda, P. O.; Cruz, D. A.; Fernández, I.; Martín, V. S.; Padrón, J. I. Iron (III)-Catalyzed Prins Cyclization Towards the Synthesis of trans-Fused Bicyclic Tetrahydropyrans. *Synthesis* **2015,** *47*, 1791–1798.
43. Zheng, K.; Liu, X.; Qin, S.; Xie, M.; Lin, L.; Hu, C.; Feng, X. Completely OH-Selective $FeCl_3$-Catalyzed Prins Cyclization: Highly Stereoselective Synthesis of 4-OH-Tetrahydropyrans. *J. Am. Chem. Soc.* **2012,** *134*, 17564–17573.
44. Pérez, S. J.; Purino, M.; Miranda, P. O.; Martín, V. S.; Fernández, I.; Padrón, J. I. Prins Cyclization Catalyzed by a FeIII/Trimethylsilyl Halide System: The Oxocarbenium Ion Pathway Versus the [2 + 2] Cycloaddition. *Chem. Eur. J.* **2015,** *21*, 15211–15217.
45. Miranda, P. O.; León, L. G.; Martín, V. S.; Padrón, J. I.; Padrón, J. M. One-Pot Synthesis and SAR Study of *cis*-2, 6-Dialkyl-4-Chloro-Tetrahydropyrans. *Bioorg. Med. Chem. Lett.* **2006,** *16*, 3135–3138.
46. Xu, T.; Yang, Q.; Li, D.; Dong, J.; Yu, Z.; Li, Y. Iron(III)-Catalyzed Cyclization of Alkynyl Aldehyde Acetals: Experimental and Computational Studies. *Chem. Eur. J.* **2010,** *16*, 9264–9272.

47. Miranda, P. O.; Díaz, D. D.; Padrón, J. I.; Ramírez, M. A.; Martín, V. S. Fe(III) Halides as Effective Catalysts in Carbon–Carbon Bond Formation: Synthesis of 1, 5-Dihalo-1, 4-Dienes, α, β-Unsaturated Ketones, and Cyclic Ethers. *J. Org. Chem.* **2005,** *70*, 57–62.
48. Calmus, L.; Corbu, A.; Cossy, J. *2H*-Chromenes Generated by an Iron(III) Complex-Catalyzed Allylic Cyclization. *Adv. Synth. Catal.* **2015,** *357*, 1381–1386.
49. Casola, K. K.; Back, D. F.; Zeni, G. Iron-Catalyzed Cyclization of Alkynols with Diorganyl Diselenides: Synthesis of 2,5-Dihydrofuran, 3,6-Dihydro-2H-pyran, and 2,5-Dihydro-1H-Pyrrole Organoselanyl Derivatives. *J. Org. Chem.* **2015,** *80*, 7702–7712.
50. Jefferies, L. R.; Weber, S. R.; Cook, S. P.; Iron-Catalyzed, C. –N. Bond Formation Via the Beckmann Rearrangement. *Synlett* **2015,** *26*, 331–334.
51. Carballo, R. M.; Ramírez, M. A.; Rodríguez, M. L.; Martín, V. S.; Padrón, J. I. Iron(III)-Promoted Aza-Prins-Cyclization: Direct Synthesis of Six-Membered Azacycles. *Org. Lett.* **2006,** *8*, 3837–3840.
52. Qiu, I. W.; Zhang, X. G.; Tang, R. Y.; Zhong, P.; Li, J. H. Iron-Catalyzed Tandem Reactions of 2-Halobenzenamines with Isothiocyanates Leading to 2-Aminobenzothiazoles. *Adv. Synth. Catal.* **2009,** *351*, 2319–2323.
53. Yang, D.; Fu, H.; Hu, L.; Jiang, Y.; Zhao, Y. Environmentally Friendly Iron-Catalyzed Cascade Synthesis of 1, 2, 4-Benzothiadiazine 1, 1-Dioxide and Quinazolinone Derivatives. *J. Comb. Chem.* **2009,** *11*, 653–657.
54. Pérez, S. J.; Purino, M. A.; Cruz, D. A.; López-Soria, J. M.; Carballo, R. M.; Ramírez, M. A.; Fernández, I.; Martín, V. S.; Padrón, J. I. Enantiodivergent Synthesis of (+)-and (−)-Pyrrolidine 197B: Synthesis of trans-2, 5-Disubstituted Pyrrolidines by Intramolecular Hydroamination. *Chem. Eur. J* **2016,** *22*, 15529–15535.
55. Carballo, R. M.; Purino, Mn; Ramírez, M. A.; Martín, Vc. S.; Padrón, J. I. Iron(III)-Catalyzed Consecutive Aza-Cope–Mannich Cyclization: Synthesis of trans-3, 5-Dialkyl Pyrrolidines and 3, 5-Dialkyl-2, 5-Dihydro-*1 H*-Pyrroles. *Org. Lett.* **2010,** *12*, 5334–5337.
56. Wang, W.; Li, G.-P.; Wang, S.-F.; Shi, Z.-F.; Cao, X.-P. Direct and Short Construction of the ACDE Ring System of Daphenylline. *Chem. Asian J.* **2015,** *10*, 377–382.
57. Wang, W.; Lu, J.-T.; Zhang, H.-L.; Shi, Z.-F.; Wen, J.; Cao, X.-P. Formal Synthesis of (±)-Cycloclavine. *J. Org. Chem.* **2014,** *79*, 122–127.
58. Stauffacher, D.; Niklaus, P.; Tscherter, H.; Weber, H.; Hofmann, A. Cycloclavin, a New Alkaloid From Ipomoea Hildebrandtii Vatke. 71. Ergot Alkaloids. *Tetrahedron* **1969,** *25*, 5879–5887.
59. Incze, M.; Dörnyei, G.; Moldvai, I.; Temesvári-Major, E.; Egyed, O.; Szántay, C. New Routes to Clavine-Type Ergot Alkaloids. Part 2: Synthesis of the Last, so far not yet Synthesized Member of the Clavine Alkaloid Family, (±)-Cycloclavine. *Tetrahedron* **2008,** *64*, 2924–2929.
60. Petronijevic, F. R.; Wipf, P. Total Synthesis of (±)-Cycloclavine and (±)-5-epi-Cycloclavine. *J. Am. Chem. Soc.* **2011,** *133*, 7704–7707.
61. Ji, S.; Xia, J.; Xu, H. Dynamic Chemistry of Selenium: Se–N and Se–Se Dynamic Covalent Bonds in Polymeric Systems. *ACS Macro Lett.* **2016,** *5*, 78–82.
62. Kildahl, N. K. Bond Energy Data Summarized. *J. Chem. Educ.* **1995,** *72*, 423–424.
63. Neto, J. S.; Iglesias, B. A.; Back, D. F.; Zeni, G. Iron-Promoted Tandem Cyclization of 1,3-Diynyl Chalcogen Derivatives with Diorganyl Dichalcogenides for the Synthesis of Benzo[b] furan-Fused Selenophenes. *Adv. Synth. Catal.* **2016,** *358*, 3572–3585.
64. Yamazaki, S.; Fujinami, K.; Maitoko, Y.; Ueda, K.; Kakiuchi, K. Lewis Acid-Promoted Cyclization Reactions of Alkenyl Ethenetricarboxylates: Stereoselective Synthesis of 2-Oxotetrahydrofurans and 2-Oxopyrrolidines. *J. Org. Chem.* **2013,** *78*, 8405–8416.

65. Zhang, Z.; Li, J.; Zhang, G.; Ma, N.; Liu, Q.; Liu, T. Iron-Catalyzed Intramolecular C (sp2)−N Cyclization of 1-(N-Arylpyrrol-2-yl)ethanone O-Acetyl Oximes Toward Pyrrolo [1,2-a]quinoxaline Derivatives. *J. Org. Chem.* **2015,** *80*, 6875−6884.
66. Elschenbroich, C. third completely revised and extended edition *Organometallics;* Wiley-VCH: Weinheim, 2006.
67. Gay, R. M.; Manarin, F.; Schneider, C. C.; Barancelli, D. A.; Costa, M. D.; Zeni, G. $FeCl_3$-Diorganyl Dichalcogenides Promoted Cyclization of 2-Alkynylanisoles to 3-Chalcogen Benzo[b]furans. *J. Org. Chem.* **2010,** *75*, 5701−5706.
68. Jiang, H.; Yao, W.; Cao, H.; Huang, H.; Cao, D. Iron-Catalyzed Domino Process for the Synthesis of α-Carbonyl Furan Derivatives via One-Pot Cyclization Reaction. *J. Org. Chem.* **2010,** *75*, 5347−5350.
69. Lin, A.; Zhang, Z.-W.; Yang, J. Iron-Catalyzed Reductive Cyclization of 1,6-Enynes. *Org. Lett.* **2014,** *16*, 386−389.
70. Mantovani, A. C.; Goulart, T. A. C.; Back, D. F.; Menezes, P. H.; Zeni, G. Iron(III) Chloride and Diorganyl Diselenides-Mediated 6-endo-dig Cyclization of Arylpropiolates and Arylpropiolamides Leading to 3-Organoselenyl-*2H*-Coumarins and 3 Organoselenyl-Quinolinones. *J. Org. Chem.* **2014,** *79*, 10526−10536.
71. Speranҫa, A.; Godoi, B.; Pinton, S.; Back, D. F.; Menezes, P. H.; Zeni, G. Regioselective Synthesis of Isochromenones by Iron(III)/PhSeSePh-Mediated Cyclization of 2-Alkynylaryl Esters. *J. Org. Chem.* **2011,** *76*, 6789−6797.
72. Speranҫa, A.; Godoi, B.; Zeni, G. Iron(III) Chloride/Diorganyl Diselenides: A Tool for Intramolecular Cyclization of Alkynone O-Methyloximes. *J. Org. Chem.* **2013,** *78*, 1630−1637.
73. Iqbal, J.; Bhatia, B.; Nayyar, N. K. Transition Metal-Promoted Free-Radical Reactions in Organic Synthesis: The Formation of Carbon−Carbon Bonds. *Chem. Rev.* **1994,** *94*, 519−564.
74. Jahn, U. Radicals in Transition Metal Catalyzed Reactions? Transition Metal Catalyzed Radical Reactions? A Fruitful Interplay Anyway: Part 1. Radical Catalysis by Group 4 to Group 7 Elements. *Top. Curr. Chem.* **2012,** *320*, 121−189.
75. Gualandi, A.; Mengozzi, L.; Cozzi, P. G. Iron-Promoted Radical Reactions: Current Status and Perspectives. *Asian J. Org. Chem.* **2017,** *6*, 1160−1179.
76. Jahn, U. Radicals in Transition Metal Catalyzed Reactions? Transition Metal Catalyzed Radical Reactions?−A Fruitful Interplay Anyway: Part 2. Radical Catalysis by Group 8 and 9 Elements. *Top. Curr. Chem.* **2011,** *320*, 191−322.
77. Haslinger, S.; Kück, J. W.; Hahn, E. M.; Cokoja, M.; Pöthig, A.; Basset, J.-M.; Kühn, F. E. Making Oxidation Potentials Predictable: Coordination of Additives Applied to the Electronic Fine Tuning of an Iron(II) Complex. *Inorg. Chem.* **2014,** *53*, 11573−11583.
78. Crichton, R. *Iron Metabolism: From Molecular Mechanisms to Clinical Consequences;* John Wiley & Sons: Hoboken, NJ, 2016.
79. Shigeoka, D.; Kamon, T.; Yoshimitsu, T. Formal Synthesis of (−)-Agelastatin A: An Iron (II)-Mediated Cyclization Strategy. *Beilstein J. Org. Chem.* **2013,** *9*, 860−865.
80. Lv, L.; Li, Z. Iron-Catalyzed Oxidative Cyclization of 1,6-Enones with Aldehydes: Synthesis of Functionalized 3,4-Dihydro-*2H*-Pyrans. *Chin. J. Chem.* **2017,** *35*, 303−306.
81. Lu, D.-F.; Zhu, C.-L.; Jia, Z.-X.; Xu, H. Iron(II)-Catalyzed Intermolecular Amino-Oxygenation of Olefins through the N−O Bond Cleavage of Functionalized Hydroxylamines. *J. Am. Chem. Soc.* **2014,** *136*, 13186−13189.

82. Liu, G.-S.; Zhang, Y.-Q.; Yuan, Y.-A.; Xu, H. Iron(II)-Catalyzed Intramolecular Aminohydroxylation of Olefins with Functionalized Hydroxylamines. *J. Am. Chem. Soc.* **2013,** *135*, 3343–3346.
83. Kim, J. G.; Son, Y. H.; Seo, J. W.; Kang, E. J. Iron-Catalyzed Tandem Cyclization and Cross-Coupling Reactions of Iodoalkanes and Aryl Grignard Reagents. *Eur J. Org. Chem.* **2015,** 1781–1789.
84. Huo, C.; Chen, F.; Yuan, Y.; Xie, H.; Wang, Y. Iron Catalyzed Dual-Oxidative Dehydrogenative (DOD) Tandem Annulation of Glycine Derivatives with Tetrahydrofurans. *Org. Lett.* **2015,** *17*, 5028–5031.
85. Kyne, S. H.; Lévêque, C.; Zheng, S.; Fensterbank, L.; Jutand, A.; Ollivier, C. Iron(II) Catalyzed Reductive Radical Cyclization Reactions of Bromoacetals in the Presence of $NaBH_4$: Optimization Studies and Mechanistic Insights. *Tetrahedron* **2016,** *72*, 7727–7737.
86. Ekomié, A.; Lefèvre, G.; Fensterbank, L.; Lacôte, E.; Malacria, M.; Ollivier, C.; Jutand, A. Iron-Catalyzed Reductive Radical Cyclization of Organic Halides in the Presence of $NaBH_4$: Evidence of an Active Hydrido Iron(I) Catalyst. *Angew. Chem. Int. Ed.* **2012,** *51*, 6942–6946.
87. Hennessy, E. T.; Betley, T. A. Complex N-Heterocycle Synthesis Via Iron-Catalyzed, Direct C–H Bond Amination. *Science* **2013,** *340*, 591–595.
88. Wang, H.; Wang, L.; Shang, J.; Li, X.; Wang, H.; Gui, J.; Lei, A. Fe-Catalysed Oxidative C–H Functionalization/C–S Bond Formation. *Chem. Commun.* **2012,** *48*, 76–78.
89. Chen, X.; Chen, T.; Ji, F.; Zhou, Y.; Yin, S.-F. Iron-Catalyzed Aerobic Oxidative Functionalization of sp3 C–H Bonds: A Versatile Strategy for the Construction of *N*-Heterocycles. *Catal. Sci. Tech.* **2015,** *5*, 2197–2202.

Chapter 6

Allene Aziridination as a Tool for the Synthesis of Complex Amines

Lu Liu and Jennifer M. Schomaker
University of Wisconsin, Madison, WI, United States

6.1 INTRODUCTION

Densely functionalized, stereochemically complex amine motifs, where the nitrogen is embedded in an array of three or more contiguous, chiral sp^3 carbons, occur frequently in bioactive natural products and pharmaceuticals (Fig. 6.1).[1] A number of alkene functionalization methods, including Sharpless asymmetric aminohydroxylation (SAA),[2] diamination,[3] and asymmetric aziridination[4] followed by ring-opening,[5] are typically used to install stereodefined amines adjacent to another heteroatom-bearing, chiral sp^3 carbon. However, issues quickly arise with these strategies when they are employed to prepare amines with three contiguous stereogenic carbons, termed "stereotriads" in this discussion. First, the required precursors must contain both a stereogenic, heteroatom-substituted sp^3 carbon at the allylic position and a specific alkene geometry; these substrates are not always trivial to synthesize.[6,7] Second, controlling the selectivity of the subsequent oxidation event is often highly dependent on the nature of the substrate, resulting in mixtures of regio- and diastereomers. This is particularly problematic in cases involving the nondirected installation of two different heteroatoms across the double bond of the chiral alkene.[8] The use of group transfer reactions, such as epoxidation and aziridination, avoids the problem of regioselectivity in the alkene oxidation event. However, depending on the substitution pattern of the original alkene, regioselectivity of the subsequent nucleophilic ring-opening can be challenging in the absence of significant steric or electronic differentiation between the two ring carbons. Although alkene functionalization is traditionally viewed as the standard strategy to access amine stereotriads, the lack of predictable catalyst or reagent control over regio- and stereoselective oxidations of a broad scope of chiral alkenes

Advances in Transition-Metal Mediated Heterocyclic Synthesis.
DOI: https://doi.org/10.1016/B978-0-12-811651-7.00006-6

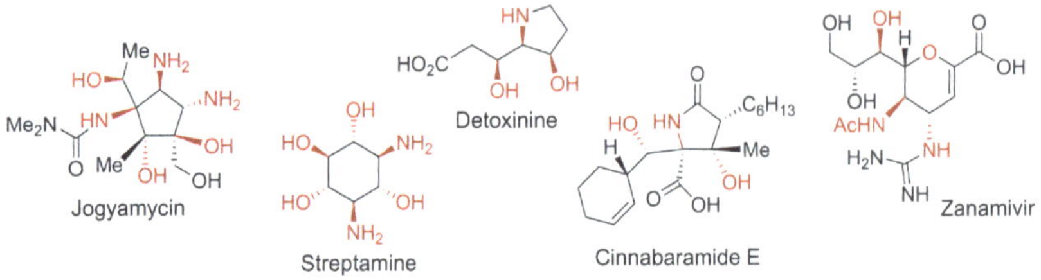

FIGURE 6.1 Bioactive molecules containing complex amine motifs.

limits flexibility in the substitution or stereochemical pattern of the amine triad products.

Allenes display structural features that can be exploited to develop powerful complementary alternatives to alkene oxidation methods for the synthesis of complex amine stereotriad building blocks (Scheme 6.1). For example, allenes are more stereochemically "flexible" than alkenes; there are no defined *E* or *Z* geometries associated with unsaturation in the allene. The appropriate reaction design could transform a single allene precursor selectively into several possible stereoisomeric products under the auspices of reagent control, rather than relying on substrates with defined alkene geometries. Allenes can also contain axial chirality (chirality arising from molecules having an axis of chirality, as opposed to a stereogenic sp^3 carbon) that might be effectively transferred to point chirality in the amine triad products, eliminating the need for chiral catalysts, and avoiding the mixtures of diastereomers that often result from oxidation of a chiral alkene precursor. Finally, the controlled introduction of an array of heteroatoms into the allene might be envisaged to transform each unsaturated carbon of the allene to a new $C(sp^3)$-heteroatom bond in a flexible manner. In this chapter, the left-most allene carbon is labeled as "1," the central carbon as "2," and the remaining carbon as "3"; this numbering convention is used throughout the text.

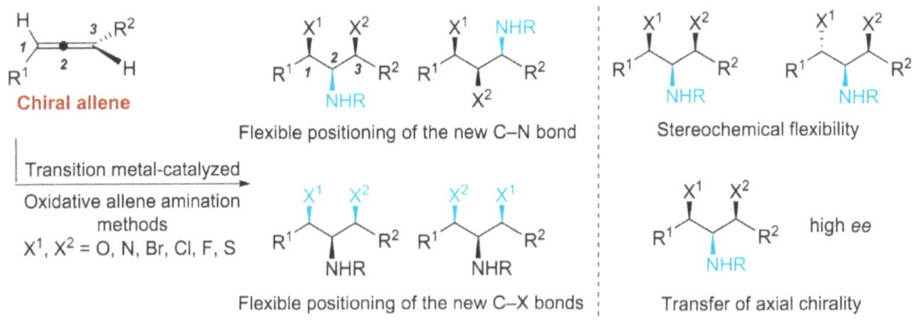

SCHEME 6.1 Goals for transition metal–catalyzed oxidative allene amination.

Contrary to the perception that allenes are exotic functional groups, they are easily accessible through simple manipulations of familiar and readily available building blocks; a selection of the most commonly used methods is

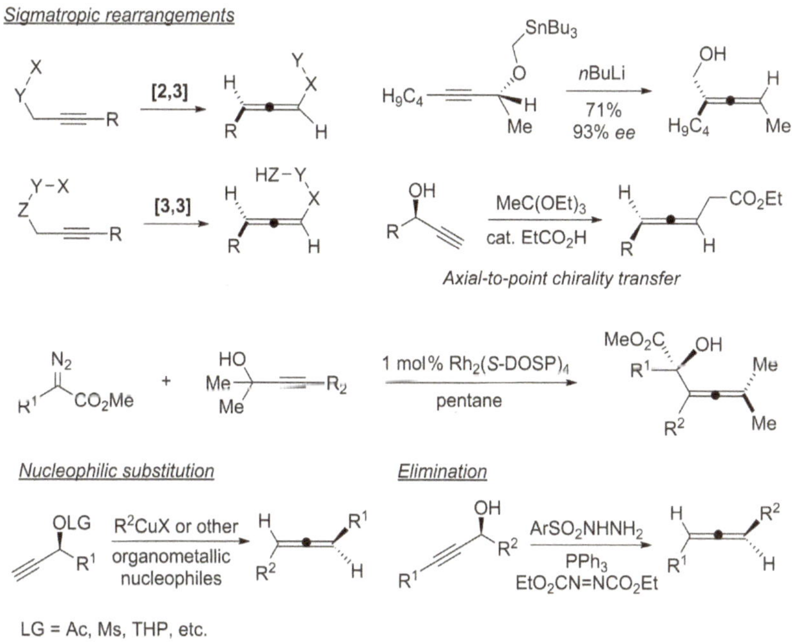

SCHEME 6.2 Selected methods for the synthesis of allenes.

illustrated in Scheme 6.2.[9,10] These precursors can be prepared in enantioenriched form and in some cases, the substitution patterns can be readily manipulated from a single key intermediate.

This chapter discusses strategies for transition metal–catalyzed oxidative allene amination that go beyond the well-reviewed metal-catalyzed allene hydroamination methods.[11] Focus is placed on allene aziridination reactions that produce unusually strained methyleneaziridine (herein, abbreviated as "MA") scaffolds with high levels of chemo-, regio-, and stereocontrol. These key intermediates show remarkable versatility in subsequent transformations to an array of densely functionalized, complex acyclic and heterocyclic amines. These amine motifs can be employed as building blocks for bioactive natural products, pharmaceuticals, unnatural amino acids, and as new scaffolds to investigate novel chemical space.

6.2 HETEROCYCLE FORMATION VIA NONTRANSITION METAL-CATALYZED ALLENE AMINATION

Early examples of allene amination employed either heat or ultraviolet light to generate the required reactive nitrogen species. For example, the first reported formal nitrogen atom transfers to an allene **1** involved photochemical decomposition of ethyl azidoformate, which underwent addition to dimethylallene to yield methylene oxazoline **2** in 47% yield as the sole product (Scheme 6.3A).[12] Two possible pathways, including the formation of MA **3** or a direct [3 + 2] cycloaddition of the allene with the carboethoxynitrene, were considered. This chemistry was later re-examined by the Gilbert group in the reaction of a nosyloxycarbamate with dimethylallene **1** (Scheme 6.3B).[13] Decomposition of the nosyloxycarbamate to a triplet nitrene furnished trace amounts of the MA **4**, as observed by MS and NMR analyses. Subsequent thermolysis yielded the oxazoline **2**, supporting the possibility of MAs as intermediates in this reaction.

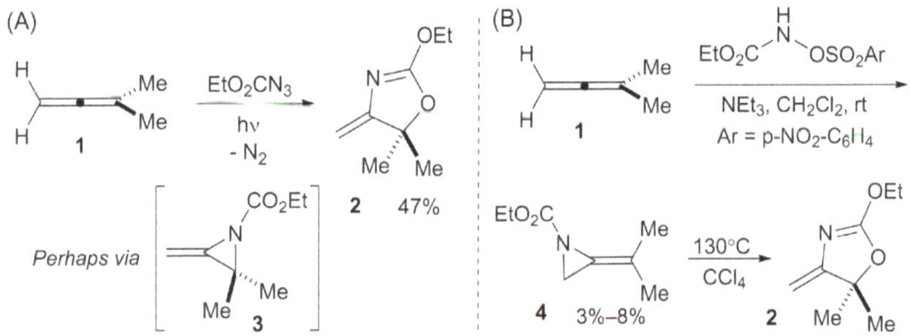

SCHEME 6.3 Early examples of non-transition metal-catalyzed nitrene additions to allenes under (A) photochemical conditions and (B) thermal conditions.

In 1975, Atkinson and Malpass described the treatment of methyl 2-methylbuta-2,3-dienonate **5** with *N*-aminophthalimide and $Pb(OAc)_4$ to deliver a strained 1,4-diazaspiro[2.2]pentane (DASP) heterocycle **6** (Scheme 6.4) in ~9:1 *dr*.[14] The further reactivity of this unusual ring system was not investigated until the work of Schomaker et al. in 2012 (*vide infra*).[15]

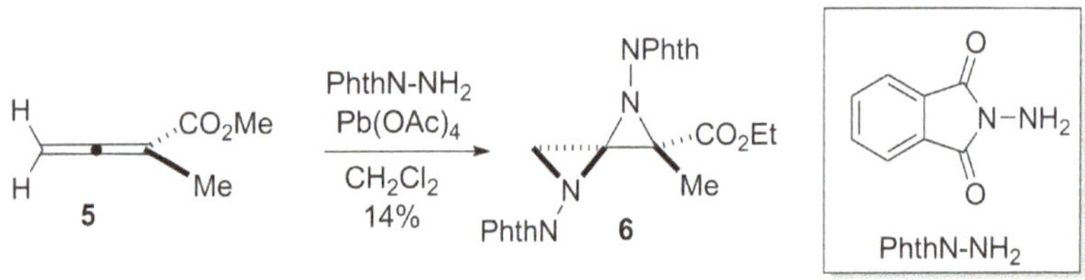

SCHEME 6.4 Intermolecular formation of a DASP.

6.3 TRANSITION METAL–CATALYZED ALLENE AMINATION NOT INVOLVING METHYLENEAZIRIDINE INTERMEDIATES

In 2010, the Blakey group reported interesting examples of intramolecular Rh-catalyzed allene amination to transform 1,1′-disubstituted allenes, such as

7 and **12**, into substituted aminocyclopropanes **11** and **13** (Scheme 6.5).[16] Imidoiodinane formation by reaction of the sulfamate of **7** with $PhI(OPiv)_2$ was followed by formation of a Rh-supported nitrene intermediate. The allene **7** was then proposed to attack the electrophilic metal nitrene to form an intermediate 2-amidoallyl cation, as represented by the resonance structures **8** and **9**. Ring closure to the iminocyclopropane **10** was followed by attack of the pivalate anion on the imine to yield the product **11**. The authors also found that a variety of Grignard reagents added to the intermediate imine formed from the reaction of allenes of the form **12** to furnish highly substituted aminocyclopropanes **13** with good diastereoselectivity.

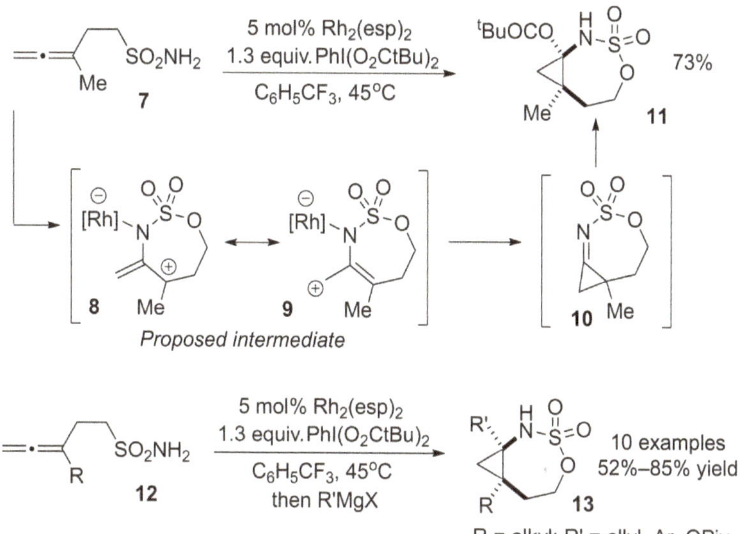

SCHEME 6.5 Rh-catalyzed rearrangement of allenes to aminocyclopropanes.

In studies to probe the mechanism of this oxidative allene amination, Blakey found support for a likely rearrangement of the proposed intermediate amidoallyl cation by showing that partial racemization of an enantioenriched allene **14** occurs under the reaction conditions (Scheme 6.6),

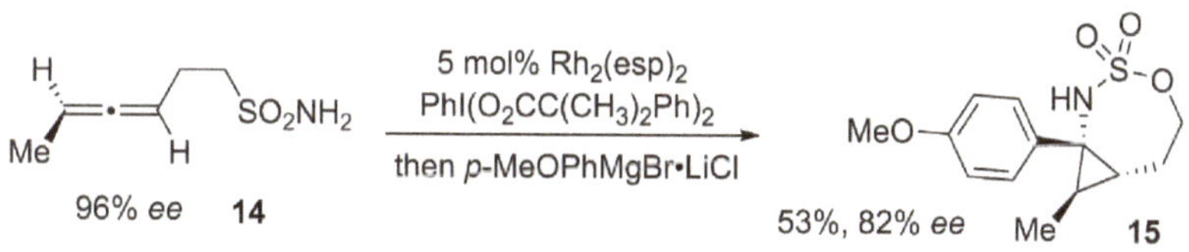

SCHEME 6.6 Transfer of axial-to-point chirality in aminocyclopropane formation.

presumably through intermediates similar to **8** and **9**. While the formation of a discrete MA was never reported in these reactions, the iminocyclopropane represents a valence tautomer and implies the aziridine might exist as a fleeting intermediate.[16]

Other coupling partners were shown to intercept the proposed 2-amidoallyl cation formed from Rh-catalyzed oxidative rearrangement of allenes. For example, the Blakey group demonstrated that addition of benzaldehyde to **16** gave the aminated tetrahydrofuran **17** in 40% yield through a [3 + 2] cycloaddition (Scheme 6.7A).[17] Regio- and diastereoselective [3 + 3] annulations of **18** using nitrone partner **19** led to the heterocycle **20** in Scheme 6.7B and was proposed to arise from attack of the nitrogen lone pair of the iminocyclopropane **22** on the electrophilic imine carbon of **19** to give **23**. Rearrangement of **23** to **24** releases the cyclopropane ring strain, while ring closure furnishes the final [3 + 3] annulation product **20**. The scope of the reaction was explored and shown to be highly selective, providing cycloadducts with excellent regio- and diastereoselectivity. Switching the substitution pattern of the allene to 1,1′,3-trisubstituted precursor **25** (Scheme 6.7C) also resulted in reaction with nitrone **19** to give the heterocyclic product **26** with high regioselectivity, albeit favoring C–O bond formation at C1 of the allene.

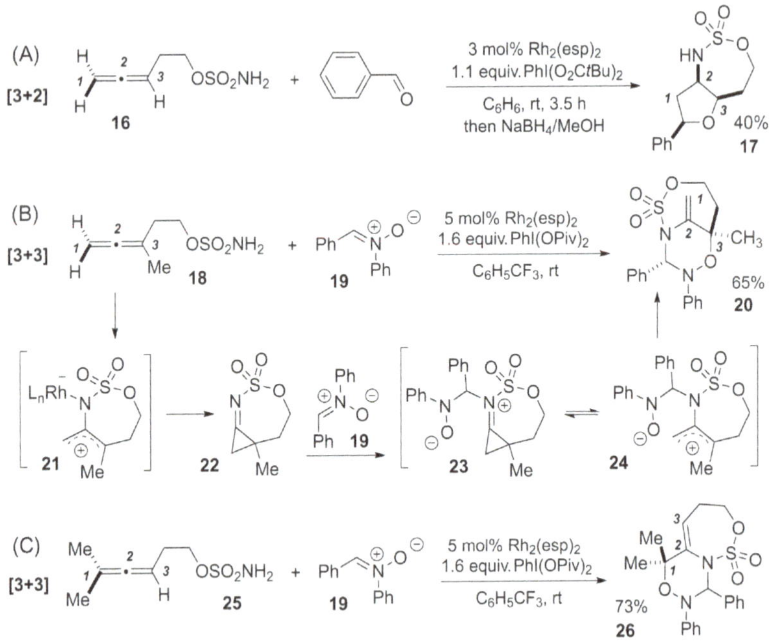

SCHEME 6.7 [3 + 2] cycloadditions of 2-amidoallyl cations formed from Rh-catalyzed allene amidation with aldehydes to furnish furans (A) and [3 + 3] reactions with nitrones (B, C).

The Feldman group has reported a series of reactions proposed to involve the addition of nitrogen radicals to allenes (selected example in Scheme 6.8).[18] The nitrogen-centered radical is generated from photolysis or thermal decomposition of the phenyl azide of **27**, with Cu(I) salts employed to improve the yield. While the complete mechanistic details are not clear, an initial intramolecular [3 + 2] reaction of the allene **27** with the azide moiety yields **28**, which is followed by rapid loss of nitrogen to generate **29**. This intermediate goes on to form the polycyclic heterocycle **30**. As the majority of chemistry involving the addition of nitrogen radicals to allenes does not involve transition metal catalysis, the reader is referred to the original papers for further details.[18–26] While these strategies illustrate beautiful examples of the formation of valuable polycyclic ring systems, the potential of transferring the allene's axial chirality to point chirality of the products is not utilized.

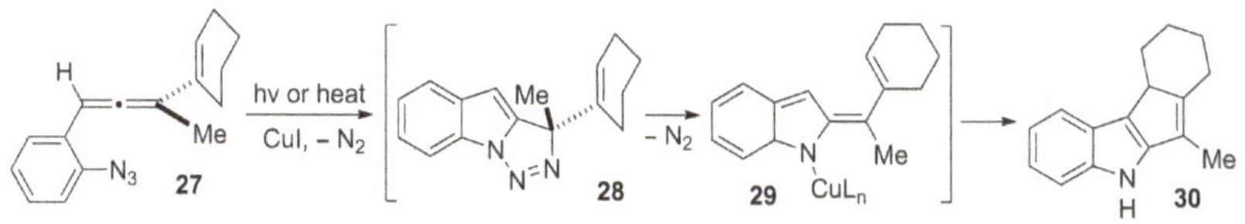

SCHEME 6.8 Addition of a nitrogen radical to an allene.

6.4 TRANSITION METAL–CATALYZED ALLENE AZIRIDINATION TO METHYLENEAZIRIDINES

6.4.1 General Features of Alkylidene or Methyleneaziridines

One of the most flexible approaches for transition metal–catalyzed intramolecular introduction of nitrogen into allenes involves an initial aziridination reaction to generate an intermediate "bicyclic MA" (Scheme 6.9). Depending on the substitution pattern of the allene and the nature of the nitrogen source (typically a carbamate, X = C in **31**, or sulfamate, X = SO), either "exocyclic" **32** or "endocyclic" **33** bicyclic MAs can be accessed. The literature on this topic often uses the terms "methylene" and "alkylidene" interchangeably, despite the fact that "methylene" refers to species containing a terminal exocyclic double bond and "alkylidene" to more substituted products.

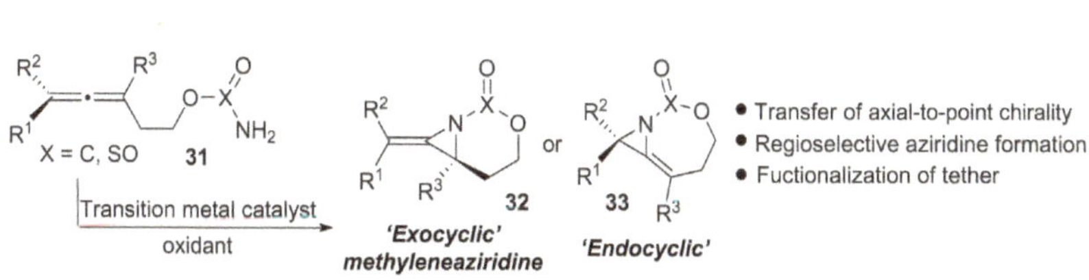

SCHEME 6.9 Modes of intramolecular transition metal–catalyzed allene aziridination.

However, in this chapter, the general convention of employing "methylene" to describe all members of these classes of compounds has been adopted.

If methods to achieve allene aziridination in a chemo-, regio-, and stereocontrolled fashion could be developed, the resulting unique, highly strained exo- and endocyclic bicyclic MAs could be envisaged as scaffolds for the flexible syntheses of complex amine stereotriads (see Scheme 6.19, *vide infra*). Several unusual features of MAs make them appealing intermediates. As one example, direct allene aziridination to an exocyclic MA should enable transfer of the axial chirality of an enantioenriched allene to the product with high fidelity.[27] Bicyclic MAs contain considerable ring strain (generally 12–13 kcal/mol greater than the parent three-membered ring),[28–30] partially due to the presence of a sp^2 carbon in the ring, as well as the additional fused ring. This means that further reactivity at either the three-membered ring or at the exocyclic double bond of the MA can be exploited for the stereocontrolled syntheses of amine triads and heterocycles.

This section of the chapter focuses on synthetic approaches that furnish MAs directly from allenes; however, the body of work on the formation and reactivity of MAs described by the Shipman group bears special mention.[31–42] In Shipman's chemistry, treatment of *N*-(2-bromoallyl)-alkylamines with alkyllithium reagents form MAs adorned with electron-donating groups on the aziridine nitrogen, as opposed to the electron-withdrawing groups on MAs discussed in this chapter. For details on the synthesis and reactivity of these types of MAs, the reader is directed to previous reviews and publications by the Shipman group.[31–42]

6.4.2 Intramolecular Allene Aziridination Using Carbamate Precursors

The development of mild and highly chemo-, regio-, and stereoselective protocols for allene aziridination is a key to engaging MAs as useful synthetic intermediates. While a number of transition metals, including Rh, Ru, Ir, Fe, Co, Mn, and Ag, are known to promote transition metal–catalyzed nitrene transfer,[43] allenes present additional challenges not observed in their alkene counterparts, as will be discussed in the following sections.

The first example of transition metal–catalyzed allene aziridination was reported by the Robertson group and involved using a $Rh_2(OAc)_4$ complex with *N*-tosyloxycarbamate-bearing allenes **34** as precursors (Scheme 6.10).[44] The "preoxidized" nature of the nitrogen source precluded the need to use hypervalent iodine sources to generate the reactive metal-nitrene intermediate. The allenes employed were achiral; thus, stereoselectivity was not an issue. While the scope of these initial studies was narrow and the yields were poor, spectroscopic experiments and single crystal X-ray structural analysis clearly confirmed that the *exo* MAs **35** were the major isolated products.

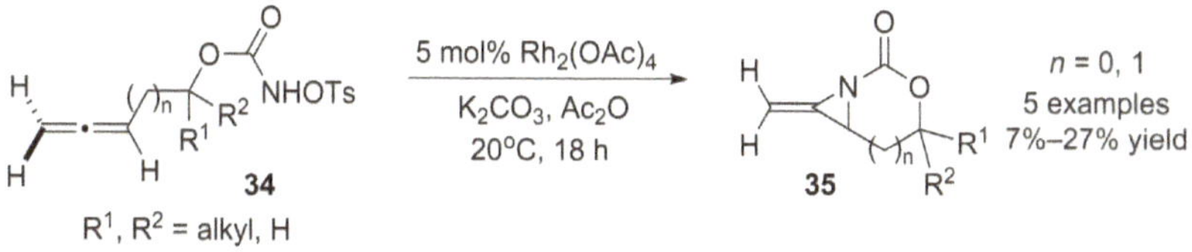

SCHEME 6.10 Rh-catalyzed aziridination of *N*-tosyloxycarbamate-bearing allenes.

Not long after this initial report, the Schomaker group detailed significantly improved protocols with broader scope and higher yields.[45] Similar to the Robertson group, they also noted poor results using *N*-tosyloxycarbamates as the nitrene precursors, but found that tethering either a carbamate (Scheme 6.11) or a sulfamate (Scheme 6.16, *vide infra*)[46] to the allene resulted in good-to-excellent yields of the desired products using a series of dinuclear Rh(II) catalysts. However, these reactions also had their drawbacks. When an allenic carbamate with no substitution in the tether (**36**, $n = 0$) was employed (Scheme 6.11, left), mixtures of *E*:*Z* isomers of the MA **38** resulted, with the *Z* isomer undergoing ring-opening to the enecarbamate **39**.[45] The substitution of an additional methyl group in the tether of **36** suppressed ring-opening and gave 87% yield of the [5.3]-fused bicyclic MA **40**; however, the *E*:*Z* ratio was still moderate. When homoallenic carbamates were used (Scheme 6.11, right, $n = 1$), ring-opening of the desired [6.3]-fused MAs **41** and **43** did not occur if the reaction was run at room temperature. However, chemoselectivity issues were problematic, with competing C–H bond amination to **42** and **44** observed. When no substitution was present in the tether between the allene and the carbamate, the use of $Rh_2(esp)_2$

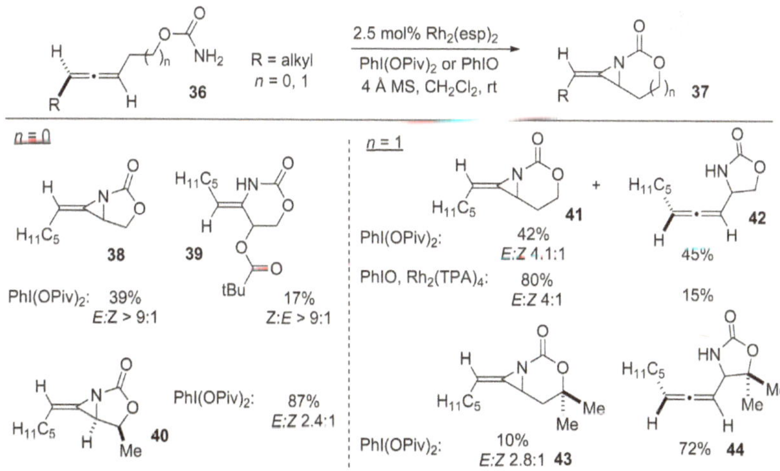

SCHEME 6.11 Rh-catalyzed aziridination of allenic and homoallenic carbamates.

(esp = α,α,α′,α′-tetramethyl-1,3-benzenedipropionic acid) and $PhI(OPiv)_2$ as the oxidant gave a poor 1:1 ratio of the aziridine **41** to the allenic amine **42**. However, some element of catalyst control could be achieved, as increasing the bulk of the catalyst to $Rh_2(TPA)_4$ (TPA = triphenylacetate) and changing to PhIO as the oxidant increased the preference for aziridination:C−H amination to 5.3:1. In the case of MA **41**, the *E:Z* ratio was similar with both catalysts. Despite this initial success, catalyst control did not hold up when substitution was introduced into the tether, giving a mixture of **43** and **44**. In this case, **44** was heavily favored, irrespective of the catalyst and oxidant identity. The *E:Z* stereochemistry in both cases was only moderate, in favor of the *E* isomer.

Due to the poor chemoselectivity observed with Rh_2L_n catalysts, an alternative catalyst system needed to be identified for the use of carbamates as nitrene precursors. Interestingly, traditional catalysts for nitrene transfer, including Fe, Mn, Cu, and Co, failed to give predictable control over the chemoselectivity.[43] However, the Schomaker group found that AgOTf, supported by bipyridine (bipy) or phenanthroline (phen) ligands, gave reasonable catalyst-control over the chemoselectivity to favor the formation of MAs.[47] Investigations of the substrate scope (selected examples in Scheme 6.12) showed that approximately equimolar amounts of AgOTf and phenanthroline ligand transformed allenes **45a−d**, containing a number of different substitution patterns and functionalities such as alkyl, phenyl, ester, and protected alcohols, selectively into MAs **46a−d**. Substitution in the tether was also tolerated in **45a** and **45c** to provide the MA products **46a** and **46c** in high yields and moderate *E:Z* ratios. The Ag(phen)OTf catalyst was also capable of providing high *E:Z* selectivity in the MA product **46d** using 1,1′,3-trisubstituted homoallenic carbamate substrates such as **45d**.

In 2013, Schomaker and coworkers further demonstrated the utility of silver-catalyzed nitrene transfer by showing that the chemoselectivity of allene amination could be tuned by altering the ratio of AgOTf and

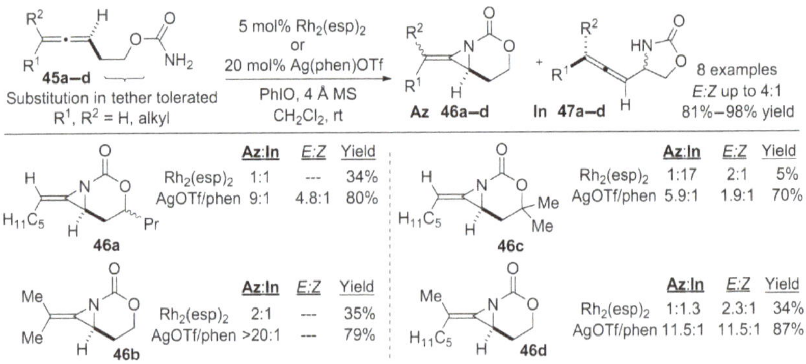

SCHEME 6.12 Ag(I)-catalyzed aziridination of homoallenic carbamates.

phenanthroline ligand.[48] When a ratio of 1:1.25 of AgOTf and phenanthroline was used, selectivity for aziridination to MAs **48a–h** over C–H insertion in ratios as high as 100:0 and yields from 57% to 87% were achieved (selected results in Scheme 6.13).

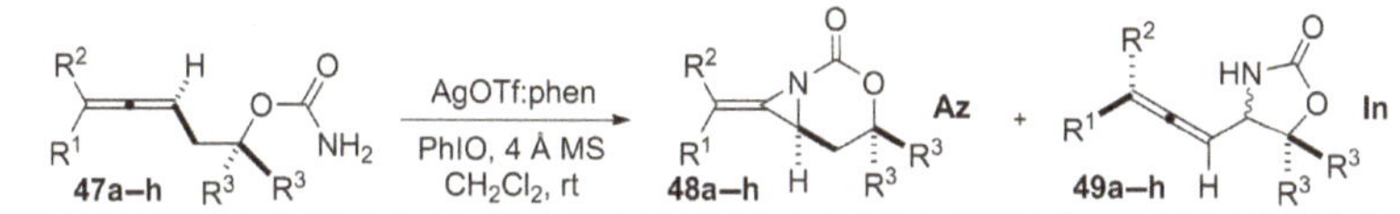

R^1	R^2	R^3	AgOTf:phen[a,b]	In:Az	Az	In
Me **47a**	Me	H	1:1.25	1:20	<4%	79%
			1:3	100:0	81%	0%
			$Rh_2(esp)_2$	1:2	17%	35%
C_5H_{11} **47b**	H	Me	1:1.25	1:5.9	13%	79%
			1:3	76:1	76%	1%
C_5H_{11} **47c**	H	H	1:1.25	1:4	18%	72%
			1:3	13:1	65%	5%
^{t}Bu **47d**	H	H	1:1.25[c]	1:3.7	18%	67%
			1:3	100:0	83%	0%
Et **47e**	Me	H	1:1.25	1:19	4%	70%
			1:3	100:0	78%	0%
C_5H_{11} **47f**	Me	H	1:1.25	1:11.5	7%	87%
			1:3	100:0	88%	0%
			$Rh_2(esp)_2$	1.3:1	44%	34%
^{t}Bu **47g**	H	H	1:1.25	1:4.8	12%	57%
			1:3	19:1	74%	4%
$(CH_2)_2Ph$ **47h**	H	H	1:1.25	0:100	0%	61%
			1:3	24:1	74%	3%

[a] **Az**: 20% AgOTf, 25% phen, 2 equiv. PhIO, 4 Å MS, CH_2Cl_2. [b] **In**: 10% AgOTf, 30% phen, 3.5 equiv. PhIO, 4 Å MS, CH_2Cl_2. [c] 2,2′-bipyridine ligand.

SCHEME 6.13 Selected scope in the tunable, Ag-catalyzed nitrene transfer of allenes.

Employing a 1:3 ratio of AgOTf to phenanthroline gave selectivity for C–H insertion against aziridination as high as >99:1 and yielded from 36% to 88%. The change in chemoselectivity was shown to be due to equilibrium between the potential catalytic species Ag(phen)OTf and $Ag(phen)_2OTf$. The origin of the chemoselectivity results from the steric congestion around the Ag center; a more sterically congested Ag center suppresses the rate of aziridination versus C–H insertion, while not significantly affecting the C–H insertion rate.[49]

A more recent development in chemoselective allene aziridination is a single example reported in 2017 showing the potential for the asymmetric aziridination of achiral allenes (Scheme 6.14).[50] Chemo- and enantioselective aziridination employing a $AgClO_4$ salt and a chiral BOX ligand **52** led to an unoptimized *ee* of 62% in the aziridination of achiral **50** to **51**. The excess $AgClO_4$ helps to break down the polymeric PhIO, ensuring complete reaction within a few hours and preventing decomposition of the MA.

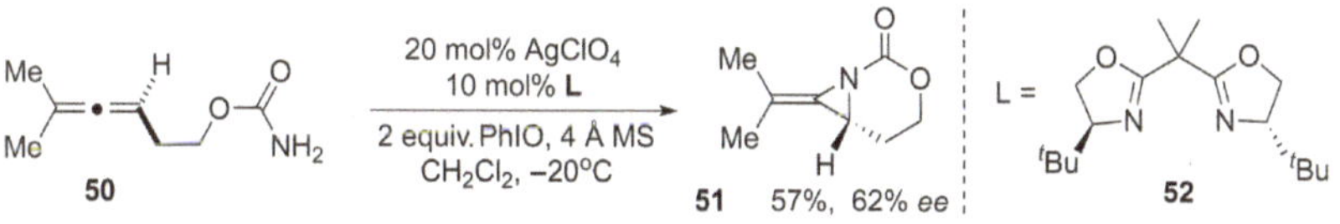

SCHEME 6.14 Ag-catalyzed enantioselective aziridination of an achiral allene.

6.4.3 Intramolecular Allene Aziridination Using Sulfamate Precursors

Not long after Blakey group reported their work in Rh-catalyzed allene amination, the Robertson group published similar findings using homoallenic sulfamates as the nitrene precursors (Scheme 6.15).[51] The substitution pattern of the allene influenced the reaction outcome, with 1,1'-disubstituted allenes furnishing the same aminocyclopropane products observed by Blakey.[16] However, a more hindered ^{t}Bu-containing allene **53** (Scheme 6.15) gave two sulfamate-derived bicyclic MAs **55** and **56** in a ratio of 5:1, favoring formation of the seven-membered ring **55**. The aminocyclopropane **57** could be accessed by treatment of **55** with NaI in DMF. Reaction of the 1,3-disubstituted allene **58** (R = H) resulted in formation of the enesulfamate **59**; the authors invoked acetate trapping of an intermediate 2-amidoallyl cation to explain this result. In contrast, a 1,3,3′-trisubstituted allene **58** (R = Et) reacted with poor regioselectivity to furnish a 1:1 mixture of the aminocyclopropane **61** and the enesulfamate **60**.

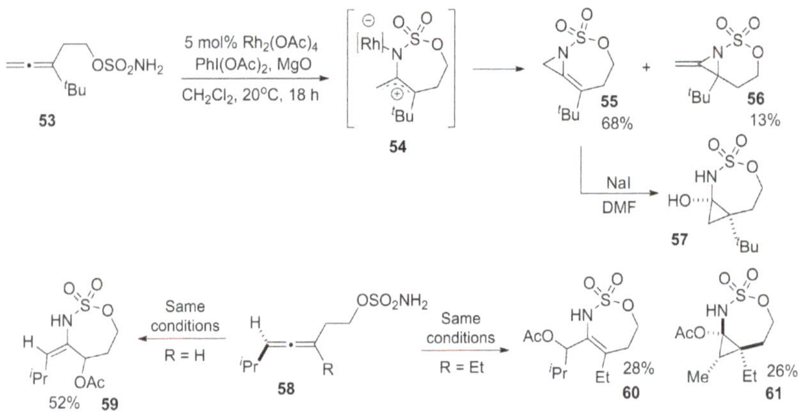

SCHEME 6.15 Rh(II)-catalyzed aziridination of homoallenic sulfamates via a 2-aminoallyl cation.

The Schomaker group was interested in identifying ways to control the regioselectivity of allene aziridination to form both exocyclic and endocyclic MAs that could be readily transformed into a diverse array of aminated stereotriads. This required the formation of an intermediate bicyclic MA that could engage with external nucleophiles prior to rearrangement to the aminocyclopropanes observed by Blakey and Robertson.[16,51] By switching the catalyst to $Rh_2(TPA)_4$ and utilizing PhIO with molecular sieves to remove water, the exocyclic MAs **63** could be formed with excellent chemo-, regio-, and stereoselectivity (Scheme 6.16).[46] These labile compounds could be observed spectroscopically, but were readily opened on treatment with water

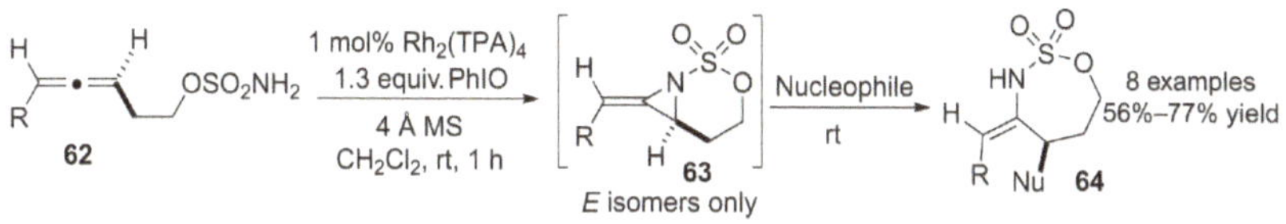

SCHEME 6.16 Rh(II)-catalyzed aziridination of homoallenic sulfamates via MAs.

or silica gel. Therefore, aziridination was followed by immediate nucleophilic ring-opening to afford enesulfamates of the form **64** in good yield and excellent stereoselectivity for the *E* isomer. Importantly, the axial chirality of an enantioenriched homoallenic sulfamate was transferred to the enesulfamate with >95% fidelity. Essentially complete retention of the stereochemical information contained in the allene precursor suggested that the 2-amidoallyl cation was not an intermediate using these reaction conditions.

To favor reaction of the distal double bond to generate endocyclic MAs that included **69a–c**, Schomaker and Burke described the use of silyl-substituted homoallenic sulfamates **68a–c** as precursors in Rh-catalyzed aziridination (Scheme 6.17).[52] Steric congestion alone was not enough to exclusively favor the endocyclic MA **66**, as even a ^{t}Bu group at C3 of the allene **65** still gave 27% yield of **67**. Nevertheless, regioselectivity was completely diverted to the distal double bond of **68a–c** via electronic effects exerted by a trialkyl silyl substituent at C3 of the allene. The excellent regioselectivity was rationalized by examining the likely nature of the transition state. The positive charge that develops on the C2 of **68c** is better stabilized by the hyperconjugative interaction with the C–Si bond that is

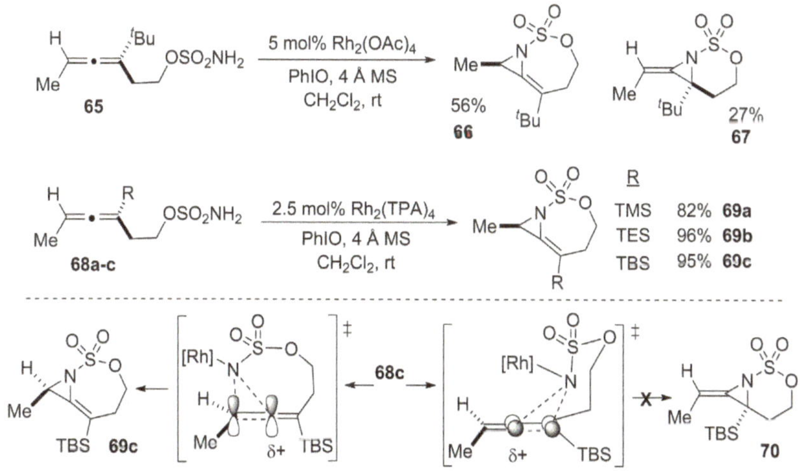

SCHEME 6.17 Steric versus electronic control of regioselectivity in the aziridination of silylated homoallenic sulfamates.

coplanar with the distal π orbitals, leading to **69c** as opposed to **70**. Monoalkyl substitution on C1 was well-tolerated, as was alkyl substitution in the tether between the allene and the sulfamate; however, 1,1′,3,3′-tetrasubstituted allenes gave only elimination products.

To summarize, this section of the chapter describes efforts to control the chemo-, regio-, and stereoselectivity of transition metal–catalyzed intramolecular allene aziridination reactions. While Rh_2L_n catalysts pair well with sulfamate nitrene precursors, the use of Ag(I) catalysts with carbamates provides opportunities to tune the chemoselectivity of allene amination, as well as enantioselectivity in the aziridination of achiral allenes. Typically, exocyclic MAs are the preferred products, but installation of a silyl substituent at C3 of the allene precursor successfully diverts the aziridination to the distal carbon to furnish endocyclic MAs.

6.4.4 Intermolecular Allene Aziridinations

In contrast to intramolecular aziridinations, where the nitrene precursor is tethered to the allene, controlling the regioselectivity of intermolecular allene aziridination would be expected to be much more challenging. In fact, the only example of such a reaction resulting in the isolation of a MA product was reported by Bingham and Gilbert in 1975.[13] Treatment of allene **71** with the nitrene precursor *N*-(*p*-nitrobenzenesulfonoxy)urethane **72** resulted in a very low yield of the MA **73** (Scheme 6.18). Due to the scarcity of intermolecular allene aziridinations, it is not known whether the stability of the product is the major problem, or whether better reaction conditions could lead to higher yields of the desired products. Efforts in this area represent future opportunities for the modular introduction of nitrogen into allene precursors.

As previously illustrated in Scheme 6.4, Atkinson and Malpass reported the addition of a nitrene formed from treatment of *N*-aminophthalimide with $Pb(OAc)_4$ to a 1,1′-disubstituted allene to form a 1,4-diazaspiro[2.2]pentane (DASP) product. The MA was proposed as a key intermediate, but was never isolated or spectroscopically characterized.[14]

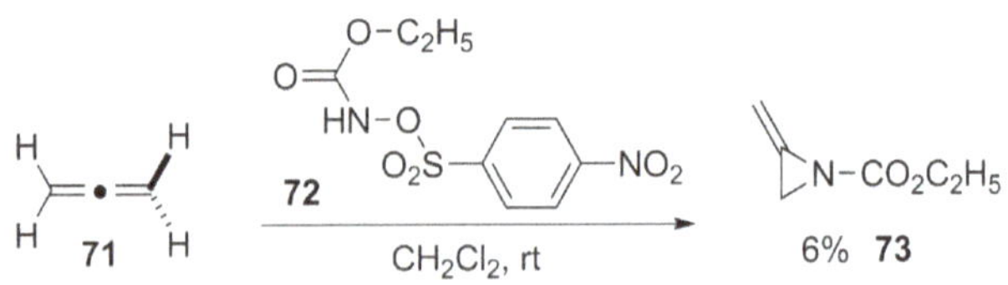

SCHEME 6.18 First reported example of intermolecular addition of a nitrene to an allene.

6.5 STRATEGIES FOR THE SYNTHESIS OF HETEROCYCLES VIA ALLENE AZIRIDINATION

The ability to synthesize bicyclic MAs in a highly chemo-, regio-, and stereoselective manner has enabled subsequent chemistry to exploit their unusual structural features to be extensively investigated. Scheme 6.19 summarizes some of the major aminated heterocycles that are accessible from allenes of the form **74** through the intermediacy of MAs **75** and **79**. The aziridine ring can be opened with excellent regioselectivity to furnish products of the form **76** with flexibility in the stereochemical relationship among C1–C3. Other modes of reactivity include formal $[n+3]$ cycloadditions through 2-amidoallyl cations **77** to give amination carbocyclic scaffolds **78** or expansion of the aziridine ring to deliver heterocycles that include **80–82**. This section of the chapter focuses on the various modes of reactivity of bicyclic MAs that give rise to these complex and densely functionalized amines.

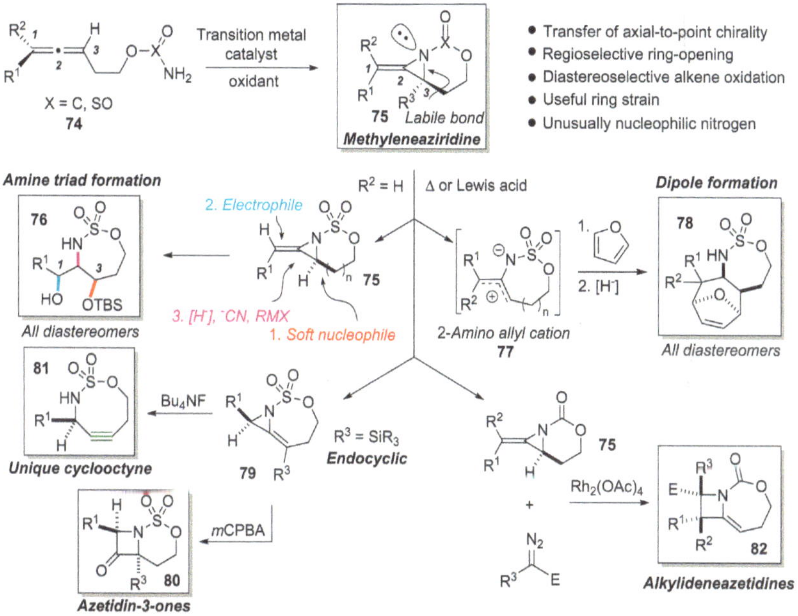

SCHEME 6.19 Transition metal–catalyzed allene aziridination and subsequent functionalization.

6.5.1 Nucleophilic Ring-Opening of Methyleneaziridines

Two advantages of utilizing MAs as reactive intermediates are their ring strain and the ability to control the regioselectivity of nucleophilic ring opening. While this section will be devoted largely to reactions involving direct nucleophilic attack on the aziridine ring, other reactivity has also been described. Robertson and coworkers reported an unexpected formal S_NV mechanism in the reaction of terminally unsubstituted MAs **84** with Grignard reagents (Scheme 6.20).[44] The $C(sp^2)$–N bond of **84** was cleaved in preference to the expected $C(sp^3)$–N bond in the presence of 3.0 equivalents of the Grignard reagent, 0.05 equivalents of CuI, and 1.5 equivalents of $BF_3 \cdot OEt_2$ to furnish allylic amines of the general structure **85**.

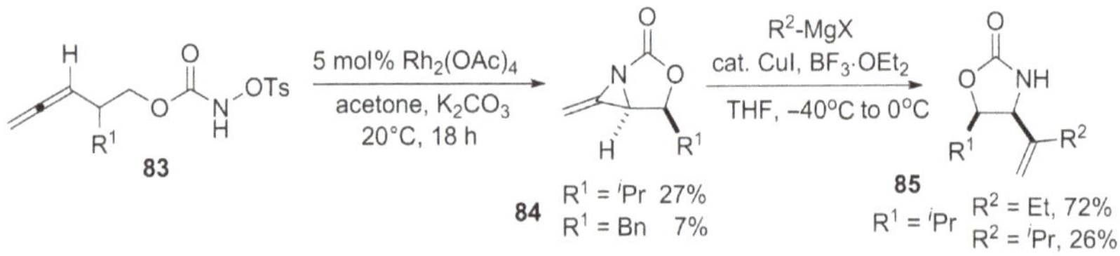

SCHEME 6.20 Ring-opening of bicyclic MAs via an S_NV mechanism.

The Schomaker group investigated the ring-opening of isolable carbamoyl-containing [6.3]-bicyclic MAs, as represented by **86** (Scheme 6.21).[45] In contrast to the reactivity observed by Robertson, **86** reacted with a variety of soft nucleophiles, including carboxylic acids, alcohols, azide, thiols, and various halides to give the desired $C(sp^3)$–N bond-cleavage products **87**. The resulting enecarbamate **87** (Nu = OAc) was sufficiently nucleophilic to react with *N*-bromosuccinimide (NBS) as the electrophilic bromine source. The intermediate iminium ion was then reduced with $NaBH_3CN$ in AcOH to generate the all-*syn* stereotriad **88**. These reactions are described in more detail in the context of sulfamoyl-derived MAs (*vide infra*).

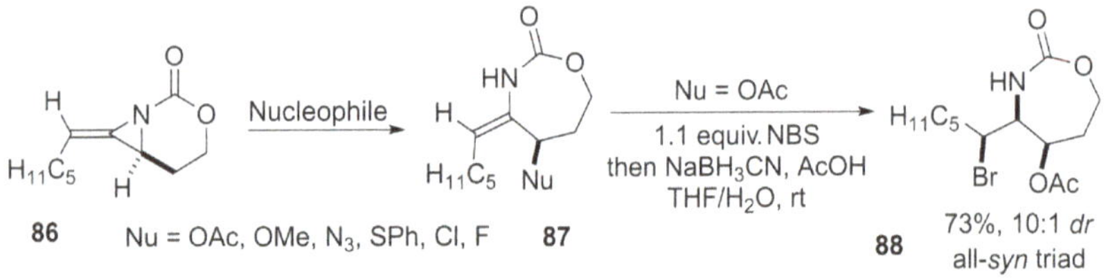

SCHEME 6.21 Ring-opening of carbamoyl-containing bicyclic methyleneaziridines.

In contrast to the carbamoyl-based MAs, the sulfamoyl-containing compounds are nonisolable under standard conditions and are significantly more reactive.[46] This enabled a convenient one-pot process that converted all three of the unsaturated carbons in the allene to motifs displaying three contiguous and stereodefined $C(sp^3)$–X or $C(sp^3)$–C bonds, termed "stereotriads." The four-step, one-pot process was accomplished through (1) allene aziridination of **89**, (2) regioselective aziridine ring-opening of **90**, (3) electrophilic

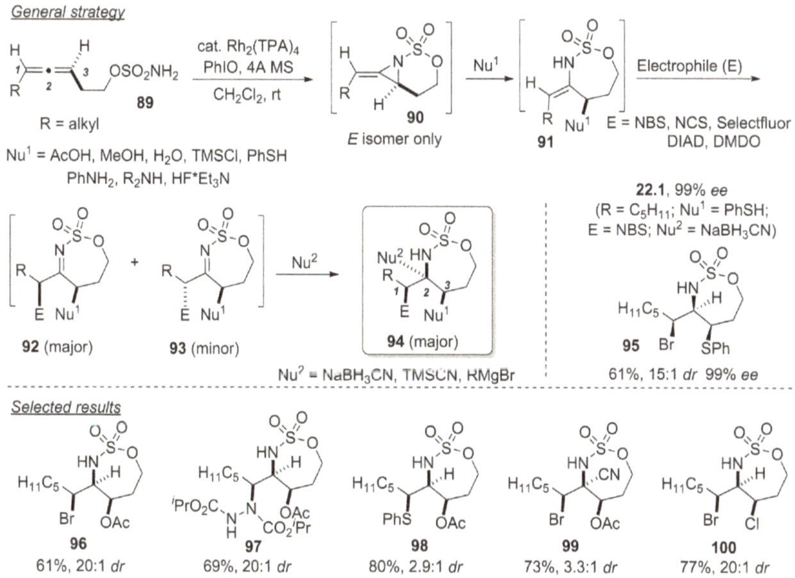

SCHEME 6.22 Ring-opening of sulfamoyl bicyclic methyleneaziridines to aminated stereotriads.

substitution of the resulting enesulfamate **91** to yield mixtures of **92** and **93** in favor of **92**, and (4) a final reduction of the imine intermediates to furnish **94** (Scheme 6.22, top). The ring strain and good leaving group capability of the sulfamoyl group enabled mild nucleophiles, including alcohols, carboxylic acids, amines, thiols, halides, and water, to open the ring of **90** at room temperature without requiring exogenous Lewis acid additives. The resulting enesulfamates **91** were sufficiently nucleophilic to further react with heteroatom electrophiles for the overall introduction of halogen, oxygen, sulfur, and nitrogen groups at each of the three original allene carbons. In the final step of the sequence, the resulting imines **92** and **93** could be reduced with hydride, cyanide, or Grignard nucleophiles to give the desired amine stereotriads **94**. In these early studies, the major stereoisomer of **94** was generally obtained in >10:1 *dr*, with the 1,2-*syn*:2,3-*syn* diastereomer noted as the major product. In this manner, the heteroatoms flanking the central amine could be tuned to give a diverse array of triads in good *dr* (selected examples in Scheme 6.22, bottom). Importantly, the axial chirality of an enantioenriched allene **89** (R = C_5H_{11}) was transferred to point chirality in the stereotriad **95** with complete fidelity, providing rapid access to complex and enantioenriched amine building blocks **96–100**.

The 1,2-*syn*:2,3-*syn* diastereoselectivity between C1 and C3 of the original unsaturated carbon atoms of the allene was explained by the model shown in Scheme 6.23. The cyclic nature of the enesulfamate **101** can impart

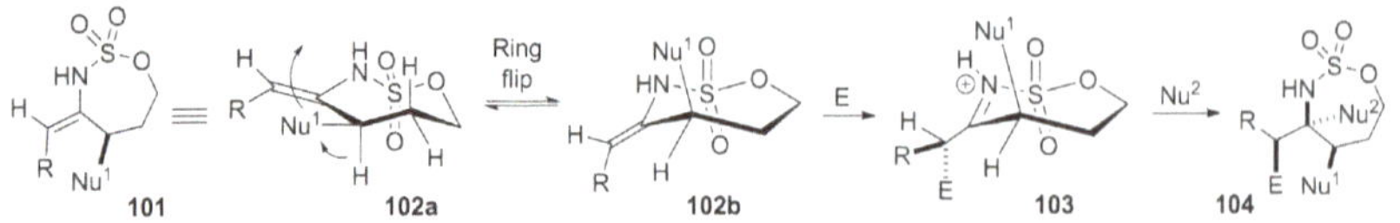

SCHEME 6.23 Stereochemical model rationalizing the 1,2-*syn*:2,3-*syn* diastereoselectivity.

good facial selectivity in its reaction with an electrophile by adopting a conformation that minimizes $A^{1,3}$ strain, as in **102b** compared to the less stable **102a**. The favored conformation of the resulting iminium ion **103** again minimizes the resultant $A^{1,3}$ strain in this intermediate. As in the enesulfamate, shielding of the top face of **103** by the Nu^1 helps to guide the stereoselectivity of the reduction event to yield the 1,2-*syn*:2,3-*syn* product **104** as the major diastereomer.

The presence of fluorine in organic scaffolds is well-known to improve the pharmacological properties of a number of drugs and bioactive molecules.[53] A number of these drugs, including antivirals, neuraminidase inhibitors, and modulators of androgen receptors, contain stereodefined $C(sp^3)-F$ bonds embedded in aminated triad motifs.[54] As these building blocks can be difficult to access using traditional alkene oxidation methodologies, the Schomaker group optimized the oxidative allene amination strategy described in Scheme 6.22 to furnish fluorine-containing triads in good-to-excellent diastereoselectivities.[55] Key to the success of this chemistry was the optimization of the conditions for the use of Selectfluor as the source for electrophilic introduction of fluorine into the intermediate enesulfamate **106** (Scheme 6.24, top). The resulting imine could be reduced with hydride, Grignard, or cyanide nucleophiles to introduce additional substitution at C2 of the original allene in the triads **108–118**. Particularly gratifying was the ability to install fluorine at quaternary carbons in **111–116**, a feat that is difficult to achieve through fluoride ring-opening of epoxides or aziridines. The diastereoselectivities were excellent in most cases, although minor amounts of isomers were often observed. These were easily separable by standard silica gel column chromatography.

In addition to achieving flexibility in the identity of the heteroatoms that can be introduced into amine triads *via* oxidative amination, allenes also offer the potential for stereochemical versatility. The Schomaker group addressed this possibility in 2014 by reporting stereodivergent syntheses of O/N/O triads from single allene precursors.[56] These particular triads were chosen as targets due to the challenges inherent in the synthesis and resulting aminohydroxylation of the analogous allylic alcohol precursors. A general roadmap of the methodology required to convert a single enantioenriched allene **119** to any one of the four possible diastereo- and enantioenriched amine stereotriads **122**, **125**, **126**, and **129** is shown in Scheme 6.25.

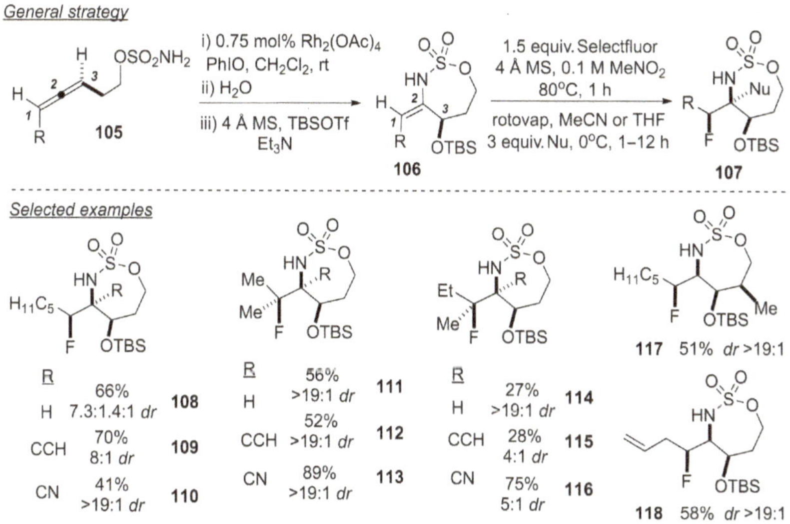

SCHEME 6.24 Introduction of fluorine into all-*syn* amine stereotriads.

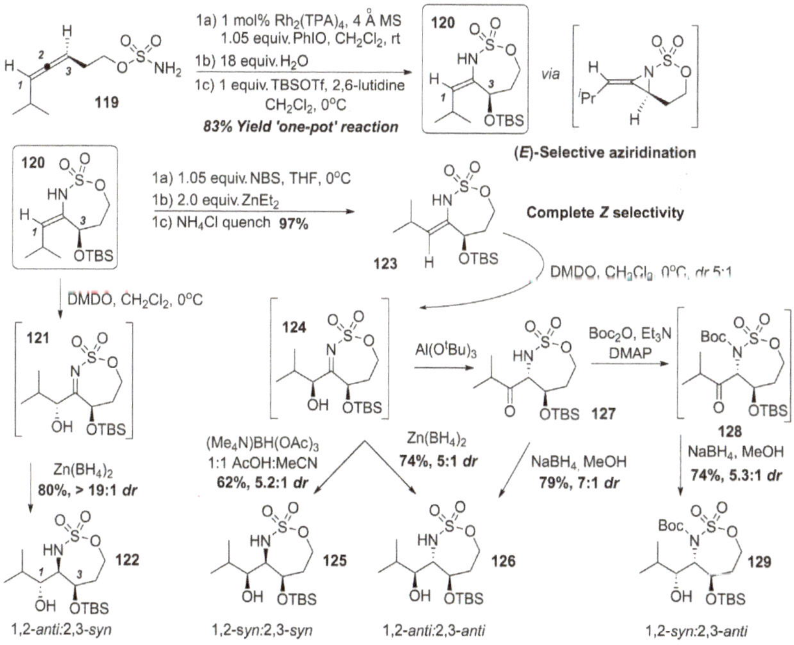

SCHEME 6.25 Stereochemical diversity in oxidative allene amination.

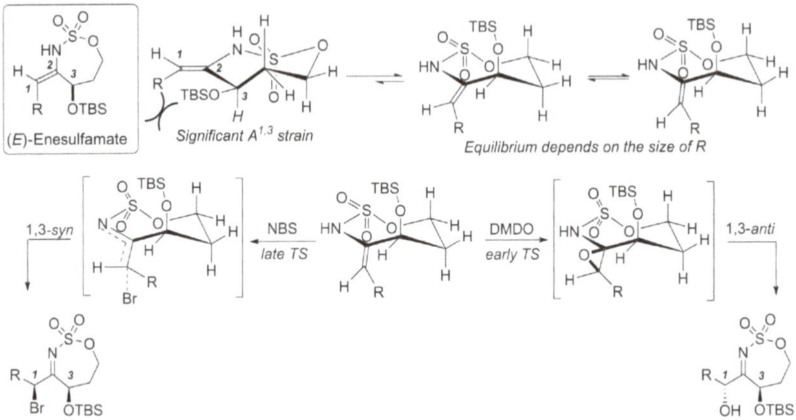

FIGURE 6.2 Model for selective oxidation of (*E*)-enesulfamates.

Key features of stereodivergent oxidative allene amination include the ability to carry out the transformation of **119** to **120** in a single pot, although a bulky OTBS group is required at C3 to ensure the diastereoselectivity in the addition of the enesulfamate to the dimethyldioxirane (DMDO) electrophile. In contrast to the majority of other electrophiles (Schemes 6.21–6.22), the relationship between C1 and C3 of the allene in **121** is *anti*. This was rationalized by proposing that the nature of the transition state in the addition of the (*E*)-enesulfamate **120** to DMDO plays an important role in determining the stereochemical outcome (Fig. 6.2). For example, if the addition of **120** to DMDO occurs in a concerted fashion with an early transition state that closely resembles the starting material, a 1,3-*anti* relationship will result in the imine intermediate **121** (see Fig. 6.2). In contrast, utilizing NBS in the reaction of **121** is likely to occur through a late transition state that contains significant C = N double bond character. The bending of the C = C bond away from a pseudoaxial orientation back to a more equatorial position may influence facial selectivity by precluding approach of NBS from the top face of the alkene due to effective shielding by the bulky OTBS group, resulting in the 1,3-*syn* imine. A second key feature is the need to isomerize the (*E*)-enesulfamate of **120** to the (*Z*)-enesulfamate **123** in order to access the remaining three stereoisomers. Fortunately, this can be achieved in nearly stoichiometric yield using a one-pot addition/elimination sequence. Finally, a combination of chelation and Felkin-Ahn control over the imine reduction event enables synthesis of **122**, **125**, **126**, and **129** in good-to-excellent diastereoselectivities.

6.5.2 Methyleneaziridine Oxidation and Rearrangement Strategies

An alternate mode of reactivity of MAs involves the functionalization of the exocyclic olefin prior to any ring-opening of the aziridine. The bicyclic nature of the ring helps to render any subsequent transformation of the double bond highly diastereoselective. For example, treatment of carbamoyl-derived bicyclic MAs **130** (Scheme 6.26) with *N*-aminophthalimide in the presence of a hypervalent iodine oxidant resulted in complete facial selectivity to give the tricyclic 1,4-diazaspiro[2.2]pentanes (DASPs) **131** in >19:1 *dr*.[15] Monoalkyl substitution at C1 of the MA was tolerated to furnish **132–134** in good yields. Both the *E* and *Z* isomers of **130** (*E:* $R^1 = C_5H_{11}$; R^3 = Me; R^2, R^4 = H) were successful in the reaction, giving diastereomeric DASP products **135** and **136**, although the greater steric hindrance presented by the (*Z*)-MA resulted in a lower yield of the DASP. The substitution pattern of the original allene, and hence the MA, could be varied to furnish DASPs **135–138** in moderate yields. One important feature of this "intra-/intermolecular" aziridination strategy is that the two rings of the DASPs are electronically differentiated, where the fused aziridine contains an electron-withdrawing group and the spiro ring an electron-donating group.

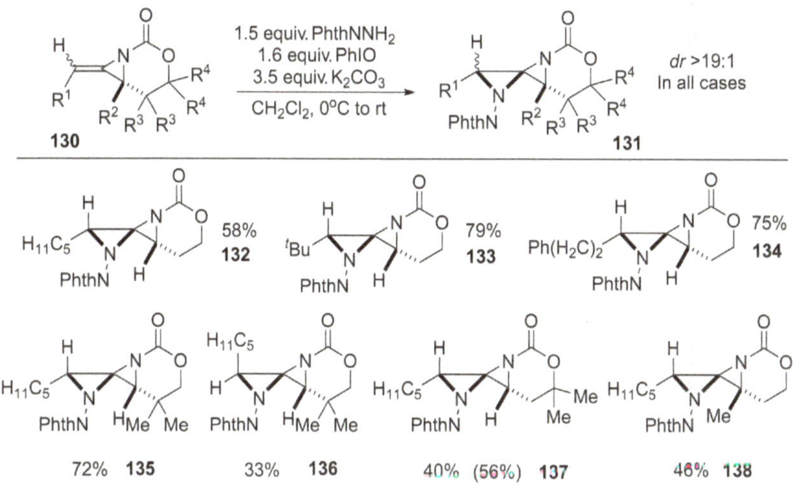

SCHEME 6.26 Intermolecular aziridination of methyleneaziridines to DASPs.

The Schomaker group briefly explored the reactivity of DASPs to determine if control over the site and regioselectivity of nucleophile addition could be achieved (Scheme 6.27).[15] The addition of oxygen, sulfur, and halogen nucleophiles to the DASP ***E*-139** typically favored ring-opening at C3 to give the *N*,*N*-aminal products **142–145** (Scheme 6.27, left) in good yields with *dr* 19:1. Allene precursors could be transformed into these stereotriads in a single reaction vessel through a sequence involving $Rh_2(OAc)_4$-catalyzed intramolecular aziridination, addition of *N*-aminophthalimide and PhIO to induce intermolecular aziridination, and final nucleophile addition. The

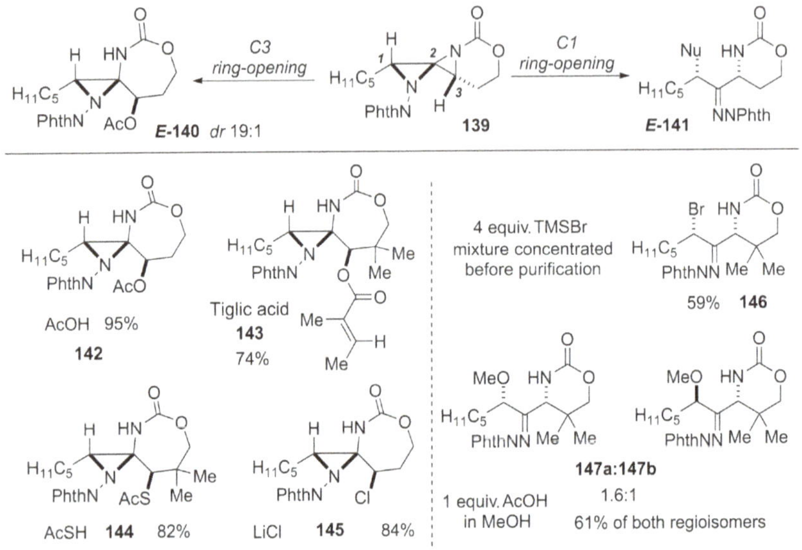

SCHEME 6.27 Regioselective ring-opening at C1 and C3 of a DASP.

stereochemical information contained in an enantioenriched allene was transferred to the products with excellent fidelity. Efforts to directly open the NPhth-substituted aziridine ring at C1 were largely unsuccessful. However, specific reaction conditions, including the use of TMSBr and 1 equivalent AcOH in MeOH, did yield products **146** and **147a–b** that initially appeared to result from the addition of the nucleophile to C1. However, as will be described later in this section, these products likely arise from an initial attack at C1 of the internal aziridine, followed by an intramolecular rearrangement reaction. Judging from the mixture of diastereomers obtained in **147a–b**, the mechanism probably proceeds through a stepwise pathway that ablates the stereochemical information in the DASP precursor.

Formal reaction at C2 of a DASP of the general structure **148** was achieved through an unusual cascade reaction involving ring-opening with acetic acid to give **149**, followed by treatment with $Bi(OTf)_3$ to deliver the 1,3-aminoketone **153** through a sequence involving the proposed intermediates **150–152** (Scheme 6.28).[57] Although the acetate was not conserved in the final product, its use was crucial to the success of the rearrangement. The stereochemistry in the DASP transferred from the bicyclic MA (*E* or Z) was maintained in the aminoketone products, as in **155** and **156**. The diastereoselectivity between the two amines at C1 and C3 was excellent in most cases, exceeding 19:1 *dr* in favor of the *anti* product employing DASPs derived from *E*-MAs (**155**, **157–159**) and *syn* when the DASP substrate was derived from a Z-MA (**156** and **160**). Diastereoselective reduction to the 1,3-diamino-2-ol was achieved using excess $NaBH(OAc)_3$, although a *gem*-dimethyl

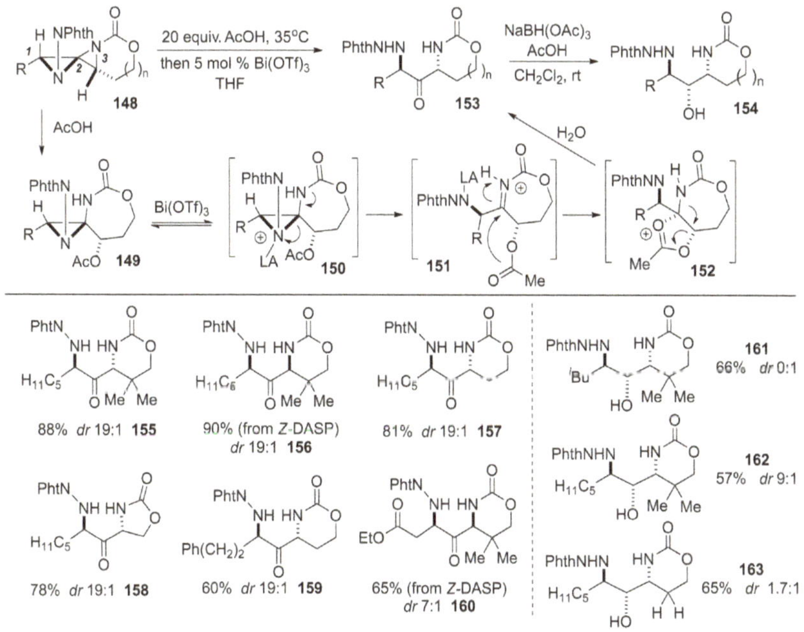

SCHEME 6.28 Formal reactions at C2 of a DASP to yield 1,3-diamino-2-ols.

motif was required to achieve high diastereoselectivities. This entire sequence could be carried out starting from the allene precursor in one pot with comparable overall yields to the two-pot process starting with the DASP.

To establish which aspects of substrate structure control the diastereoselectivity of the resulting 1,3-diamino-2-one products, the reduction of 1,3-diaminoketones with varied alkyl substitution was carried out (Scheme 6.29).[57] The presence of the *gem*-dimethyl groups in the carbamate ring of **155** increases the steric differentiation between the quaternary carbon and

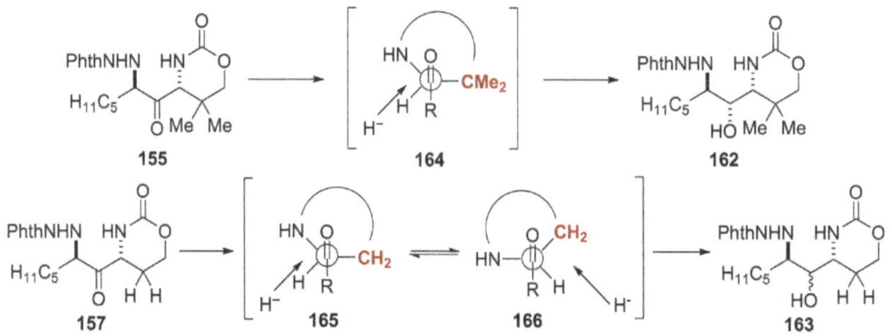

SCHEME 6.29 Felkin-Ahn model applied to the reduction of 1,3-diamino-2-ones.

the nitrogen of the carbamate, likely favoring conformation **164**, leading to high *dr* in **162**. In contrast, when the substrate **157** lacks the *gem*-dimethyl groups, a *dr* of only 1.7:1 was observed in **163**. This likely results from the lack of steric differentiation between the two substituents at C1 and C3; thus, conformations **165** and **166** are similar in energy and little selectivity is observed.

Shorter routes to 1,2-dihydroxy-3-amino and 1,3-diamino-2-ol stereotriads were developed *via* transition metal–catalyzed dihydroxylation and aminohydroxylation of MA scaffolds (Schemes 6.30 and 6.31).[58,59] In both cases, significant reaction optimization was required to minimize product decomposition; nonetheless, these approaches accessed valuable O/O/N and N/O/N stereotriads in short order and excellent *dr*.

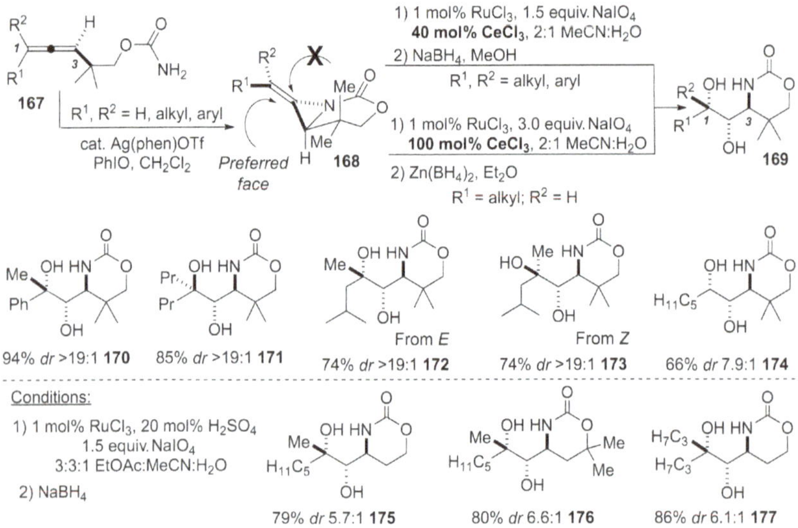

SCHEME 6.30 Stereocontrolled dihydroxylation of bicyclic methyleneaziridines.

Silver-catalyzed aziridination of homoallenic carbamates **167** (Scheme 6.30) was followed by a $RuCl_3$-catalyzed dihydroxylation and reduction to yield products of the form **169**.[58] Depending on whether the allene was 1,1′,3-trisubstituted or 1,3-disubstituted, differing amounts of $CeCl_3$ as an additive were necessary for optimal yields. However, the caveat was that a *gem*-dimethyl group needed to be present at the carbon adjacent to C3 of the allene to prevent competitive aziridine ring-opening. The identity of the best reagent for the reduction of the intermediate ketone also depended on the substitution pattern at C1 of the allene precursor **167**. The stereochemical relationship in the 1,2-dihydroxy-3-amine motifs **170–172** and **174** was 1,2-*syn*:2,3-*anti* when the MA precursor **168** displayed an *E* stereochemistry at

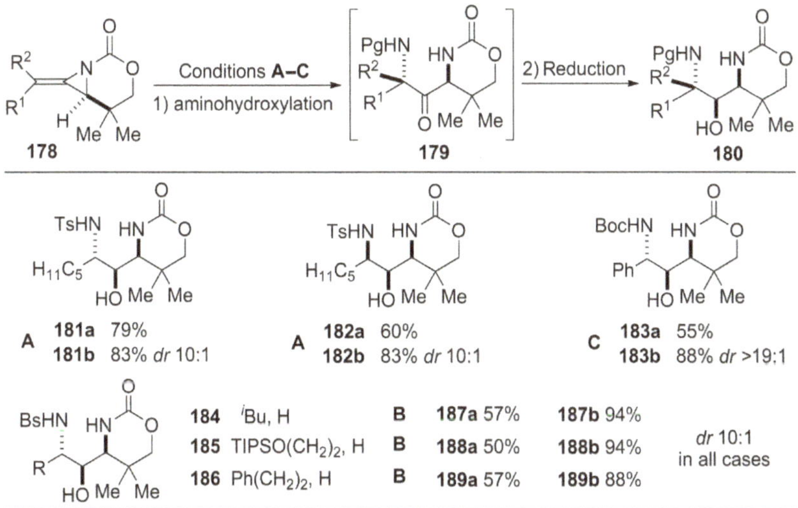

Condition **A**: Aminohydroxylation: 7.5 mol% $K_2OsO_2(OH)_4$, chloramine T, 33°C, reduction: $NaBH_4$ in CH_2Cl_2, rt; Condition **B**: Aminohydroxylation: 10 mol% OsO_4, Chloramine B, 37°C, reduction: $NaBH_4$ in CH_2Cl_2; Condition **C**: Aminohydroxylation: 7.5 mol% OsO_4, BocNNaCl, 30°C, reduction: $NaBH_4$ in 1:1 $CHCl_3$:H_2O with sonication.

SCHEME 6.31 Aminohydroxylation/reduction of bicyclic MAs.

the exocyclic alkene. In contrast, the use of a *Z*-MA yielded the opposite configuration at C1 of the 1,2-*anti*:2,3-*anti* product **173**.

Removal of the *gem*-dimethyl group in the substrate **167** required a modification of the dihydroxylation conditions to prevent competitive ring-opening of the aziridine by $CeCl_3$. Substitution of $CeCl_3$ with H_2SO_4 as the additive (Scheme 6.30, bottom) successfully furnished the 1,2-*syn*:2,3-*anti* amine triads **175–177**; however, the *dr* in the subsequent reduction was lowered to 5.7–6.6:1. The addition of carbon-based nucleophiles to the intermediate ketones was not reported, but has potential to deliver adjacent, heteroatom-bearing quaternary stereocenters.

Aminohydroxylation of MAs of the form **178** furnished 1,3-diamino-2-ols **180** in good *dr* as long as a *gem*-dimethyl group was installed adjacent to C3 of the allene precursor.[59] Successful reaction required extensive modifications to the standard Sharpless Os-catalyzed aminohydroxylation conditions.[2] The initial oxidation event gave direct access to the 1,3-diaminoketone products **181a–189a**, with the exocyclic nitrogen protected with a readily cleavable sulfonyl or Boc group (Scheme 6.31). Selective reduction was achieved by the use of $NaBH_4$ in either CH_2Cl_2 or a 1:1 mixture of $CHCl_3$:H_2O to give **181b–189b**. The process was general for a variety of alkyl and aryl-substituted MAs derived from 1,3-disubstituted allenes to give triads, with the identity of the chloramine crucial to obtaining good yields. For example, sulfonamides (Conditions A and B) worked best for alkyl-substituted allenes, while BocNNaCl was a better nitrogen source when the allene was aryl-substituted. In contrast to MA dihydroxylation, trisubstituted MAs were not tolerated in this chemistry and *gem*-dimethyl group was critical to achieving good *dr* in the reduction.

In 2015, the Schomaker group explored the formation and subsequent oxidation of the endocyclic MAs, which were generated via aziridination of the distal allene double bond of an allene of the form **190** (Scheme 6.17, *vide supra*).[52] Following the formation and purification of the column-stable MAs **191**, oxidation of the double bond could be achieved using *m*CPBA as the electrophilic oxygen source to deliver the fused azetidin-3-one **194** in excellent yields and diastereoselectivities (Scheme 6.32). The formation of the strained ring **194** was proposed to occur through the rearrangement of a highly reactive 1,4-oxaza-[2.2]spiropentane **193**, although this intermediate was not observed spectroscopically. In terms of the substrate scope, 1,3,3′-trisubstituted allenes were well-tolerated to furnish heterocycles **195b–198b**. The size of the silyl group at R^2 (TBS vs TES vs TMS) did not affect the regioselectivity or the subsequent transformation of the endocyclic MA to the azetidine-3-one. When diastereomeric mixtures of allene precursors were employed in the reaction, two diastereomeric azetidine-3-ones were formed, as in **199b–c**; these could be readily separated by column chromatography. Further manipulations on azetidin-3-ones could be carried out by either opening the sulfamate ring of TBS-substituted rings, while TMS-substituted products reacted preferentially at the carbonyl group (C2).

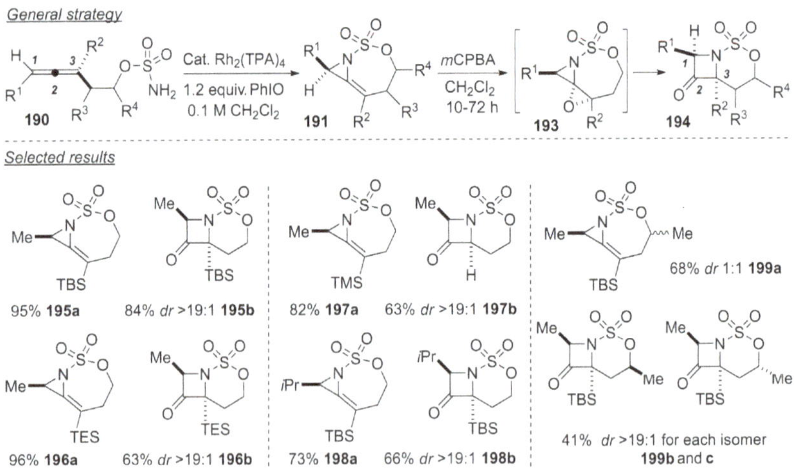

SCHEME 6.32 Selected substrate scope of the azetidin-3-one formation from endocyclic MAs.

6.5.3 Formation of 2-Amidoallyl Cation Intermediates

In 2004, the Shipman group reported Lewis acid–catalyzed intramolecular [4 + 3] cycloadditions between MAs and a furan or 1,3-diene moiety appended to the substrate.[34] While the methods to synthesize MAs of the form **200** (Scheme 6.33) do not involve allene aziridination, their chemistry could inform similar efforts utilizing bicyclic MAs. Treatment of **200** with a Lewis acid resulted in the formation of polycyclic cycloheptene-containing ring systems **203** and **204**, depending on whether the intermediate imine **203** was hydrolyzed to the ketone **204**. In the case of a diene-tethered MA **205**, treatment with $BF_3{*}OEt_2$ and EDC (1-ethyl-3-(3-dimethylaminopropyl)carbodiimide) gave the cycloheptenone **206** as a 56:44 mixture of diastereomers. The proposed mechanism involves formation of a 2-aminoallyl cation, such as **202**, upon activation of the MA by the Lewis acid. Subsequent attack by the tethered furan or 1,3-diene on the aminoallyl cation delivered the imine product, which could be converted to its ketone counterpart if an acidic workup was employed.

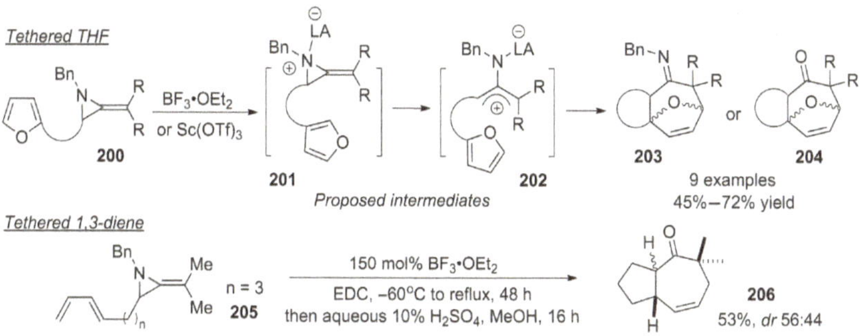

SCHEME 6.33 Shipman's formal [4 + 3] cycloadditions of methyleneaziridines.

Another example of an intramolecular [4 + 3] cycloaddition was demonstrated using an allene substrate containing a tethered furan **207** (Scheme 6.34), analogous to work reported by Shipman on the reactivity of monocyclic MAs.[44] Although the yield of **208** was a modest 22%, **207** represented the first report of the synthesis of a carbamoyl-derived MAs since the work of Gilbert and coworkers in 1975 (Scheme 6.5, *vide supra*).[13]

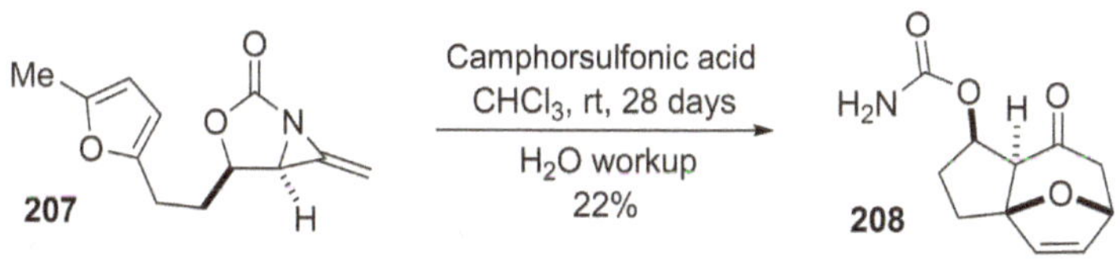

SCHEME 6.34 Intramolecular [4 + 3] cycloaddition involving a bicyclic MA.

Blakey's previous work in the Rh-catalyzed transformations of alkynes and allenes suggested the intermediacy of 2-amidoallyl cations.[16,17] In contrast to the analogous 2-oxaallyl cations, the nitrogenated versions have received much less attention, largely due to the limited methods to cleanly generate these reactive species.[60] In Blakey's work, the purported 2-amidoallyl cations followed one of two pathways: (1) capture by dipolarophiles, such as benzaldehyde, to form substituted tetrahydrofurans (Scheme 6.7A)[16] or (2) rearrangement to cyclopropylimines, which were trapped by Grignard reagents (Schemes 6.5 and 6.6)[16] or nitrones (Scheme 6.7B)[17] to form unusual polycyclic systems. However, these approaches relied on substrate control to determine the stereochemical outcome in the products. The Schomaker group was interested in demonstrating examples of intermolecular [4 + 3] reactions of 2-amidoallyl cations **211** formed *via* aziridination of **209** to **210**, followed by treatment with either heat or a Lewis acid. It was hypothesized that the stereochemical outcome of the [4 + 3] cycloaddition to **212** might be altered through simple changes in the reaction conditions. Likewise, changing the nature of the reagent in the reduction of **212** to **213** might offer an additional opportunity to vary the relative stereochemistry between C1 and C3 of the original allene precursor. A 2016 report by Schomaker described the first examples of intermolecular [4 + 3] reactions between a furan dipolarophile and 2-amidoallyl cation intermediates generated from the treatment of bicyclic MAs with either heat or a Lewis acid (Scheme 6.35).[61]

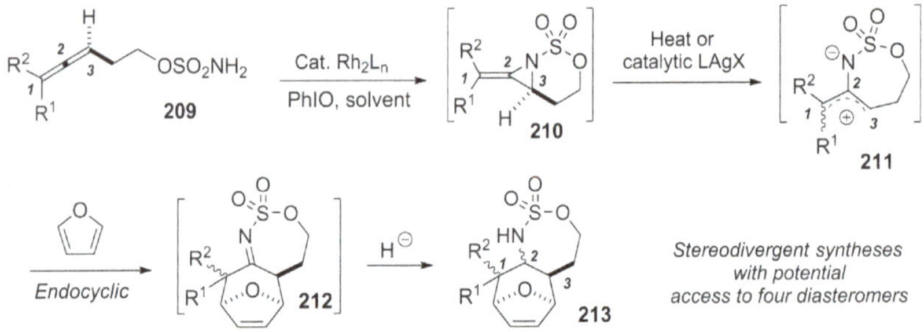

SCHEME 6.35 Intermolecular [4 + 3] cycloadditions of furan with bicyclic MAs.

The Schomaker group also showed that the stereochemistry of the 2-amidoallyl cation intermediate could be controlled by using different co-solvents (Scheme 6.36).[61] Experimental results showed that a mixture of 1:1 furan/THF furnished the 1,3-*syn* heterocycle **215** from **214**, whereas a 1:1 furan/$MeNO_2$ mixture delivered the 1,3-*anti* heterocycle **216**. Imine reduction of **215** with small hydride sources, such as $NaBH_3CN$, favors the approach of the nucleophile from the top face to give the stereoisomer **217** as the preferred product. On the other hand, a bulky $LiBHEt_3$ reducing agent

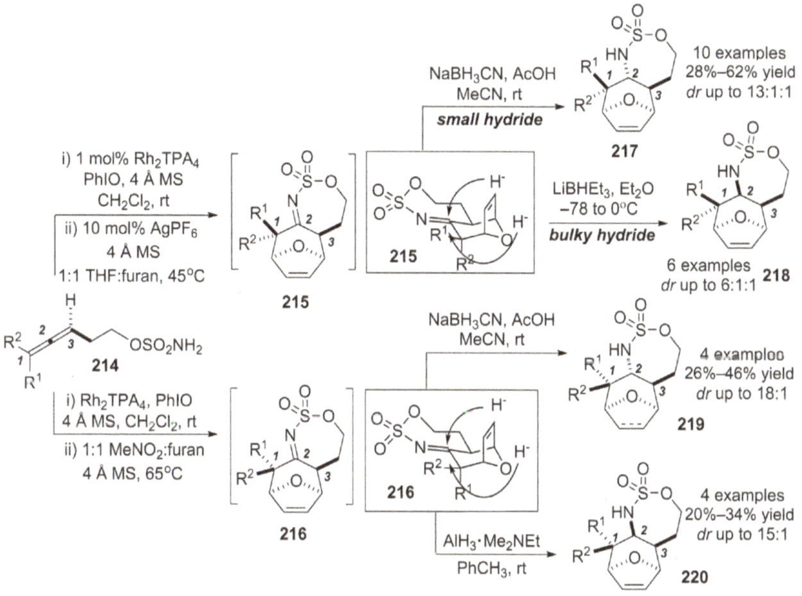

SCHEME 6.36 Stereodivergent intermolecular [4 + 3] cycloadditions of furan with bicyclic MAs.

delivered **218**, provided R^2 was H. For the 1,3-*anti* heterocycle **216**, reduction with $NaBH_3CN$ furnished **219** in good *dr*, while in selected cases, $AlH_3 \cdot Me_2NEt$ provided access to the final remaining diastereomer **220**. As expected, the formation of an achiral 2-amidoallyl cation intermediate precluded transfer of the axial chirality of the allene to the products; however, the presence of a chiral center in the side chain R^1 or R^2 of a substrate such as **214** enables a highly diastereoselective [4 + 3] reaction.[61]

6.5.4 Other Reactivity of Bicyclic Methyleneaziridines

6.5.4.1 *Synthesis of Strained Cycloalkynes*

Readily accessible silylated allenes, represented by **221**, undergo regio-, chemo-, and stereocontrolled Rh-catalyzed aziridinations to form endocyclic MAs of the form **222** (Scheme 6.17, *vide supra*). The Schomaker group discovered that treatment of **222** with tetrabutylammonium fluoride triggered elimination of the silyl group, followed by ring-opening of the aziridine, to yield stable strained alkynes **223** (Scheme 6.37).[62] These novel strained cyclooctynes were termed "SNO-OCTs," due to the S, N, and O heteroatoms incorporated into the ring both at the propargylic and homopropargylic positions. Groups at the C1 position could be varied through the choice of the starting homoallenic alcohol; R^1 groups with a range of steric bulk were

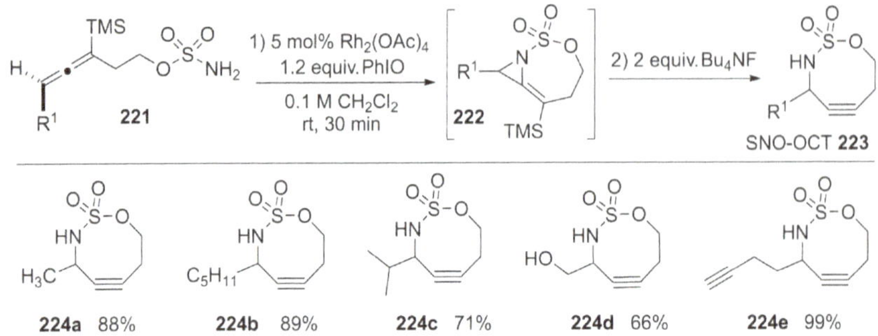

SCHEME 6.37 Synthesis of strained cyclooctynes (SNO-OCTs) from endocyclic MAs.

tolerated in the synthesis, giving SNO-OCTs represented by **224a–e**. Changing the nature of the R[1] group in the SNO-OCT could be used to tune various features of subsequent reactions, including the rates of reactivity with a variety of 1,3-dipoles (methyl-substituted **224a)** and the regioselectivity of the [3 + 2] dipolar cycloaddition products. SNO-OCTs **224d** and **224e** with a tethered alcohol or a terminal alkyne offer opportunities to act as linkers when adding desired functionality to the scaffold.[63]

SNO-OCTs and their derivatives react quickly with a diverse set of 1,3-dipoles, including azides, diazos, nitrile oxides, and tetrazines (Scheme 6.38).[62,63] The presence of heteroatoms in SNO-OCTs enables tunable polarization of the alkyne, depending on the substitution pattern of the heterocycle. This influences both the chemo- and regioselectivity observed in the [3 + 2] products. Regioselectivity is controlled largely by the substrate that is utilized; for example, SNO-OCT **225a** furnished a 2.1:1 mixture of triazoles **226a:226b** in reactions with benzyl azide. Increasing the size of the R group from C_5H_{11} in SNO-OCT **225a** to an isopropyl group in SNO-OCT **225b** resulted in an improvement in the ratio of the triazole regioisomers

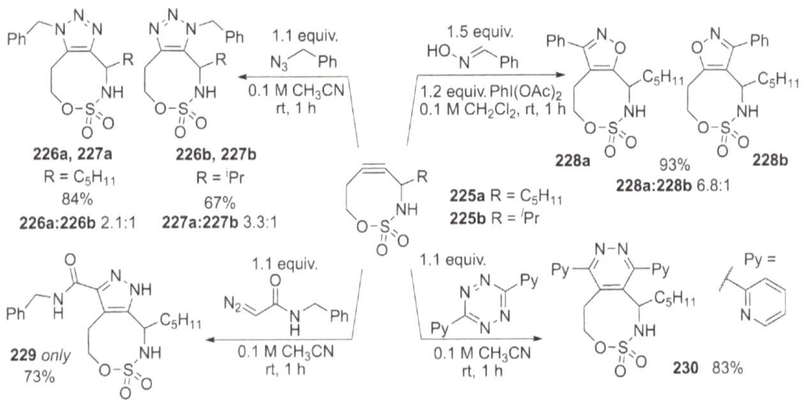

SCHEME 6.38 Use of SNO-OCTs in regioselective 1,3-dipolar cycloadditions.

227a:227b to 3.3:1. There is also a significant electronic component to the regioselectivity, as switching to more polarized 1,3-dipoles, including benzaldehyde oxime or diazoacetate, results in improved regioselectivities. For example, oxime addition to **225a** led to the formation of the oxazole regioisomers **228a:228b** in a 7:1 ratio. Diazoacetate addition to **225a** gave only the product **229** as a single regioisomer. A tetrazine also functioned as a good coupling partner, delivering **230** in 83% yield.

6.5.4.2 Synthesis of Highly Substituted Methyleneazetidines and Azetidines

Azetidines are four-membered saturated nitrogen heterocycles that have received significantly less attention than their aziridine, pyrrolidine, and piperidine counterparts.[64] Recent interest in utilizing azetidines to rigidify alkyl-substituted amines and as synthetic building blocks have increased efforts to identify new stereocontrolled methods for the preparation of these scaffolds.[65–68] Transition metal–catalyzed allene aziridination to **231** (Scheme 6.39), followed by a stereoselective one-carbon ring expansion reaction using a metal-supported carbene **232**, offers one solution for obtaining highly substituted and densely functionalized methylene azetidines of the form **234**.[69]

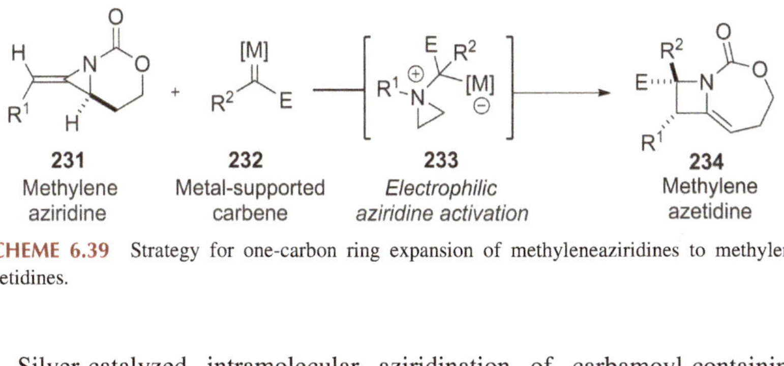

SCHEME 6.39 Strategy for one-carbon ring expansion of methyleneaziridines to methylene azetidines.

Silver-catalyzed intramolecular aziridination of carbamoyl-containing allenes of the form **235** yields MAs **236**. The bicyclic nature of **236** aligns the nitrogen lone pair orthogonal to the carbonyl, rendering it more nucleophilic than might otherwise be predicted. The additional ring strain provides an excellent driving force for ring-opening, while the presence of the carbamate tether helps to ensure that any subsequent ring-closure occurs in a highly regioselective fashion. Treating an aziridine of the form **236** with an electrophilic one-carbon source, such as a Rh-bound carbene, triggered the ring-opening/ring-closing cascade proposed in Scheme 6.39 to yield methyleneazetidines with good-to-excellent diastereoselectivity.

Schomaker and coworkers reported the scope of Rh-catalyzed ring expansions employing a variety of donor/acceptor carbene precursors and MAs

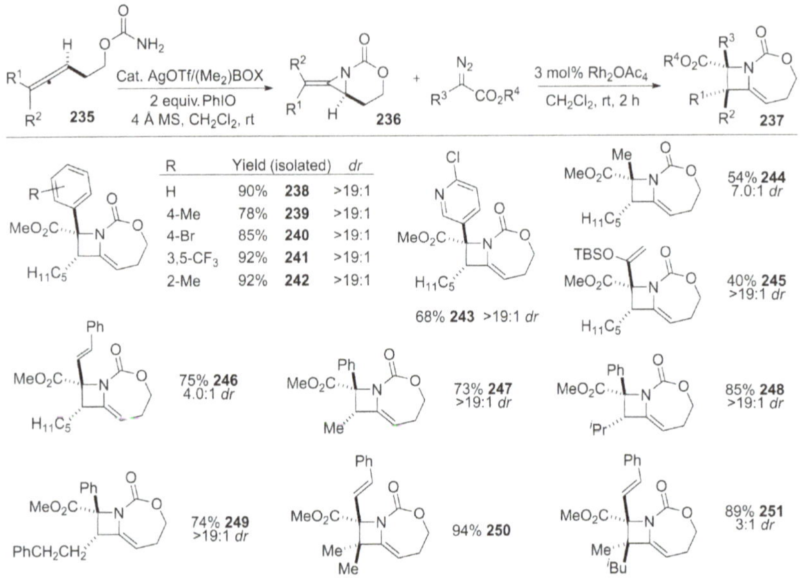

Standard conditions: Slow addition of the diazoester (1.3 equiv.) to aziridine (1.0 equiv.) over 2 h.

SCHEME 6.40 Rh-catalyzed one-carbon ring expansions of methyleneaziridines.

(Scheme 6.40).[69] Phenyl diazoacetate, as well as electron-rich and electron-poor aryl diazoacetates, were successful, furnishing the azetidines **238–242** in 78%–92% yield and with diastereoselectivities typically >19:1 *dr*. Ortho substitution on the aryl ring of the carbene was tolerated, as were heterocycles, as represented by **243**.

The [3 + 1] ring expansion reaction did not require an aryl diazoester precursor, as the aryl group could be replaced with an alkyl or ester group, as in **244**; alternatively, diazoacetates with vinyl substitution could be employed to yield exclusive [3 + 1] addition, delivering azetidine **245** in 40% yield as a single diastereomer. Styrenyl derivatives cleanly gave azetidine **246**, albeit in lower *dr*; this is attributed to the smaller steric differentiation between a styrene and methyl ester, as compared to the phenyl variant.

The *E* isomers (> 10:1 *E*:*Z*) of a variety of MAs were successful in the reaction. Precursors containing methyl, isopropyl, and remote phenyl substituents were all well-tolerated, giving azetidines **247–249** in good yields and excellent *dr*. Trisubstituted allene precursors generated bicyclic MAs that reacted slowly with bulky carbene partners; however, changing to a less-hindered styrenyl diazo coupling partner delivered the fully substituted methyleneazetidine **250** in 94% yield. A highlight of the method was the conversion of a nonsymmetric MA precursor, consisting of a 4:1 mixture of inseparable *E*/*Z* isomers, to **251** in 3:1 *dr* and 89% combined yield. The two diastereomers were easily separable, yielding the *syn*-Me, $-CO_2Me$ isomer of **251** in 54% yield and 15:1 *dr*.

The alkene of the methyleneazetidine **252** could be further functionalized through hydrogenation to give the azetidine **254** in high stereoselectivity (Scheme 6.41). Alternatively, Baran's iron-catalyzed C–C coupling successfully yielded **253** in good *dr*, forging a second quaternary carbon on the azetidine ring.[70]

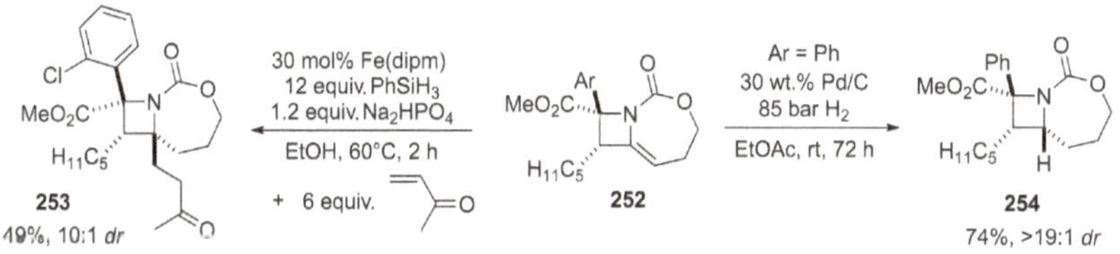

SCHEME 6.41 Further transformations of bicyclic methyleneazetidines.

Mechanistic studies of the Rh-catalyzed one-carbon ring expansion of MAs showed that chirality was transferred from the enantioenriched MA **255** to the azetidine **256** with excellent fidelity (Scheme 6.42). A tentative mechanism proposes that aziridinium ylide formation occurs *via* attack of the nitrogen of the MA on the Rh-bound carbene to form **257**. Ring-opening to **258** is followed by a bond rotation and ring closure to yield **260** in an overall ring-opening/ring-closing cascade. The proposed ylide intermediate likely retains its stereochemistry by existing as a C-bound Rh enolate that does not

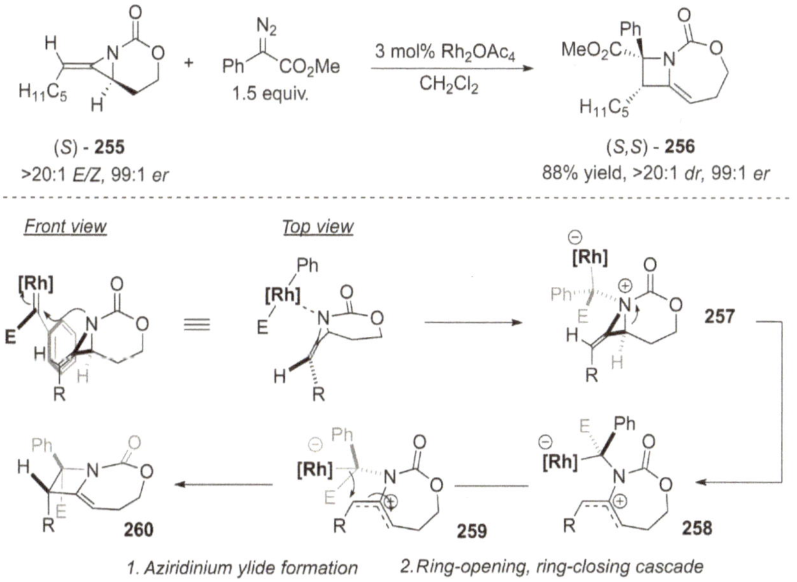

SCHEME 6.42 Mechanistic insight into Rh-catalyzed [3 + 1] ring expansions of MAs.

racemize *via* isomerization through an O-bound Rh species.[71] This assumption, coupled with transfer of chirality experiment, suggests that aziridinium formation occurs with high *dr* in the case of aryl diazoacetates. The relative configuration among the substituents on the azetidine can be rationalized *via* a steric-based model, wherein the approaching Rh-bound carbene orients its larger R group (aryl, styrenyl, etc. vs the ester) in the back, convex pocket of the bowl-shaped methyleneaziridine to avoid steric interactions with the alkene.

6.6 SYNTHETIC APPLICATIONS OF ALLENE AZIRIDINATION TO COMPLEX MOLECULE SYNTHESIS

6.6.1 Aminosugars

Aminosugars occur in an array of bioactive compounds, including mono- and polysaccharides, glycolipids, nucleotides, anthracycline antitumor agents, and antibiotics (Fig. 6.3).[72] Neuraminic acid and glucosamine are commonly found in nature, but new modular, flexible *de novo* approaches to construct libraries of novel sugars for incorporation into glycoconjugates are highly desirable. The ability to flexibly install three contiguous heteroatom-bearing stereocenters at each of the unsaturated carbons in the allene, while controlling the relative stereochemistry, could give useful building blocks for aminosugar synthesis through axial-to-point chirality transfer.

A general strategy for aminosugar synthesis from allene precursors was described by Schomaker and coworkers (Scheme 6.43).[73] Enantioenriched allenes of the form **261** are easily prepared from D-mannitol; a one-pot aziridination to **262**, followed by water ring-opening and alcohol protection, furnishes an enesulfamate of the form **263**. Treatment of **263** with DMDO and a reductant was expected to lead to the key 2-amino-1,3-diol motif **264** that can be elaborated to either known or novel aminosugar scaffolds. Since the allene precursors can be prepared in enantioenriched form and transfer of axial-to-point chirality is achieved with excellent fidelity, the major challenge is to control the diastereoselectivity in the transformations of **262** to **263**, as well as the reduction step to **264**.

Interestingly, aziridination of a dimethylketal-substituted allene **265** gave a mixture of the (*E*)-MA **266**, along with the cyclopropane **267** (Scheme 6.43, top). The diastereomeric allene **268** gave only the cyclopropylimine

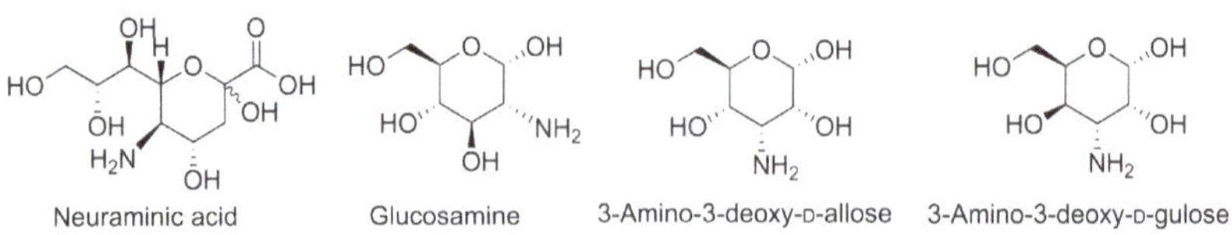

FIGURE 6.3 Selected examples of aminosugars.

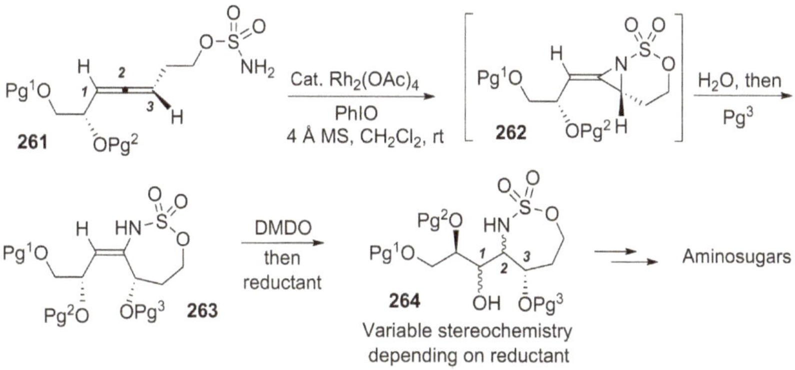

SCHEME 6.43 Allene aziridination strategy for the synthesis of aminosugars.

269, indicating that the relative stereochemistry between the allene and the secondary alcohol plays an important role in determining the reaction outcome. One possible explanation is that the stereochemistry in **268** aligns the oxygen with the aziridine ring of **270a** to readily ring-open the sensitive bicyclic MA giving **271** as the intermediate (Scheme 6.43, bottom) that undergoes ring closure to **269**. In contrast, unfavorable steric interactions in **270b** do not allow it to readily adopt a conformation that permits nucleophilic attack on the MA. As a result, less rearrangement occurs, although some **269** is still observed (Scheme 6.44).

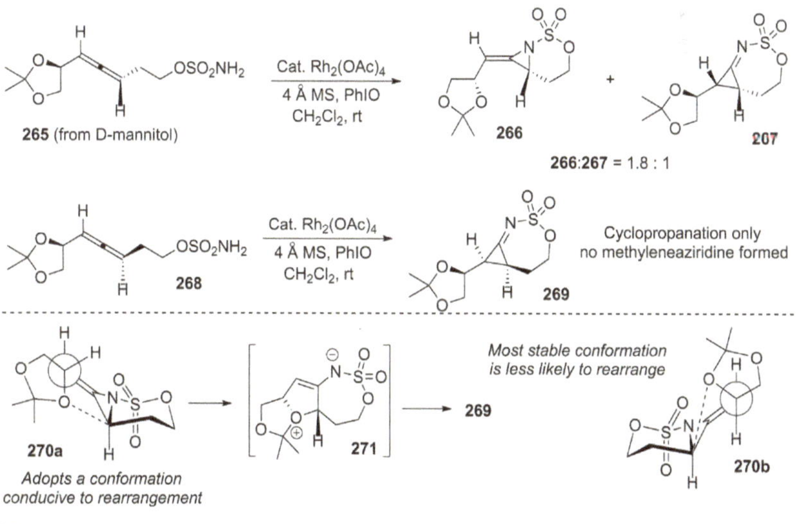

SCHEME 6.44 Competing cyclopropane formation in Rh-catalyzed allene aziridination.

The competing rearrangement of MAs to the cyclopropylimine was minimized by installing a diethylketal in **272** (Scheme 6.45). This enabled preparation of the enesulfamate **274** (Scheme 6.45, top) in 32% yield over the three steps and >19:1 *dr*. DMDO oxidation gave the sensitive α-hydroxyimine **275** in greater than 10:1 *dr* and was followed by immediate reduction with $Zn(BH_4)_2$ to furnish **276a** and **276b** as a 4:1 mixture of easily separable diastereomers. A one-pot removal of the sulfamoyl group and cleavage of the resulting olefin *via* ozonolysis gave a protected derivative of 3-deoxy-3-amino-D-allose **278**.

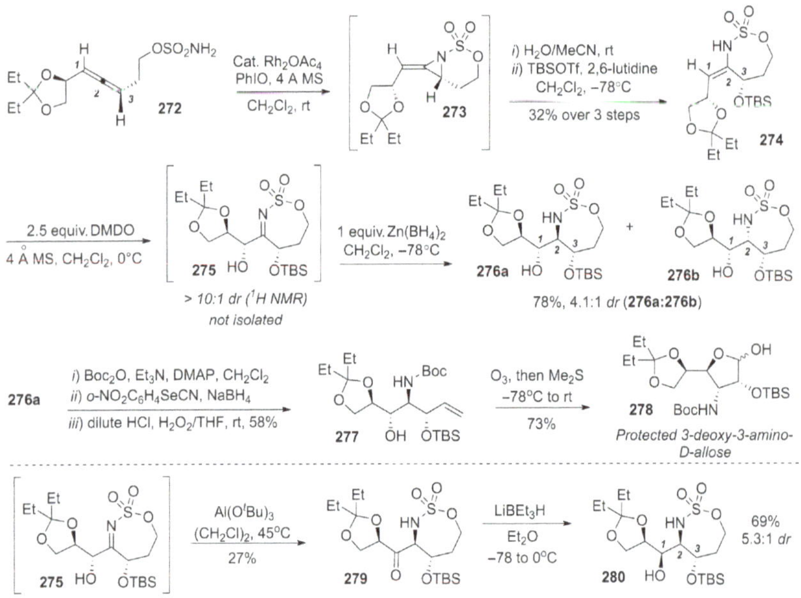

SCHEME 6.45 Synthesis of aminosugars using an oxidative allene aziridination strategy.

Stereochemical flexibility was not thoroughly explored in these studies; however, Amadori-type rearrangement of the α-hydroxy imine **275** to the α-aminoketone **279** (Scheme 6.45, bottom), followed by reduction with $LiBEt_3H$, gave a stereoisomer of **276a,b**, **280**, with the stereochemical configuration of 3-amino-3-deoxy-D-gulose. With this strategy in hand, the introduction of other heteroatoms into the amine triads can be envisaged to lead to novel aminosugars.

6.6.2 Total Synthesis of (±)-Detoxinine Methyl Ester and Its Stereoisomers

The natural product detoxinine is an unusual bis-hydroxylated α-amino acid that comprises the core of the detoxifying agent detoxin, which is coadministered with blasticidin S for the treatment of rice blast disease.[74] The stereochemical flexibility of the oxidative allene amination chemistry developed by the Schomaker group permits a single homoallenic sulfamate precursor to be transformed to all four possible diastereomers of the methyl ester of detoxinine (Scheme 6.46).

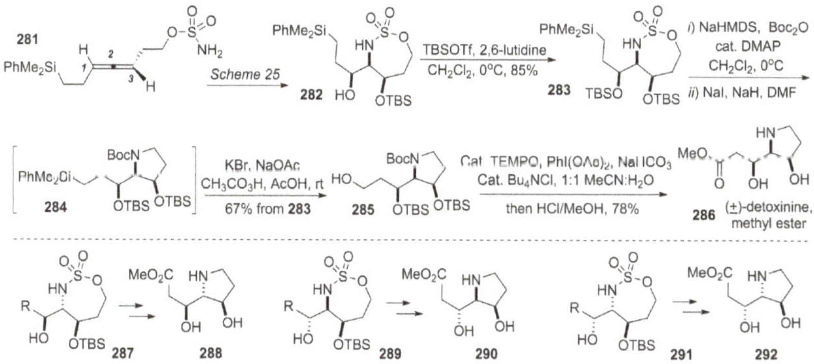

SCHEME 6.46 Total synthesis of (±)-detoxinine methyl ester and stereoisomers.

A silylated homoallenic sulfamate **281** was transformed into four diastereomeric O/N/O stereotriads **282** and stereoisomers using the chemistry described in Scheme 6.25.[56] Each of the four diastereomers was transformed into a detoxinine analog through the sequence illustrated in Scheme 6.46. Protection of the free secondary alcohol of all-*syn* stereotriad **282** gave **283** in 85% yield. Installation of a Boc group on the amine was followed by NaI-mediated ring contraction to give intermediate **284**, which was subjected to Tamao-Fleming oxidation to furnish **285** in 67% yield from **283**. A two-step oxidation sequence, followed by an acidic workup afforded (±)-detoxinine methyl ester **287**. The same sequence of reactions was applied to the three remaining diastereomers of **282** (**287**, **289**, and **291**, Scheme 6.46, bottom) to yield the nonnatural stereoisomers of detoxinine, **288**, **290**, and **292** in comparable overall yields and in excellent diastereoselectivity.

6.6.3 Fluorinated Pyrrolidines and Unnatural Amino Acids

Fluorine is often introduced into drugs and other bioactive scaffolds to improve a host of physical properties that include lipophilicity, bioavailability, hydrophobic interactions between the drug and binding sites on receptors or enzymes, and higher oxidative and thermal stability as compared to the analogous C–H bonds.[53] In 2017, the Schomaker group reported the use of oxidative allene amination (Scheme 6.24, *vide supra*) as a strategy to prepare fluorine-containing amine triads.[55] To demonstrate the utility of the method, it was applied to the synthesis of fluorinated pyrrolidines found in powerful glucosidase inhibitors, such as 2,5-dideoxy-2,5-imino-D-mannitol,[75] as well as inhibitors of purine nucleoside phosphorylase (Scheme 6.47).[76] Oxidative allene amination of **293** to furnish **294** was followed by N-Boc protection of the nitrogen and NaI-mediated ring contraction to deliver the pyrrolidine **295** in good yield and *dr* >19:1.

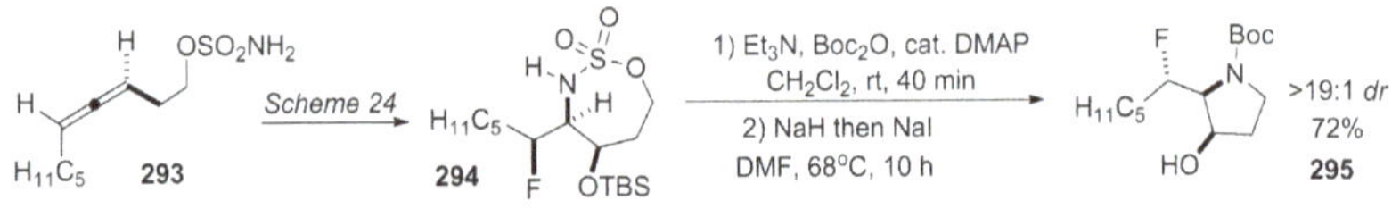

SCHEME 6.47 Synthesis of fluorinated pyrrolidine scaffolds.

Fluorinated, unnatural amino acids play key roles in directing protein—-protein interactions, are utilized in the design of hyperstable protein folds,[77] and display antitumor and antimicrobial activities.[78] Fluorinated amines **297** and **298** from oxidative allene amination of **296** can be readily transformed in a few steps to α-, β-, and γ-amino acids using simple chemistry (Scheme 6.48).[55] For example, ozonolysis of **297**, followed by Pinnick oxidation, furnished the α-aminoacid **299** in good yield and excellent diastereoselectivity. Both β- and γ-amino acids could be synthesized from key intermediate **300**; the former through ozonolysis, Pinnick oxidation, and esterification to deliver **301**. The protected γ-amino acid **302** was obtained through a sequence involving hydroboration and $RuCl_3$-catalyzed oxidation of the alcohol to the carboxylic acid. Again, the ability to transfer axial-to-point chirality in **296** enables access to either enantiomer of the unnatural aminoacid scaffolds.

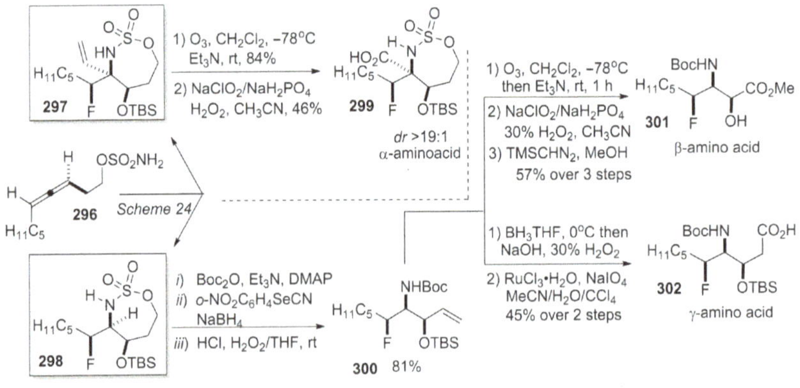

SCHEME 6.48 Transformations of fluorinated amine triads to unnatural amino acids.

6.6.4 Aminocyclopentitol Core of Jogyamycin

The natural product jogyamycin was first isolated in 2012 from a culture broth of *Streptomyces* sp. a-WM-JG-16.2 (Fig. 6.4). This complex aminocyclopentitol is one of a handful of other natural products with a similar scaffold, including pactamycin, cranomycin, and TM-026.[79] Jogyamycin itself exhibits potent antiprotozoal activity against the organisms that cause diseases including malaria and African sleeping sickness, while pactamycin and its analogs have been demonstrated to possess anticancer, antiviral, and antimicrobial activities in addition to their antiprotozoal activities.[80] Reported crystal structures of pactamycin binding to the ribosomal 30 S subunit indicate that it is a universal inhibitor of translocation in a highly conserved region of the ribosome, explaining the wide range of biological activity exhibited by this class of molecules.[81] Nonetheless, subtle structural changes in the aminocyclitol motif have been shown to alter the activity of these natural products significantly for reasons that are not well understood.[80] All members of this class of compounds exhibit a fully substituted cyclopentane ring system with a heteroatom present on each carbon. Three contiguous quaternary carbons, as well as the dense functionalization around the ring that includes sensitive urea and aniline moieties, have inspired various strategies to achieve the syntheses of these challenging motifs.[82] However, none of these efforts has led to methods flexible enough for deep-seated changes to the core of jogyamycin, rendering explorations of the structure–activity relationship of this family of molecules difficult.

Oxidative allene aziridination and subsequent functionalization of the resulting MAs have been demonstrated to be a powerful and versatile method for the synthesis of diverse amine stereotriads that could comprise key building blocks for the preparation of jogyamycin and its analogs. The Schomaker group reported the application of this methodology to the synthesis of the core of jogyamycin (Scheme 6.49).[83] A three-step, one-pot conversion of allene **303**–**305** was followed by another three-step, one-pot sequence to transform enesulfamate **305**–**306** in good *dr*. The *dr* of the aminoketone was increased to >20:1 after purification. N-Boc protection of **306** furnished

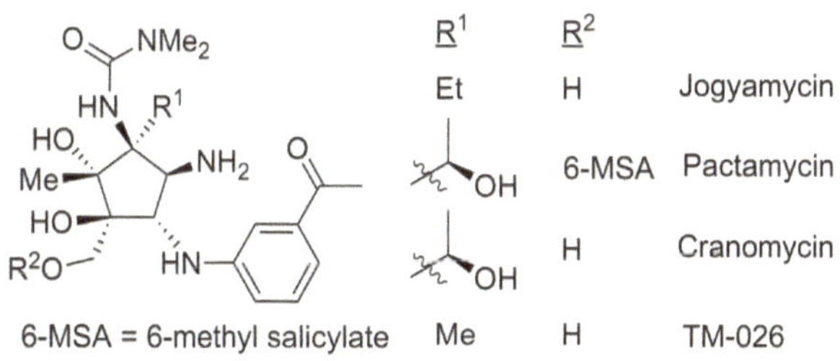

FIGURE 6.4 Aminocyclopentitol-containing, bioactive natural products.

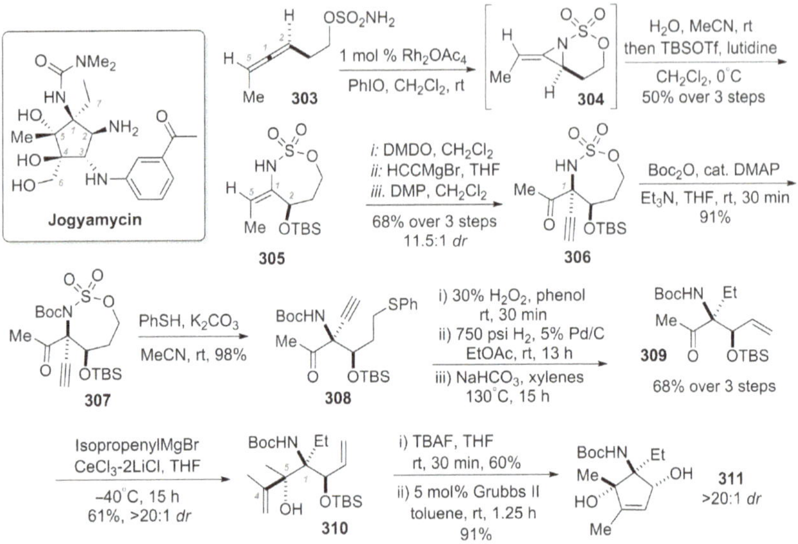

SCHEME 6.49 Synthesis of the core of jogyamycin via oxidative allene amination.

307; ring-opening of the sulfamate with a thiophenol nucleophile to give **308** was followed by a one-pot, three-step sequence to install the alkene in **309**. Chelation-controlled addition with the isopropenyl Grignard reagent to **309** gave the amine triad **310** in >20:1 *dr*. Removal of the TBS-protecting group from the secondary alcohol of **310** was required for successful ring-closing metathesis catalyzed by Grubbs II to deliver the core cyclopentene **311** in 91% yield. In addition to accessing the core of jogyamycin, this approach offers significant flexibility in terms of the introduction of other heteroatoms into the core, as well as versatility in the relative stereochemistries of the heteroatom substituents on the ring.

6.7 CONCLUSION AND OUTLOOK

Transition metal–catalyzed aminations of allenes offer advantages and complementary strategies to conventional alkene oxidation methods for the synthesis of complex amine triad motifs. The potential to transfer the readily accessible axial chirality of allenes to point chirality in the products represents a major advantage over alkene oxidation, both by simplifying substrate preparation and enabling flexibility in the stereoisomeric outcome of the reaction.

As oxidative allene amination is a relatively new area, this chapter focuses mainly on highly chemo-, regio-, and stereocontrolled aziridination as the first step in transforming each of the three sp and sp^2 carbons of an allene precursor to three new sp^3 stereocenters. The development of new silver catalysts for chemoselective, intramolecular allene aziridination was reviewed, methods described to control the regioselectivity of the aziridination to favor either the proximal or the distal double bond of an allene precursor and the differences between employing carbamoyl versus sulfamoyl nitrene precursors discussed.

Several modes of unusual reactivity of bicyclic MAs generated from allene amination were presented. Ring-opening of the sp^3 carbon of the aziridine is the first step in a powerful, one-pot reaction sequence that delivers $X^1/N/X^2$ amine stereotriads, where the identities of the two flanking heteroatoms can be easily varied. Applications of this strategy to the syntheses of aminosugars, detoxinine and its stereoisomers, fluorinated pyrrolidines, unnatural α-, β-, and γ-amino acids, and the core of a bioactive aminocyclopentitol, jogyamycin, were demonstrated.

Alternative modes of reactivity of bicyclic MAs included functionalization of the exocyclic double bond of the MA to furnish X/O/N stereotriads, engaging the nitrogen lone pair of the MA with Rh-supported carbenes to achieve ring expansion to densely substituted azetidines and treatment of MAs with heat or Lewis acid to generate 2-amidoallyl cations that could be trapped in a stereodivergent manner to deliver [4 + 3] products. Modifications to the substitution pattern of the allene enabled aziridination of the distal double bond of the allene, leading to endocyclic MAs. These exhibited complementary reactivity to the exocyclic MAs and were readily converted to rare azetidin-3-one scaffolds and novel-strained cyclooctynes that could be employed for bioorthogonal labeling.

Despite recent advances in transition metal–catalyzed allene aminations, the field is relatively new. The mechanistic details of some of the methodologies described herein are not completely understood and may comprise the basis for future studies. Many other synthetic challenges remain, particularly in the context of intermolecular allene amination reactions and other methods to introduce nitrogen directly into the allene that do not rely on hydroamination or aziridination. Transition metal–catalyzed strategies capable of converting racemic allenes to enantioenriched triads through catalyst control are also highly desirable and are likely to be the target of further studies in the near future. In addition, the scope of the reactivity displayed by MAs has not been thoroughly explored. While each of the unsaturated carbons of the allene can be readily converted to a new sp^3 carbon-heteroatom bond, incorporating new C–C bonds into amine triad has not been investigated, but would greatly expand the scope and utility of amine triads obtained from allene amination.

The ease, scope, and versatility of this chemistry, especially in gaining access to enantioenriched and densely substituted amines, render it ideal for the efficient exploration of structure–activity relationships in bioactive molecules with complex molecular architectures. It is expected that oxidative allene amination methods will be particularly useful for the synthesis and study of novel aminosugars, unnatural amino acids, analogs of aminocyclitols, such as jogyamycin and pactamycin, as well as other bioactive natural and unnatural product scaffolds.

REFERENCES

1. For selected references, see: (a) Pasco, M.; Moumne, R.; Lecourt, T.; Micouin, L. Stereoselective Synthesis of Fluorinated 1,3-cis-Diaminocyclopentanes. *J. Org. Chem.* **2011,** *76*, 5137–5142. (b) King, S. B.; Ganem, B. Synthetic Studies on Mannostatin A and Its Derivatives: A New Family of Glycoprotein Processing Inhibitors. *J. Am. Chem. Soc.* **1994,** *116*, 562–570. (c) Magerlein, B. J.; Kagan, F. Lincomycin. VIII. 4′-Alkyl-1′-Demethyl-4′-Depropylclindamycins, Potent Antibacterial and Antimalarial Agents. *J. Med. Chem.* **1969,** *12*, 780–784. (d) Zhang, Y.; Liu, S.; Che, Y.; Liu, X. Epicoccins A–D, Epipolythiodioxopiperazines From a *Cordyceps*-Colonizing Isolate of *Epicoccum nigrum*. *J. Nat. Prod.* **2007,** *70*, 1522–1525. (e) Lewis, J. G.; Anandan, S. K.; O'Dowd, H.; Gordeev, M. F. Novel Lincomycin Derivatives Possessing Antimicrobial Activity. *PCT Int. Appl.* **2005**. WO 2005012320 A2 20050210. (f) Ghosh, A. K.; Dawson, Z. L.; Mitsuya, H. Darunavir, a Conceptually New HIV-1 Protease Inhibitor for the Treatment of Drug-Resistant HIV. *Bioorg. Med. Chem.* **2007,** *15*, 7576–7580. (g) He, H.; Williamson, R. T.; Shen, B.; Graziani, E. I.; Yang, H. Y.; Sakya, S. M.; Petersen, P. J.; Carter, G. T. Mannopeptimycins, Novel Antibacterial Glycopeptides From *Streptomyces hygroscopicus*, LL-AC98. *J. Am. Chem. Soc.* **2002,** *124*, 9729–9736. (h) Prakesch, M.; Denisov, A. Y.; Naim, M.; Gehring, K.; Arya, P. The Discovery of Small Molecule Chemical Probes of Bcl-X_L and Mcl-1. *Bioorg. Med. Chem.* **2008,** *16*, 7443–7449. (i) Iserloh, U.; Wu, Y.; Cumming, J. N.; Pan, J.; Wang, L. Y.; Stamford, A. W.; Kennedy, M. E.; Kuvelkar, R.; Chen, X.; Parker, E. M.; Strickland, C.; Voight, J. Potent Pyrrolidine- and Piperidine-Based BACE-1 Inhibitors. *Bioorg. Med. Chem. Lett.* **2008,** *18*, 414–417. (j) Hanessian, S.; Soma, U.; Dorich, S.; Deschenes-Simard, B. Total Synthesis of (+)-*ent*-Cyclizidine: Absolute Configurational Confirmation of Antibiotic M146791. *Org. Lett.* **2011,** *13*, 1048–1051. (k) Ohsaki, A.; Ishiyama, H.; Yonedac, K.; Kobayashi, J. Secu'amamine A, a Novel Indolizidine Alkaloid From *Securinega suffruticosa* var. *amamiensis*. *Tetrahedron Lett.* **2003,** *44*, 3097–3099. (l) Kim, H.; Kang, S. Total Synthesis of (+)-*ent*-Cyclizidine: Absolute Configurational Confirmation of Antibiotic M146791. *Org. Lett.* **2011,** *13*, 1048–1051. (m) Kim, H.; Kang, S.; Beaulieu, P.; Lavallee, P.; Abraham, A.; Anderson, P. C.; Boucher, C.; Bousquet, Y.; Duceppe, J.; Gillard, J.; Gorys, V.; Grand-Maitre, C.; Grenier, L.; Guindon, Y.; Guse, I.; Plamondon, L.; Soucy, F.; Valois, S.; Wernic, D.; Yoakim, C. Practical, Stereoselective Synthesis of Palinavir, a Potent HIV Protease Inhibitor. *J. Org. Chem.* **1997,** *62*, 3440–3448. (n) Shinada, T.; Ikebe, E.; Oe, K.; Namba, K.; Kawasaki, M.; Ohfune, Y. Synthesis and Absolute Structure of Manzacidin B. *Org. Lett.* **2010,** *12*, 2170. (o) Arya, D. P., Ed. *Aminoglycoside Antibiotics;* John Wiley & Sons: Hoboken, NJ, 2007. (p) He, H.; Williamson, R. T.; Shen, B.; Graziani, E. I.; Yang, H. Y.; Sakya, S. M.; Petersen, P. J.; Carter, G. T. Mannopeptimycins, Novel Antibacterial Glycopeptides From *Streptomyces hygroscopicus*, LL-AC98. *J. Am. Chem. Soc.* **2002,** *124*, 9729–9736. (q) Gan, M.; Zheng, X.; Gan, L.; Guan, Y.; Hao, X.; Liu, Y.; Si, S.; Zhang, Y.; Yu, L.; Xiao, C. Streptothricin Derivatives From *Streptomyces* sp. I08A 1776. *J. Nat. Prod.* **2011,** *74*, 1142–1147. (r) He, H.; Williamson, R. T.; Shen, B.; Graziani, E. I.; Yang, H. Y.; Sakya, S. M.; Petersen, P. J.; Carter, G. T. Streptothricin Derivatives From *Streptomyces* sp. I08A 1776. *J. Am. Chem. Soc.* **2002,** *124*, 9729. (s) Shibasaki, M.; Kanai, M.; Yamatsugu, K. Recent Development in Synthetic Strategies for Oseltamivir Phosphate. *Isr. J. Chem.* **2011,** *51*, 316–328. (t) Magano, J. Synthetic Approaches to the Neuraminidase Inhibitors Zanamivir (Relenza) and Oseltamivir Phosphate (Tamiflu) for the Treatment of Influenza. *Chem. Rev.* **2009,** *109*, 4398–4438. (u) Kobayashi, J.; Kanda, F.;

Ishibashi, M.; Shigemori, H. Manzacidins A-C, Novel Tetrahydropyrimidine Alkaloids From the Okinawan Marine Sponge *Hymeniacidon* sp. *J. Org. Chem.* **1991,** *56*, 4574–4576. (v) Magano, J. Recent Synthetic Approaches to Oseltamivir Phosphate (Tamiflu™) for the Treatment of Influenza. *Tetrahedron* **2011,** *67*, 7875–7899. (w) Iwatsuki, M.; Nishihara-Tsukashima, A.; Ishiyama, A.; Namatame, M.; Watanabe, Y.; Handasah, S.; Pranamuda, H.; Marwoto, B.; Matsumoto, A.; Takahashi, Y.; Otoguro, K.; Omura, S. Jogyamycin, a New Antiprotozoal Aminocyclopentitol Antibiotic, Produced by *Streptomyces* sp. a-WM-JG-16.2. *J. Antibiot.* **2012,** *65*, 169–171.

2. For selected examples of Sharpless asymmetric aminohydroxylation, see: (a) Li, G.; Chang, H.-T.; Sharpless, K. B. Catalytic Asymmetric Aminohydroxylation (AA) of Olefins. *Angew. Chem. Int. Ed.* **1996,** *35*, 451–454. (b) O'Brien, P. Sharpless Asymmetric Aminohydroxylation: Scope, Limitations, and Use in Synthesis. *Angew. Chem. Int. Ed.* **1999,** *38*, 326–329. (c) Li, G.; Angert, H. H.; Sharpless, K. B. *N*-Halocarbamate Salts Lead to More Efficient Catalytic Asymmetric Aminohydroxylation. *Angew. Chem. Int. Ed.* **1996,** *35*, 2813–2817. (d) Tao, B.; Schlingloff, G.; Sharpless, K. B. Reversal of Regioselection in the Asymmetric Aminohydroxylation of Cinnamates. *Tetrahedron Lett.* **1998,** *39*, 2507–2510. (e) Bruncko, M.; Schlingloff, G.; Sharpless, K. B. *N*-Bromoacetamide—A New Nitrogen Source for the Catalytic Asymmetric Aminohydroxylation of Olefins. *Angew. Chem. Int. Ed.* **1997,** *36*, 1483–1486. (f) Rudolph, J.; Sennhenn, P. C.; Vlaar, C. P.; Sharpless, K. B. Smaller Substituents on Nitrogen Facilitate the Osmium-Catalyzed Asymmetric Aminohydroxylation. *Angew. Chem. Int. Ed.* **1996,** *35*, 2810–2813.
3. For selected recent examples on diamination of alkenes, see: (a) Bar, G. L. J.; Lloyd-Jones, G. C.; Booker-Milburn, K. I. Pd(II)-Catalyzed Intermolecular 1,2-Diamination of Conjugated Dienes. *J. Am. Chem. Soc.* **2005,** *127*, 7308–7309. (b) Muñiz, K.; Streuff, J.; Hövelmann, C. H.; Núñez, A. Exploring the Nickel-Catalyzed Oxidation of Alkenes: A Diamination by Sulfamide Transfer. *Angew. Chem. Int. Ed.* **2007,** *46*, 7125–7127. (c) Ingalls, E. L.; Sibbald, P. A.; Kaminsky, W.; Michael, F. E. Enantioselective Palladium-Catalyzed Diamination of Alkenes Using *N*-Fluorobenzenesulfonimide. *J. Am. Chem. Soc.* **2013,** *135*, 8854–8856. (d) Zhang, H.; Pu, W.; Xiong, T.; Li, Y.; Zhou, X.; Sun, K.; Liu, Q.; Zhang, Q. Copper-Catalyzed Intermolecular Aminocyanation and Diamination of Alkenes. *Angew. Chem. Int. Ed.* **2013,** *52*, 2529–2533. (e) Sequeira, F. C.; Turnpenny, B. W.; Chemler, S. R. Copper-Promoted and Copper-Catalyzed Intermolecular Alkene Deamination. *Angew. Chem.* **2010,** *122*, 6509–6512. (f) Sibbald, P. A.; Michael, F. E. Palladium-Catalyzed Diamination of Unactivated Alkenes Using *N*-Fluorobenzenesulfonimide as Source of Electrophilic Nitrogen. *Org. Lett.* **2009,** *11*, 1147–1149.
4. For selected reviews of asymmetric aziridination of alkenes, see: (a) Yudin, A. K., Ed. *Aziridines and Epoxides in Organic Synthesis;* Wiley-VCH, Verlag GmbH & Co. KGaA, Weinheim, 2006. (b) Müller, P.; Fruit, C. Enantioselective Catalytic Aziridinations and Asymmetric Nitrene Insertions Into CH Bonds. *Chem. Rev.* **2003,** *103*, 2905–2920. (c) Zhu, Y.; Wang, Q.; Cornwall, R. G.; Shi, Y. Organocatalytic Asymmetric Epoxidation and Aziridination of Olefins and Their Synthetic Applications. *Chem. Rev.* **2014,** *114*, 8199. (d) Pellissier, H. Recent Developments in Asymmetric Aziridination. *Tetrahedron* **2010,** *66*, 1509–1555.

5. For selected examples and reviews of ring opening of aziridines, see: (a) Sweeney, J. B. Aziridines: Epoxides' Ugly Cousins? *Chem. Soc. Rev.* **2002,** *31*, 247–258. (b) Hu, X. E. Nucleophilic Ring Opening of Aziridines. *Tetrahedron* **2004,** *60*, 2701–2743. (c) Lu, P. Recent Developments in Regioselective Ring Opening of Aziridines. *Tetrahedron* **2010,** *66*, 2549–2560. (d) Fukuta, Y.; Mita, T.; Fukuda, N.; Kanai, M.; Shibasaki, M. De Novo Synthesis of Tamiflu Via a Catalytic Asymmetric Ring-Opening of Meso-Aziridines With $TMSN_3$. *J. Am. Chem. Soc.* **2006,** *128*, 6312–6313. (e) Tanner, D. Chiral Aziridines—Their Synthesis and Use in Stereoselective Transformations. *Angew. Chem. Int. Ed.* **1994,** *33*, 599–619. (f) Stanković, S.; D'hooghe, M.; Catak, S.; Eum, H.; Waroquier, M.; Speybroeck, V.; De Kimpe, N.; Ha, H. J. Regioselectivity in the Ring Opening of Non-Activated Aziridines. *Chem. Soc. Rev.* **2012,** *41*, 643–665.
6. For selected reviews on allylic amination and allylic oxidation, see: (a) Johannsen, M.; Jørgensen, K. A. Allylic Amination. *Chem. Rev.* **1998,** *9*, 1689–1708. (b) Grangea, R. L.; Clizbeb, E. A.; Evans, P. A. Recent Developments in Asymmetric Allylic Amination Reactions. *Synthesis* **2016,** *48*, 2911–2968. (c) Butta, N. A.; Zhang, W. Transition Metal-Catalyzed Allylic Substitution Reactions With Unactivated Allylic Substrates. *Chem. Soc. Rev.* **2015,** *44*, 7929–7967. (d) Nakamura, A.; Nakada, M. Allylic Oxidations in Natural Product Synthesis. *Synthesis* **2013,** *45*, 1421–1451.
7. For selected examples on allylic halogenation, see: (a) Katcher, M. H.; Sha, A.; Doyle, A. G. Palladium-Catalyzed Regio- and Enantioselective Fluorination of Acyclic Allylic Halides. *J. Am. Chem. Soc.* **2011,** *133*, 15902–15905. (b) Hollingworth, C.; Hazari, A.; Hopkinson, M. N.; Tredwell, M.; Benedetto, E.; Huiban, M.; Gee, A. D.; Brown, J. M.; Gouverneur, V. Palladium-Catalyzed Allylic Fluorination. *Angew. Chem. Int. Ed.* **2011,** *50*, 2613–2617. (c) Tunge, J. A.; Mellegaard, S. R. Selective Selenocatalytic Allylic Chlorination. *Org. Lett.* **2004,** *6*, 1205–1207. (d) Moreno-Dorado, F. J.; Guerra, F. M.; Manzano, F. L.; Aladro, F. J.; Jorge, Z. D.; Massanet, G. M. $CeCl_3/NaClO$: A Safe and Efficient Reagent for the Allylic Chlorination of Terminal Olefins. *Tetrahedron Lett.* **2003,** *44*, 6691–6693.
8. For selected examples and reviews on difunctionalizing alkenes with heteroatoms, see: (a) Romero, R. M.; Wçste, T. H.; Muçiz, K. Vicinal Difunctionalization of Alkenes With Iodine(III) Reagents and Catalysts. *Chem. Asian J.* **2014,** *9*, 972–983. (b) Jacques, B.; Muñiz, K. In *"Palladium Catalysis for Oxidative 1,2-Difunctionalization of Alkenes" in Catalyzed Carbon-Heteroatom Bond Formation;* Yudin, A. K., Ed.; Wiley-VCH, Verlag GmbH & Co. KGaA, Weinheim, 2010.
9. For reviews on synthesis of allenes, see: (a) Krause, N.; Hashmi, S., Eds. *Modern Allene Chemistry*, Vol. 1–2; Wiley-VCH, Verlag GmbH & Co. KGaA, Weinheim, 2004. (b) Sydnes, L. K. Allenes From Cyclopropanes and Their Use in Organic Synthesis—Recent Developments. *Chem. Rev.* **2003,** *103*, 1133–1150. (c) Taylor, D. R. The Chemistry of Allenes. *Chem. Rev.* **1967,** *67*, 317–359. (d) Krause, N.; Hoffmann-Röder, A. Synthesis of Allenes With Organometallic Reagents. *Tetrahedron* **2004,** *60*, 11671–11694. (e) Ogasawara, M. Catalytic Enantioselective Synthesis of Axially Chiral Allenes. *Tetrahedron: Asymm.* **2009,** *20*, 259–271. (f) Schuster, J.; Coppola, B., Eds. *Allenes in Organic Synthesis;* Wiley, New York, 1984.
10. For other selected examples of allene synthesis, see: (a) Marshall, J. A.; Robinson, E. D.; Zapata, Z. A. [2, 3] Wittig Rearrangement of Nonracemic Propargyl Ethers Leading to Allenes of High Stereochemical Integrity. *J. Org. Chem.* **1989,** *54*, 5854–5855. (b) Li, Z.; Boyarskikh, V.; Hansen, J. H.; Autschbach, J.; Musaev, D. G.; Davies, H. M. L. Scope and Mechanistic Analysis of the Enantioselective Synthesis of Allenes by Rhodium-Catalyzed Tandem Ylide Formation/[2,3]-Sigmatropic Rearrangement Between Donor/Acceptor Carbenoids and Propargylic Alcohols. *J. Am. Chem. Soc.* **2012,**

15497–15504. (c) Leclère, M.; Fallis, A. G. Asymmetric Allenophanes: Synthesis of a Tris-*meta*-Allenophane and Tetrakis-*meta*-Allenophane by Sequential Cross-Coupling. *Angew. Chem. Int. Ed.* **2008,** *47*, 568–572. (d) Crimmins, M. T.; Emmitte, K. A. Asymmetric Total Synthesis of (−)-Isolaurallene. *J. Am. Chem. Soc.* **2001,** *123*, 1533–1534. (e) Claesson, A.; Olsson, L.-I. Allenes and Acetylenes. 22. Mechanistic Aspects of the Allene-Forming Reductions (SN_2' Reaction) of Chiral Propargylic Derivatives With Hydride Reagents. *J. Am. Chem. Soc.* **1979,** *101*, 7302–7311. (f) Torres, E.; Larson, G. L.; McGarvey, G. J. An Asymmetric Hydrogen Equivalent: Use of the 1-Naphthylphenylmethylsilyl Group in the Preparation of Optically Active Allyl Alcohols and (S) 1-Phenyl-1,2-Butadiene. *Tetrahedron Lett.* **1988,** *29*, 1355–1358. (g) Tsubouchi, A.; Kira, T.; Takeda, T. Peterson Allenation Using (*Z*)-(1-Lithio-1-Alkenyl) Trimethylsilanes. *Synlett* **2006,** *2006*, 2577–2580. (h) Henderson, M. A.; Heathcock, C. H. Acyclic Stereoselection. 44. Diastereoselectivity in the Ortho Ester Claisen Rearrangement of Chiral Propargylic Alcohols. Use of β-Allenic Esters as Chiral Methylmalonaldehyde Synthons. *J. Org. Chem.* **1988,** *53*, 4736–4745. (i) Myers, A. G.; Zheng, B. New and Stereospecific Synthesis of Allenes in a Single Step From Propargylic Alcohols. *J. Am. Chem. Soc.* **1996,** *118*, 4492–4493.

11. For selected examples of transition metal-catalyzed hydroamination of allenes, see: (a) Zimmer, R.; Dinesh, C. U.; Nandanan, E.; Khan, F. Z. Palladium-Catalyzed Reactions of Allenes. *Chem Rev.* **2000,** *100*, 3067–3126. (b) Chemler, S. The Enantioselective Intramolecular Aminative Functionalization of Unactivated Alkenes, Dienes, Allenes and Alkynes for the Synthesis of Chiral Nitrogen Heterocycles. *Org. Biomol. Chem.* **2009,** *7*, 3009–3019. (c) Trost, B. M.; Pinkerton, A. B.; Kremzow, D. A Ruthenium-Catalyzed Pyrrolidine and Piperidine Synthesis. *J. Am. Chem. Soc.* **2000,** *122*, 12007–12008. (d) Nakano, S.; Inoue, N.; Hamada, Y.; Nemoto, T. Pd-Catalyzed Cascade Cyclization by Intramolecular Heck Insertion of an Allene–Allylic Amination Sequence: Application to the Synthesis of 3,4-Fused Tricyclic Indoles. *Org. Lett.* **2015,** *17*, 2622–2625. (e) Rigamonti, M.; Prestat, G.; Broggini, G.; Poli, G. Synthesis of 1,4-Benzodiazepinones Via Palladium-Catalysed Allene Carbopalladation/Amination Domino Sequence. *J. Organometal. Chem.* **2014,** *760*, 149–155. (f) Beccalli, E. M.; Broggini, G.; Clerici, F.; Galli, S.; Kammerer, C.; Rigamonti, M.; Sottocornola, S. Palladium-Catalyzed Domino Carbopalladation/5-exo-Allylic Amination of α-Amino Allenamides: An Efficient Entry to Enantiopure Imidazolidinones. *Org. Lett.* **2009,** *11*, 1563–1566. (g) Ohno, H.; Hamaguchi, H.; Tanaka, T. Stereoselective Synthesis of Chiral 2,3-cis-2-Ethynylaziridines by Base-Mediated Intramolecular Amination of Bromoallenes. *Org. Lett.* **2001,** *3*, 2269–2271. (h) Xu, T.; Mu, X.; Peng, H.; Liu, G. Silver-Catalyzed Intramolecular Aminofluorination of Activated Allenes. *Angew. Chem.* **2011,** *123*, 8326–8329.
12. Bleiholder, R. F.; Schechter, H. Addition of Electronegatively Substituted Azides to Allenes. *J. Am. Chem. Soc.* **1968,** *90*, 2131–2137.
13. Bingham, E. M.; Gilbert, J. C. Reaction of Carbethoxynitrene With Allenes. *J. Org. Chem.* **1975,** *40*, 224–228.
14. Atkinson, R. S.; Malpass, J. R. Nitrene Addition to Allenes: 1, 4-Diazaspiro [2.2] Pentanes. *Tetrahedron Lett.* **1975,** *16*, 4305–4306.
15. Rigoli, J. W.; Boralsky, L. A.; Hershberger, J. C.; Meis, A. R.; Guzei, I. A.; Schomaker, J. M. 1, 4-Diazaspiro [2.2] Pentanes as a Flexible Platform for the Synthesis of Diamine-Bearing Stereotriads. *J. Org. Chem.* **2012,** *77*, 2446–2455.
16. Stoll, A. H.; Blakey, S. B. Rhodium Catalyzed Allene Amination: Diastereoselective Synthesis of Aminocyclopropanes Via a 2-Amidoallylcation Intermediate. *J. Am. Chem. Soc.* **2010,** *132*, 2108–2109.

17. Stoll, A. H.; Blakey, S. B. Rhodium Catalyzed Allene Amidation: A Facile Entry Into 2-Amidoallylcations for Unusual [3 + 3] Annulation Reactions. *Chem. Sci.* **2011,** *2*, 112–116.
18. Feldman, K. S.; Hester, D. K.; López, C. S.; Faza, O. N. Allenyl Azide Cycloaddition Chemistry. Photochemical Initiation and CuI Mediation Leads to Improved Regioselectivity. *Org. Lett.* **2008,** *10*, 1665–1668.
19. Feldman, K. S.; Iyer, M. R. Allenyl Azide Cycloaddition Chemistry. Synthesis of Pyrrolidine-Containing Bicycles and Tricycles Via the Possible Intermediacy of Azatrimethylenemethane Species. *J. Am. Chem. Soc.* **2005,** *127*, 4590–4591.
20. Feldman, K. S.; Iyer, M. R.; Hester, D. K. Allenyl azide Cycloaddition Chemistry. Synthesis of Annelated Indoles From 2-(Allenyl)Phenyl Azide Substrates. *Org. Lett.* **2006,** *8*, 3113–3116.
21. López, C. S.; Faza, O. N.; Feldman, K. S.; Iyer, M. R.; Hester, D. K. Cyclization Cascade of Allenyl Azides: A Dual Mechanism. *J. Am. Chem. Soc.* **2007,** *129*, 7638–7646.
22. Feldman, K. S.; Iyer, M. R.; López, C. S.; Faza, O. N. Allenyl azide Cycloaddition Chemistry: Exploration of the Scope and Mechanism of Cyclopentennelated Dihydropyrrole Synthesis Through Azatrimethylenemethane Intermediates. *J. Org. Chem.* **2008,** *73*, 5090–5099.
23. Feldman, K. S.; Hester, D. K.; Iyer, M. R.; Munson, P. J.; López, C. S.; Faza, O. N. Allenyl Azide Cycloaddition Chemistry. 2,3-Cyclopentennelated Indole Synthesis Through Indolidene Intermediates. *J. Org. Chem.* **2009,** *74*, 4958–4974.
24. Feldman, K. S.; Antoline, J. F. Allenyl Azide Cycloaddition Chemistry: Application to the Total Synthesis of (±)-Meloscine. *Org. Lett.* **2012,** *14*, 934–937.
25. Feldman, K. S.; Gonzalez, I. Y.; Glinkerman, C. M. Intramolecular [3 + 2] Cyclocondensations of Alkenes With Indolidenes and Indolidenium Cations. *J. Am. Chem. Soc.* **2014,** *136*, 15138–15141.
26. Feldman, K. S.; Gonzalez, I. Y.; Glinkerman, C. M. Indolidenes and Indolidenium Intermediates in the Synthesis of Cyclopent[*b*]indoles: Mechanistic Studies on Intramolecular Cyclizations. *J. Org. Chem.* **2015,** *80*, 11849–11862.
27. For selected references on allenes and examples of the transfer of the axial chirality of an allene to point chirality, see: (a) Ogasawara, M. Catalytic Enantioselective Synthesis of Axially Chiral Allenes. *Tetrahedron-Asymm.* **2009,** *20*, 259–271. (b) Kim, H.; Williams, L. J. Recent Developments in Allene-Based Synthetic Methods. *Curr. Opin. Drug Disc.* **2008,** *11*, 870–894. (c) Ma, S. M. Transition-Metal-Catalyzed Reactions of Allenes. *Pure Appl. Chem.* **2006,** *78*, 197–208. (d) Krause, N.; Hoffmann-Röder, A. Synthesis of Allenes With Organometallic Reagents. *Tetrahedron* **2004,** *60*, 11671–11694. (e) Brawn, R. A.; Panek, J. S. Synthesis of Enantioenriched Homopropargylic Sulfonamides by a Three Component Reaction of Aldehydes, Sulfonamides, and Chiral Allenylsilanes. *Org. Lett.* **2009,** *11*, 4362–4365. (f) Yue, Z.; Williams, L. J. Spirodiepoxides: Synthesis of Epoxomicinoids. *J. Org. Chem.* **2009,** *74*, 7707–7714 and references therein.
28. Adams, C. S.; Weatherly, C. D.; Burke, E. G.; Schomaker, J. M. The Conversion of Allenes to Strained Three-Membered Heterocycles. Invited Review for. *Chem. Soc. Rev.* **2014,** *43*, 3136–3163.
29. Tehrani, K. A.; Kimpe, N. Syntheses and Transformations of 2-Methyleneaziridines and 2-Methyleneazetidines. *Curr. Org. Chem.* **2009,** *13*, 854–877.
30. Shipman, M. Methyleneaziridines: Unusual Vehicles for Organic Synthesis. *Synlett* **2006,** *19*, 3205–3217.

31. Cariou, C. C. A.; Clarkson, G. J.; Shipman, M. Rapid Synthesis of 1, 3, 4, 4-Tetrasubstituted β-Lactams From Methyleneaziridines Using a Four-Component Reaction. *J. Org. Chem.* **2008,** *73*, 9762–9764.
32. Ennis, D. S.; Ince, J.; Rahman, S.; Shipman, M. Observations on the Ring Opening Reactions of 2-Methyleneaziridines With Acid Chlorides and Alkyl Chloroformates. *J. Chem. Soc. Perk. Trans. 1* **2000,** *13*, 2047–2053.
33. Shiers, J. J.; Clarkson, G. J.; Shipman, M.; Hayesh, J. F. Rapid Generation of Molecular Complexity Using Hybrid Multi-Component Reactions (MCRs): Application to the Synthesis of α-Amino Nitriles and 1,2-Diamines. *Chem. Commun.* **2006,** *6*, 649–651.
34. Prie, G.; Prevost, N.; Twin, H.; Fernandes, S. A.; Hayes, J. F.; Shipman, M. A Lewis Acid Catalyzed Intramolecular [4 + 3] Route to Polycyclic Systems Containing a Seven Membered Ring. *Angew. Chem. Int. Edit.* **2004,** *43*, 6517–6519.
35. Montagne, C.; Prevost, N.; Shiers, J. J.; Prie, G.; Rahman, S.; Hayes, J. F.; Shipman, M. Generation and Electrophilic Substitution Reactions of 3-Lithio-2-Methyleneaziridines. *Tetrahedron* **2006,** *62*, 8447–8457.
36. Montagne, C.; Shiers, J. J.; Shipman, M. Rapid Generation of Molecular Complexity Using Sequenced Multi-Component Reactions. One-Pot Synthesis of 5,5′-Disubstituted Hydantoins From Methyleneaziridines. . *Tetrahedron Lett.* **2006,** *47*, 9207–9209.
37. Hayes, J. F.; Shipman, M.; Slawin, A. M. Z.; Twin, H. Multi-Component Reactions of 2-Isopropylideneaziridines: Application to the Synthesis of Enantiopure Neopentylamines. *Heterocycles* **2002,** *58*, 243–250.
38. Hayes, J. F.; Shipman, M.; Twin, H. Asymmetric Synthesis of 2-Substituted Piperidines Using a Multi-Component Coupling Reaction: Rapid Assembly of (*S*)-(+)-Coniine From (S)-1-Phenylethyl-2-Methyleneaziridine. *Chem. Commun.* **2001,** *18*, 1784–1785.
39. Hayes, J. F.; Shipman, M.; Twin, H. Multi-Component Reactions Involving 2-Methyleneaziridines: Rapid Synthesis of 1,3-Disubstituted Propanones. *J. Org. Chem.* **2002,** *67*, 935–942.
40. Mumford, P. M.; Shiers, J. J.; Tarver, G. J.; Hayes, J. F.; Shipman, M. Synthesis of 1,1-Disubstituted Tetrahydro-β-Carbolines From 2-Methyleneaziridines. *Tetrahedron Lett.* **2008,** *49*, 3489–3491.
41. Ince, J.; Ross, T. M.; Shipman, M.; Slawin, A. M. Z.; Ennis, D. S. Observations on the Synthesis of Functionalised Methyleneaziridines. *Tetrahedron* **1996,** *52*, 7037–7044.
42. Ince, J.; Shipman, M.; Ennis, D. S. Novel Ring Opening Reactions of Methyleneaziridines. *Tetrahedron Lett.* **1997,** *38*, 5887–5890.
43. For selected reviews on metal-catalyzed nitrene transfer, see: (a) Zalatan, D. N.; Du Bois, J. Metal-Catalyzed Oxidations of C-H to C-N bonds. *Top. Curr. Chem.* **2010,** *292*, 347. (b) Roizen, J. L.; Harvey, M. E.; Du Bois, J. Metal-Catalyzed Nitrogen-Atom Transfer Methods for the Oxidation of Aliphatic C–H Bonds. *Accts. Chem. Res.* **2012,** *45*, 911–922. (c) Muller, P.; Fruit, C. Enantioselective Catalytic Aziridinations and Asymmetric Nitrene Insertions Into CH Bonds. *Chem. Rev.* **2003,** *103*, 2905–2920. (d) Li, Z. G.; He, C. Recent Advances in Silver-Catalyzed Nitrene, Carbene, and Silylene-Transfer Reactions. *Eur. J. Org. Chem.* **2006,** *19*, 4313–4322. (e) Halfen, J. A. Recent Advances in Metal-Mediated Carbon-Nitrogen Bond Formation Reactions: Aziridination and Amidation. *Curr. Org. Chem.* **2005,** *9*, 657–669. (f) Espino, C. G.; Du Bois, J. *Modern Rhodium-Catalyzed Organic Reactions* **2005,** 379–416. (g) Dequirez, G.; Pons, V.; Dauban, P. Nitrene Chemistry in Organic Synthesis: Still in Its Infancy? *Angew. Chem. Int. Ed.* **2012,** *51*, 7384–7395. (h) Davies, H. M. L.; Du Bois, J.; Yu, J. Q. C–H Functionalization in Organic Synthesis. *Chem. Soc. Rev.* **2011,** *40*, 1855–1856. (i) Collet,

F.; Dodd, R. H.; Dauban, P. Catalytic C–H Amination: Recent Progress and Future Directions. *Chem. Commun.* **2009,** *34*, 5061–5074.

44. (a) Robertson, J.; Feast, G. C.; White, L. V.; Steadman, V. A.; Claridge, T. D. W. Structure and Reactivity of Bicyclic Methylene Aziridines Prepared by Intramolecular Aziridination of Allenes. *Org. Biomol. Chem.* **2010,** *8*, 3060–3063. (b) Zhurakovskyi, O.; Ellis, S. R.; Thompson, A. L.; Robertson, J. Access to a Guanacastepene and Cortistatin-Related Skeleton Via Ethynyl Lactone Ireland–Claisen Rearrangement and Transannular (4 + 3)-Cycloaddition of an Azatrimethylenemethane Diyl. *Org. Lett.* **2017,** *19*, 2174–2177.
45. Boralsky, L. A.; Marston, D.; Grigg, R. D.; Hershberger, J. C.; Schomaker, J. M. Allene Functionalization Via Bicyclic Methylene Aziridines. *Org. Lett.* **2011,** *13*, 1924–1927.
46. Adams, C. S.; Boralsky, L. A.; Guzei, I. A.; Schomaker, J. M. Modular Functionalization of Allenes to Aminated Stereotriads. *J. Am. Chem. Soc.* **2012,** *134*, 10807–10810.
47. Rigoli, J. W.; Weatherly, C. D.; Vo, B. T.; Neale, S.; Meis, A. R.; Schomaker, J. M. Chemoselective Allene Aziridination Via Ag (I) Catalysis. *Org. Lett.* **2013,** *15*, 290–293.
48. Rigoli, J. W.; Weatherly, C. D.; Alderson, J.; Vo, B. T.; Schomaker, J. M. Tunable, Chemoselective Amination Via Silver Catalysis. *J. Am. Chem. Soc.* **2013,** *135*, 1723–1726.
49. Weatherly, C. D.; Alderson, J. M.; Hein, J.; Berry, J. F.; Schomaker, J. M. Catalyst-Controlled Nitrene Transfer by Tuning Metal: Ligand Ratios: Insight Into the Mechanisms of Chemoselectivity. *Organometallics* **2017,** *36*, 1649–1661.
50. Ju, M.; Weatherly, C. D.; Guzei, I. A.; Schomaker, J. M. Chemo-and Enantioselective Intramolecular Silver-Catalyzed Aziridination. *Angew. Chem. Int. Ed.* **2017,** *56*, 9944–9948.
51. Feast, G. C.; Page, L. W.; Robertson, J. The Intramolecular Amination of Allenes. *Chem. Commun.* **2010,** *46*, 2835–2837.
52. Burke, E. G.; Schomaker, J. M. Oxidative Allene Amination for the Synthesis of Azetidin-3-Ones. *Angew. Chem. Int. Ed.* **2015,** *54*, 12097–12101.
53. For selected references, see: (a) Sorochinsky, A. E.; Fustero, S.; Soloshonok, V. A.; Liu, H. Fluorine in Pharmaceutical Industry: Fluorine-Containing Drugs Introduced to the Market in the Last Decade (2001–2011). *Chem. Rev.* **2014,** *114*, 2432–2506. (b) Vitale, A.; Bongiovanni, R.; Ameduri, B. Fluorinated Oligomers and Polymers in Photopolymerization. *Chem. Rev.* **2015,** *115*, 8835–8866. (c) Ilardi, E. A.; Vitaku, E.; Njardarson, J. T. Data-Mining for Sulfur and Fluorine: An Evaluation of Pharmaceuticals to Reveal Opportunities for Drug Design and Discovery: Miniperspective. *J. Med. Chem.* **2014,** *57*, 2832–2842. (d) Jeschke, P. The Unique Role of Fluorine in the Design of Active Ingredients for Modern Crop Protection. *ChemBioChem.* **2004,** *5*, 570–589. (e) Purser, S.; Moore, P. R.; Swallow, S.; Gouverneur, V. Fluorination in Medicinal Chemistry. *Chem. Soc. Rev.* **2008,** *37*, 320–330. (f) Isanbor, C.; O'Hagan, D. Fluorine in Medicinal Chemistry: A Review of Anti-Cancer Agents. *J. Fluor. Chem.* **2006,** *127*, 303–319. (g) O'Hagen, D. Understanding Organofluorine Chemistry. An Introduction to the C–F bond. *Chem. Soc. Rev.* **2008,** *37*, 308–319. (h) Müller, K.; Faeh, C.; Diederich, F. Fluorine in Pharmaceuticals: Looking Beyond Intuition. *Science* **2007,** 1881–1886.
54. For selected references, see: (a) Liang, T.; Neumann, C. N.; Ritter, T. Introduction of Fluorine and Fluorine-Containing Functional Groups. *Angew. Chem. Int. Ed.* **2013,** *52*, 8214–8264. (b) Kirk, K. L. Fluorination in Medicinal Chemistry: Methods, Strategies, and Recent Developments. *Org. Process Res. Dev.* **2008,** *12*, 305–321. (c) Withers, S.;

Watts, A.G.; Kim, J.-H.; Wennekes, T. 2,3-Fluorinated Glycosides as Neuramidase Inhibitors and Their Use as Anti-Virals. U.S. Patent 8,815,941, **2011**. (d) Ullrich, T.; Sasmal, S.; Boorgu, V.; Pasagadi, S.; Cheera, S.; Rajagopalan, S.; Bhumireddy, A.; Shashikumar, D.; Chelur, S.; Belliappa, C.; Pandit, C.; Krishnamurthy, N.; Mukherjee, S.; Ramanathan, A.; Ghadiyaram, C.; Ramachandra, M.; Santos, P. G.; Lagu, B.; Bock, M. G.; Perrone, M. H.; Weiler, S.; Keller, H. 3-Alkoxy-Pyrrolo [1,2-b] Pyrazolines as Selective Androgen Receptor Modulators With Ideal Physicochemical Properties for Transdermal Administration. *J. Med. Chem.* **2014,** *57*, 7396–7411.

55. Liu, L.; Gerstner, N. C.; Oxtoby, L. J.; Guzei, I. A.; Schomaker, J. M. Fluorinated Amine Stereotriads Via Allene Amination. *Org. Lett.* **2017,** *19*, 3239–3242.
56. Adams, C. S.; Grigg, R. D.; Schomaker, J. M. Complete Stereodivergence in the Synthesis of 2-Amino-1, 3-Diols From Allenes. *Chem. Sci.* **2014,** *5*, 3046–3056.
57. Weatherly, C. D.; Rigoli, J. W.; Schomaker, J. M. Synthesis of 1, 3-Diaminated Stereotriads Via Rearrangement of 1, 4-Diazaspiro [2.2] Pentanes. *Org. Lett.* **2012,** *14*, 1704–1707.
58. Rigoli, J. W.; Guzei, I. A.; Schomaker, J. M. Aminodiols Via Stereocontrolled Oxidation of Methyleneaziridines. *Org. Lett.* **2014,** *16*, 1696–1699.
59. Weatherly, C. D.; Guzei, I. A.; Schomaker, J. M. Stereocontrolled Synthesis of 1, 3-Diamino-2-Ols by Aminohydroxylation of Bicyclic Methylene-Aziridines. *Eur. J. Org. Chem.* **2013,** 3667–3670.
60. For selected examples, see: (a) Schlegel, M.; Schneider, C. Lewis Acid-Catalyzed Nucleophilic Addition of Indoles to In Situ Generated 2-Amidoallyl Cations. *J. Org. Chem.* **2017,** *82*, 5986–5992. (b) Saputra, M. A.; Dange, N. ,S.; Cleveland, A. H.; Malone, J. A.; Fronczek, F. R.; Kartika, R. Regioselective Functionalization of Enamides at the α-Carbon Via Unsymmetrical 2-Amidoallyl Cations. *Org. Lett.* **2017,** *19*, 2414–2417. (c) Kende, A. S.; He, H. Asymmetric [4 + 3] Cycloadditions From Chiral α-Chloro Imines. *Tetrahedron Lett.* **1997,** *38*, 3353–3356. (d) Griffin, K.; Montagne, C.; Hoang, C. T.; Clarkson, G. J.; Shipman, M. Lewis Acid Promoted Intramolecular (3 + 2) 'Cycloadditions' of Methyleneaziridines With Alkene and Alkyne Acceptors. *Org. Biomol. Chem.* **2012,** *10*, 1032–1039.
61. Gerstner, N.; Adams, C. S.; Schomaker, J. M. Stereocontrolled Syntheses of Seven-Membered Carbocycles by Tandem Allene Aziridination/[4 + 3] Reaction. *Angew. Chem. Int. Ed.* **2016,** *55*, 13240–13243.
62. Burke, E. G.; Gold, B.; Hoang, T. T.; Raines, R. T.; Schomaker, J. M. Fine-Tuning Strain and Electronic Activation of Strain-Promoted 1,3-Dipolar Cycloadditions With Endocyclic Sulfamates in SNO-OCTs. *J. Am. Chem. Soc.* **2017,** *139*, 8029–8037.
63. Burke, E. G.; Schomaker, J. M. Synthetic Applications of Flexible SNO-OCT Strained Alkynes and Their Use in Post-Polymerization Modifications. *J. Org. Chem.* **2017**. in process.
64. For selected references, see: (a) Singh, G. S.; D'hooge, M.; De Kimpe, N. In *Comprehensive Heterocyclic Chemistry III;* Katritzky, A., Ramsden, C., Taylor, E. R., Eds.; , Vol. 2; Elsevier Science: York, 2008; pp 1–110. (b) Rousseau, G.; Robin, S. In *Modern Heterocyclic Chemistry;* Álvarez-Builla, J., Vaquero, J. J., Barluenga, J., Eds.; , Vol. 4; Wiley-VCH: Weinheim, 2011; pp 163–268.
65. (a) Brown, A.; Ellis, D.; Wallace, O.; Ralph, M. Identification of a Urea Bioisostere of a Triazole Oxytocin Antagonist. *Bioorg. Med. Chem. Lett.* **2010,** *20*, 1851–1853. (b) Han, Y.; Han, M.; Shin, D.; Song, C.; Hahn, H.-G. Exploration of Novel 3-Substituted Azetidine Derivatives as Triple Reuptake Inhibitors. *J. Med. Chem.* **2012,** *55*, 8188–8192. (c) Han,

M.; Song, C.; Jeong, N.; Hahn, H.-G. Exploration of 3-Aminoazetidines as Triple Reuptake Inhibitors by Bioisosteric Modification of 3-α-Oxyazetidine. *ACS Med. Chem. Lett.* **2014,** *5*, 999–1004.

66. (a) Carreira, E. M.; Fessard, T. C. Four-Membered Ring-Containing Spirocycles: Synthetic Strategies and Opportunities. *Chem. Rev.* **2014,** *114*, 8257–8322. (b) Marson, C. M. New and Unusual Scaffolds in Medicinal Chemistry. *Chem. Soc. Rev.* **2011,** *40*, 5514–5533.
67. (a) Brandi, A.; Cicchi, S.; Cordero, F. M. Novel Syntheses of Azetidines and Azetidinones. *Chem. Rev.* **2008,** *108*, 3988–4035. (b) Couty, F.; Evano, G. Azetidines: New Tools for the Synthesis of Nitrogen Heterocycles. *Synlett* **2009,** 3053–3064. (c) Bott, T. M.; West, F. G. Preparation and Synthetic Applications of Azetidines. *Heterocycles* **2012,** *84*, 223. (d) Antermite, D.; Degennaro, L.; Luisi, R. Recent Advances in the Chemistry of Metallated Azetidines. *Org. Biomol. Chem.* **2017,** *15*, 34–50.
68. Kato, N., et al. Diversity-Oriented Synthesis Yields Novel Multistage Antimalarial Inhibitors. *Nature* **2016,** *538*, 344–349.
69. Schmid, S. C.; Schomaker, J. M. A Stereoselective [3 + 1] Ring Expansion for the Synthesis of Highly-Substituted Methylene Azetidines. *Angew. Chem. Int. Ed.* **2017**. in process.
70. Lo, J. C.; Kim, D.; Pan, C.-M.; Edwards, J. T.; Yabe, Y.; Gui, J.; Qin, T.; Gutierrez, S.; Giacoboni, J.; Smith, M. W.; Hollan, P. L.; Baran, P. S. Fe-Catalyzed C–C Bond Construction From Olefins Via Radicals. *J. Am. Chem. Soc.* **2017,** *139*, 2484–2503.
71. Computations detailing the nature of rhodium ylides as carbon bound in protic mechanisms see: (a) Liang, Y.; Zhou, H.; Yu, Z.-X. Why Is Copper (I) Complex More Competent Than Dirhodium (II) Complex in Catalytic Asymmetric O – H Insertion Reactions? A Computational Study of the Metal Carbenoid O – H Insertion Into Water. *J. Am. Chem. Soc.* **2009,** *131*, 17783–17785. (b) Li, Z.; Boyarskikh, V.; Hansen, J. H.; Autschbach, J.; Museav, D. G.; Davies, H. M. L. Scope and Mechanistic Analysis of the Enantioselective Synthesis of Allenes by Rhodium-Catalyzed Tandem Ylide Formation/[2, 3]-Sigmatropic Rearrangement Between Donor/Acceptor Carbenoids and Propargylic Alcohols. *J. Am. Chem. Soc.* **2012,** *134*, 15497–15540. (c) Li, Z.; Parr, B. T.; Davies, H. M. L. "Highly Stereoselective C–C Bond Formation by Rhodium-Catalyzed Tandem Ylide Formation/[2,3]-Sigmatropic Rearrangement Between Donor/Acceptor Carbenoids and Chiral Allylic Alcohols. *J. Am. Chem. Soc.* **2012,** *134*, 10942–10946. (d) Xie, Q.; Song, X.-S.; Qu, D.; Guo, L.-P.; Xie, Z.-Z. DFT Study on the Rhodium (II)-Catalyzed C–H Functionalization of Indoles: Enol Versus Oxocarbenium Ylide. *Organometallics* **2015,** *34*, 3112–3119. (e) Pospech, J.; Lennox, A. J.; Beller, M. Rhodium-Catalysed Alkoxylation/Acetalization of Diazo Compounds: One-Step Synthesis of Highly Functionalised Quaternary Carbon Centres. *Chem. Commun.* **2015,** *51*, 14505–14508. (f) Liu, Y.; Luo, Z.; Zhang, J. Z.; Xia, F. DFT Calculations on the Mechanism of Transition-Metal-Catalyzed Reaction of Diazo Compounds With Phenols: O–H Insertion Versus C–H Insertion. *J. Phys. Chem. A.* **2016,** *120*, 6485–6492.
72. For selected references, see: (a) Johnson, D. A.; Liu, H.-W. Mechanisms and Pathways From Recent Deoxysugar Biosynthesis Research. *Curr. Opin. Chem. Biol.* **1998,** *2*, 642–649. (b) Weymouth-Wilson, A. C. The Role of Carbohydrates in Biologically Active Natural Products. *Nat. Prod. Rep.* **1997,** *14*, 99–110. (c) Hooper, I. R. *Aminoglycoside Antibiotics;* Springer-Verlag: New York, 1982. (d) Ogura, H. Development of Miracle Medicines From Sialic Acids. *Proc. Jpn Acad. Ser. B* **2011,** *87*, 328–361. (e) Wong, C.-H., Ed. *Carbohydrate-Based Drug Discovery;* Wiley-VCH Verlag GmbH & C. KGaA:

Weinheim, 2003. (f) Wang, J.; Chang, C.-W. T. In *Glycobiology and Drug Design;* Klyosov, A. A., Ed.; ACS Publications, Washington, DC, 2012; pp 265–286. (g) Rai, R.; McAlexander, I.; Chang, C.-W. T. Synthetic Glycodiversification. From Aminosugars to Aminoglycoside Antibiotics. A Review. *Org. Prep. Proc. Int.* **2005,** *37*, 337–375. (h) Hanessian, S., Ed. *Preparative Carbohydrate Chemistry;* Marcel Dekker: New York, 1997. (j) Boons, G.-J.; Hale, K. J. *Organic Synthesis With Carbohydrates;* Sheffield Academic Press: Sheffield, 2000.

73. Adams, C. S.; Grigg, R. D.; Schomaker, J. M. Aminosugar Motifs Via an Allene Aziridination Strategy. *Tetrahedron* **2014,** *70*, 4128–4134.
74. (a) Joullié, M. M.; Li, W.-R.; Han, S.-Y. The Detoxin Complex: A Naturally Occurring Safener. *Heterocycles* **1993,** *36*, 359–388. (b) Yonehara, H.; Seto, H.; Aizawa, S.; Hidaka, T.; Shimazu, A.; Otake, N. The Detoxin Complex, Selective Antagonists of Blasticidin S. *J. Antibiot.* **1968,** *21*, 369–370. (c) Yonehara, H.; Seto, H.; Shimazu, A.; Aizawa, S.; Hidaka, T.; Kakinuma, K.; Otake, N. The Production, Isolation and Biological Properties of Detoxin Complex. *Agricul. Biol. Chem.* **1973,** *37*, 2771–2776.
75. (a) Legler, G. Glycoside Hydrolases: Mechanistic Information From Studies With Reversible and Irreversible Inhibitors. *Adv. Carbohydr. Chem. Biochem.* **1990,** *48*, 319–384. (b) Wrodnigg, T. From Lianas to Glycobiology Tools: Twenty-Five Years of 2, 5-Dideoxy-2, 5-Imino-D-mannitol. *Monatsh. Chem.* **2002,** *133*, 393–426. (c) Andersen, S. M.; Ebner, M.; Ekhart, C. W.; Gradnig, G.; Legler, G.; Lundt, I.; Stütz, A. E.; Withers, S. G.; Wrodnigg, T. Two Isosteric Fluorinated Derivatives of the Powerful Glucosidase Inhibitors, 1-Deoxynojirimycin and 2, 5-Dideoxy-2, 5-Imino-d-Mannitol: Syntheses and Glucosidase-Inhibitory Activities of 1, 2, 5-Trideoxy-2-Fluoro-1, 5-Imino-d-Glucitol and of 1, 2, 5-Trideoxy-1-Fluoro-2, 5-Imino-d-Mannitol. *Carbohydr. Res.* **1997,** *301*, 155–166. (d) Wrodnigg, T. M.; Diness, F.; Gruber, C.; Häusler, H.; Lundt, I.; Rupitz, K.; Steiner, A. J.; Stütz, A. E.; Tarling, C. A.; Withers, S. G.; Wölfler, H. Probing the Aglycon Binding Site of a β-Glucosidase: A Collection of C-1-Modified 2, 5-Dideoxy-2, 5-Imino-d-Mannitol Derivatives and Their Structure–Activity Relationships as Competitive Inhibitors. *Bioorg. Med. Chem.* **2004,** *12*, 3485–3495.
76. (a) Krenitsky, T. A. Purine Nucleoside Phosphorylase: Kinetics, Mechanism, and Specificity. *Mol. Pharmacol.* **1967,** *3*, 526–536. (b) Mason, J. M.; Murkin, A. S.; Li, L.; Schramm, V. L.; Gainsford, G. J.; Skelton, B. W. A β-Fluoroamine Inhibitor of Purine Nucleoside Phosphorylase. *J. Med. Chem.* **2008,** *51*, 5880–5884. (c) Evans, G. B.; Furneaux, R. H.; Lewandowicz, A.; Schramm, V. L.; Tyler, P. C. Exploring Structure – Activity Relationships of Transition State Analogues of Human Purine Nucleoside Phosphorylase. *J. Med. Chem.* **2003,** *46*, 3412–3423.
77. For reviews and selected examples of fluorinated amino acids used in protein design and engineering, see: (a) Yoder, N. C.; Kumar, K. Fluorinated Amino Acids in Protein Design and Engineering. *Chem. Soc. Rev.* **2002,** *31*, 335–341. (b) Odar, C.; Winkler, M.; Wiltschi, B. Fluoro Amino Acids: A Rarity in Nature, Yet a Prospect for Protein Engineering. *Biotechnol. J.* **2015,** *10*, 427–446. (c) Wang, P.; Tang, Y.; Tirrell, D. A. Incorporation of Trifluoroisoleucine Into Proteins *In Vivo*. *J. Am. Chem. Soc.* **2003,** *125*, 6900–6906. (d) Niemz, A.; Tirrell, D. A. Self-Association and Membrane-Binding Behavior of Melittins Containing Trifluoroleucine. *J. Am. Chem. Soc.* **2001,** *123*, 7407–7413. (e) Bilgiçer, B.; Fichera, A.; Kumar, K. A Coiled Coil With a Fluorous Core. *J. Am. Chem. Soc.* **2001,** *123*, 4393–4399.
78. For selected examples of antitumor and antimicrobial activities of fluorinated amino acids, see: (a) Gottler, L. M.; Lee, H.-Y.; Shelburne, C. E.; Ramamoorthy, A.; Marsh, E. N. G.

Using Fluorous Amino Acids to Modulate the Biological Activity of an Antimicrobial Peptide. *ChemBioChem.* **2008,** *9*, 370–373. (b) Meng, H.; Kumar, K. Antimicrobial Activity and Protease Stability of Peptides Containing Fluorinated Amino Acids. *J. Am. Chem. Soc.* **2007,** *129*, 15615–15622. (c) Tsushima, T.; Kawada, K.; Ishihara, S.; Uchida, N.; Shiratori, O.; Higaki, J.; Hirata, M. Fluorine Containing Amino Acids and Their Derivatives. 7. Synthesis and Antitumor Activity of α-and γ-Substituted Methotrexate Analogs. *Tetrahedron* **1988,** *44*, 5375–5387. (d) Tsushima, T.; Kawada, K. Fluorine-Containing Amino Acids and Their Derivatives. 6. An Efficient Synthesis of β-Fluorinated Alanines Via Fluorohalomethylation of Aminomalonates. *Tetrahedron Lett.* **1985,** *26*, 2445–2448.

79. Iwatsuki, M.; Nishihara-Tsukashima, A.; Ishiyama, A.; Namatame, M.; Watanabe, Y.; Handasah, S.; Pranamuda, H.; Marwoto, B.; Matsumoto, A.; Takahashi, Y.; Otoguro, K.; Ōmura, S. Jogyamycin, a New Antiprotozoal Aminocyclopentitol Antibiotic, Produced by *Streptomyces* sp. a-WM-JG-16.2. *J. Antibiot.* **2012,** *65*, 169–171.

80. (a) White, F. R. Pactamycin. *Cancer Chemother. Rep.* **1962,** *24*, 75–78. (b) Ito, T.; Roongsawang, N.; Shirasaka, N.; Lu, W.; Flatt, P. M.; Kasanah, N.; Miranda, C.; Mahmud, T. Deciphering Pactamycin Biosynthesis and Engineered Production of New Pactamycin Analogues. *ChemBioChem.* **2009,** *10*, 2253–2265. (c) Lu, W.; Roongsawang, N.; Mahmud, T. Biosynthetic Studies and Genetic Engineering of Pactamycin Analogs With Improved Selectivity Toward Malarial Parasites. *Chem. Biol.* **2011,** *18*, 425–431. (d) Otoguro, K.; Iwatsuki, M.; Ishiyama, A.; Namatame, M.; Nishihara-Tukashima, A.; Shibahara, S.; Kondo, S.; Yamada, H.; Ōmura, S. Promising Lead Compounds for Novel Antiprotozoals. *J. Antibiot.* **2010,** *63*, 381–384. (e) Hanessian, S.; Vakiti, R. R.; Chattopadhyay, A. K.; Dorich, S.; Lavallée, C. Probing Functional Diversity in Pactamycin Toward Antibiotic, Antitumor, and Antiprotozoal Activity. *Bioorg. Med. Chem.* **2013,** *21*, 1775–1786. (f) Sharpe, R. J.; Malinowski, J. T.; Sorana, F.; Luft, J. C.; Bowerman, C. J.; DeSimone, J. M.; Johnson, J. S. Preparation and Biological Evaluation of Synthetic and Polymer-Encapsulated Congeners of the Antitumor Agent Pactamycin: Insight Into Functional Group Effects and Biological Activity. *Bioorg. Med. Chem.* **2015,** *23*, 1849–1857. (g) Almabruk, K. H.; Lu, W.; Li, X.; Abugreen, M.; Kelly, J. X.; Mahmud, T. Mutasynthesis of Fluorinated Pactamycin Analogues and Their Antimalarial Activity. *Org. Lett.* **2013,** *15*, 1678–1681.

81. (a) Egebjerg, J.; Garrett, R. A. Binding Sites of the Antibiotics Pactamycin and Celesticetin on Ribosomal RNAs. *Biochimie* **1991,** *73*, 1145–1149. (b) Carter, A. P.; Clemons, W. M.; Brodersen, D. E.; Morgan-Warren, R. J.; Wimberly, B. T.; Ramakrishnan, V. Functional Insights From the Structure of the 30S Ribosomal Subunit and Its Interactions With Antibiotics. *Nature* **2000,** *407*, 340–348. (c) Dinos, G.; Wilson, D. N.; Teraoka, Y.; Szaflarski, W.; Fucini, P.; Kalpaxis, D.; Nierhaus, K. H. Dissecting the Ribosomal Inhibition Mechanisms of Edeine and Pactamycin: The Universally Conserved Residues G693 and C795 Regulate P-Site RNA Binding. *Mol. Cell.* **2004,** *13*, 113–124. (d) Tourigny, D. S.; Fernández, I. S.; Kelley, A. C.; Vakiti, R. R.; Chattopadhyay, A. K.; Dorich, S.; Hanessian, S.; Ramakrishnan, V. Crystal Structure of a Bioactive Pactamycin Analog Bound to the 30S Ribosomal Subunit. *J. Mol. Biol.* **2013,** *425*, 3907–3910.

82. (a) Hanessian, S.; Vakiti, R. R.; Dorich, S.; Banerjee, S.; Lecomte, F.; DelValle, J. R.; Zhang, J.; Deschênes-Simard, B. Total Synthesis of Pactamycin. *Angew. Chem. Int. Ed.* **2011,** *50*, 3497–3500. (b) Malinowski, J. T.; Sharpe, R. J.; Johnson, J. S. Enantioselective Synthesis of Pactamycin, a Complex Antitumor Antibiotic. *Science* **2013,** *340*,

180–182. (c) Malinowski, J. T.; McCarver, S. J.; Johnson, J. S. Diastereocontrolled Construction of Pactamycin's Complex Ureido Triol Functional Array. *Org. Lett.* **2012,** *14*, 2878–2881. (d) Loertscher, B. M.; Young, P. R.; Evans, P. R.; Castle, S. L. Diastereoselective Synthesis of Vicinal Tertiary Diols. *Org. Lett.* **2013,** *15*, 1930–1933. (e) Haussner, T. J.; Looper, R. E. Access to the Pactamycin Core Via an Epoxide Opening Cascade. *Org. Lett.* **2012,** *14*, 3632–3635. (f) Knapp, S.; Yu, Y. Synthesis of the Oxygenated Pactamycin Core. *Org. Lett.* **2007,** *9*, 1359–1362.

83. Gerstner, N. C.; Adams, C. S.; Grigg, R. D.; Tretbar, M.; Rigoli, J. W.; Schomaker, J. M. Diastereoselective Synthesis of the Aminocyclitol Core of Jogyamycin Via an Allene Aziridination Strategy. *Org. Lett.* **2016,** *18*, 284–287.

Chapter 7

Zinc-Mediated Synthesis of Heterocycles

Rubén Vicente and Sergio Mata
Universidad de Oviedo, Oviedo, Spain

7.1 INTRODUCTION

Heterocyclic compounds are likely the largest family of organic compounds. As a consequence, heterocycles are prevalent in chemistry having influence in virtually all research areas from medicinal chemistry, agrochemistry, and natural products to advanced or functional materials. These features make the synthesis of heterocycles an everlasting field of research. In this sense modern trends of organic synthesis, in which sustainability becomes as relevant as efficiency and selectivity, consider the use of metal-catalysis as an important tool. This fact is particularly accurate when inexpensive and low-toxic metal-based catalysts are employed. For this reason, during the past years, important research efforts have been devoted to the use of these type of metals as replacement to those traditionally employed (Pd, Rh, Ir, Ru, Pt, Au, among others).

The use of zinc-based catalyst for the synthesis of heterocycles is attractive since zinc meets most of the requirements mentioned earlier.[1] Thus metallic zinc and most of its salts are readily available in high purity at a low price and toxicity of zinc compounds is generally low. Moreover, for most of the synthetic purposes, zinc catalysts often show remarkable functional group tolerances and they are relatively insensitive towards air and moisture.

A comprehensive review is beyond the aim of this compilation. Instead, this chapter summarizes recent progresses (up to the beginning of 2017) on zinc-catalyzed transformations involving the formation of heterocycles. The large variety of heterocyclic classes and methodologies for their synthesis is a difficult rational to describe; however, we tried to organize the contents by reaction types, even though they could lead to different heterocyclic skeletons. Apart from catalytic methods, some selected relevant examples of stoichiometric reactions have been also included.

Advances in Transition-Metal Mediated Heterocyclic Synthesis.
DOI: https://doi.org/10.1016/B978-0-12-811651-7.00007-8

7.2 ZINC-CATALYZED SYNTHESIS OF HETEROCYCLES THROUGH ADDITION REACTIONS TO ALKENES AND ALKYNES

7.2.1 Syntheses of *N*-Heterocycles by Zinc-Catalyzed Hydroamination Reactions

Nitrogen-containing heterocycles are arguably in a dominant position among the huge variety of heterocyclic compounds. As a result, several methodologies have been developed for their syntheses. One of the most attractive approaches for the synthesis of *N*-containing compounds (cyclic or acyclic) is the hydroamination reaction, namely the direct addition of an N–H bond to an unsaturated substrate. Several metals are capable of promoting this reaction. Zinc-based catalysts can be considered as privileged complexes to accomplish hydroaminations since they show remarkable functional group tolerances, low sensitivity to moisture and air (in contrast to early transition metals and *f*-elements), and an adequate activity/lifetime (comparable or superior to late transition metals). These valuable features are key to explain the number of reports appeared in the last decade dealing with homogeneous zinc-catalyzed hydroamination reactions.[2] A selection of representative examples is shown below.

At the end of the last century, early examples of zinc-catalyzed intramolecular hydroamination reactions were reported by Müller and coworkers.[3] They found that simple $Zn(OTf)_2$ was a very efficient catalyst to promote the addition of amines to either terminal or internal alkynes (Scheme 7.1).

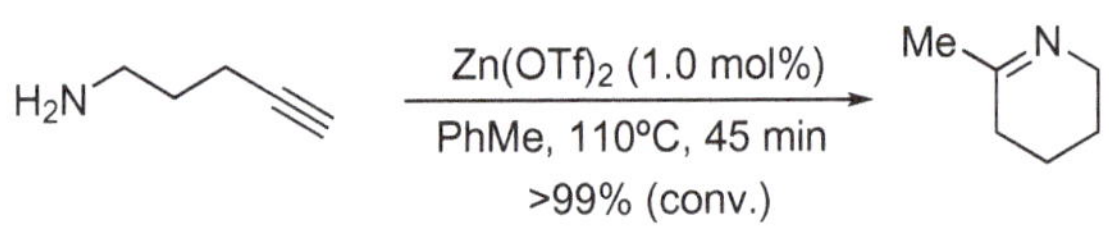

SCHEME 7.1 Early example of a zinc-catalyzed hydroamination reaction.

Subsequently, Roesky, Blechert, and coworkers rationally developed a series of organozinc complexes containing aminotroponiminate (ATI) ligands as (pre)catalysts for the hydroamination of alkynes and alkenes (Scheme 7.2). These catalytic systems showed high efficiencies and operated, in some examples, at low-catalyst loadings.[2a,4] The activity of these complexes can be modulated by controlling the substitution pattern of the ligand and the activator, normally a Brønsted acid. Immobilized versions of these complexes showed comparable activity and, more importantly, recyclability.[5] In contrast, results concerning asymmetric hydroaminations are still modest and need further improvement.[6] Other ligands containing amino groups, such as phenalenyl[7] or β-diketiminates[8] ligands have also been successfully employed in the intramolecular hydroamination reactions of alkenes and alkynes. Importantly, mechanistic studies pointed to zinc activation of the alkene or alkyne, rather than of the amine.

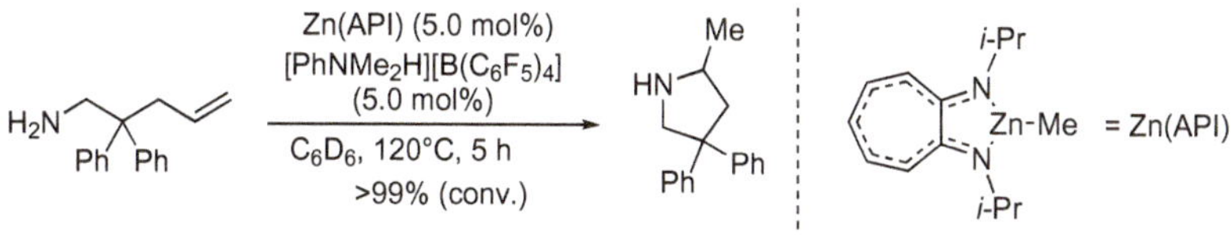

SCHEME 7.2 Catalytic hydroamination reaction using a Zn(API) complex.

Later, the ability of organozinc compounds as catalysts for hydroamination reactions was reported. For instance, a combination of simple diethylzinc (Et_2Zn) with acid activator has been successfully applied to intramolecular alkene hydroaminations. Noteworthy, this catalytic system worked at a loading of 2.5 mol% at ambient temperature (Scheme 7.3).[9,10] In addition, zinc metallocenes ($Cp^*_2Zn_2$ and Cp^*_2Zn) also proved to be applicable in hydroamination reactions.[11]

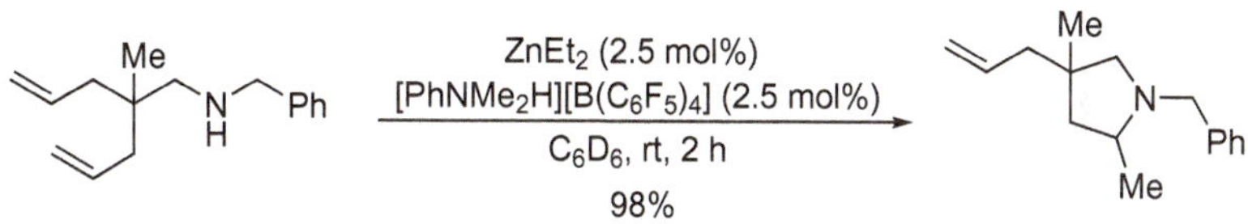

SCHEME 7.3 $ZnEt_2$-catalyzed hydroamination reaction.

By choosing adequately functionalized substrates, intramolecular hydroaminations serve as a platform for preparing several *N*-heterocyclic skeletons. Thus *ortho*-alkynyl anilines could be converted into 2-substituted indoles. Both $ZnBr_2$ and ZnI_2 could be used as catalysts providing N–H-indoles in excellent yields (Scheme 7.4A).[12] A $Zn(OTf)_2$-mediated modular approach based on a sequence intermolecular hydroamination of alkynes with hydrazines followed by a Fischer indole cyclization was reported by Beller and coworkers for the synthesis of indoles (Scheme 7.4B).[13]

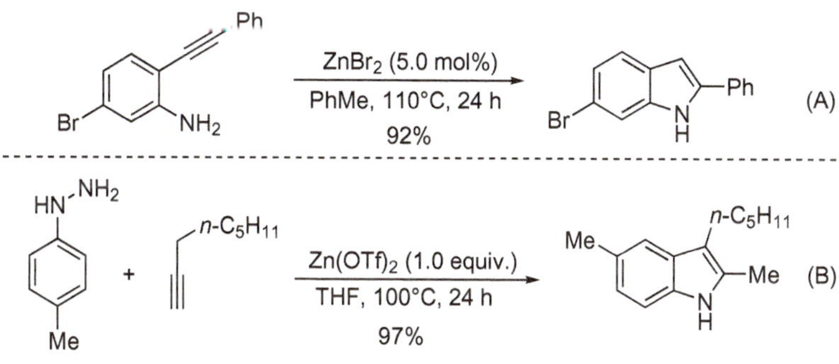

SCHEME 7.4 Zn-mediated indole synthesis using alkynes.

Syntheses of other heterocyclic scaffolds such as imidazoles and pyrroles via zinc-catalyzed hydroaminations were also reported using propargyl amides or enamines, respectively (Scheme 7.5A–B).[14] A different strategy for pyrrole syntheses was disclosed by Zhan and coworkers.[15] In this case, the aminoalkyne intermediate for the intramolecular hydroamination was prepared through a multicomponent coupling, providing additional synthetic versatility (Scheme 7.5C). Moreover, hydroquinolines were prepared using anilines and alkynes in a three-component process involving a hydroamination-alkyne addition sequence.[16] This process took place efficiently when using a β-ketiminate zinc complex as precatalyst (Scheme 7.5D).[17]

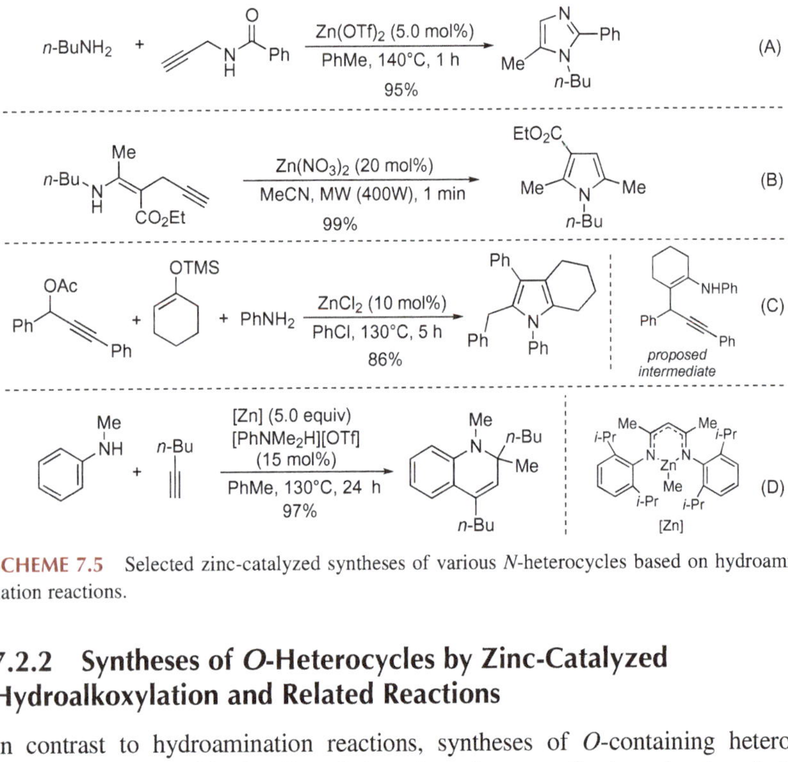

SCHEME 7.5 Selected zinc-catalyzed syntheses of various *N*-heterocycles based on hydroamination reactions.

7.2.2 Syntheses of *O*-Heterocycles by Zinc-Catalyzed Hydroalkoxylation and Related Reactions

In contrast to hydroamination reactions, syntheses of *O*-containing heterocycles by means of hydroalkoxylations have been so far less developed. In 2000, Carreira and coworkers described the ZnI_2-catalyzed hydroalkoxylation of propargyl *N*-hydroxylamines for the synthesis of 2,3-dihydroisoxazoles. The reaction was accomplished efficiently under mild conditions by using catalytic 4-dimethylaminopyridine (DMAP) as the base (Scheme 7.6).[18]

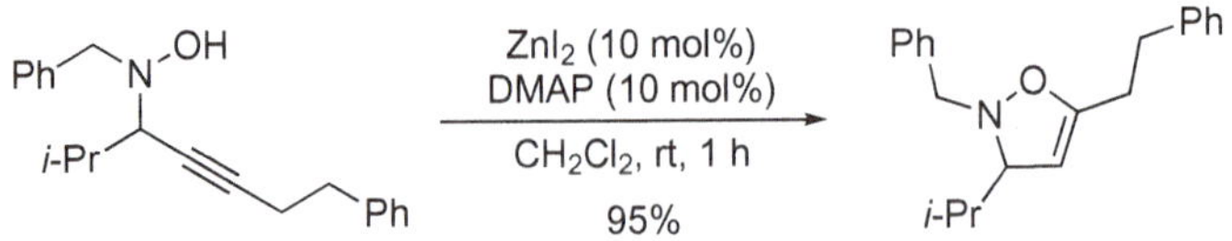

SCHEME 7.6 ZnI_2-catalyzed hydroalkoxylation reaction with *N*-hydroxylamines.

Related to indole synthesis, zinc-mediated intramolecular hydroalkoxylation reactions with *ortho*-alkynylphenols to prepare benzofurans were reported by Nakamura and coworkers (Scheme 7.7).[19] Albeit stoichiometric amounts of $ZnCl_2$ were required, the formation of an organozinc intermediate could be exploited for either reaction with electrophiles or cross-coupling reactions.

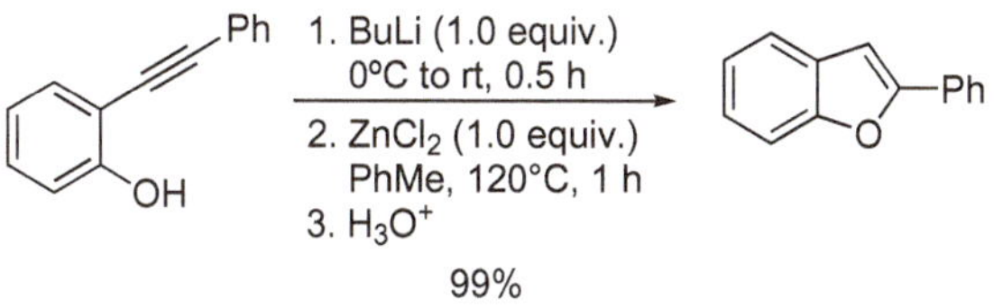

SCHEME 7.7 $ZnCl_2$-mediated synthesis of benzofurans.

The reaction of allenyl *N*-hydroxylamines in the presence of nitrosoarenes affording 1,2-oxazine derivatives was reported by Liu and coworkers (Scheme 7.8).[20] $Zn(OTf)_2$ showed better ability to promote this transformation when compared with other transition metals. The process was initiated with an intramolecular hydroalkoxylation with a concomitant addition of the nitrosoarene and a sequence of rearrangements, resulting in a formal [4 + 2]-annulation.

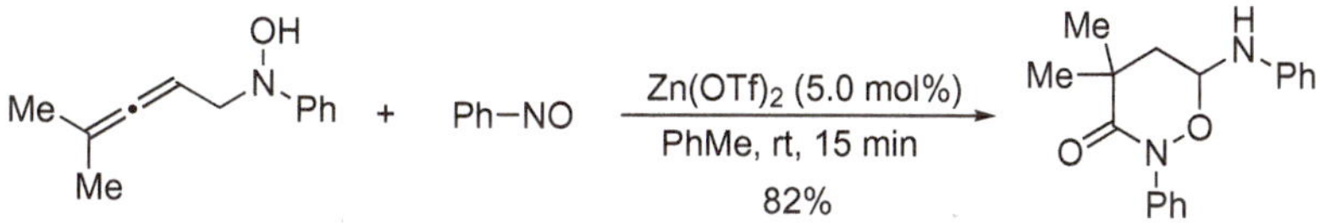

SCHEME 7.8 $Zn(OTf)_2$-catalyzed synthesis of 1,2-oxazine derivatives.

Besides carbonyl compounds can be used as internal nucleophiles to prepare furan derivatives. Dembinski and coworkers described the synthesis of furans by 5-*endo*-dig cyclization of α-alkynyl ketones using $ZnCl_2$ as the catalyst.[21] The utility of this reaction was highlighted by an elegant application for the synthesis of biologically relevant furopyrimidine nucleosides, which required the use of a zinc cluster as the catalyst to obtain higher efficiencies (Scheme 7.9).[22]

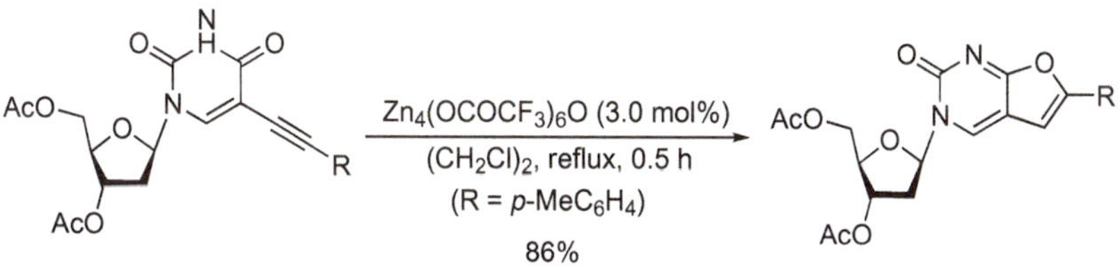

SCHEME 7.9 Zinc-catalyzed cyclization of alkynylketone to furans: application to relevant nucleoside derivatives.

By using a conceptually similar reaction, Vicente, López, and coworkers designed a method for the catalytic generation of zinc carbenes from conjugated enynones. In this case, $ZnCl_2$ promoted a 5-*exo*-dig cyclization which led to a 2-furyl zinc carbene, which could be then trapped by a number of nucleophilic reagents, including alkenes for the synthesis of furyl cyclopropanes (Scheme 7.10).[23]

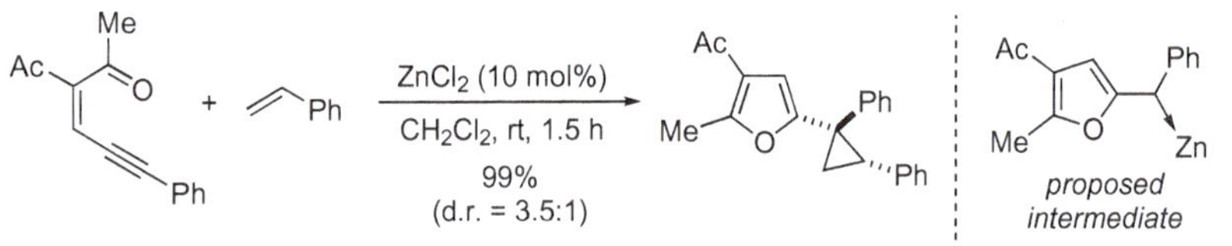

SCHEME 7.10 $ZnCl_2$-catalyzed cyclization of enynones to furans via zinc carbene intermediate.

The same authors reported the synthesis of furfuryl ethers by using the same strategy with alcohols (Scheme 7.11).[24] However, a different mechanistic scenario was proposed. Hence $ZnCl_2$ accomplished a dual activation by enhancing the acidity of the alcohol through coordination. This fact likely favored the cyclization step. Then, the process finished with the nucleophilic addition of the zinc alkoxide.

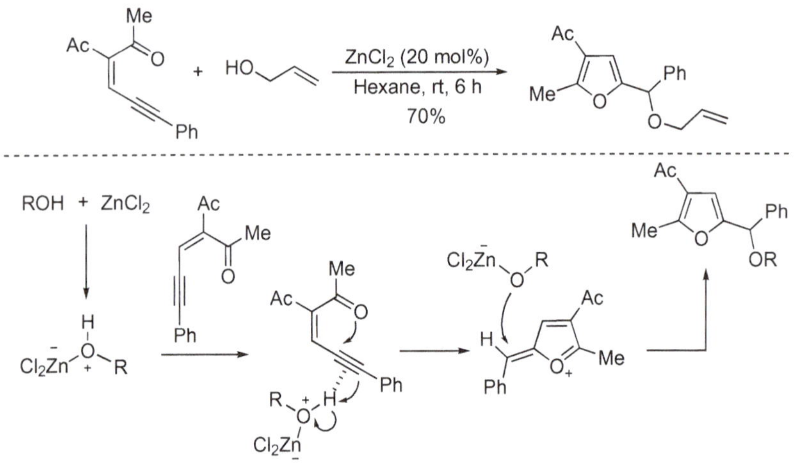

SCHEME 7.11 $ZnCl_2$-catalyzed cyclization of enynones to furans via alcohol activation.

The synthesis of oxazoles through a zinc-catalyzed cyclization of propargyl amides followed by an allylic alkylation was reported by Wang, Xu, and coworkers.[25] In this process, zinc catalyst had again two roles, first as π-acid for the cyclization and, second, as σ-acid for the alkylation of the intermediate with allylic alcohols (Scheme 7.12).

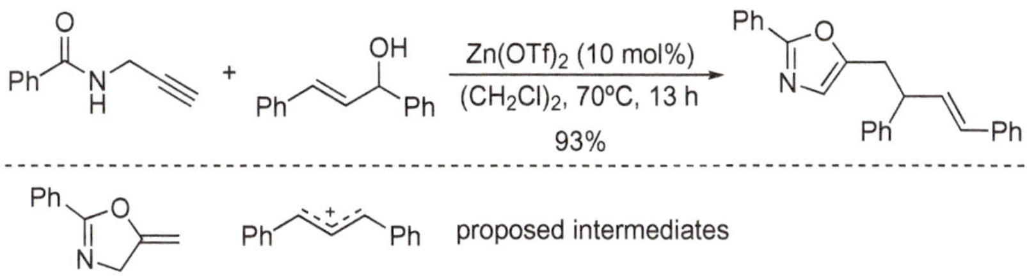

SCHEME 7.12 Zn(OTF)$_2$-catalyzed synthesis of oxazoles by a tandem reaction.

7.2.3 Syntheses of Heterocycles Involving Zinc-Catalyzed Addition Reactions of C-Nucleophiles to Alkynes

Addition of C-nucleophiles to alkynes via Conia-ene-type cyclizations have been used as a strategy to access heterocycles. In particular, the combination of Michael addition with subsequent cyclizations allowed modular and straightforward syntheses of heterocyclic scaffolds. For instance the addition of propargylic alcohols to alkylidene malonate derivatives led to the formation of highly substituted furan (Scheme 7.13A) or pyrrolidine derivatives.[26] Similarly, substituted piperidines were prepared using propargylamines and 1,1-cyclopropane diesters as C-nucleophiles (Scheme 7.13B).[27]

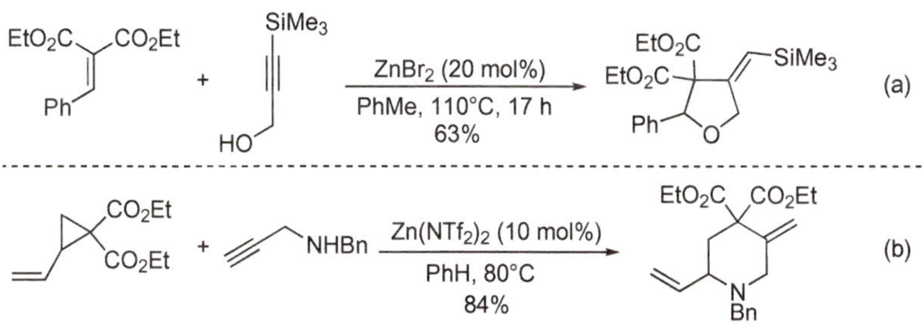

SCHEME 7.13 Zinc-catalyzed synthesis of heterocycles via *C*-addition to alkynes.

The addition of nucleophilic electron-rich arenes to propiolate derivatives was employed for the synthesis of coumarines and neoflavones. This transformation worked using $ZnCl_2$ as the catalyst in the absence of solvents providing the corresponding products in an efficient and rapid manner via alkyne hydroarylation–cyclization sequence (Scheme 7.14).[28]

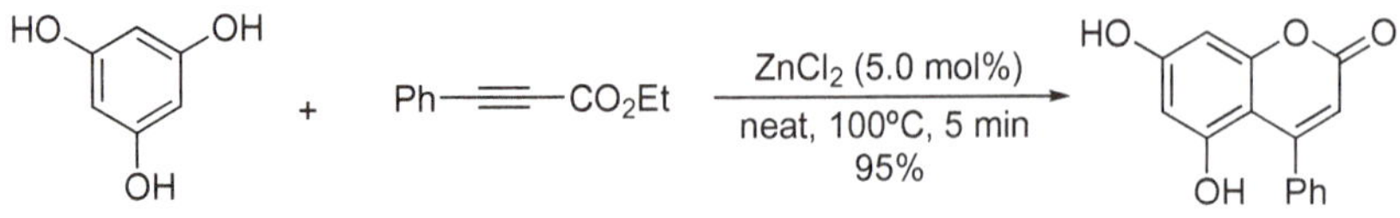

SCHEME 7.14 Zinc-catalyzed synthesis of neoflavones.

An interesting zinc-catalyzed sequence comprising an alkyne hydroarylation of propargylic anilines followed by a cross-dehydrogenative coupling with indoles was reported by Nakamura and coworker.[29] In this manner, tetrahydroquinoline derivatives were prepared by means of $Zn(OAc)_2$ catalysis (Scheme 7.15). A plausible mechanism involves a zinc-promoted intramolecular hydroarylation. An equilibration among dihydroquinolines and an iminium cation set the stage for the nucleophilic addition of indole and final cleavage of the Zn–C bond.

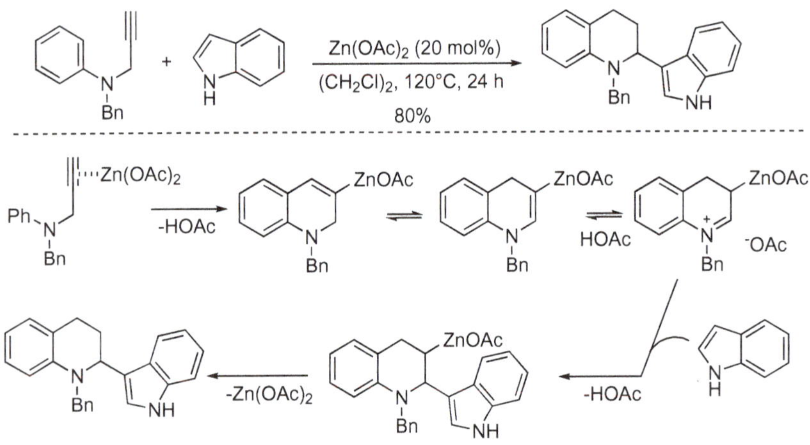

SCHEME 7.15 Zinc-catalyzed synthesis of tetrahydroquinolines from propargyl anilines and proposed mechanism.

7.2.4 Other Zinc-Catalyzed Syntheses of Heterocycles Involving Transformations with Unsaturated Substrates

In some cases, the use of azides as *N*-source led to the formation of *N*-heterocycles, in processes that resemble hydroamination reactions. Thus Driver and coworkers described the synthesis of pyrroles from dienyl azides by means of ZnI_2 catalyst under mild reaction conditions (Scheme 7.16A).[30] Similarly, Dembinski and coworkers described the synthesis of pyrroles using alkynyl azides. In this case, $ZnCl_2$ was the catalyst employed and harsher reaction conditions were necessary (Scheme 7.16B).[31]

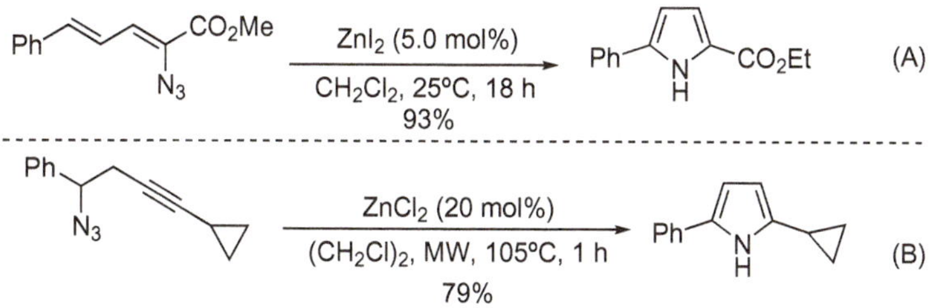

SCHEME 7.16 Formal selected zinc-catalyzed syntheses of various *N*-heterocycles based on hydroamination reactions.

In the recent years, ynamides have become relevant compounds in organic synthesis by means of metal catalysis. In this sense, some interesting applications using zinc-based catalysts have been recently reported. Ye, Lu, and coworkers described the synthesis of isoquinolone derivatives through the cyclization of benzyl ynamides in the presence of *N*-oxides as *O*-source using $Zn(OTf)_2$ as the catalyst (Scheme 7.17A).[32] On the contrary, Cao and coworkers reported the synthesis of amidobenzofuran derivatives through the reaction of *ortho*-anisole-substituted ynamides with alcohols (Scheme 7.17B).[33] Interestingly, a similar intermolecular reaction of ynamides with anisole derivatives led to the formation of amidochromane derivatives (Scheme 7.17C).[34] Unfortunately, both reactions required stoichiometric amounts of $ZnBr_2$, yet proceeded under mild conditions.

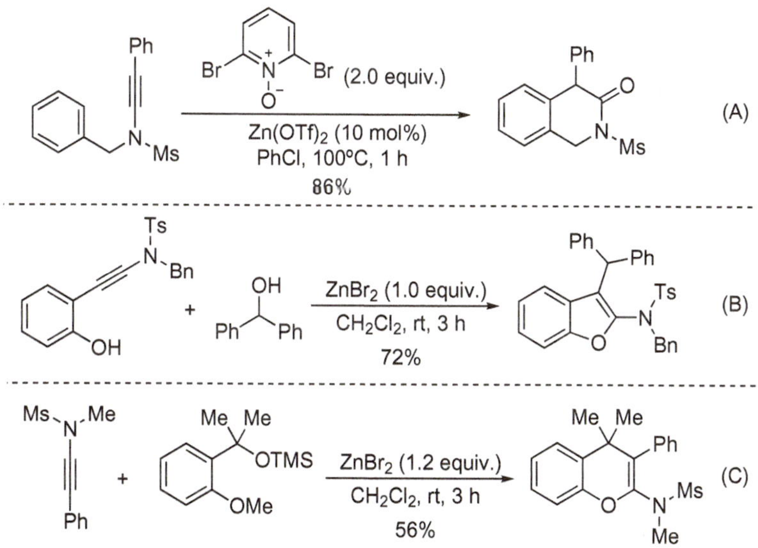

SCHEME 7.17 Zinc-promoted synthesis of heterocycles using ynamides.

7.3 ZINC-CATALYZED SYNTHESIS OF HETEROCYCLES THROUGH FORMAL CYCLOADDITION REACTIONS

Cycloaddition reactions are indispensable for organic synthesis. Lewis acids are commonly employed to facilitate cycloaddition reactions. Again, the use of zinc provides some advantages over other metals due to the relative inertness of most of the zinc complexes and their superior tolerance towards sensitive functional groups with respect to other Lewis acids. This section covers recent examples of zinc-catalyzed formal cycloaddition reactions for the synthesis of heterocycles.

7.3.1 Zinc-Catalyzed Formal [4 + 2]-Cycloadditions

Hetero[4 + 2]-cycloadditions comprise a well-stablished strategy for the synthesis of six-membered ring heterocycles. Early works by Danishefsky et al. showed the ability of zinc to promote this type of reactions under stoichiometric conditions.[35] Further research has led to the development of catalytic versions, which enabled convenient asymmetric applications. Thus Ding and coworkers reported the synthesis of pyrone derivatives through an enantioselective cycloaddition of Danishefsky's diene with aldehydes (Scheme 7.18A).[36] The utility of this reaction was highlighted by its application by Waters and coworkers as the key step in the synthesis of cytotoxic compound (+)-aspergillide C using an easily accessible enantiopure aldehyde (Scheme 7.18B).[37]

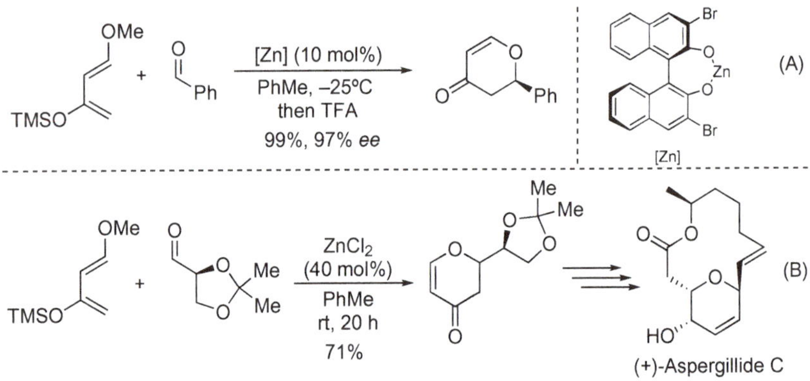

SCHEME 7.18 Zinc-catalyzed [4 + 2]-cycloadditions of Danishefsky diene with aldehydes.

An inverse-electron-demand enantioselective imino-[4 + 2]-cycloaddtion was reported by Kumar, Waldmann, and coworkers.[38] This reaction took place between imines and chromone-dienes to afford intricate aza-heterocycles, namely indoloquinolizines, which show interesting activities in mitosis modulation. While $ZnCl_2$ could be employed in the racemic version of the reaction, a catalytic system comprising $ZnEt_2$ and enantiopure 1,1′-bi-2-naphtol (BINOL)-type ligand were used for the asymmetric versions (Scheme 7.19). Further implementations might solve the high catalyst loadings required.

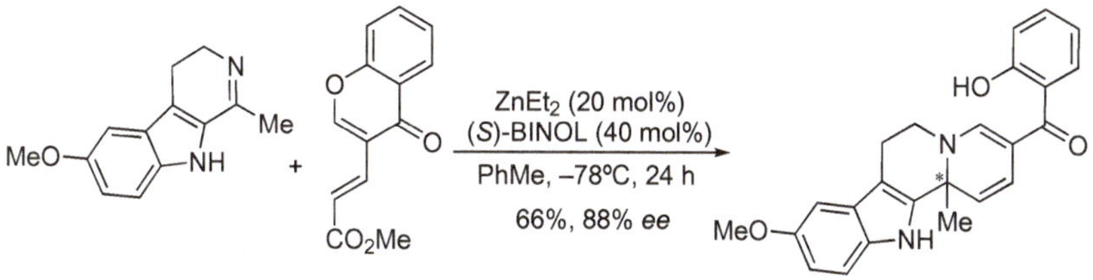

SCHEME 7.19 Zinc-catalyzed imino-[4 + 2]-cycloaddition for the synthesis of indoloquinolizines.

Doyle and coworkers employed imines for $Zn(OTf)_2$-catalyzed [4 + 2]-cycloaddition with 1-azadienes.[39] In this manner, a variety of densely substituted tetrahydropyrimidine derivatives were prepared in high yields, albeit long reaction times at low temperature were necessary (Scheme 7.20).

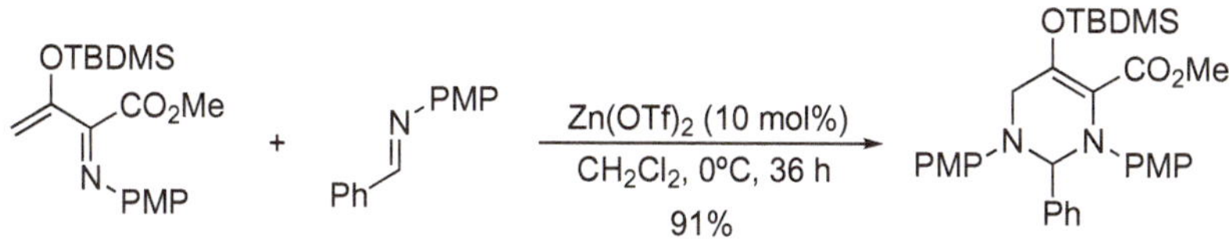

SCHEME 7.20 Zinc-catalyzed [4 + 2]-cycloaddition between imines and 1-azadienes (PMP = 4-MeO-C_6H_4).

A deep study on challenging [4 + 2]-cycloadditions involving two electron-deficient partners was performed by Rovis and coworkers.[40] A high-throughput screening of several metals revealed that zinc salts were able to promote the cycloaddition between 1-azadienes and nitroalkenes to afford piperidines after a subsequent reduction of the enamine intermediate. The use of $Zn(OTf)_2$ along with chiral nonracemic bis(oxazolinylphenyl)amide ligands led to the formation of highly substituted piperidines with promising levels of stereoinduction (Scheme 7.21).

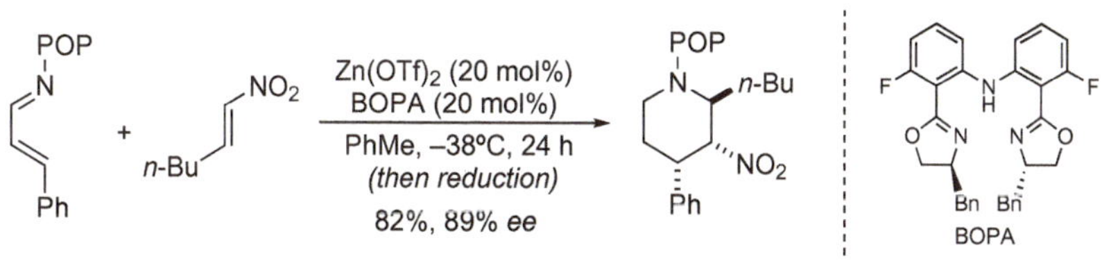

SCHEME 7.21 Zinc-catalyzed [4 + 2]-cycloaddition between two electron-deficient partners (POP = PhO-C_6H_4).

7.3.2 Zinc-Promoted [3 + 2]-Cycloadditions

Five-membered ring containing heteroatoms can be prepared through [3 + 2]-cycloadditions using 1,3-dipoles. Again, zinc complexes have been fruitfully used for the synthesis of five-membered heterocycles. For instance, a catalytic system comprising $Zn(OTf)_2$ and bisoxazolidine ligands showed better selectivities than other metals in the synthesis of optically active proline derivatives by asymmetric cycloaddition of azomethine ylides with acrylates (Scheme 7.22).[41]

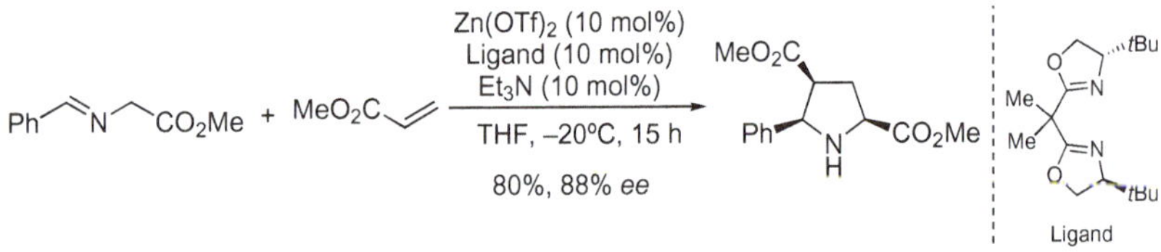

SCHEME 7.22 Zinc-catalyzed asymmetric [3 + 2]-cycloaddition with azomethine ylides and acrylates.

Aziridines can be used as *N*-containing three-membered partner for [3 + 2]-cycloadditions through Lewis acid promoted C–C cleavages. In this sense, an efficient and highly diastereoselective $Zn(OTf)_2$-catalyzed synthesis of oxazolidines by reaction of aziridines with aldehydes was described by Lu, Wang, and coworkers (Scheme 7.23A).[42] The reaction with imines proved also feasible, yet yields and diastereoselectivities were modest. Besides, Stoltz and coworkers reported [3 + 2]-cycloadditions of aziridines with heterocumulenes such as isothiocyanates (Scheme 7.23B) or carbodiimides.[43] These reactions proceeded with complete regioselectivity, albeit stoichiometric amounts of zinc salts were employed.

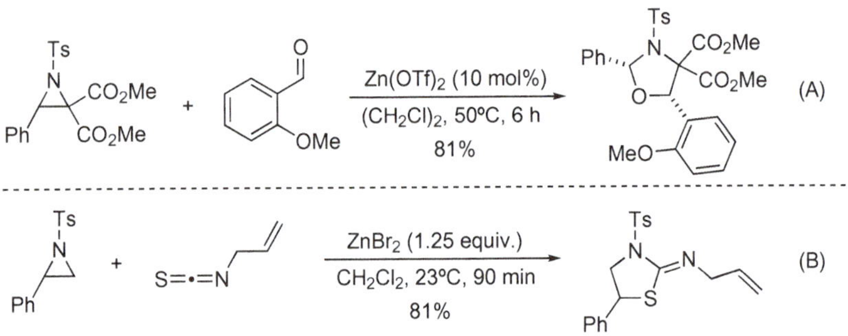

SCHEME 7.23 Zinc-promoted [3 + 2]-cycloadditions using aziridines as three-membered partner.

Liang and coworkers reported regioselective [3 + 2]-cycloadditions of stabilized diazo compounds with terminal alkynes to yield pyrazoles using $Zn(OTf)_2$ (Scheme 7.24).[44] This reaction takes advantage of the inability of zinc salts to promote N_2 extrusion of the diazo compound to allow the synthesis of the pyrazole.

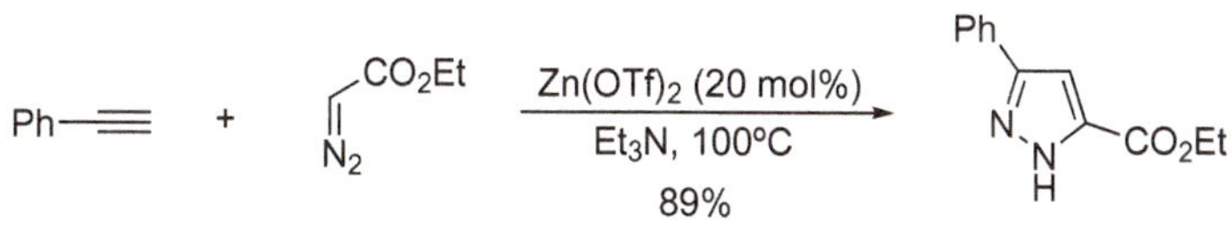

SCHEME 7.24 Zinc-catalyzed [3 + 2] cycloadditions between alkynes and diazo compounds.

The development of click-type reactions is among the most influential achievements of chemistry. In spite of the dominion of copper as the catalyst for these transformations, relevant examples based on zinc have also been reported. At the beginning of the century, Sharpless and coworkers described the [3 + 2]-cycloaddition of sodium azides to nitriles in the presence of substoichiometric amounts of zinc salts.[45] This process was cleverly applied to the synthesis of tetrazole analogues of α-amino acids (Scheme 7.25A). Later, applications of this reaction to thiocyanates were also reported.[46] Interestingly, Greaney and coworkers showed that a stoichiometric amount of $ZnEt_2$ was able to promote the [3 + 2]-cycloaddition of organic azides and alkynes in the presence of *N*-methylimidazole (NMI) (Scheme 7.25B).[47] Remarkable features of this transformation are the complementary regioselectivity with respect to copper-catalyzed reactions and the possibility to functionalize the triazolyl zinc intermediate.

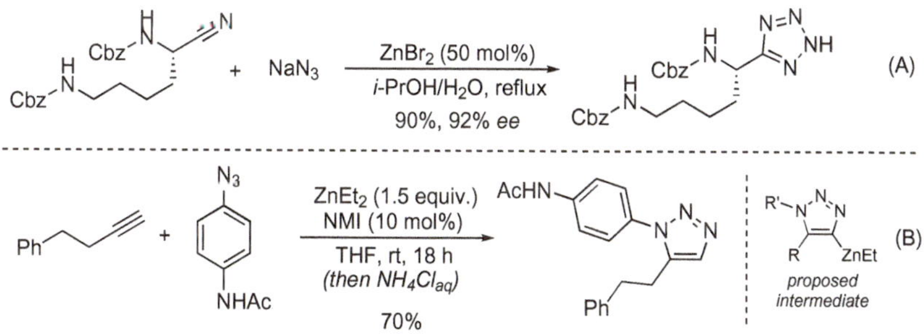

SCHEME 7.25 Zinc-mediated click-type reactions with azides.

7.3.3 Other Zinc-Promoted Cycloadditions

A [3 + 3]-cycloaddition involving azomethine imines with in-situ generated azomethine ylides was reported by Liu, Na, and coworkers.[48] $Zn(OTf)_2$ was used as the catalyst for the generation of the nonstabilized azomethine ylides from the appropriate 1-methoxy-methylamine by releasing methanol, followed by the subsequent cycloaddition with the azomethine imine to afford 1,2,4-triazine derivatives (Scheme 7.26).

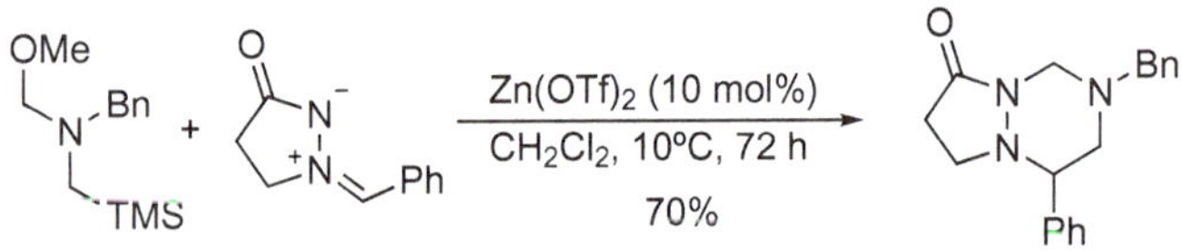

SCHEME 7.26 $Zn(OTf)_2$-catalyzed [3 + 3]-cycloaddition with azomethine imines and ylides.

Wang and coworkers reported a new synthesis of multisubstituted pyridines from isonitriles and *N*-formylmethyl-substituted enamides through a formal [1 + 5]-cycloaddition.[49] The transformation comprises a sequence of reactions starting with a zinc-promoted addition of the isonitrile to the enamine, which triggers the cyclization. Then, a series of H-transfer/isomerization events followed by an intramolecular acyl shift afforded the corresponding pyridine (Scheme 7.27). Subsequent modifications of the reaction events enabled the synthesis of tetra- or dihydropyridine derivatives.[50]

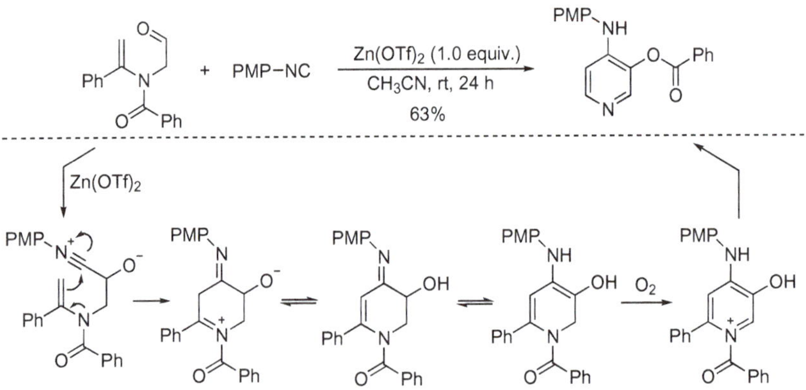

SCHEME 7.27 Synthesis of pyridines via $Zn(OTf)_2$-mediated [1 + 5]-cycloadditions.

7.4 ZINC-PROMOTED SYNTHESIS OF HETEROCYCLES THROUGH CONDENSATION–CYCLIZATION REACTION SEQUENCES

Condensations of amines with carbonyl compounds are among the most important reactions in chemistry and biology. These reactions have been fruitfully used in the synthesis of heterocycles. As an example, pyrroles can be prepared easily by a zinc-catalyzed Paal–Knorr-like synthesis (Scheme 7.28A).[51] The reaction took place in short reaction times and without the use of solvent. Similarly, condensation reactions of amino-substituted indoles with aldehydes were employed for the synthesis of polycyclic *N*-containing heterocycles using zinc halides as catalysts (Scheme 7.28B).[52] It should be noticed that the presence of benzotriazole was required to facilitate the Friedel–Crafts-like cyclization. A $ZnCl_2$-catalyzed Knoevenagel condensation followed by 1,5-hydride shift cyclization sequence led to the formation of tetrahydroquinolines bearing a spirane core. The reaction took place using *ortho*-pyrrolidine benzaldehydes with isoxazolones under relatively mild reaction conditions (Scheme 7.28C).[53]

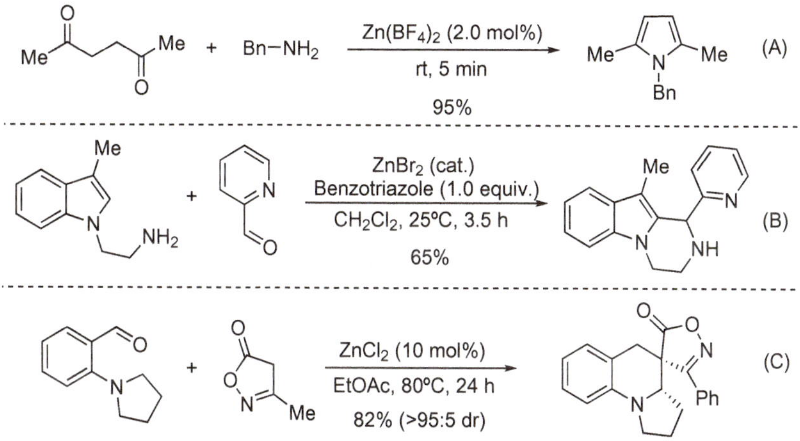

SCHEME 7.28 Zinc-catalyzed synthesis of heterocycles using condensation reactions of aldehyde and amine derivatives.

By using zinc catalysis, carboxylic acid derivatives were also employed in the synthesis of heterocycles via condensation reaction. For instance, a formal [5 + 1]-annulation of enamidines with orthoesters for the synthesis of pyrimidines was reported (Scheme 7.29A).[54] Noteworthy, this $ZnBr_2$-catalyzed reaction enabled the synthesis of fully substituted pyrimidines. An interesting condensation reaction for the synthesis of organic materials based on a benzimidazole skeleton was described by Anzenbacher and coworkers. In this case, $Zn(OAc)_2$ was employed as the catalyst for the reaction of carboxylic acid anhydrides with *ortho*-phenylenediamines in a process which took place in the solid state (Scheme 7.29B).[55] Polysubstituted pyridone derivatives were prepared by $ZnBr_2$-catalyzed self-condensation of β-ketoamides. P_2O_5 was required as dehydrating agent (Scheme 7.29C).[56]

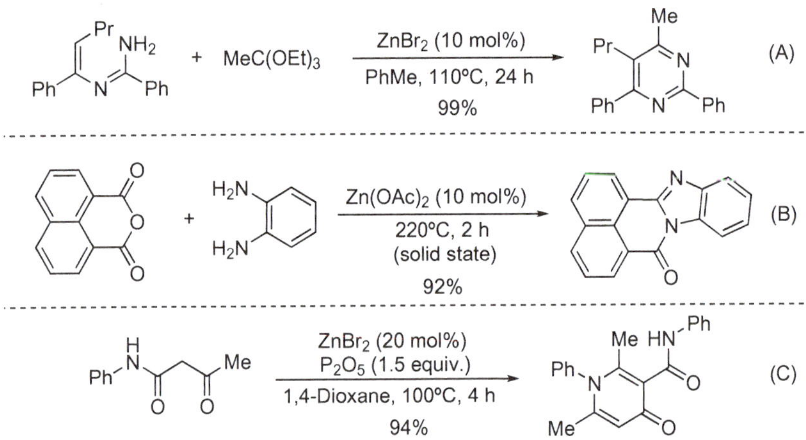

SCHEME 7.29 Zinc-catalyzed synthesis of heterocycles using condensation reactions of carboxylic acid and amine derivatives.

7.5 ZINC-PROMOTED SYNTHESIS OF HETEROCYCLES THROUGH MICHAEL ADDITION–CYCLIZATION REACTION SEQUENCES

As important as the condensation of amines and carbonylic compounds, Michael additions comprise a group of reactions with undoubted relevance in organic synthesis. Indeed, Michael addition–cyclization sequences can be used as a tool for the synthesis of heterocycles. Some selected examples of this strategy are described in this section. For instance, Trost and coworkers reported a $ZnEt_2$-catalyzed synthesis of lactones through a Michael addition–intramolecular transesterification sequence.[57] Making use of 3-hydroxyoxindoles and α,β-unsaturated esters, spirocyclic lactones were obtained in high yield, diastereo- and enantioselectivity (Scheme 7.30A). A similar strategy was used by Chen, Xiao, and coworkers for the synthesis of spirooxindoles. In this case, they used nitrostyrenes as Michael acceptors and isothiocyanate oxindoles as the nucleophiles (Scheme 7.30B).[58] A catalytic system comprising $Zn(OTf)_2$ along with an enatiopure bisoxazolidine ligand provided good yields and excellent enantioselectivities.

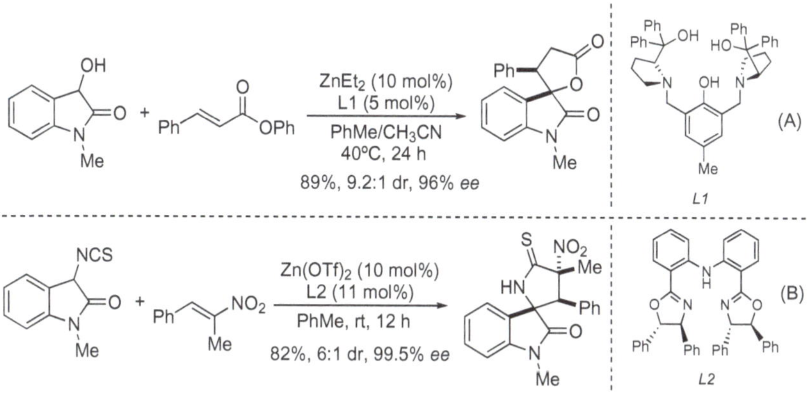

SCHEME 7.30 Zinc-catalyzed synthesis of spirooxindoles via sequential Michael addition–cyclization.

Multicomponent syntheses of heterocycles involving Michael reactions along the process have been recently reported. Thus quinoline derivatives were prepared by $ZnCl_2$-catalyzed three-component reaction of 4-hydroxiquinolones, aldehydes, and nitroketene *N,S*-acetals (Scheme 7.31A).[59] This transformation relies on the ability of the nitroketene acetal to lose a SMe fragment as nucleofuge. Besides, amino-substituted pyridines, a class of compounds with interesting antibacterial activity, were synthesized by multicomponent coupling of malononitrile, aldehydes, and amines (Scheme 7.31B). In this case, stoichiometric amounts of $ZnCl_2$ were used along with ethanol as solvent.[60]

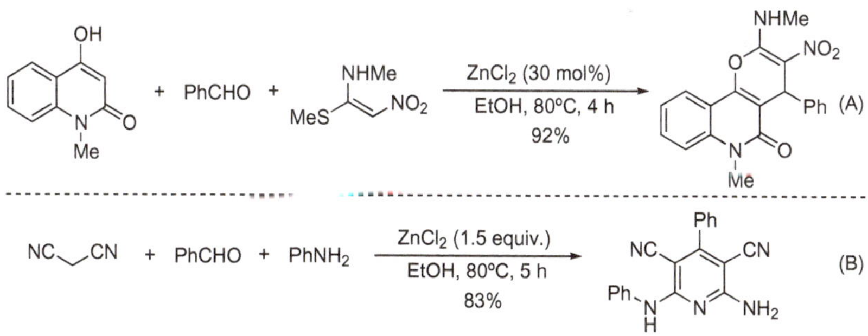

SCHEME 7.31 Multicomponent zinc-promoted synthesis of heterocycles.

7.6 ZINC-CATALYZED SYNTHESIS OF CYCLIC CARBONATE DERIVATIVES USING CO_2

Cyclic carbonates are relevant compounds in the industry due to their applications as polymer precursors, fuel additives or electrolytes in batteries, and their uses as aprotic high-boiling point solvents.[61] One of the most attractive routes for their syntheses comprises the reaction of epoxides with CO_2 by means of Lewis acid catalysis. Zinc-based catalysts, along with Lewis base cocatalysts, have demonstrated an impressive ability to promote this transformation, as very recently gathered by Enthaler.[62] This research field is currently very active and a number of new catalytic systems comprising zinc complexes were reported recently.[63] Indeed, zinc-catalyzed synthesis of carbonates working at low-catalyst loadings or low CO_2 pressures and easily recyclable systems are available at the moment. Two representative examples of this type of transformation are briefly described. On the one hand, He and coworkers described a catalytic system based on structurally simple (OPO)-zinc complex and tetrabutylammonium iodide. This catalytic system could be recycled at least five times without decreasing the activity in the reaction of 1,2-epoxyhexane with CO_2. Importantly, this catalyst operated at ambient CO_2 pressure (Scheme 7.32A).[63g] On the other hand, Ema, Maeda, and coworkers developed a bridged porphyrin trimer, which worked as bifunctional catalysis for the synthesis of carbonates.[63e] Remarkable turnover number (TON) number of 310,000 was described in the reaction with 1,2-epoxyhexane, showing the robustness of the catalyst (Scheme 7.32B).

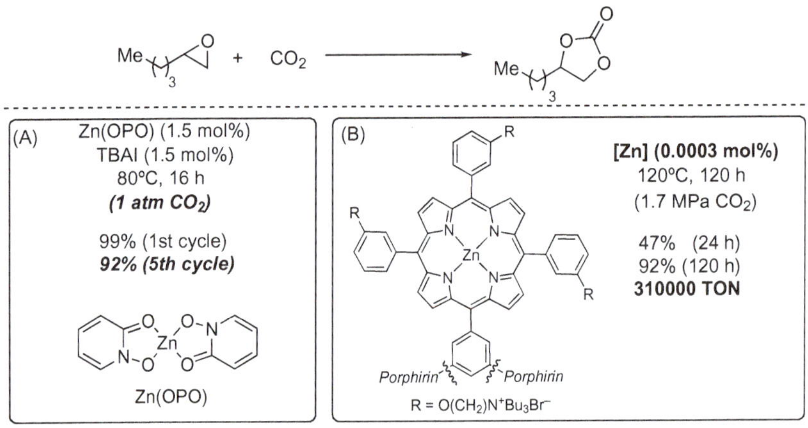

SCHEME 7.32 Selected examples of zinc-catalyzed synthesis of cyclic carbonates using CO_2.

As an alternative to epoxides, propargylic alcohols could be used as substrates for the synthesis of cyclic carbonates in their reaction with CO_2. In this sense, Ma, Han, and coworkers described the synthesis of α-alkylidene carbonates from tertiary propargylic alcohols using ZnI_2 as the catalyst, trimethylamine as the base, and 1 MPa pressure of CO_2 (Scheme 7.33A).[64] Similarly, He and coworkers reported a zinc-catalyzed synthesis of alkylidene 2-oxazolidinones when using propargylamines.[65] In this case, a zinc-triazabicyclodecene complex provided high yields even at ambient pressure of CO_2 (Scheme 7.33B).

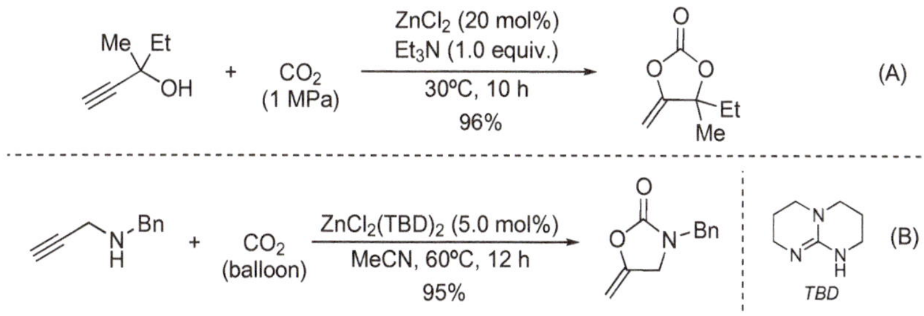

SCHEME 7.33 Zinc-catalyzed synthesis of cyclic carbonates and carbamates using CO_2.

7.7 SUMMARY AND OUTLOOK

The importance of heterocycles in chemistry and their beneficial influence in our society demand new synthetic methods to prepare common scaffolds or new structures. Efficiency and selectivity are major issues to consider when preparing heterocycles, making very appealing the use of metal catalysis as a synthetic tool. While precious metals still dominate the field of transition metal catalysis, the use of inexpensive and low-toxic metal-based catalysts is currently claimed. Zinc is among these metals and the use of zinc-based catalyst is becoming relevant. However, additional efforts for the development of new zinc catalysts showing similar efficiencies, in terms of TON or turnover frequency (TOF) numbers, and scope as precious metals are still required to turn them into a true alternative. For this reason, we strongly believe that future studies should provide exciting advances in the zinc-catalyzed synthesis of heterocycles.

REFERENCES

1. For general reviews on zinc catalysis in organic synthesis, see: (a) Wu, X.-F.; Neumann, H. Zinc-Catalyzed Organic Synthesis: C–C, C–N, C–O Bond Formation Reactions. *Adv. Synth. Catal.* **2012,** *354*, 3141–3160. (b) Wu, X.-F. Non-Redox-Metal-Catalyzed Redox Reactions: Zinc Catalysts. *Chem. Asian J.* **2012,** *7*, 2502–2509. (c) Enthaler, S. Rise of the Zinc Age in Homogeneous Catalysis? *ACS Catal* **2013,** *3*, 150–158. (d) Enthaler, S.; Wu, X.-F., Eds. *Zinc Catalysis, Applications in Organic Synthesis;* Wiley-VCH: Weinheim, 2015.
2. For some revisions, see: (a) Li, T.; Wiecko, J.; Roesky, P. W. Zinc-Catalyzed Hydroamination Reactions. In *Zinc Catalysis, Applications in Organic Synthesis;* Enthaler, S., Wu, X.-F., Eds.; Wiley-VCH: Weinheim, 2015. (b) López, L. A.; González, J. Zinc-Catalyzed C–N and C–O Bond Formation. In *Zinc Catalysis, Applications in Organic Synthesis;* Enthaler, S., Wu, X.-F., Eds.; Wiley-VCH: Weinheim, 2015.
3. (a) Müller, T. E.; Plier, A.-K. Intramolecular Hydroamination of Alkynes Catalysed by Late Transition Metals. *J. Chem. Soc., Dalton Trans.* **1999,** *4*, 583–588. (b) Müller, T. E.; Grosche, M.; Herdtwerk, E.; Plier, A.-K.; Walter, E.; Yan, Y.-K. Developing Transition-Metal Catalysts for the Intramolecular Hydroamination of Alkynes. *Organometallics* **1999,** *19*, 170–183.

4. Selected references: (a) Zulys, A.; Dochnahl, M.; Hollmann, D.; Löhnwitz, K.; Herrmann, J.-S.; Roesky, P. W.; Blechert, S. Intramolecular Hydroamination of Functionalized Alkenes and Alkynes with a Homogenous Zinc Catalyst. *Angew. Chem. Int. Ed.* **2005,** *44*, 7794–7798. (b) Dochnahl, M.; Pissarek, J.-W.; Blechert, S.; Löhnwitz, K.; Roesky, P. W. A New Homogeneous Zinc Complex with Increased Reactivity for the Intramolecular Hydroamination of Alkenes. *Chem. Commun.* **2006,** 3405–3407. (c) Dochnahl, M.; Löhnwitz, K.; Pissarek, J.-W.; Biyikal, M.; Schulz, S. R.; Schön, S.; Meyer, N.; Roesky, P. W.; Blechert, S. Intramolecular Hydroamination with Homogeneous Zinc Catalysts: Evaluation of Substituent Effects in N,N′-Disubstituted Aminotroponiminate Zinc Complexes. *Chem. Eur. J* **2007,** *13*, 6654–6666. For a review, see: (d) Jenter, J.; Lühl, A.; Roesky, P. W.; Blechert, S. Aminotroponiminate Zinc Complexes as Catalysts for the Intramolecular Hydroamination. *J. Organomet. Chem.* **2011,** *696*, 406–418.
5. (a) Duncan, C. T.; Flitsch, S.; Asefa, T. Aminotroponiminate–Zinc Complex-Functionalized Mesoporous Materials: Efficient and Recyclable Intramolecular Hydroamination Catalysts. *ChemCatChem* **2009,** *1*, 365–368. (b) Duncan, C.; Biradar, A. V.; Asefa, T. Aminotroponate/Aminotroponiminate Zinc Complexes Functionalized Mesoporous Silica Catalysts for Intramolecular Hydroamination of Non-Activated Alkenes with Varied Steric and Electronic Properties. *ACS Catal.* **2011,** *1*, 736–750.
6. Meyer, N.; Roesky, P. W. Chiral Aminotroponiminate Zinc Complexes. *Organometallics* **2009,** *28*, 306–311.
7. Mukherjee, A.; Sen, T. K.; Ghorai, P. Kr; Samuel, P. P.; Schulzke, C.; Mandal, S. K. Phenalenyl-Based Organozinc Catalysts for Intramolecular Hydroamination Reactions: A Combined Catalytic, Kinetic, and Mechanistic Investigation of the Catalytic Cycle. *Chem. Eur. J.* **2012,** *18*, 10530–10545.
8. Biyikal, M.; Löhnwitz, K.; Meyer, N.; Dochnahl, M.; Roesky, P. W.; Blechert, S. β-Diketiminate Zinc Complexes for the Hydroamination of Alkynes. *Eur. J. Inorg. Chem.* **2010,** *7*, 1070–1081.
9. (a) Yin, Y.; Ma, W.; Chai, Z.; Zhao, G. Et_2Zn-Catalyzed Intramolecular Hydroamination of Alkynyl Sulfonamides and the Related Tandem Cyclization/Addition Reaction. *J. Org. Chem.* **2007,** *72*, 5731–5736. (b) Pissarek, J.-W.; Schlesiger, D.; Roesky, P. W.; Blechert, S. Diethylzinc: A Simple and Efficient Catalyst for the Swift Hydroamination at Room Temperature. *Adv. Synth. Catal.* **2009,** *351*, 2081–2085. (c) Chilleck, M. A.; Hartenstein, L.; Braun, T.; Roesky, P. W.; Braun, B. Cationic Zinc Organyls as Precatalysts for Hydroamination Reactions. *Chem. Eur. J.* **2015,** *21*, 2594–2602.
10. For a relevant example of a stoichiometric reaction and subsequent use of the organozinc intermediate in cross-coupling reaction, see: Sunsdahl, B.; Smith, A. R.; Livinghouse, T. Intramolecular Metalloamination of *N,N*-Dimethylhydrazinoalkenes: A Versatile Method to Access Functionalized Piperidines and Pyrrolidines. *Angew. Chem. Int. Ed.* **2014,** *53*, 14352–14356.
11. (a) Luhl, A.; Pada Nayek, H.; Blechert, S.; Roesky, P. W. Zinc-Zinc Bonded Decamethyldizincocene $Zn_2(\eta^5\text{-}C_5Me_5)_2$ as Catalyst for the Inter- and Intramolecular Hydroamination Reaction. *Chem. Commun.* **2011,** *47*, 8280–8282. (b) Lühl, A.; Hartenstein, L.; Blechert, S.; Roesky, P. W. Comparison of Decamethyldizincocene $[(\eta^5\text{-}Cp^*)_2Zn_2]$ versus Decamethylzincocene $[Cp^*{}_2Zn]$ and Diethylzinc Et_2Zn as Precatalysts for the Intermolecular Hydroamination Reaction. *Organometallics* **2012,** *31*, 7109–7116 See also Ref. 9b.
12. Okuma, K.; Seto, J.-i; Sakaguchi, K.-i; Ozaki, S.; Nagahora, N.; Shioji, K. Palladium-Free Zinc-Mediated Hydroamination of Alkynes: Efficient Synthesis of Indoles from 2-Akynylaniline Derivatives. *Tetrahedron Lett.* **2009,** *50*, 2943–2945.

13. (a) Alex, K.; Tillack, A.; Schwarz, N.; Beller, M. Zinc-Promoted Hydrohydrazination of Terminal Alkynes: An Efficient Domino Synthesis of Indoles. *Angew. Chem. Int. Ed.* **2008,** *47*, 2304–2307. (b) Pews-Davtyan, A.; Beller, M. Efficient and Simple Zinc-Mediated synthesis of 3-Amidoindoles. *Org. Biomol. Chem.* **2011,** *9*, 6331–6334. (c) Ahmed, A.; Ghosh, M.; Sarkar, P.; Ray, J. K. $ZnCl_2$ and Pd/C Catalyzed Synthesis of 2-Substituted Indoles. *Tetrahedron Lett.* **2013,** *54*, 6691–6694 For related processes for the synthesis of pyrazolines, see. (d) Patil, N. T.; Singh, V. Synthesis of 1,3,5-Trisubstituted Pyrazolines via Zn(II)-Catalyzed Double Hydroamination of Enynes with Aryl Hydrazines. *Chem. Commun.* **2011,** 11116–11118. (e) Alex, K.; Tillack, A.; Schwarz, N.; Beller, M. Zinc-Catalyzed Synthesis of Pyrazolines and Pyrazoles via Hydrohydrazination. *Org. Lett.* **2008,** *10*, 2377–2379.
14. (a) Pews-Davtyan, A.; Beller, M. A Novel Zn-Catalyzed Hydroamination of Propargylamides: A General Synthesis of Di- and Tri-Substituted Imidazoles. *Chem. Commun.* **2011,** *47*, 2152–2154. (b) Prior, A. M.; Robinson, R. S. An Assessment of Late Transition Metals as Hydroamination Catalysts in the Cyclization of C-Propargyl Vinylogous Amides into Pyrroles. *Tetrahedron Lett.* **2008,** *49*, 411–414.
15. Liu, X.-T.; Hao, L.; Lin, M.; Chen, L.; Zhan, Z.-P. One-Pot Highly Efficient Synthesis of Substituted Pyrroles and *N*-Bridgehead Pyrroles by Zinc-Catalyzed Multicomponent Reaction. *Org. Biomol. Chem.* **2010,** *8*, 3064–3072.
16. Purkait, N.; Blechert, S. Synthesis of Bi- and Tricyclic 1,2-Dihydroquinoline Derivatives from Arylamines and Alkynes by a Consecutive Zinc-Ammonium Salt Catalysis. *Adv. Synth. Catal.* **2012,** *354*, 2079–2083.
17. For additional recent examples, see: (a) Viji, M.; Nagarajan, R. Zinc Triflate Catalyzed Regioselective Synthesis of Pyrrolo[2,3-c]carbazoles via Heteroannulation. *RSC Adv.* **2012,** *2*, 10544–10549. (b) Reddy, V.; Jadhav, A. S.; Anand, R. V. Catalyst-Controlled Regioselective Approach to 1-Aminoisoquinolines and/or 1-Aminoisoindolines through Aminative Domino Cyclization of 2-Alkynylbenzonitriles. *Eur. J. Org. Chem.* **2016,** 453–458.
18. (a) Aschwanden, P.; Frantz, D. E.; Carreira, E. M. Synthesis of 2,3-Dihydroisoxazoles from Propargylic N-Hydroxylamines via Zn(II)-Catalyzed Ring-Closure Reaction. *Org. Lett.* **2000,** *2*, 2331–2333. For the synthesis of the same heterocycles via zinc-mediated reaction of nitrones with alkynes, see: (b) Xiao, Z.-F.; Ding, T.-H.; Mao, S.-W.; Ning, X.-S.; Kang, Y.-B. Zinc Iodide-Mediated Direct Synthesis of 2,3-Dihydroisoxazoles from Alkynes and Nitrones. *Adv. Synth. Catal.* **2016,** *358*, 1859–1863.
19. (a) Nakamura, M.; Ilies, L.; Otsubo, S.; Nakamura, E. 3-Zinciobenzofuran and 3-Zincioindole: Versatile Tools for the Construction of Conjugated Structures Containing Multiple Benzoheterole Units. *Angew. Chem. Int. Ed.* **2006,** *45*, 944–947. (b) Nakamura, M.; Ilies, L.; Otsubo, S.; Nakamura, E. 2,3-Disubstituted Benzofuran and Indole by Copper-Mediated C – C Bond Extension Reaction of 3-Zinciobenzoheterole. *Org. Lett.* **2006,** *8*, 2803–2805. For a revision, see: (c) Tsuji, H.; Ilies, L.; Nakamura, E. Synthetic Strategy for Multisubstituted Fused Furan Compounds Using Main-Group Metal Reagents. *Synlett* **2014,** 2099–2110.
20. Sharma, P.; Liu, R.-S. Skeletal Rearrangement in the Zn(II)-Catalyzed [4 + 2]-Annulation of Disubstituted N-Hydroxy Allenylamines with Nitrosoarenes to Yield Substituted 1,2-Oxazinan-3-one Derivatives. *Chem. Eur. J* **2016,** *22*, 15881–15887.
21. Sniady, A.; Durham, A.; Morreale, M. S.; Wheeler, K. A.; Dembinski, R. Room Temperature Zinc Chloride-Catalyzed Cycloisomerization of Alk-3-yn-1-ones: Synthesis of Substituted Furans. *Org. Lett.* **2007,** *9*, 1175–1178.

22. Sniady, A.; Durham, A.; Morreale, M. S.; Marcinek, A.; Szafert, S.; Lis, T.; Brzezinska, K. R.; Iwasaki, T.; Ohshima, T.; Mashima, K.; Dembinski, R. Zinc-Catalyzed Cycloisomerizations. Synthesis of Substituted Furans and Furopyrimidine Nucleosides. *J. Org. Chem.* **2008,** *73*, 5881–5889.
23. (a) Vicente, R.; González, J.; Riesgo, L.; González, J.; López, L. A. Catalytic Generation of Zinc Carbenes from Alkynes: Zinc-Catalyzed Cyclopropanation and Si–H Bond Insertion Reactions. *Angew. Chem. Int. Ed.* **2012,** *51*, 8063–8067. (b) González, J.; López, L. A.; Vicente, R. Zinc-Catalyzed Synthesis of 2-Alkenylfurans via Cross-Coupling of Enynones and Diazo Compounds. *Chem. Commun.* **2014,** *50*, 8536–8538. (c) González, M. J.; López, L. A.; Vicente, R. Zinc-Catalyzed Cyclopropenation of Alkynes via 2-Furylcarbenoids. *Org. Lett.* **2014,** *16*, 5780–5783. (d) Mata, S.; López, L. A.; Vicente, R. Zinc-Catalyzed Functionalization of Si–H Bonds with 2-Furyl Carbenoids through Three-Component Coupling. *Chem. Eur. J* **2015,** *21*, 8998–9002. (e) Mata, S.; López, L. A.; Vicente, R. Synthesis of Silylcyclopropanes through the Catalytic Generation of Zinc Silylcarbenoids from Enynones. *Synlett* **2015,** 2685–2689. (f) Mata, S.; González, J.; Vicente, R.; López, L. A. Zinc-Catalyzed Multicomponent Reactions: Easy Access to Furyl-Substituted Cyclopropane and 1,2-Dioxolane Derivatives. *Eur. J. Org. Chem.* **2016,** 2681–2687. (g) Mata, S.; González, M. J.; González, J.; López, L. A.; Vicente, R. Zinc-Catalyzed Synthesis of Conjugated Dienoates through Unusual Cross-Couplings of Zinc Carbenes with Diazo Compounds. *Chem. Eur. J* **2017,** *23*, 1013–1017. See also: (h) Song, B.; Li, L.-H.; Song, X.-R.; Qiu, Y.-F.; Zhong, M.-J.; Zhou, P.-X.; Liang, Y.-M. Zinc-Catalyzed [4 + 3] Cycloaddition with Concomitant Furan Annulation: Formation of Cyclohepta[b]Furans. *Chem. Eur. J.* **2014,** *20*, 5910–5913.
24. González, J.; González, J.; Pérez-Calleja, C.; López, L. A.; Vicente, R. Zinc-Catalyzed Synthesis of Functionalized Furans and Triarylmethanes from Enynones and Alcohols or Azoles: Dual X–H Bond Activation by Zinc. *Angew. Chem. Int. Ed.* **2013,** *52*, 5853–5857.
25. Wang, B.; Chen, Y.; Zhou, L.; Wang, J.; Tung, C.-H.; Xu, Z. Synthesis of Oxazoles by Tandem Cycloisomerization/Allylic Alkylation of Propargyl Amides with Allylic Alcohols: $Zn(OTf)_2$ as π Acid and σ Acid Catalyst. *J. Org. Chem.* **2015,** *80*, 12718–12724.
26. (a) Nakamura, M.; Liang, C.; Nakamura, E. Zn(II)/Amine-Catalyzed Coupling Reaction of Alkylidenemalonates with Propargyl Alcohol: A One-Pot Synthesis of Methylenetetrahydrofurans. *Org. Lett.* **2004,** *6*, 2015–2017. (b) Morikawa, S.; Yamazaki, S.; Tsukada, M.; Izuhara, S.; Morimoto, T.; Kakiuchi, K. Lewis Acid-Catalyzed Conjugate Addition – Cyclization Reactions of Ethenetricarboxylates with Substituted Propargyl Alcohols: Stereoselectivity in the Efficient One-Pot Synthesis of Methylenetetrahydrofurans. *J. Org. Chem.* **2007,** *72*, 6459–6463. (c) Yamazaki, S.; Morikawa, S.; Miyazaki, K.; Takebayashi, M.; Yamamoto, Y.; Morimoto, T.; Kakiuchi, K.; Mikata, Y. Zinc-Catalyzed Reactions of Ethenetricarboxylates with 2-(Trimethylsilylethynyl)anilines Leading to Bridged Quinoline Derivatives. *Org. Lett.* **2009,** *11*, 2796–2799. (d) Deng, C.-L.; Song, R.-J.; Liu, Y.-L.; Li, J.-H. General Zinc-Catalyzed Conia-Ene Reactions of 1,3-Dicarbonyl Compounds with Alkynes Including the Classically Challenging Substrates under Neat Conditions. *Adv. Synth. Catal.* **2009,** *351*, 3096–3100. (e) Hess, W.; Burton, J. W. A Zinc(II) Catalyst System for the Conia-Ene Reaction of Alkynyl-Aminomalonates Applicable to 5-*Endo*-Dig Reactions. *Adv. Synth. Catal.* **2011,** *353*, 2966–2970. (f) Morikawa, S.; Yamazaki, S.; Furusaki, Y.; Amano, N.; Zenke, K.; Kakiuchi, K. Zinc- and Indium-Promoted Conjugate Addition – Cyclization

Reactions of Ethenetricarboxylates with Propargylamines and Alcohol: Novel Methylenepyrrolidine and Methylenetetrahydrofuran Syntheses. *J. Org. Chem.* **2006,** *71*, 3540–3544. (g) Urabe, F.; Miyamoto, S.; Takahashi, K.; Ishihara, J.; Hatakeyama, S. Formal [4 + 1]-Cycloaddition of Homopropargyl Alcohols to Diazo Dicarbonyl Compounds Giving Substituted Tetrahydrofurans. *Org. Lett.* **2014,** *16*, 1004–1007.

27. Lebold, T. P.; Leduc, A. B.; Kerr, M. A. Zn(II)-Catalyzed Synthesis of Piperidines from Propargyl Amines and Cyclopropanes. *Org. Lett.* **2009,** *11*, 3770–3772.
28. Leão, R. A. C.; de Moraes, P. de F.; Pedro, M. C. B. C.; Costa, P. R. R. Synthesis of Coumarins and Neoflavones through Zinc Chloride Catalyzed Hydroarylation of Acetylenic Esters with Phenols. *Synthesis* **2011,** 3692–3696.
29. Li, G.; Nakamura, H. Synthesis of 2-Indolyltetrahydroquinolines by Zinc(II)-Catalyzed Intramolecular Hydroarylation-Redox Cross-Dehydrogenative Coupling of N-Propargylanilines with Indoles. *Angew. Chem. Int. Ed.* **2016,** *55*, 6758–6761.
30. Dong, H.; Shen, M.; Redford, J. E.; Stokes, B. J.; Pumphrey, A. L.; Driver, T. G. Transition Metal-Catalyzed Synthesis of Pyrroles from Dienyl Azides. *Org. Lett.* **2007,** *9*, 5191–5194.
31. Wyrebek, P.; Sniady, A.; Bewick, N.; Li, Y.; Mikus, A.; Wheeler, K. A.; Dembinski, R. Microwave-Assisted Zinc Chloride-Catalyzed Synthesis of Substituted Pyrroles from Homopropargyl Azides. *Tetrahedron* **2009,** *65*, 1268–1275.
32. Li, L.; Zhou, B.; Wang, Y.-H.; Shu, C.; Pan, Y.-F.; Lu, X.; Ye, L.-W. Zinc-Catalyzed Alkyne Oxidation/C–H Functionalization: Highly Site-Selective Synthesis of Versatile Isoquinolones and β-Carbolines. *Angew. Chem. Int. Ed.* **2015,** *54*, 8245–8249.
33. Kong, Y.; Jiang, K.; Cao, J.; Fu, L.; Yu, L.; Lai, G.; Cui, Y.; Hu, Z.; Wang, G. Synthesis of 3-Alkyl- or 3-Allenyl-2-amidobenzofurans via Electrophilic Cyclization of *o*-Anisole-Substituted Ynamides with Carbocations. *Org. Lett.* **2013,** *15*, 422–425.
34. (a) Chen, L.; Yu, L.; Deng, Y.; Zheng, Z.-J.; Xu, Z.; Cao, J.; Xu, L.-W. C–H Functionalization/C–O Bond Cleavage of Benzyl Silyl Ethers with Ynamides for the Chemoselective Synthesis of Skeletally Diverse Compounds. *Adv. Synth. Catal.* **2016,** *358*, 480–485. See also: (b) Chen, L.; Cui, Y.-M.; Xu, Z.; Cao, J.; Zheng, Z.-J.; Xu, L.-W. An Efficient Approach Toward Formation of Polycyclic Coumarin Derivatives via Carbocation-Initiated [4 + 2] Cycloaddition and Atom-Economical Photo-Irradiated Cyclization. *Chem. Commun.* **2016,** *52*, 11131–11134.
35. Danishefsky, S.; Kerwin, J. F.; Kobayashi, S. Lewis Acid Catalyzed Cyclocondensations of Functionalized Dienes with Aldehydes. *J. Am. Chem. Soc.* **1982,** *104*, 358–360.
36. (a) Du, H.; Long, J.; Hu, J.; Li, X.; Ding, K. 3,3'-Br_2-BINOL-Zn Complex: A Highly Efficient Catalyst for the Enantioselective Hetero-Diels – Alder Reaction. *Org. Lett.* **2002,** *4*, 4349–4352. See also: (b) Du, H.; Zhang, X.; Wang, Z.; Ding, K. One Catalyst for Two Distinct Reactions: Sequential Asymmetric Hetero Diels–Alder Reaction and Diethylzinc Addition. *Tetrahedron* **2005,** *61*, 9465–9477.
37. Panarese, J. D.; Waters, S. P. Enantioselective Formal Total Synthesis of (+)-Aspergillide C. *Org. Lett.* **2009,** *11*, 5086–5088.
38. Eschenbrenner-Lux, V.; Kuechler, P.; Ziegler, S.; Kumar, K.; Waldmann, H. An Enantioselective Inverse-Electron-Demand Imino Diels-Alder Reaction. *Angew. Chem. Int. Ed.* **2014,** *53*, 2134–2137.
39. Mandler, M. D.; Truong, P. M.; Zavalij, P. Y.; Doyle, M. P. Catalytic Conversion of Diazocarbonyl Compounds to Imines: Applications to the Synthesis of Tetrahydropyrimidines and β-Lactams. *Org. Lett.* **2014,** *16*, 740–743.

40. Chu, J. C. K.; Dalton, D. M.; Rovis, T. Zn-Catalyzed Enantio- and Diastereoselective Formal [4 + 2] Cycloaddition Involving Two Electron-Deficient Partners: Asymmetric Synthesis of Piperidines from 1-Azadienes and Nitro-Alkenes. *J. Am. Chem. Soc.* **2015,** *137*, 4445–4452.
41. (a) Gothelf, A. S.; Gothelf, K. V.; Hazell, R. G.; Jørgensen, K. A. Catalytic Asymmetric 1,3-Dipolar Cycloaddition Reactions of Azomethine Ylides—A Simple Approach to Optically Active Highly Functionalized Proline Derivatives. *Angew. Chem. Int. Ed.* **2002,** *41*, 4236–4238. (b) Dogan, Ö.; Koyuncu, H.; Garner, P.; Bulut, A.; Youngs, W. J.; Panzner, M. New Zinc(II)-Based Catalyst for Asymmetric Azomethine Ylide Cycloaddition Reactions. *Org. Lett.* **2006,** *8*, 4687–4690. See also: (c) Yamashita, Y.; Minami, K.; Saito, Y.; Kobayashi, S. Development of a Simple Adjustable Zinc Acid/Base Hybrid Catalyst for C–C and C–O Bond-Forming and C–C Bond-Cleavage Reactions. *Chem. Asian J.* **2016,** *11*, 2372–2376.
42. Jiang, Z.; Wang, J.; Lu, P.; Wang, Y. Diasteroselective Synthesis of Oxazolidines and Imidazolidines via the Lewis Acid Catalyzed C C Cleavage of Aziridines. *Tetrahedron* **2011,** *67*, 9609–9617.
43. Craig II, R. A.; O'Connor, N. R.; Goldberg, A. F. G.; Stoltz, B. M. Stereoselective Lewis Acid Mediated (3 + 2) Cycloadditions of N–H- and N-sulfonylaziridines with Heterocumulenes. *Chem. Eur. J* **2014,** *20*, 4806–4813.
44. He, S.; Chen, L.; Niu, Y.-N.; Wu, L.-Y.; Liang, Y.-M. 1,3-Dipolar Cycloaddition of Diazoacetate Compounds to Terminal Alkynes Promoted by $Zn(OTf)_2$: An Efficient Way to the Preparation of Pyrazoles. *Tetrahedron Lett.* **2009,** *50*, 2443–2445.
45. (a) Demko, Z. P.; Sharpless, K. B. Preparation of 5-Substituted 1*H*-Tetrazoles from Nitriles in Water. *J. Org. Chem.* **2001,** *66*, 7945–7950. (b) Demko, Z. P.; Sharpless, K. B. An Expedient Route to the Tetrazole Analogues of α-Amino Acids. *Org. Lett.* **2002,** *4*, 2525–2527. (c) Himo, F.; Demko, Z. P.; Noodleman, L.; Sharpless, K. B. Why Is Tetrazole Formation by Addition of Azide to Organic Nitriles Catalyzed by Zinc(II) Salts? *J. Am. Chem. Soc.* **2003,** *125*, 9983–9987. See also: (d) Cantillo, D.; Gutmann, B.; Kappe, C. O. An Experimental and Computational Assessment of Acid-Catalyzed Azide-Nitrile Cycloadditions. *J. Org. Chem.* **2012,** *77*, 10882–10890.
46. (a) Vorona, S.; Artamonova, T.; Zevatskii, Y.; Myznikov, L. An Improved Protocol for the Preparation of 5-Substituted Tetrazoles from Organic Thiocyanates and Nitriles. *Synthesis* **2014,** *46*, 781–786. (b) Kuhn, B. L.; Fortes, M. P.; Kaufman, T. S.; Silveira, C. C. Facile, Efficient and Eco-friendly Synthesis of 5-Sulfenyl Tetrazole Derivatives of Indoles and Pyrroles. *Tetrahedron Lett.* **2014,** *55*, 1648–1652.
47. (a) Smith, C. D.; Greaney, M. F. Zinc Mediated Azide–Alkyne Ligation to 1,5- and 1,4,5-Substituted 1,2,3-Triazoles. *Org. Lett.* **2013,** *15*, 4826–4829. See also: (b) Jana, S.; Thomas, J.; Dehaen, W. A One-Pot Procedure for the Synthesis of "Click-Ready" Triazoles from Ketones. *J. Org. Chem.* **2016,** *81*, 12426–12432.
48. Li, S.-M.; Yu, B.; Liu, J.; Li, H.-L.; Na, R. Bronsted Acid or Lewis Acid Catalyzed [3 + 3] Cycloaddition of Azomethine Imines with N-Benzyl Azomethine Ylide: A Facile Access to Bicyclic N-Heterocycles. *Synlett* **2016,** *27*, 282–286.
49. Lei, C.-H.; Wang, D.-X.; Zhao, L.; Zhu, J.; Wang, M.-X. Synthesis of Substituted Pyridines from Cascade [1 + 5] Cycloaddition of Isonitriles to *N*-Formylmethyl-Substituted Enamides, Aerobic Oxidative Aromatization, and Acyl Transfer Reaction. *J. Am. Chem. Soc.* **2013,** *135*, 4708–4711.
50. Lei, C.-H.; Wang, D.-X.; Zhao, L.; Zhu, J.; Wang, M.-X. Synthesis of Multifunctionalized 1,2,3,4-Tetrahydropyridines, 2,3-Dihydro-4(1*H*)-pyridinones and Pyridines from Tandem

Reactions Initiated by [5 + 1]-Cycloaddition of *N*-Formylmethyl-Substituted Enamides to Isocyanides: Mechanistic Insight and Synthetic Application. *Chem. Eur. J.* **2013,** *19*, 16981–16987.

51. Ranu, B. C.; Gosh, S.; Das, A. Zinc Tetrafluoroborate-Catalysed Synthesis of Highly Substituted Pyrroles by a Solvent-Free Reaction. *Mendeleev Commun.* **2006,** *16*, 220–221.
52. Tiwari, R. K.; Singh, J.; Singh, D.; Verma, A. K.; Chandra, R. Highly Efficient One-Pot Synthesis of 1-Substituted-1,2,3,4-tetrahydropyrazino[1,2-a]indoles. *Tetrahedron* **2005,** *61*, 9513–9518.
53. (a) Zhao, T.; Zhang, H.; Cui, L.; Wang, C.; Qu, J.; Wang, B. A $ZnCl_2$-Catalyzed Knoevenagel Condensation/1,5-Hydride Shift/Cyclization Sequence: Synthesis of Novel Spiroisoxazol-5-one Tetrahydroquinolines. *ChemistrySelect* **2016,** *1*, 3713–3717 See also. (b) Zhao, T.; Zhang, H.; Cui, L.; Qu, J.; Wang, B. Zinc Chloride Catalyzed Stereoselective Construction of Spiropyrazolone Tetrahydroquinolines via Tandem [1,5]-hydride Shift/Cyclization Sequence. *RSC Adv.* **2015,** *5*, 86056–86060.
54. Sasada, T.; Aoki, Y.; Ikeda, R.; Sakai, N.; Konakahara, T. Synthesis of Tri- or Tetrasubstituted Pyrimidine Derivatives through the [5 + 1] Annulation of Enamidines with either N,N-Dimethylformamide Dialkyl Acetals or Orthoesters and Their Application in a Ring Transformation of Pyrimidines to Pyrido[2,3-d]pyrimidin-5-one Derivatives. *Chem. Eur. J.* **2011,** *17*, 9385–9394.
55. Mamada, M.; Pérez-Bolívar, C.; Anzenbacher, P. Green Synthesis of Polycyclic Benzimidazole Derivatives and Organic Semiconductors. *Org. Lett.* **2011,** *13*, 4882–4885.
56. Tan, L.; Zhou, P.; Chen, C.; Liu, W. An Efficient Method for the Construction of Polysubstituted 4-Pyridones via Self-Condensation of β-Keto Amides Mediated by P_2O_5 and Catalyzed by Zinc Bromide. *Beilstein J. Org. Chem.* **2013,** *9*, 2681–2687.
57. Trost, B. M.; Hirano, K. Dinuclear Zinc Catalyzed Asymmetric Spirannulation Reaction: An Umpolung Strategy for Formation of α-Alkylated-α-Hydroxyoxindoles. *Org. Lett.* **2012,** *14*, 2446–2449.
58. (a) Tan, F.; Lu, L.-Q.; Yang, Q.-Q.; Guo, W.; Bian, Q.; Chen, J.-R.; Xiao, W.-J. Enantioselective Cascade Michael Addition/Cyclization Reactions of 3-Nitro-2*H*-chromenes with 3-Isothiocyanato Oxindoles: Efficient Synthesis of Functionalized Polycyclic Spirooxindoles. *Chem. Eur. J.* **2014,** *20*, 3415–3420 See also. (b) Zhao, J.-Q.; Wu, Z.-J.; Zhou, M.-Q.; Xu, X.-Y.; Zhang, X.-M.; Yuan, W.-C. Zn-Catalyzed Diastereo- and Enantioselective Cascade Reaction of 3-Isothiocyanato Oxindoles and 3-Nitroindoles: Stereocontrolled Syntheses of Polycyclic Spirooxindoles. *Org. Lett.* **2015,** *17*, 5020–5023.
59. Gunasekaran, P.; Prasanna, P.; Perumal, S.; Almansour, A. I. $ZnCl_2$-Catalyzed Three-Component Domino Reactions for the Synthesis of Pyrano[3,2-c]quinolin-5(6*H*)-ones. *Tetrahedron Lett.* **2013,** *54*, 3248–3252.
60. Huang, J.; Zhou, J.; Song, S.; Song, H.; Chen, Z.; Yi, W. A New and Efficient $ZnCl_2$-Catalyzed Synthesis and Biological Evaluation of Novel 2 Amino-3,5-dicyano-4-aryl-6-aryl-aminopyridines as Potent Antibacterial Agents against *Helicobacter pylori* (HP). *Tetrahedron* **2015,** *71*, 8628–8636.
61. Sakakura, T.; Kohno, K. The Synthesis of Organic Carbonates from Carbon Dioxide. *Chem. Commun.* **2009,** 1312–1330.
62. For an excellent revision, see: Enthaler, S. Zinc-Catalyzed Transformations of Carbon Dioxide. In *Zinc Catalysis, Applications in Organic Synthesis;* Enthaler, S., Wu, X.-F., Eds.; Wiley-VCH: Weinheim, 2015.
63. (a) Seo, U. R.; Chung, Y. K. Poly(4-vinylimidazolium)s/diazabicyclo[5.4.0]undec-7-ene/zinc(II) Bromide-Catalyzed Cycloaddition of Carbon Dioxide to Epoxides. *Adv. Synth.*

Catal. **2014,** *356*, 1955–1961. (b) Martin, C.; Whiteoak, C. J.; Martin, E.; Martinez Belmonte, M.; Escudero-Adan, E. C.; Kleij, A. W. Easily Accessible Bifunctional Zn(salpyr) Catalysts for the Formation of Organic Carbonates. *Catal. Sci. Technol.* **2014,** *4*, 1615–1621. (c) Luo, R.; Zhou, X.; Zhang, W.; Liang, Z.; Jiang, J.; Ji, H. New Bi-Functional Zinc Catalysts Based on Robust and Easy-to-Handle N-chelating Ligands for the Synthesis of Cyclic Carbonates from Epoxides and CO_2 Under Mild Conditions. *Green Chem.* **2014,** *16*, 4179–4189. (d) Liu, M.; Liang, L.; Liang, T.; Lin, X.; Shi, L.; Wang, F.; Sun, J. Cycloaddition of CO_2 and Epoxides Catalyzed by Dicationic Ionic Liquids Mediated Metal Halide: Influence of the Dication on Catalytic Activity. *J. Mol. Catal. A: Chem.* **2015,** *408*, 242–249. (d) Ren, Y.; Chen, J.; Qi, C.; Jiang, H. A New Type of Lewis Acid-Base Bifunctional M(salphen) (M = Zn, Cu and Ni) Catalysts for CO_2 Fixation. *ChemCatChem* **2015,** *7*, 1535–1538. (e) Maeda, C.; Taniguchi, T.; Ogawa, K.; Ema, T. Bifunctional Catalysts Based on *m*-Phenylene-Bridged Porphyrin Dimer and Trimer Platforms: Synthesis of Cyclic Carbonates from Carbon Dioxide and Epoxides. *Angew. Chem. Int. Ed.* **2015,** *54*, 134–138. (f) Desens, W.; Kohrt, C.; Spannenberg, A.; Werner, T. A Novel Zinc Based Binary Catalytic System for CO_2 Utilization Under Mild Conditions. *Org. Chem. Front.* **2016,** *3*, 156–164. (g) Ma, R.; He, L.-N.; Zhou, Y.-B. An Efficient and Recyclable Tetraoxo-Coordinated Zinc Catalyst for the Cycloaddition of Epoxides with Carbon Dioxide at Atmospheric Pressure. *Green Chem.* **2016,** *18*, 226–231. (h) He, S.; Wang, F.; Tong, W.-L.; Yiu, S.-M.; Chan, M. C. W. Topologically Diverse Shape-Persistent Bis-(Zn-salphen) Catalysts: Efficient Cyclic Carbonate Formation under Mild Conditions. *Chem. Commun.* **2016,** *52*, 1017–1020. (i) Mercade, E.; Zangrando, E.; Claver, C.; Godard, C. Robust Zinc Complexes that Contain Pyrrolidine-Based Ligands as Recyclable Catalysts for the Synthesis of Cyclic Carbonates from Carbon Dioxide and Epoxides. *ChemCatChem* **2016,** *8*, 234–243 For examples in heterogenous catalysis, see. (j) Bhunia, S.; Molla, R. A.; Kumari, V.; Islam, S. M.; Bhaumik, A. Zn(II) Assisted Synthesis of Porous Salen as an Efficient Heterogeneous Scaffold for Capture and Conversion of CO_2. *Chem. Commun.* **2015,** *51*, 15732–15735. (k) Wang, W.; Li, C.; Yan, L.; Wang, Y.; Jiang, M.; Ding, Y. Ionic Liquid/Zn-PPh_3 Integrated Porous Organic Polymers Featuring Multifunctional Sites: Highly Active Heterogeneous Catalyst for Cooperative Conversion of CO_2 to Cyclic Carbonates. *ACS Catal.* **2016,** *6*, 6091–6100.

64. Hu, J.; Ma, J.; Zhu, Q.; Qian, Q.; Han, H.; Mei, Q.; Han, B. Zinc(II)-Catalyzed Reactions of Carbon Dioxide and Propargylic Alcohols to Carbonates at Room Temperature. *Green Chem.* **2016,** *18*, 382–385.
65. Liu, X.; Wang, M.-Y.; Wang, S.-Y.; Wang, Q.; He, L.-N. In-Situ Generated Zinc(II) Catalyst for Incorporation of CO_2 into 2-Oxazolidinones with Propargylic Amines at Atmospheric Pressure. *ChemSusChem* **2017,** *10*, 1210–1216.

Chapter 8

Controlling Selectivities in Palladium-Catalyzed Cyclization Reactions Leading to Heterocycles: From Ambiphilic Reactions of Arylpalladium Species to Carbene Insertions

Daniel Solé[1] and Israel Fernández[2]
[1]*Universitat de Barcelona, Barcelona, Spain,* [2]*Universidad Complutense de Madrid, Madrid, Spain*

8.1 INTRODUCTION

The use of divergent catalytic reactions to access structurally diverse compounds from a common precursor through controlled reaction pathways is a highly attractive tool for the development of new drugs and functional materials.[1] Catalyst-controlled selectivity is currently well established in enantioselective transformations.[2] In contrast, the three other main types of selectivities, recognized and described in a seminal review by Trost as diastereoselectivity, regioselectivity, and chemoselectivity,[3] have been less targeted in this field. Despite that, today there is a growing recognition that all classes of selectivity can be exquisitely controlled by the appropriate choice of catalyst, ligands, and reaction conditions.[4]

Due to its catalytic effectiveness and versatility, palladium is a pivotal element in organic synthesis. Indeed, no other metal acts as the catalyst of so many named reactions,[5] which make palladium the most promising metal for divergent catalytic selective synthesis. The outcome of the reaction of a given substrate in the presence of palladium can be finely tuned depending on the catalyst type, surrounding ligands,[6] and reaction conditions.[7]

The extremely diverse chemistry of palladium in C–C bond-forming processes is mainly based on the electrophilic nature of σ-organopalladium

Advances in Transition-Metal Mediated Heterocyclic Synthesis.
DOI: https://doi.org/10.1016/B978-0-12-811651-7.00008-X

intermediates.[5] However, recent research has demonstrated that in organopalladium complexes, switching the usual reactivity from electrophilic to nucleophilic is relatively simple, since the same palladium species can also react with carbon-heteroatom multiple bonds in a nucleophilic manner.[8] Despite the inherent interest of being able to control the potentially ambiphilic character of the organopalladium intermediates involved in these reactions, little effort has been dedicated to the selective promotion of either their electrophilic or nucleophilic reactivity from the same starting material.[9]

The aim of this chapter is to provide an overview of the research on the control of the ambiphilic character of σ-arylpalladium species in intramolecular palladium-catalyzed coupling reactions with carbonyl derivatives leading to heterocyclic compounds (Scheme 8.1A). Also discussed is the use of palladium catalysis for modeling the selectivity of intramolecular metal-carbenoid C–H insertion reactions of α-diazo carbonyl derivatives in aza-heterocyclic synthesis (Scheme 8.1B). Throughout, emphasis has been placed on how the combination of experimental and computational tools can be extremely helpful to better understand the reaction mechanisms involved in these transformations.

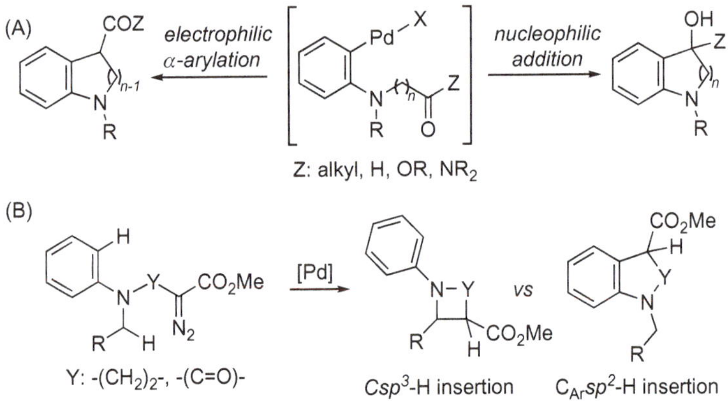

SCHEME 8.1 Controlling selectivities in palladium-catalyzed cyclization reactions leading to heterocycles.

8.2 CHEMOSELECTIVITY CONTROL IN INTRAMOLECULAR Pd-CATALYZED REACTIONS OF AMINO-TETHERED ARYL HALIDES AND KETONES

The palladium-catalyzed intramolecular coupling reactions between aryl halides and carbonyl compounds constitute a challenging field of research due to the potentially ambiphilic character of the transient σ-arylpalladium intermediates.[8,10] In particular the palladium-catalyzed α-arylation of carbonyls, which exploits the electrophilic nature of the palladium atom in the σ-organopalladium intermediates, has received considerable attention.[11] In this context the intramolecular α-arylation of ketones has emerged as an extremely powerful methodology for the synthesis of complex natural products.[12] In parallel with the progress in this field, interest in the direct addition of aryl palladium species to the electrophilic ketone carbonyl group has also grown significantly.[13,14] Although both reactions involve similar σ-C-bonded palladium(II) intermediates, the two envisaged pathways seem to operate quite independently of each other, and almost no competition between them has been described.[15,16]

Solé and coworkers have reported that two different and competitive reaction pathways, involving either enolate arylation or a nucleophilic attack at the carbonyl group, can be promoted by palladium when starting from (2-haloanilino) ketones, which show a structure-dependent behavior when undergoing palladium-catalyzed cyclization.[17,18] Thus whereas treatment of γ-(anilino)ketone **1** with $PdCl_2(PPh_3)_2$ and Cs_2CO_3 led to the α-arylation compound **2**, under the same reaction conditions α-(anilino)ketone **3** and β-(anilino)ketone **5** exclusively afforded alcohols **4** and **6**, respectively, as a result of the addition of the corresponding palladium intermediate to the ketone carbonyl group (Scheme 8.2).

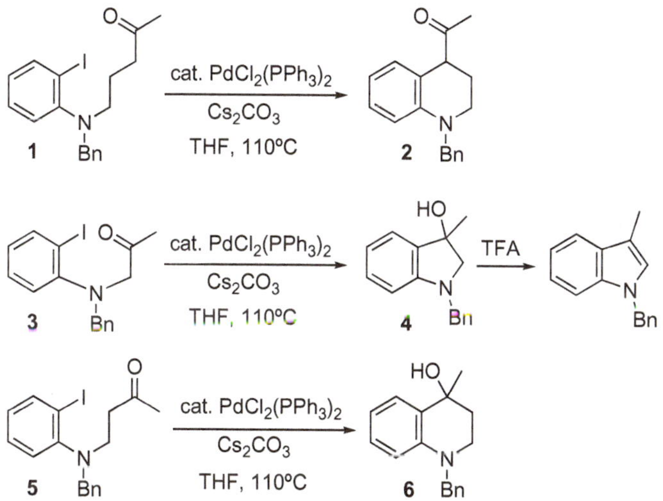

SCHEME 8.2 Pd(0)-catalyzed reactions of (2-haloanilino) ketones.

The competition between both reaction pathways, i.e., nucleophilic addition vs α-arylation, was analyzed computationally by means of Density Functional Theory (DFT) calculations.[19] For the process involving β-(anilino)ketone **5**, which exclusively produces alcohol **6** (see above), it was found

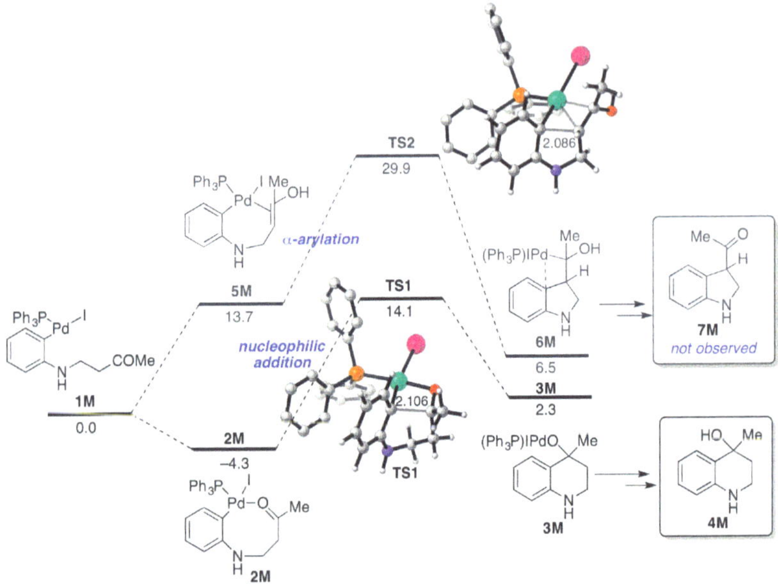

FIGURE 8.1 Computed reaction profiles [PCM(tetrahydrofuran)-B3LYP/def2-SVP level] for the competition between nucleophilic addition and α-arylation reactions involving complex **1M**. Relative energies and bond distances are given in kcal/mol and angstroms, respectively.

that the nucleophilic addition pathway is clearly favored over the possible α-arylation reaction. As shown in Fig. 8.1, in the model system where the benzyl substituent was replaced by a hydrogen atom, the process begins with complex **1M**, the intermediate formed upon the initial oxidative addition step. From this common intermediate, either the nucleophilic addition pathway via **2M**→**TS1**→**3M** or the α-arylation reaction via **5M**→**TS2**→**6M** may occur. However, the latter pathway is clearly less favored in view of the high energetic cost associated with the endergonic enolization process necessary to produce intermediate **5M** and the higher activation barrier of the subsequent α-arylation step.

Interestingly, in β-(anilino)ketones, changing the substituent at the nitrogen from alkyl to alkoxycarbonyl provoked competition between the nucleophilic addition to the carbonyl and the α-arylation reaction.[18] This is a direct consequence of the electron-withdrawing effect of the alkoxycarbonyl group, which avoids the π-delocalization of the lone pair of the nitrogen atom into the aryl ring. As a result, the nucleophilicity of the reactive aryl carbon atom is dramatically reduced, rendering the nucleophilic addition reaction (via **TS1**) far more difficult (i.e., with a much higher activation barrier), which is translated into the observed mixture of reaction products.

The chemoselectivity of the palladium-catalyzed reaction of β-(anilino) ketones was switched from nucleophilic addition to electrophilic α-arylation by a slight modification of the catalyst and reaction conditions.[20] Thus the reaction of **5** using $Pd(PPh_3)_4$ as the precatalyst and PhOK as the base, which is generated in situ by reaction of PhOH with KO^tBu, afforded a 1:2 mixture of alcohol **6** and ketone **7**, the latter arising from the aromatization of the initially formed α-arylation product. When PPh_3 was replaced by 4,5-bis(diphenylphosphino)-9,9-dimethylxanthene (xantphos), ketone **7** was obtained in 80% yield in a highly chemoselective transformation (i.e., **6/7** = 1:14.5) (Scheme 8.3).

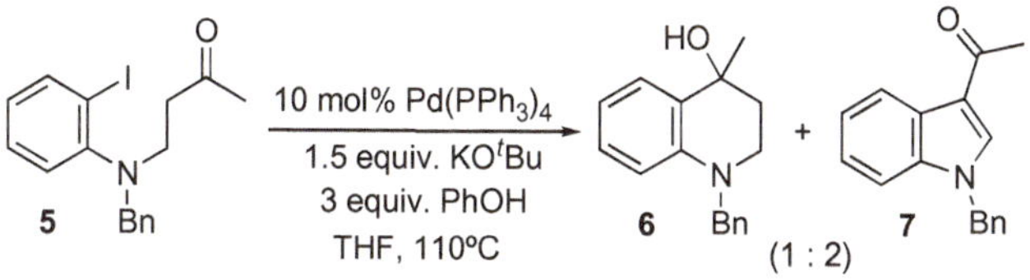

SCHEME 8.3 Switching from nucleophilic addition to electrophilic α-arylation.

According to the DFT calculations, an alternative reaction pathway leading to the α-arylation products, different from that described above, is found in the presence of phenoxide. This process involves the initial enolization reaction and subsequent exergonic substitution of the iodide ligand by PhO^- to produce **1M-OPh-enol**. Although this species may undergo a similar α-arylation reaction to that described above (via **TS2-OPh**), it experiences a highly exergonic intramolecular proton-transfer from the OH moiety to the phenoxy ligand to produce intermediate **8M**. As a consequence of the low donor ability of the PhOH ligand, the Pd—O interaction in **8M** is relatively weak, which permits the easy dissociation of PhOH ($\Delta G_{298} = -8.4$ kcal/mol) to form intermediate **9M**. This species is finally transformed into indoline **7M** via the transition state **TS3** with an activation barrier of only 13.9 kcal/mol (Fig. 8.2).

The competition between α-arylation and the nucleophilic attack on the ketone carbonyl was also observed in the palladium-catalyzed cyclization of (3-iodoindolyl) ketones.[21] Thus for example, the use of 2,2′-bis(diphenylphosphino)-1,1′-binaphthalene (binap) as the ligand for the palladium-catalyzed reaction of **8** resulted in the formation of a nearly equimolar mixture of ketone **9** and alcohol **10**. While the nucleophilic addition to the carbonyl was the only apparent reaction when using 1,2-bis(diphenylphosphino)ethane (dppe) as the ligand, it was not possible to increase the ratio of the α-arylation product **9** by changing the phosphine ligand (Scheme 8.4).

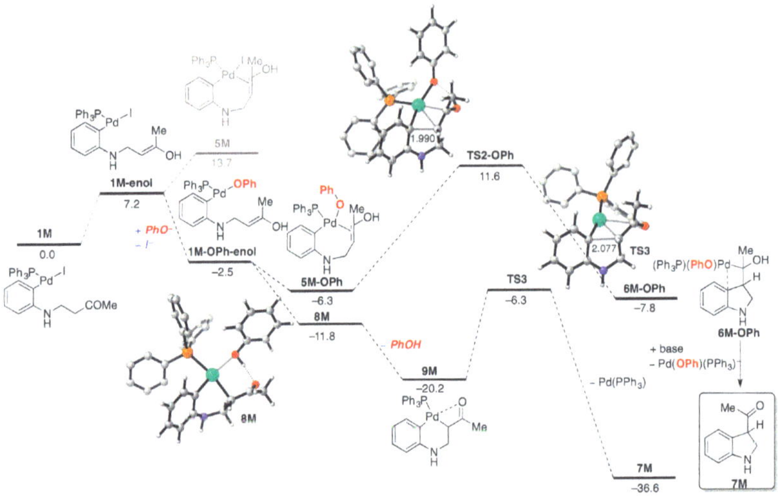

FIGURE 8.2 Computed reaction profiles for model ketone **1M** in the presence of PhO^-. See caption of Fig. 8.1 for additional details.

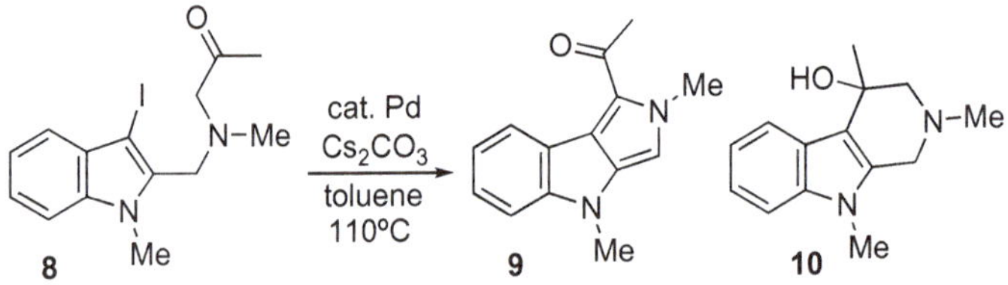

cat. Pd: $Pd_2(dba)_3$/binap, **9** (31% yield), **10** (35% yield)
cat. Pd: $Pd(OAc)_2$/dppe, **10** (48% yield)

SCHEME 8.4 Pd(0)-catalyzed reactions of 3-iodoindole **8.**

8.3 CHEMOSELECTIVITY CONTROL IN INTRAMOLECULAR Pd-CATALYZED REACTIONS OF AMINO-TETHERED ARYL HALIDES AND ALDEHYDES

Since the seminal paper by Heck and coworkers describing a single example of palladium-catalyzed annulation of *o*-iodobenzaldehyde and alkynes to give indenones,[22] the number of reports on the intramolecular nucleophilic addition of σ-vinylpalladium species to aldehydes has steadily increased.[23] In sharp contrast, there are only a few reported examples of intramolecular palladium-catalyzed reactions of aryl halides with the formyl moiety, a process that has been successfully used to prepare benzocyclobutenones[24] and a variety of azaheterocycles.[25] The intramolecular α-arylation of aldehydes has also been reported in the carbocyclic series.[26] Interestingly, the competition between the intramolecular aldehyde α-arylation and the reaction of the σ-arylpalladium species with the formyl group had only previously been observed by Muratake and coworkers.[26a]

In this context, we explored the control of the ambiphilic character of the σ-arylpalladium species in their reactions with amino-tethered aldehydes.[25] It was found that starting from β-(anilino)aldehyde **11**, the acylation process was the preferred reaction pathway when using bulky monodentate phosphines as the ligand, with a direct correlation between the phosphine cone angle and the observed chemoselectivity. Interestingly, the best acylation-to-α-arylation ratio was obtained with 1,1′-bis(di-tert-butylphosphino)ferrocene (dtpf), which afforded ketone **13** in 65% yield in a highly chemoselective transformation (i.e., **12**:**13** = 1:26). Although this ligand contains two phosphorous donors, dtpf probably ligates to the metal in a k^1-fashion, thus also acting as a sterically hindered monophosphine (Scheme 8.5).[27]

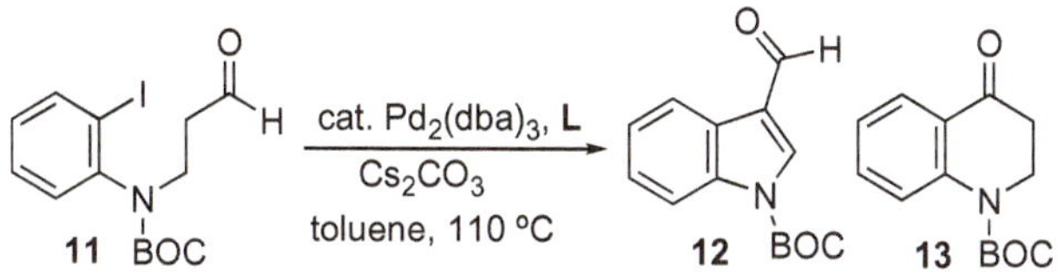

Ligand (**L**)	cone angle	α-arylation:acylation ratio
$(Cy)_3P$	170°	**12**:**13** = 1:2.3
$(^tBu)_3P$	182°	**12**:**13** = 1:9.1
(*o*-tolyl)$_3$P	194°	**12**:**13** = 1:11.5
dtpf	–	**12**:**13** = 1:26

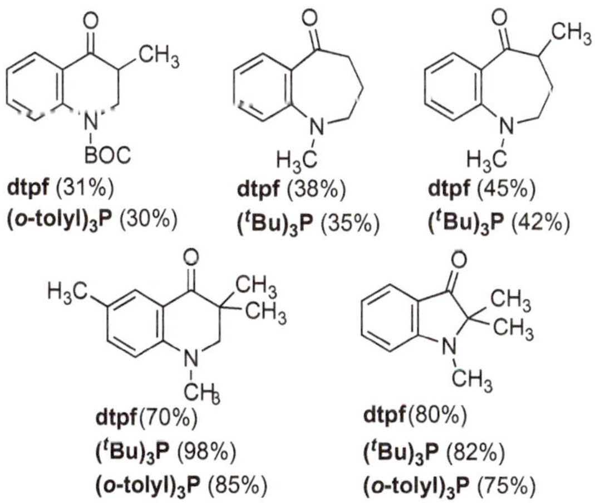

SCHEME 8.5 Pd(0)-catalyzed acylation of (2-iodoanilino)aldehydes.

The palladium-catalyzed acylation of aryl halides with aldehydes was successfully promoted starting from α-, β-, or γ-(anilino) aldehydes. As expected, the aminoaldehydes without α-carbonyl hydrogen atoms smoothly underwent acylation to give the corresponding ketones in high yields (Scheme 8.5).

While we were able to efficiently effect the acylation from a variety of aldehydes, the possibility of switching the chemoselectivity to promote the enolate arylation was in general a more challenging task.[25] As an exception, the α-arylation of aldehyde **14** proceeded with high yield to afford aldehyde **15** when using xantphos as the ligand in the presence of the base PhOK (Scheme 8.6).

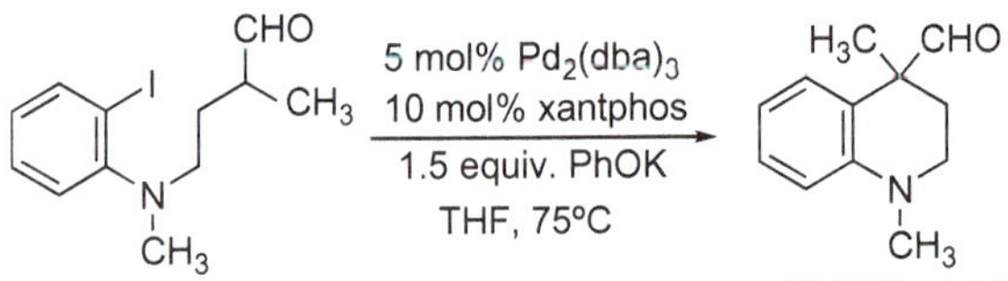

14 **15** (93% yield)

SCHEME 8.6 Pd(0)-catalyzed α-arylation of aldehyde **14**.

Based on the experimental results, we proposed a plausible reaction mechanism for the palladium-catalyzed intramolecular acylation with aldehydes. A carbopalladation reaction between the σ-arylpalladium(II) moiety and the carbonyl group would give a Pd(II) alkoxide. Then, a β-hydride elimination from the latter intermediate would afford ketone **13** and regenerate the Pd(0) catalyst (pathway A). This nucleophilic addition mechanism is supported by the formation of minor amounts of alcohol **16** in some of the cyclization reactions of aldehyde **11**. This alcohol would be formed by the competitive protonation of the palladium(II) alkoxide intermediate. The alternative mechanism for the acylation involving the insertion of the σ-aryl palladium(II) species into the formyl C – H bond (pathway B), which has been proposed to explain some related processes,[24] does not appear to be operative in this case (Scheme 8.7).

Our DFT computational studies suggest that the nucleophilic addition reaction is favored over the pathway leading to the α-arylation products.[25] This is ascribed to the computed lower barrier for the acylation and to the energetic cost associated with the endergonic enolization process (see above). In addition, the competition between the nucleophilic addition and the alternative C–H insertion was also computed starting from the common oxidative addition intermediate **10M**. From the data in Fig. 8.3, it becomes

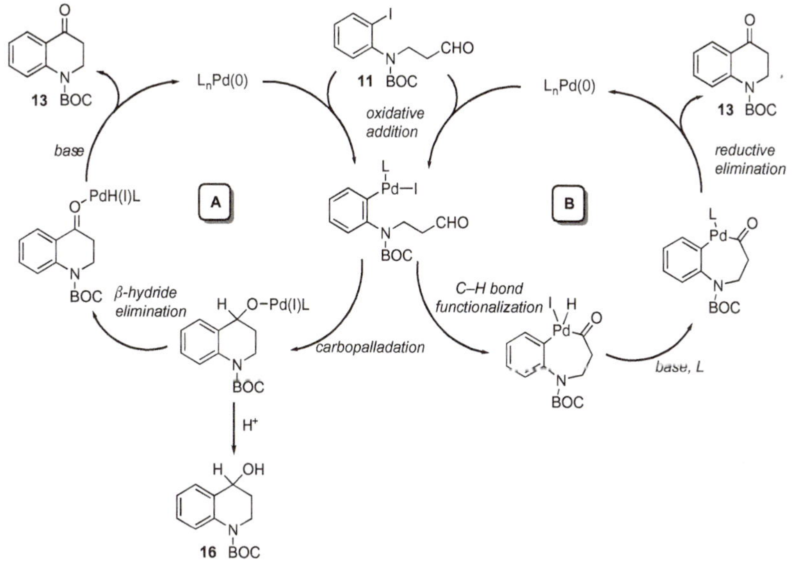

SCHEME 8.7 Proposed catalytic cycles for the intramolecular Pd(0)-catalyzed acylation with aldehydes.

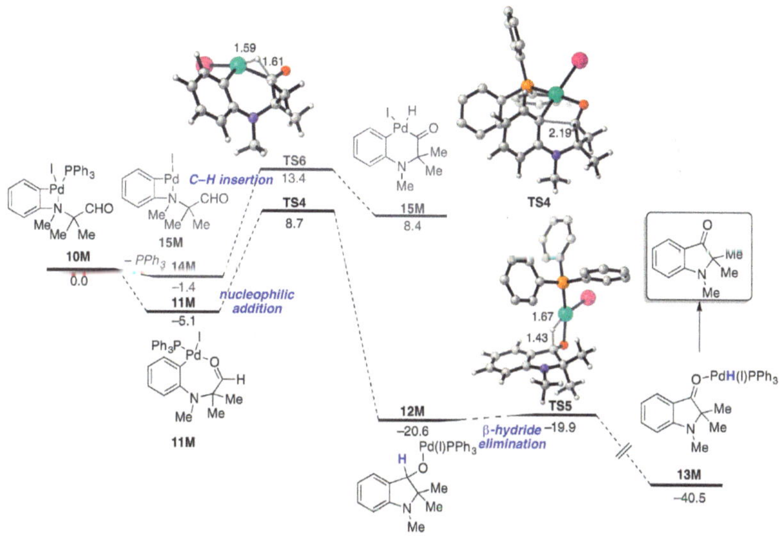

FIGURE 8.3 Computed reaction profile for complex **10M**. See caption of Fig. 8.1 for additional details. All data have been computed at the PCM(toluene)-B3LYP/def2-SVP level.

clear that the C−H insertion is not competitive, in view of the higher activation barrier and endergonicity associated with the formation of the Pd(IV) intermediate **15M**. In contrast, the nucleophilic addition occurs in a highly exergonic reaction step, which is followed by an easy β-elimination reaction (activation barrier of only 0.7 kcal/mol) to produce the precursor of the observed 3-indolinone.

The scope of the acylation with aldehydes was expanded by preparing a series of dibenzo[*b*,*e*]azepin-11-ones.[28] The functional group tolerance of the reaction is well illustrated by the fact that halides, ethers, esters, aldehydes, and free hydroxyl as well as dimethylamino groups were all perfectly accommodated (Scheme 8.8).

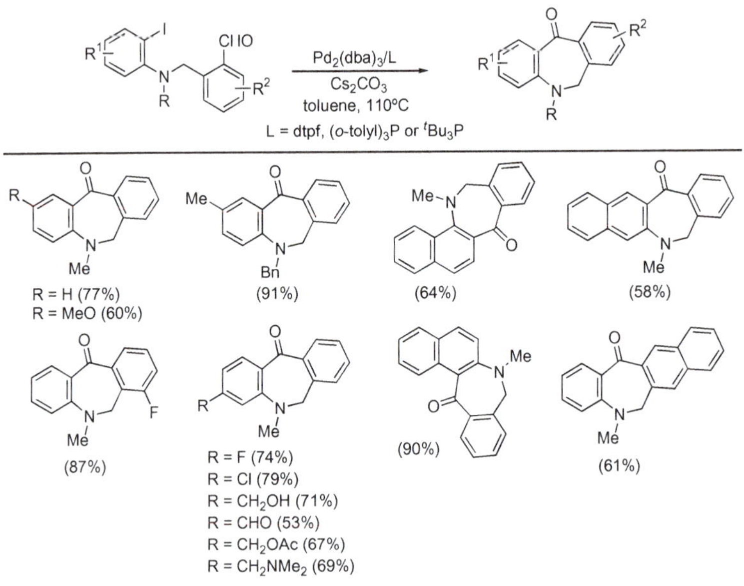

SCHEME 8.8 Synthesis of dibenzo[*b*,*e*]azepin-11-ones.

We also reported that when starting from (2-iodobenzyl)amino aldehydes, the nucleophilic addition to the formyl group can be modulated to selectively afford the corresponding alcohols.[29] Thus treatment of (2-iodobenzyl)amino aldehydes with a catalytic amount of $Pd_2(dba)_3$ and PPh_3, in the presence of Et_3N as the base, gave the corresponding tetrahydroisoquinolin-4-ols in good yields (Scheme 8.9).

In this case, our DFT calculations suggest that the carbopalladation reaction of the formyl group in complex **16M** leads, via **TS7**, to a new intermediate (**17M**) in which the coordination of the nitrogen atom to the transition metal hinders the *cisoid* conformation essential for β-hydride elimination. In

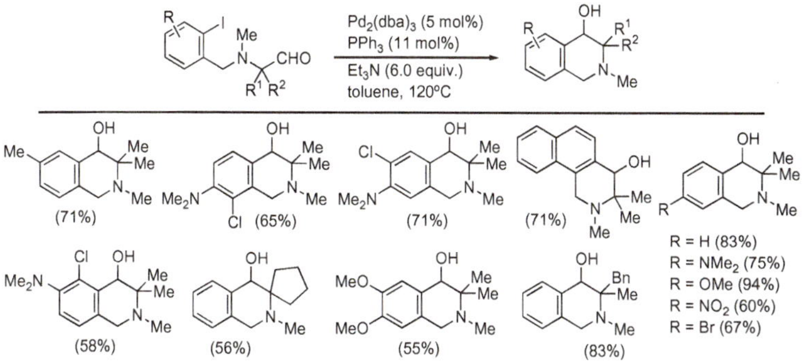

SCHEME 8.9 Synthesis of isoquinolin-4-ols

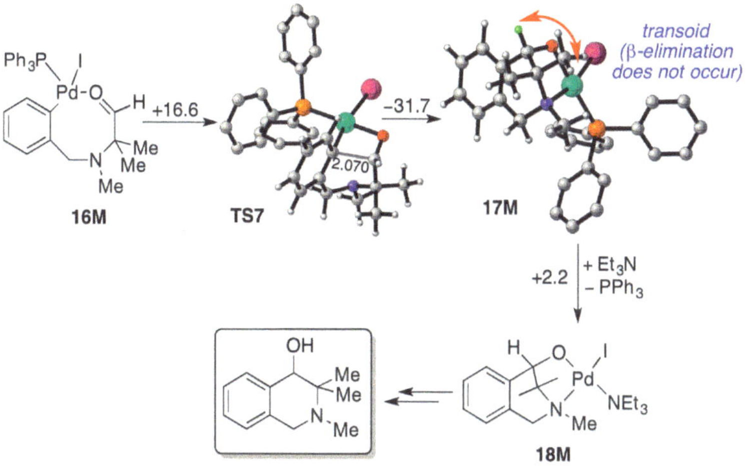

FIGURE 8.4 Computed reaction profile for complex **16M**. Free energy values (ΔG_{393}, computed at 120°C) and bond distances are given in kcal/mol and angstroms, respectively. All data have been computed at the PCM(toluene)-B3LYP/def2-TZVP//B3LYP/def2-SVP level.

the presence of Et_3N, the observed alcohol is very likely formed via a Et_3N-mediated reductive protonation process (Fig. 8.4).

The influence of the tethering heteroatom (changing from nitrogen to oxygen and sulfur) on the course of the palladium-catalyzed intramolecular reactions of aryl iodides with aldehydes was also studied.[30] In general, although some differences were found as a consequence of either the coordination ability of the heteroatom or the delocalization of the heteroatom lone-pair into the aryl fragment, the aldehydes showed a similar chemical behavior regardless of the nature of the heteroatom.

8.4 CHEMOSELECTIVITY CONTROL IN INTRAMOLECULAR Pd-CATALYZED REACTIONS OF β-(2-IODOANILINO)ESTERS AND β-(2-IODOANILINO)CARBOXAMIDES

The ester and carboxamide groups have been traditionally considered inert toward organopalladium reagents. The scarcely reported palladium-catalyzed reactions resulting in the modification of an ester functionality do not involve the carbopalladation of the ester carbonyl group but the attack of the oxygen atom at the carbon-palladium bond.[22,31] On the other hand, until recently, transition metal-catalyzed transformations of amides by N−C bond activation have been overlooked.[32]

We have studied the possibility of controlling the ambiphilic character of the σ-arylpalladium species derived from 2-haloanilines in their reaction with amino-tethered esters. It was found that the nucleophilic substitution at the alkoxycarbonyl group on ester **17** can be successfully promoted by using a catalytic amount of $Pd(PPh_3)_4$ in the presence of K_3PO_4 in toluene at high temperature (Scheme 8.10).[33]

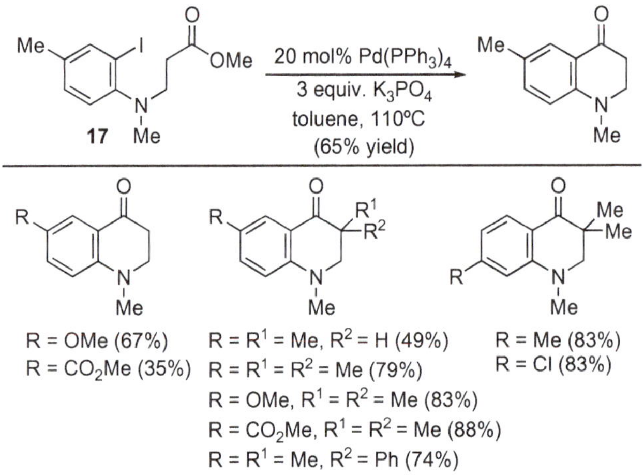

SCHEME 8.10 Pd(0)-catalyzed nucleophilic substitution at the alkoxycarbonyl group.

A range of differently substituted dihydroquinolin-4-ones were synthesized by means of this acylation reaction. In general, the acylation was largely unaffected by the aromatic ring substituents, whose electronic properties do not significantly influence the nucleophilicity of the aryl palladium species. The reaction of amino esters without hydrogen atoms placed α to the carbonyl group proceeded smoothly to give the corresponding ketones in high yields.

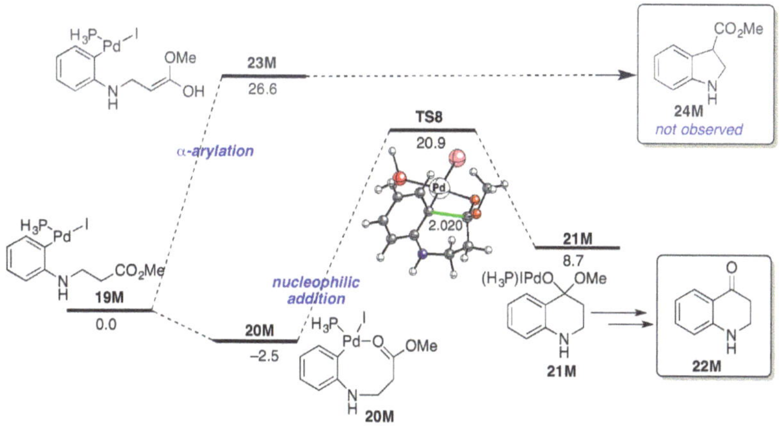

FIGURE 8.5 Computed reaction profile for the transformation of ester **19M**. Relative free energies (ΔG_{298} in kcal/mol) were computed at the PCM(toluene)-B3LYP/def2-SVP level.

The resulting experimental evidence together with subsequent computational studies[20] suggest that the acylation reaction with esters once again proceeds through a mechanism involving a nucleophilic addition to the carbonyl group. Thus a carbopalladation between the σ-arylpalladium moiety and the alkoxycarbonyl group would give a Pd(II) alkoxide. A β-alkoxide elimination would then afford the ketone and a Pd(II) methoxide complex, which would finally undergo β-hydride elimination to regenerate the palladium-catalyst with simultaneous release of formaldehyde. As expected, considering that an ester group is less electrophilic than a ketone, the computed barrier for the nucleophilic addition at the alkoxycarbonyl group is higher than at the ketone ($\Delta G_a = 23.4$ kcal/mol).[20] Interestingly, no competition between the nucleophilic attack at the carbonyl and the α-arylation was observed in the reactions of esters containing hydrogen atoms at the α-position to the carbonyl group. Similar to the transformation involving ketones (see above), our calculations on model compounds once again indicate that this is due to the required highly endergonic (ca. 27 kcal/mol) enolization process (Fig. 8.5).[20]

Further investigation of the palladium-catalyzed reactions of **17** revealed that switching their reactivity from nucleophilic to electrophilic is relatively easy. Thus the nucleophilic attack at the alkoxycarbonyl group was avoided simply by changing the base to potassium phenoxide, which is generated in situ by reaction of PhOH with KO^{t}Bu. Under these conditions, α-arylation took place to give indoline **18**. We also observed that the use of the couple K_3PO_4/PhOH in *N,N*-dimethylformamide (DMF) resulted in the direct formation of indole **19**.[34] These α-arylation reactions were successfully

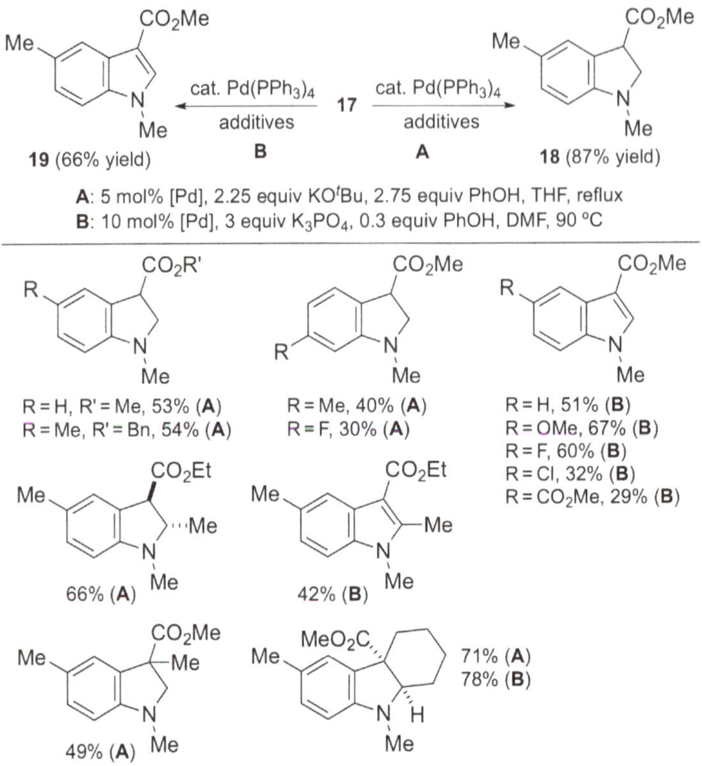

SCHEME 8.11 Pd(0)-catalyzed α-arylation of β-(2-iodoanilino)esters.

applied to the preparation of differently substituted indoles and indolines (Scheme 8.11).

To complete this study, we also explored the control of the ambiphilic character of σ-arylpalladium species towards carboxamides.[35] Using the same reaction conditions employed in the α-arylation or acylation reactions of esters, we found that amide **20** can also undergo either enolate arylation to give indoline **21** or acylation to form dihydroquinolin-4-one **22**. Both reactions were also extended to Weinreb amides. Notably, under the reaction conditions of the enolate arylation, Weinreb amide **23** afforded indole **24**, which results from the demethoxylation and aromatization of the initially formed indoline (Scheme 8.12).[35]

A variety of differently substituted dihydroquinolin-4-ones were prepared under the acylation reaction conditions starting from either amides or Weinreb amides. These acylation reactions very likely proceed through a nucleophilic addition mechanism similar to that proposed for the palladium-catalyzed acylation with esters. It should be noted that Weinreb amides

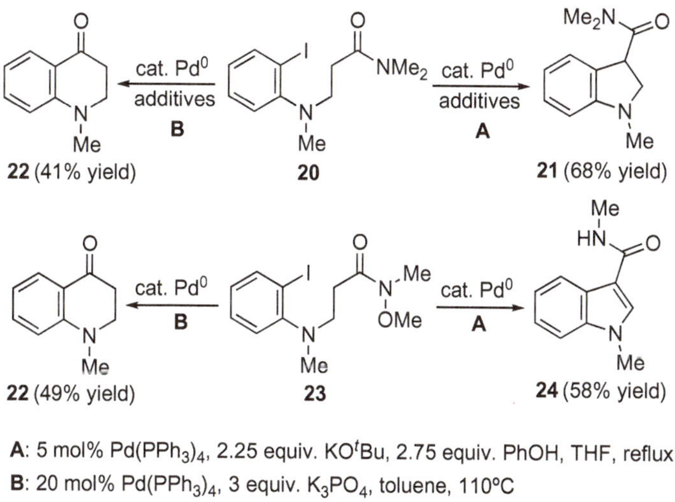

SCHEME 8.12 Ambiphilic reactions from β-(2-iodoanilino)carboxamides.

afforded better results than *N,N*-dimethyl amides, probably due to the chelation ability of the second oxygen atom, which would both facilitate the carbopalladation step and stabilize the resulting Pd(II) alkoxide intermediate (Scheme 8.13).[20]

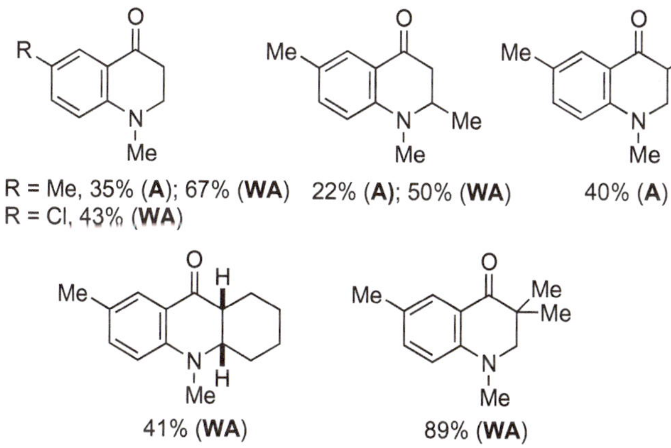

SCHEME 8.13 Scope of the acylation reaction with amides (**A**) and Weinreb amides (**WA**).

The scope of the α-arylation reaction with amides and Weinreb amides was also demonstrated by the synthesis of a variety of diversely substituted indoles and indolines (Scheme 8.14).

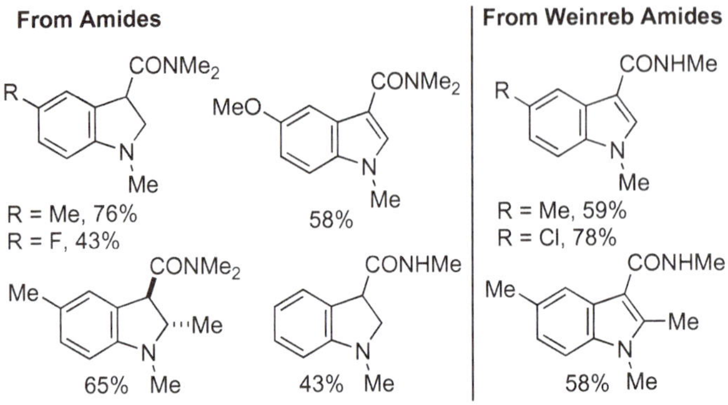

SCHEME 8.14 Scope of the α-arylation reaction with amides and Weinreb amides.

8.5 CATALYST-CONTROLLED SELECTIVE INTRAMOLECULAR C–H INSERTION OF α-DIAZO CARBONYL DERIVATIVES

The transition metal-catalyzed intramolecular carbenoid C–H insertion by decomposition of α-diazocarbonyl compounds constitutes a powerful methodology for the construction of carbocyclic and heterocyclic ring systems.[36] For a given substrate, several C–H insertion pathways may be available, the chemoselectivity, regioselectivity, and stereoselectivity being dependent on the nature of both the substrate and the catalyst.[37] In these reactions, the electron-rich C–H bonds generally exhibit higher reactivity toward the carbenoid center and show an activation order of tertiary > secondary ≫ primary Csp^3–H. As a consequence, it is not unexpected that while the insertion into tertiary and secondary Csp^3–H bonds has been thoroughly studied, the analogous process involving primary Csp^3–H and $C(Ar)sp^2$–H bonds remains comparatively underdeveloped. The use of dirhodium(II)[38] and copper catalysts[39] has traditionally monopolized this area of research, although in recent years different transition metals have emerged as useful alternatives.[40] Palladium, one of the most commonly employed metals in homogeneous catalysis, has been scarcely applied to these processes, its use being restricted to a couple of examples of α-diazo-β-ketoester insertion into Csp^2–H bonds.[41] This fact is highly surprising considering the great success of palladium catalysis in cross-coupling reactions of diazo compounds with either organic halides, pseudohalides, or arylboronic acids.[42]

We examined the feasibility of palladium as a catalyst for the carbenoid Csp^3–H insertion reaction of α-diazoesters to provide pyrrolidines through Csp^3–Csp^3 bond formation.[43] During the optimization studies with α-diazoester **25**, it was found that the carbenoid C–H insertion can be promoted by both Pd(0) and Pd(II) catalysts. These reactions afforded mixtures of pyrrolidine **26** and tetrahydroquinoline **27**, arising from the activation of the Csp^3–H and aryl $C(Ar)sp^2$–H bonds, respectively. The best chemoselectivity (i.e., a **26**:**27** ratio of ≈2.7:1) was obtained when using a combination of $Pd_2(dba)_3$ with a bidentate phosphine [1,3-bis(diphenylphosphino)propane (dppp), 1,1-bis(diphenylphosphino)ferrocene (dppf), or xantphos] as the catalyst (Scheme 8.15).

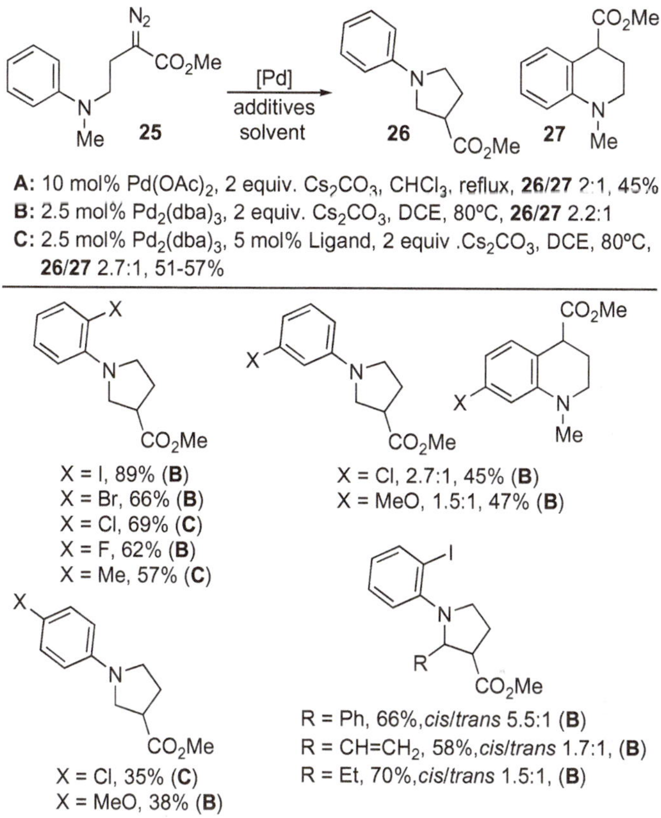

SCHEME 8.15 Pd-catalyzed cyclization of α-diazoesters to give pyrrolidines.

The scope and limitations of the C–H insertion reaction were established by preparing a series of *N*-arylpyrrolidines. The Csp^3–H insertion was the only reaction observed when *ortho*-substituted anilines were subjected to the optimized reaction conditions. In contrast, competition between the Csp^3–H and C(Ar)sp^2–H insertions was evident in the reactions involving *meta*- and *para*-substituted substrates, the electronic properties of the substituent affecting the cyclization reactions to a different extent according to its position at the aromatic ring. The carbenoid C–H insertion was not limited to *N*-methylanilines but also proved suitable for substituted Csp^3–H bonds. It should be noted that no product resulting from the competitive Pd-catalyzed reaction of the aryl iodide with the α-diazoester moiety[42] was observed in the reactions of anilines bearing the 2-iodo substituent. Similarly, no competition between allylic insertion and cyclopropanation was apparent in the reactions of the aniline with the *N*-allyl substituent, which exclusively afforded the corresponding pyrrolidine in good yield.

DFT calculations starting from palladacarbene **25M** suggest that the Csp^3–H insertion proceeds through a two-step reaction pathway. Thus **25M** evolves into intermediate **26M** through an unprecedented Pd-mediated 1,5-H migration from the N–CH_3 moiety to the carbenoid carbon atom, which results in the formal oxidation of the transition metal via **TS9**. Then, a reductive elimination reaction via **TS10** affords the observed pyrrolidine with concomitant regeneration of the Pd(dppp) catalyst. The reaction profile computed for the analogous process involving a Pd(II)-catalyst is essentially different, since the first step of the transformation consists of a concerted metalation-deprotonation (CMD) C–H activation assisted by carbonate, followed by insertion of the carbene carbon atom into the σ-Pd–C bond (Fig. 8.6).[43]

We also explored the palladium-catalyzed carbenoid C–H insertion of α-diazo-α-(methoxycarbonyl)acetanilides.[44] The use of transition metals to catalyze the insertion reactions of these substrates has been extensively studied,[45] since the insertion products, namely β- and γ-lactams as well as

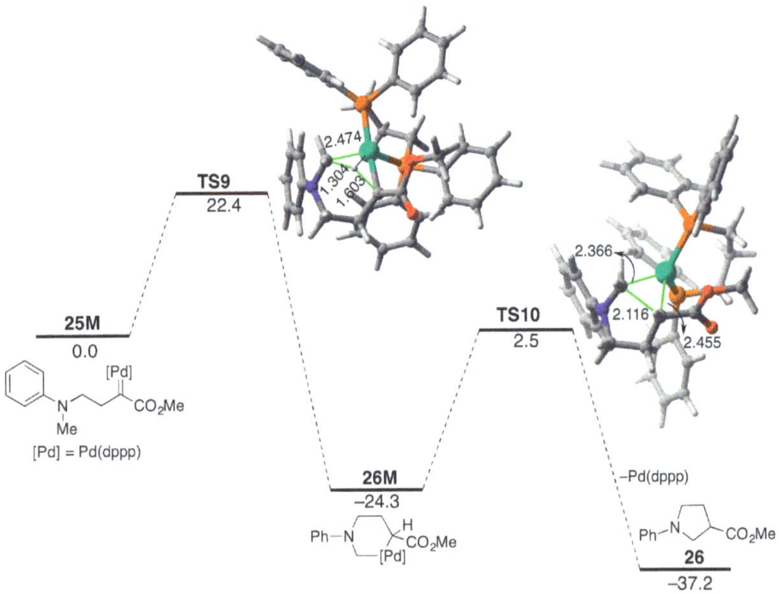

FIGURE 8.6 Computed reaction profiles for the formation of pyrrolidines. Relative free energies (ΔG_{298}, at 298 K) and bond distances are given in kcal/mol and angstroms, respectively. All data were computed at the PCM($CHCl_3$)-M06L/def2-TZVP//RI-BP86-D3/def2-SVP level.

2-oxindoles, are common scaffolds found in numerous natural products. Thus for instance, Rh(II) carboxylate and, in particular, carboxamide catalysts have been used to develop highly chemo-, regio-, and stereoselective transformations with a variety of reaction modes.[46] More recently, the commercially available $[Ru(p\text{-cymene})Cl_2]_2$ has been used as a catalyst in diverse methodologies for oxindole synthesis based on α-diazoamide carbenoid insertion.[47]

In line with the results previously reported for related substrates, the Rh (II) acetate-catalyzed decomposition of **28** resulted in the intramolecular carbenoid insertion into the benzylic C–H bond to give the *trans*-β-lactam **29**. In contrast, treatment of **28** with a catalytic amount of $Pd_2(dba)_3$ in toluene at reflux afforded oxindole **30**, arising from the arylic C–H bond insertion, while when using dichloroethane as the solvent, oxindole **31** was directly obtained from a sequential carbenoid insertion/alkylation process. The scope of the two-step sequence (Method **C**) was demonstrated by preparing a variety of 3-(2-chloroethyl)-3-(methoxycarbonyl)oxindoles. The Pd-catalyzed

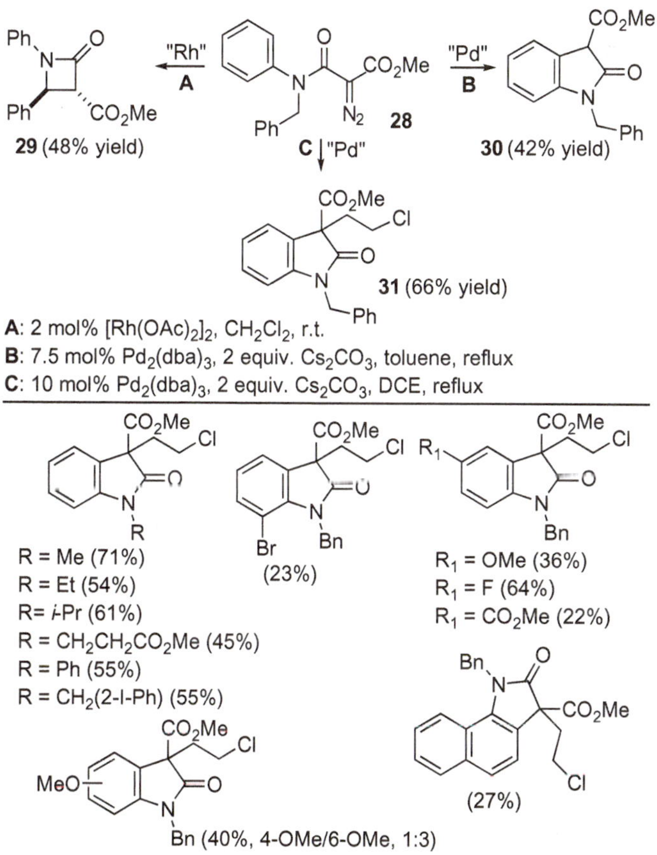

SCHEME 8.16 Transition metal-catalyzed cyclization of α-diazoamides.

insertion reaction took place selectively into the arylic C–H bond, in the presence of primary, secondary, as well as tertiary Csp^3–H bonds. The oxindoles bearing a 2-(methoxycarbonyl)ethyl, phenyl, and 2-iodobenzyl group at the nitrogen atom were also obtained in acceptable yields. Strikingly, no product resulting from the competitive Pd-catalyzed reaction of the aryl iodide with the α-diazoamide was observed in the reactions of the latter substrate. In contrast, it was found that the stereoelectronic properties of the substituents on the aniline ring considerably affect the course of the cyclization reaction (Scheme 8.16).

DFT calculations indicate that the observed oxindoles are formed through a rather similar reaction pathway to that computed for the formation of pyrrolidines (see above). Thus the initially formed palladacarbene **27M** undergoes a Pd-mediated 1,5-H migration from the C(aryl)–H moiety to the carbenoid carbon atom (via **TS11-A**), followed by a reductive elimination reaction (via **TS12-A**). From the data in Fig. 8.7, it is evident that the alternative Csp^3–H insertion that should generate the β-lactam is not feasible. This is due to the much higher activation barriers computed for the analogous 1,4-H migration and especially for the subsequent reductive elimination (via **TS12-B**) that would lead to the β-lactam reaction product.

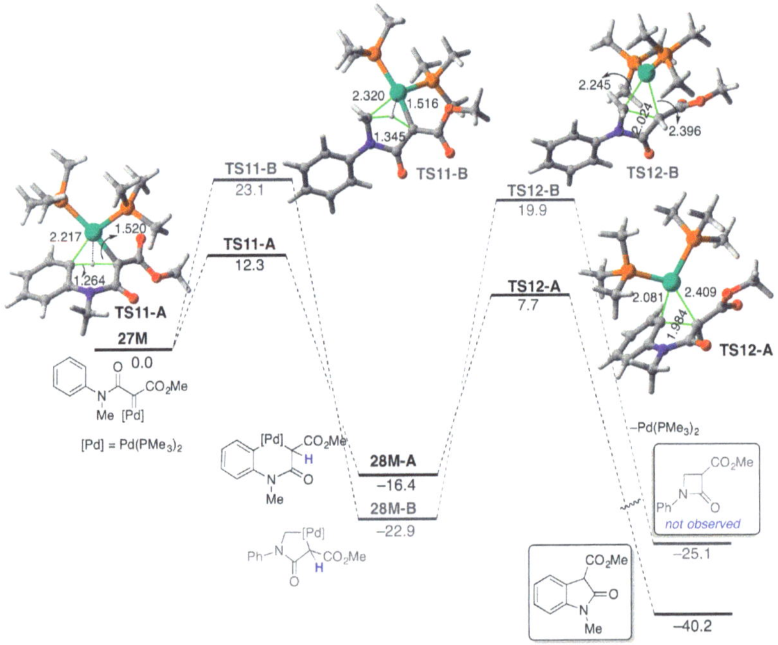

FIGURE 8.7 Computed reaction profiles for the Pd-catalyzed insertion favorably leading to oxindoles over β-lactams. All data were computed at the PCM(dichloroethane)-M06L/def2-TZVPP//PCM(dichloroethane)-B3LYP-D3/def2-SVP level.

8.6 SUMMARY AND CONCLUSIONS

The final goal of catalysis is to selectively transform a given substrate into the desired product. Today the principles of catalyst-controlled selectivity are well-established and widely applied in organopalladium chemistry. However, within the extremely diverse chemistry of transition metals, there are still many fundamental transformations that could benefit from the application of palladium catalysis to control the selectivities in C–C bond-forming reactions.

In this chapter we have first summarized our studies on the control of the ambiphilic character of σ-organopalladium species in intramolecular coupling reactions with different carbonyl moieties leading to heterocycles. These palladium intermediates act as electrophiles in α-arylation reactions or as nucleophiles in the transformations that involve the direct attack of the C–Pd bond at the different carbonyl groups. By subtle modification of the catalyst, ligands, or reaction conditions, we have been able to switch from one reaction mode to the other starting from the same precursor, therefore leading to divergent products, often with high selectivity.

In addition, we have shown that palladium catalysts also constitute a useful alternative to other transition metals commonly used for selectivity control in the intramolecular carbenoid C–H insertion reactions of α-diazocarbonyl compounds.

The above findings are the result of the synergy between experimental and computational studies, the latter used not only to explore and rationalize the different reaction pathways involved in the transformations, but also to further enhance the selectivity during the practical application of the reactions to the synthesis of nitrogen heterocycles.

REFERENCES

1. For a review, see Mahatthananchai, J.; Dumas, A. M.; Bode, J. W. Catalytic Selective Synthesis. *Angew. Chem. Int. Ed.* **2012,** *51*, 10954–10990.
2. For selective reviews, see: (a) Tietze, L. F.; Ila, H.; Bell, H. P. Enantioselective Palladium-Catalyzed Transformations. *Chem. Rev.* **2004,** *104*, 3453–3516. (b) Zheng, C.; You, S.-L. Recent Development of Direct Asymmetric Functionalization of Inert C–H bonds. *RSC Adv.* **2014,** *4*, 6173–6214. (c) Cherney, A. H.; Kadunce, N. T.; Reisman, S. E. Enantioselective and Enantiospecific Transition-Metal-Catalyzed Cross-Coupling Reactions of Organometallic Reagents to Construct C–C Bonds. *Chem. Rev.* **2015,** *115*, 9587–9652.
3. Trost, B. M. Selectivity: A Key to Synthetic Efficiency. *Science* **1983,** *219*, 245–250.
4. For selected examples, see: (a) Plietker, B.; Dieskau, A.; Möws, K.; Jatsch, A. Ligand-Dependent Mechanistic Dichotomy in Iron-Catalyzed Allylic Substitutions: σ-Allyl versus π-Allyl Mechanism. *Angew. Chem. Int. Ed.* **2008,** *47*, 198–201. (b) Yamauchi, M.; Morimoto, M.; Miura, T.; Murakami, M. Enantioselective Synthesis of 3,4-Dihydroisoquinolin-1(2*H*)-ones by Nickel-Catalyzed Denitrogenative Annulation of 1,2,3-Benzotriazin-4(3*H*)-ones with Allenes. *J. Am. Chem. Soc.* **2010,** *132*, 54–55. (c) Gao, F.;

Hoveyda, A. H. α-Selective Ni-Catalyzed Hydroalumination of Aryl- and Alkyl-Substituted Terminal Alkynes: Practical Syntheses of Internal Vinyl Aluminums, Halides, or Boronates. *J. Am. Chem. Soc.* **2010,** *132*, 10961–10963. (d) Di Giuseppe, A.; Castarlenas, R.; Pérez-Torrente, J. J.; Crucianelli, M.; Polo, V.; Sancho, R.; Lahoz, F. J.; Oro, L. A. Ligand-Controlled Regioselectivity in the Hydrothiolation of Alkynes by Rhodium N-Heterocyclic Carbene Catalysts. *J. Am. Chem. Soc.* **2012,** *134*, 8171–8183. (e) Chen, P.-H.; Xu, T.; Dong, G. Divergent Syntheses of Fused β-Naphthol and Indene Scaffolds by Rhodium-Catalyzed Direct and Decarbonylative Alkyne–Benzocyclobutenone Couplings. *Angew. Chem. Int. Ed.* **2014,** *53*, 1674–1678. (f) Li, B.; Park, Y.; Chang, S. Regiodivergent Access to Five- and Six-Membered Benzo-Fused Lactams: Ru-Catalyzed Olefin Hydrocarbamoylation. *J. Am. Chem. Soc.* **2014,** *136*, 1125–1131.

5. (a) Negishi, E., Ed. *Handbook of Organopalladium Chemistry for Organic Synthesis*, Vols. I and II; Wiley-VCH: Weinheim, 2002. (b) Palladium in Organic Synthesis; In *Topics in Organometallis Chemistry;* Tsuji, J. *Topics in Organometallis Chemistry;* Springer-Verlag: Berlin, 2005.
6. Engle, K. M.; Yu, J.-Q. Developing Ligands for Palladium(II)-Catalyzed C – H Functionalization: Intimate Dialogue between Ligand and Substrate. *J. Org. Chem.* **2013,** *78*, 8927–8955.
7. For recent selected examples of controlled divergent palladium-catalyzed reactions, see: (a) Tsvelikhovsky, D.; Buchwald, S. L. Synthesis of Heterocycles via Pd-Ligand Controlled Cyclization of 2-Chloro-*N*-(2-vinyl)aniline: Preparation of Carbazoles, Indoles, Dibenzazepines, and Acridines. *J. Am. Chem. Soc.* **2010,** *132*, 14048–14051. (b) Neufeldt, S. R.; Sanford, M. S. Controlling Site Selectivity in Palladium-Catalyzed C–H Bond Functionalization. *Acc. Chem. Res.* **2012,** *45*, 936–946. (c) Yang, Y.; Buchwald, S. L. Ligand-Controlled Palladium-Catalyzed Regiodivergent Suzuki – Miyaura Cross-Coupling of Allylboronates and Aryl Halides. *J. Am. Chem. Soc.* **2013,** *135*, 10642–10645. (d) Fujino, D.; Yorimitsu, H.; Osuka, A. Regiocontrolled Palladium-Catalyzed Arylative Cyclizations of Alkynols. *J. Am. Chem. Soc.* **2014,** *136*, 6255–6258. (e) Kim, J. H.; Bouffard, J.; Lee, S.-G. Formation of Four Different Aromatic Scaffolds from Nitriles through Tandem Divergent Catalysis. *Angew. Chem. Int. Ed.* **2014,** *53*, 6435–6438. (f) Smith, S. M.; Buchwald, S. L. Regioselective 2-Amination of Polychloropyrimidines. *Org. Lett.* **2016,** *18*, 2180–2183. (g) Wu, C.; Zhao, F.; Du, Y.; Zhao, L.; Chen, L.; Wang, J.; Liu, H. Highly Selective Intramolecular Addition of C–N and S–N Bonds to Alkynes Catalyzed by Palladium: A Practical Access to Two Distinct Functional Indoles. *RSC Adv.* **2016,** *6*, 70682–70690.
8. Yamamoto, Y.; Nakamura, I. Nucleophilic Attack by Palladium Species. In *Palladium in Organic Synthesis. Topics in Organometallic Chemistry;* Tsuji, J., Ed.; , Vol. 14; Springer-Verlag: Berlin, 2005; pp 211–239.
9. (a) Zhao, Y.-B.; Mariampillai, B.; Candito, D. A.; Laleu, B.; Li, M.; Lautens, M. Exploiting the Divergent Reactivity of Aryl-Palladium Intermediates for the Rapid Assembly of Fluorene and Phenanthrene Derivatives. *Angew. Chem. Int. Ed.* **2009,** *48*, 1849–1852. (b) Nareddy, P.; Mazet, C. α-Arylation, α-Arylative Esterification, or Acylation: A Stoichiometry-Dependent Trichotomy in the Pd-Catalyzed Cross-Coupling between Aldehydes and Aryl Bromides. *Chem. Asian J.* **2013,** *8*, 2579–2583.
10. Solé, D.; Fernández, I. Controlling the Ambiphilic Nature of σ-Arylpalladium Intermediates in Intramolecular Cyclization Reactions. *Acc. Chem. Res.* **2014,** *47*, 168–179.
11. For reviews, see: (a) Culkin, D. A.; Hartwig, J. F. Palladium-Catalyzed α-Arylation of Carbonyl Compounds and Nitriles. *Acc. Chem. Res.* **2003,** *36*, 234–245. (b) Bellina, F.; Rossi, R. Transition Metal-Catalyzed Direct Arylation of Substrates with Activated sp^3-

Hybridized C–H Bonds and Some of Their Synthetic Equivalents with Aryl Halides and Pseudohalides. *Chem. Rev.* **2010,** *110*, 1082–1146. (c) Johansson, C. C. C.; Colacot, T. J. Metal-Catalyzed α-Arylation of Carbonyl and Related Molecules: Novel Trends in C–C Bond Formation by C–H Bond Functionalization. *Angew. Chem. Int. Ed.* **2010,** *49*, 676–707.

12. Sivanandan, S. T.; Shaji, A.; Ibnusaud, I.; Johansson-Seechurn, C. C. C.; Colacot, T. J. Palladium-Catalyzed α-Arylation Reactions in Total Synthesis. *Eur. J. Org. Chem.* **2015,** 38–49.
13. For a review, see Burke, A. J. Recent Advances in Metal Catalyzed Intramolecular Aryl Additions—Accessing Bioactive Molecules. *Tetrahedron Lett.* **2016,** *57*, 1197–1204.
14. (a) Quan, L. G.; Lamrani, M.; Yamamoto, Y. Intramolecular Nucleophilic Addition of Aryl Bromides to Ketones Catalyzed by Palladium. *J. Am. Chem. Soc.* **2000,** *122*, 4827–4828. (b) Liu, G.; Lu, X. Cationic Palladium Complex Catalyzed Highly Enantioselective Intramolecular Addition of Arylboronic Acids to Ketones. A Convenient Synthesis of Optically Active Cycloalkanols. *J. Am. Chem. Soc.* **2006,** *128*, 16504–16505. (c) Liu, G.; Lu, X. Palladium(II)-catalyzed Intramolecular Addition of Arylboronic Acids to Ketones. *Tetrahedron* **2008,** *64*, 7324–7330. (d) Jia, Y.-X.-; Katayev, D.; Kündig, E. P. Synthesis of 3-Hydroxyoxindoles by Pd-Catalysed Intramolecular Nucleophilic Addition of Aryl Halides to α-Ketoamides. *Chem. Commun.* **2010,** *46*, 130–132. (e) Yin, L.; Kanai, M.; Shibasaki, M. A Facile Pathway to Enantiomerically Enriched 3-Hydroxy-2-Oxindoles: Asymmetric Intramolecular Arylation of α-Keto Amides Catalyzed by a Palladium-DifluorPhos Complex. *Angew. Chem. Int. Ed.* **2011,** *50*, 7620–7623. (f) Hu, J.-X.; Wu, H.; Li, C.-Y.; Sheng, W.-J.; Jia, Y.-X.; Gao, J.-R. Nickel-Catalyzed Intramolecular Nucleophilic Addition of Aryl or Vinyl Chlorides to α-Ketoamides Through C–Cl Bond Activation. *Chem. Eur. J.* **2011,** *17*, 5234–5237. (g) He, J.-Q.; Chen, C.; Yu, W.-B.; Liu, R.-R.; Xu, M.; Li, Y.-J.; Gao, J.-R.; Jia, Y.-X. Nickel-Catalyzed Intramolecular Addition of Vinyl or Aryl Bromides to Ketoamides. *Tetrahedron Lett.* **2014,** *55*, 2805–2808. (h) Shin, I.; Ramgren, S. D.; Krische, M. J. Reductive Cyclization of Halo-Ketones to form 3-Hydroxy-2-oxindoles via Palladium Catalyzed Hydrogenation: A Hydrogen-Mediated Grignard Addition. *Tetrahedron* **2015,** *71*, 5776–5780.
15. Giorgi, G.; Maiti, S.; López-Alvarado, P.; Menéndez, J. C. Synthesis of Benzo- and Naphtho-fused Bicyclo[n.3.1]alkane Frameworks with a Bridgehead Nitrogen Function by Palladium-Catalyzed Intramolecular α-Arylation of α-Nitroketones. *Org. Biomol. Chem.* **2011,** *9*, 2722–2730.
16. Saadi, J.; Bentz, C.; Redies, K.; Lentz, D.; Zimmer, R.; Reissig, H.-U. Stereoselective Synthesis of Tricyclic Compounds by Intramolecular Palladium-Catalyzed Addition of Aryl Iodides to Carbonyl Groups. *Beilstein J. Org. Chem.* **2016,** *12*, 1236–1242.
17. Solé, D.; Vallverdú, L.; Peidró, E.; Bonjoch, J. Palladium-Catalysed Intramolecular Annulation of 2-Haloanilines and Ketones: Enolate Arylation *vs.* Nucleophilic Addition to the Carbonyl Group. *Chem. Commun.* **2001,** 1888–1889.
18. Solé, D.; Vallverdú, L.; Solans, X.; Font-Bardia, M.; Bonjoch, J. Intramolecular Pd-Mediated Processes of Amino-Tethered Aryl Halides and Ketones: Insight into the Ketone α-Arylation and Carbonyl-Addition Dichotomy. A New Class of Four-Membered Azapalladacycles. *J. Am. Chem. Soc.* **2003,** *125*, 1587–1594.
19. Fernández, I.; Solé, D.; Sierra, M. A. A DFT Study of the Ambiphilic Nature of Arylpalladium Species in Intramolecular Cyclization Reactions. *J. Org. Chem.* **2011,** *76*, 1592–1598.

20. Solé, D.; Fernández, I.; Sierra, M. A. Control over the Chemoselectivity of Pd-Catalyzed Cyclization Reactions of (2-Iodoanilino)carbonyl Compounds. *Chem. Eur. J.* **2012,** *18*, 6950–6958.
21. Solé, D.; Bennasar, M.-L.; Jiménez, I. Intramolecular Palladium-Catalysed Enolate Arylation of 2- and 3-Iodoindole Derivatives for the Synthesis of β-Carbolines, γ-Carbolines, and Pyrrolo[3,4-*b*]indoles. *Org. Biomol. Chem.* **2011,** *9*, 4535–4544.
22. Tao, W.; Silverberg, L. J.; Rheingold, A. L.; Heck, R. F. Alkyne Reactions with Arylpalladium Compounds. *Organometallics* **1989,** *8*, 2550–2559.
23. (a) Larock, R. C.; Doty, M. J.; Cacchi, S. Synthesis of Indenones via Palladium-Catalyzed Annulation of Internal Alkynes. *J. Org. Chem.* **1993,** *58*, 4579–4583. (b) Vicente, J.; Abad, J.-A.; Gil-Rubio, J. Palladium-Assisted Formation of Carbon-Carbon Bonds. 6. Study of the Reactivity of (*o*-Formylaryl)- or (*o*-Acetylaryl)palladium Complexes with Alkynes. Synthesis of Indenones and Indenols. *Organometallics* **1996,** *15*, 3509–3519. (c) Gevorgyan, V.; Quan, L. G.; Yamamoto, Y. Synthesis of Indenols and Indanones via Catalytic Cyclic Vinylpalladation of Aromatic Aldehydes. *Tetrahedron Lett.* **1999,** *40*, 4089–4092. (d) Zhao, L.; Lu, X. Pd^{II}-Catalyzed Cyclization of Alkynes Containing Aldehyde, Ketone, or Nitrile Groups Initiated by the Acetoxypalladation of Alkynes. *Angew. Chem. Int. Ed.* **2002,** *41*, 4343–4345. (e) Yang, M.; Zhang, X.; Lu, X. Cationic Palladium(II)-Catalyzed Highly Enantioselective [3 + 2] Annulation of 2-Acylarylboronic Acids with Substituted Alkynes. *Org. Lett.* **2007,** *9*, 5131–5133. (f) Han, X.; Lu, X. Control of Chemoselectivity by Counteranions of Cationic Palladium Complexes: A Convenient Enantioselective Synthesis of Dihydrocoumarins. *Org. Lett.* **2010,** *12*, 108–111.
24. (a) Álvarez-Bercedo, P.; Flores-Gaspar, A.; Correa, A.; Martin, R. Pd-Catalyzed Intramolecular Acylation of Aryl Bromides via C–H Functionalization: A Highly Efficient Synthesis of Benzocyclobutenones. *J. Am. Chem. Soc.* **2010,** *132*, 466–467. (b) Flores-Gaspar, A.; Gutiérrez-Bonet, A.; Martin, R. N-Heterocyclic Carbene Dichotomy in Pd-Catalyzed Acylation of Aryl Chlorides via C–H Bond Functionalization. *Org. Lett.* **2012,** *14*, 5234–5237.
25. Solé, D.; Mariani, F.; Fernández, I.; Sierra, M. A. Intramolecular Pd(0)-Catalyzed Reactions of (2-Iodoanilino)-aldehydes: A Joint Experimental-Computational Study. *J. Org. Chem.* **2012,** *77*, 10272–10284.
26. (a) Muratake, H.; Nakai, H. Intramolecular Cyclization Using Palladium-Catalyzed Arylation toward Formyl and Nitro Groups. *Tetrahedron Lett.* **1999,** *40*, 2355–2358. (b) García-Fontanet, J.; Buchwald, S. L. Asymmetric Palladium-Catalyzed Intramolecular α-Arylation of Aldehydes. *Angew. Chem. Int. Ed.* **2008,** *47*, 8108–8111. (c) Nareddy, P.; Mantilli, L.; Guénée, L.; Mazet, C. Atropoisomeric (P,N) Ligands for the Highly Enantioselective Pd-Catalyzed Intramolecular Asymmetric α-Arylation of α-Branched Aldehydes. *Angew. Chem. Int. Ed.* **2012,** *51*, 3826–3831.
27. Kawatsura, M.; Hartwig, J. F. Simple, Highly Active Palladium Catalysts for Ketone and Malonate Arylation: Dissecting the Importance of Chelation and Steric Hindrance. *J. Am. Chem. Soc.* **1999,** *121*, 1473–1478.
28. Solé, D.; Mariani, F. Synthesis of Dibenzo[*b,e*]azepin-11-ones by Intramolecular Palladium-Catalyzed Acylation of Aryliodides with Aldehydes. *J. Org. Chem.* **2013,** *78*, 8136–8142.
29. Solé, D.; Mariani, F.; Fernández, I. Synthesis of Isoquinolin-4-ols by Palladium-Catalysed Nucleophilic Addition of Aryl Iodides to Aldehydes. *Adv. Synth. Catal.* **2014,** *356*, 3237–3243.

30. Solé, D.; Mariani, F.; Fernández, I. A Joint Experimental-Computational Comparative Study of the Pd^0-Catalysed Reactions of Aryl Iodides and Aldehydes with N, O, and S Tethers. *Eur. J. Org. Chem.* **2015,** 3925–3942.
31. (a) Larock, R. C.; Yunn, E. U.; Doty, M. J.; Sham, K. K. C. Synthesis of Aromatic Heterocycles via Palladium-Catalyzed Annulation of Internal Alkynes. *J. Org. Chem.* **1995,** *60*, 3270–3271. (b) Larock, R. C.; Han, X.; Doty, M. J. Synthesis of α-Pyrones via Palladium-Catalyzed Annulation of Internal Alkynes. *Tetrahedron Lett.* **1998,** *39*, 5713–5716.
32. Meng, G.; Shi, S.; Szostak, M. Cross-Coupling of Amides by N–C Bond Activation. *Synlett* **2016,** *27*, 2530–2540.
33. Solé, D.; Serrano, O. Palladium-Catalyzed Intramolecular Nucleophilic Substitution at the Alkoxycarbonyl Group. *Angew. Chem. Int. Ed.* **2007,** *46*, 7270–7272.
34. Solé, D.; Serrano, O. Synthesis of Indole-3-carboxylic Acid Derivatives by Pd(0)-Catalyzed Intramolecular α-Arylation of β-(2-Iodoanilino) Esters. *J. Org. Chem.* **2008,** *73*, 2476–2479.
35. Solé, D.; Serrano, O. Intramolecular Pd(0)-Catalyzed Reactions of β-(2-Iodoanilino) Carboxamides: Enolate Arylation and Nucleophilic Substitution at the Carboxamide Group. *J. Org. Chem.* **2008,** *73*, 9372–9378.
36. For recent reviews, see: (a) Ford, A.; Miel, H.; Ring, A.; Slattery, C. N.; Maguire, A. R.; McKervey, M. A. Modern Organic Synthesis with α-Diazocarbonyl Compounds. *Chem. Rev.* **2015,** *115*, 9981–10080. (b) Burtoloso, A. C. B.; Dias, R. M. P.; Bernardim, B. α,β-Unsaturated Diazoketones as Useful Platforms in the Synthesis of Nitrogen Heterocycles. *Acc. Chem. Res.* **2015,** *48*, 921–934.
37. (a) Davies, H. M. L.; Beckwith, R. E. J. Catalytic Enantioselective C–H Activation by Means of Metal-Carbenoid-Induced C–H Insertion. *Chem. Rev.* **2003,** *103*, 2861–2903. (b) Merlic, C. A.; Zechman, A. L. Selectivity in Rhodium(II) Catalyzed Reactions of Diazo Compounds: Effects of Catalyst Electrophilicity, Diazo Substitution, and Substrate Substitution. From Chemoselectivity to Enantioselectivity. *Synthesis* **2003,** 1137–1156. (c) Davies, H. M. L.; Morton, D. Guiding Principles for Site Selective and Stereoselective Intermolecular C–H Functionalization by Donor/Acceptor Rhodium Carbenes. *Chem. Soc. Rev.* **2011,** *40*, 1857–1869. (d) Zheng, C.; You, S.-L. Recent Development of Direct Asymmetric Functionalization of Inert C–H Bonds. *RSC Adv.* **2014,** *4*, 6173–6214. (e) DeAngelis, A.; Panish, R.; Fox, J. M. Rh-Catalyzed Intermolecular Reactions of α-Alkyl-α-Diazo Carbonyl Compounds with Selectivity over β-Hydride Migration. *Acc. Chem. Res.* **2016,** *49*, 115–127.
38. Doyle, M. P.; Duffy, R.; Ratnikov, M.; Zhou, L. Catalytic Carbene Insertion into C–H Bonds. *Chem. Rev.* **2010,** *110*, 704–724.
39. (a) Díaz-Requejo, M. M.; Pérez, P. J. Coinage Metal Catalyzed C–H Bond Functionalization of Hydrocarbons. *Chem. Rev.* **2008,** *108*, 3379–3394. (b) Zhao, X.; Zhang, Y.; Wang, J. Recent Developments in Copper-Catalyzed Reactions of Diazo Compounds. *Chem. Commun.* **2012,** *48*, 10162–10173.
40. (a) Cai, Y.; Zhu, S.-F.; Wang, G.-P.; Zhou, Q.-L. Iron-Catalyzed C–H Fuctionalization of Indoles. *Adv. Synth. Catal.* **2011,** *353*, 2939–2944. (b) Yao, T.; Hirano, K.; Satoh, T.; Miura, M. Nickel- and Cobalt-Catalyzed Direct Alkylation of Azoles with N-Tosylhydrazones Bearing Unactivated Alkyl Groups. *Angew. Chem. Int. Ed.* **2012,** *51*, 775–779. (c) Yu, Z.; Ma, B.; Chen, M.; Wu, H.-H.; Liu, L.; Zhang, J. Highly Site-Selective Direct C–H Bond Functionalization of Phenols with α-Aryl-α-diazoacetates and Diazooxindoles via Gold Catalysis. *J. Am. Chem. Soc.* **2014,** *136*, 6904–6907. (d) Liu,

X.-G.; Zhang, S.-S.; Wu, J.-Q.; Li, Q.; Wang, H. Cp*Co(III)-Catalyzed Direct Functionalization of Aromatic C–H Bonds with α-Diazomalonates. *Tetrahedron Lett.* **2015,** *56*, 4093–4095. (e) Zhao, D.; Kim, J. H.; Stegemann, L.; Strassert, C. A.; Glorius, F. Cobalt(III)-Catalyzed Directed C–H Coupling with Diazo Compounds: Straightforward Access towards Extended π-Systems. *Angew. Chem. Int. Ed.* **2015,** *54*, 4508–4511. (f) Fructos, M. R.; Díaz-Requejo, M. M.; Pérez, P. J. Gold and Diazo Reagents: A Fruitful Tool for Developing Molecular Complexity. *Chem. Commun.* **2016,** *52*, 7326–7335. (g) Liu, L.; Zhang, J. Gold-Catalyzed Transformations of α-Diazocarbonyl Compounds: Selectivity and Diversity. *Chem. Soc. Rev.* **2016,** *45*, 506–516. (h) Conde, A.; Sabenya, G.; Rodríguez, M.; Postils, V.; Luis, J. M.; Díaz-Requejo, M. M.; Costas, M.; Pérez, P. J. Iron and Manganese Catalysts for the Selective Functionalization of Arene $C(sp^2)$–H Bonds by Carbene Insertion. *Angew. Chem. Int. Ed.* **2016,** *55*, 6530–6534.

41. (a) Taber, D. F.; Amedio, J. C., Jr.; Sherill, R. G. Palladium-Mediated Diazo Insertions: Preparation of 3-Alkyl-2-carbomethoxycyclopentenones. *J. Org. Chem.* **1986,** *51*, 3382–3384. (b) Rosenberg, M. L.; Aasheim, J. H. F.; Trebbin, M.; Uggerud, E.; Hansen, T. Synthesis of a 1,3,4,5-tetrahydrobenzindole β-ketoester. *Tetrahedron Lett.* **2009,** *50*, 6506–6508.
42. For reviews, see: (a) Barluenga, J.; Valdés, C. Tosylhydrazones: New Uses for Classic Reagents in Palladium-Catalyzed Cross-Coupling and Metal-Free Reactions. *Angew. Chem. Int. Ed.* **2011,** *50*, 7486–7500. (b) Xiao, Q.; Zhang, Y.; Wang, J. Diazo Compounds and N-Tosylhydrazones: Novel Cross-Coupling Partners in Transition-Metal-Catalyzed Reactions. *Acc. Chem. Res.* **2013,** *46*, 236–247.
43. Solé, D.; Mariani, F.; Bennasar, M.-L.; Fernández, I. Palladium-Catalyzed Intramolecular Carbene Insertion into $C(sp^3)$–H Bonds. *Angew. Chem. Int. Ed.* **2016,** *55* (6467–6470).
44. Solé, D.; Pérez-Janer, F.; Fernández, I. Palladium-Catalysed Intramolecular Carbenoid Insertion of α-Diazo-α-(methoxycarbonyl)acetanilides for Oxindole Synthesis. *Chem. Commun.* **2017,** *53*, 3110–3113.
45. (a) Gois, P. M. P.; Afonso, C. A. M. Stereo- and Regiocontrol in the Formation of Lactams by Rhodium-Carbenoid C–H Insertion of α-Diazoacetamides. *Eur. J. Org. Chem.* **2004,** 3773–3788. (b) Ring, A.; Ford, A.; Maguire, A. R. Substrate and Catalyst Effects in C–H Insertion Reactions of α-Diazoacetamides. *Tetrahedron Lett.* **2016,** *57*, 5399–5406.
46. (a) Doyle, M. P.; Shanklin, M. S.; Pho, H. Q.; Mahapatro, S. N. Rhodium(II) Acetate and Nafion-H Catalyzed Decomposition of *N*-Aryldiazoamides. An Efficient Synthesis of 2 (3*H*)-Indolinones. *J. Org. Chem.* **1988,** *53*, 1017–1022. (b) Wee, A. G. H.; Liu, B.; Zhang, L. Dirhodium Tetraacetate Catalyzed Carbon-Hydrogen Insertion Reaction in N-Substituted α-Carbomethoxy-α-diazoacetanilides and Structural Analogues. Substituent and Conformational Effects. *J. Org. Chem.* **1992,** *57*, 4404–4414. (c) Brown, D. S.; Elliott, M. C.; Moody, C. J.; Mowlem, T. J.; Marino, J. P.; Padwa, A. Ligand Effects in the Rhodium(II)-Catalyzed Reactions of α-Diazoamides. Oxindole Formation is Promoted by the Use of Rhodium(II) Perfluorocarboxamide Catalysts. *J. Org. Chem.* **1994,** *59*, 2447–2455. (d) Miah, S.; Slawin, A. M. Z.; Moody, C. J.; Sheedan, S. M.; Marino, J. P.; Semones, M. A.; Padwa, A.; Richards, I. C. Ligand Effects in the Rhodium(II) Catalysed Reactions of Diazoamides and Diazoimides. *Tetrahedron* **1996,** *52*, 248–2514. (e) Qiu, H.; Li, M.; Jiang, L.-Q.; Lv, F.-P.; Zan, l; Zhai, C.-W.; Doyle, M. P.; Hu, W.-H. Highly Enantioselective Trapping of Zwitterionic Intermediates by Imines. *Nat. Chem.* **2012,** *4*, 733–738.

47. (a) Chan, W.-W.; Kwong, T.-L.; Yu, W.-Y. Ruthenium-Catalyzed Intramolecular Cyclization of Diazo-β-ketoanilides for the Synthesis of 3-Alkylideneoxindoles. *Org. Biomol. Chem.* **2012,** *10*, 3749–3755. (b) Liu, N.; Tian, Q.-P.; Yang, Q.; Yang, S.-D. Ruthenium-Catalyzed Intramolecular Cyclization and Fluorination to form 3-Fluorooxindoles. *Synlett* **2016,** 2621–2625. (c) Yamamoto, K.; Qureshi, Z.; Tsoung, J.; Pisella, G.; Lautens, M. Combining Ru-Catalyzed C–H Fuctionalization with Pd-Catalyzed Asymmetric Allylic Alkylation: Synthesis of 3-Allyl-3-aryl Oxindole Derivatives from Aryl α-Diazoamides. *Org. Lett.* **2016,** *18*, 4954–4957.

Index

Note: Page numbers followed by "*f*" refer to figures.

A

C

D

E

H

N

O

P

Q

R

S

T

U

V

W

X

Y

Z

Kolibri 360 GmbH
Ferdinand-Jühlke-Straße 7
99095 Erfurt, Deutschland
produktsicherheit@kolibri360.de